ENCYCLOPEDIA OF
CHEMISTRY

DON RITTNER
AND
RONALD A. BAILEY, PH.D.

Facts On File, Inc.

To Nancy, Christopher, Kevin, Jackson, Jennifer, and Jason

Encyclopedia of Chemistry

For information contact:

Facts On File, Inc.
132 West 31st Street
New York NY 10001

Library of Congress Cataloging-in-Publication Data
Rittner, Don.
Encyclopedia of chemistry / Don Rittner and Ronald A. Bailey.
p. cm.
Includes bibliographical references and index.
ISBN 0-8160-4894-0
1. Chemistry—Encyclopedias. I. Bailey, R. A. (Ronald Albert), 1933 –
II. Title.
QD4.R57 2005
540′.3—dc22 2004011242

Facts On File books are available at special discounts when purchased in bulk quantities for businesses, associations, institutions, or sales promotions. Please call our Special Sales Department in New York at (212) 967-8800 or (800) 322-8755.

You can find Facts On File on the World Wide Web at http://www.factsonfile.com.

Text design by Joan M. Toro
Cover design by Cathy Rincon
Illustrations by Richard Garratt

Printed in the United States of America

VB Hermitage 10 9 8 7 6 5 4 3 2 1

This book is printed on acid-free paper.

CONTENTS

ACKNOWLEDGMENTS

I would like to thank the following for their generosity in helping to make this book as complete as possible, especially in the use of images, biographies, definitions, essays, and encouragement: Kristina Fallenias, Nobel Foundation; Fabienne Meyers, International Union of Pure and Applied Chemistry; Daryl Leja, NHGRI, and National Institutes of Health; essayists Harry K. Garber, Karl F. Moschner, and Theresa Beaty; the many chemistry Webmasters; and Nancy, Chris, Kevin, Jack, Jennifer, and Jason.

I also wish to thank Frank K. Darmstadt, executive editor, Sara Hov and Melissa Cullen-DuPont, editorial assistants, and the rest of the staff at Facts On File, Inc., and my colleague Ron Bailey. I apologize to anyone left out due to error.

PREFACE

Students beginning their study of chemistry are faced with understanding many terms that are puzzling and unrelated to contexts that make them understandable. Others may seem familiar, but in chemistry they have meanings that are not quite the same as when used in popular discourse. In science, terms need to have definite and specific meanings. One of the purposes of the *Encyclopedia of Chemistry* is to provide definitions for many of these terms in a manner and at a level that will make their meanings clear to those with limited backgrounds in chemistry, and to those in other fields who need to deal with chemistry. The International Union of Pure and Applied Chemistry (IUPAC), an international organization of chemists and national chemistry societies, makes the final determination of terminology and nomenclature in chemistry. Among other things, this organization decides the names for new elements and sets up systematic rules for naming compounds so that a given structure can be defined uniquely. Compounds are frequently called by common or trade names, often because their IUPAC names may be long and complex, but the IUPAC name permits a chemist to know the structure of any compound based on the rules of the terminology, while the common name requires remembering what structure goes with what name.

Chemistry has been called "the central science" because it relates to and bridges all of the physical and biological sciences. For example, biology, as it focuses more and more on processes at the cellular and molecular level, depends heavily on chemistry. There is great overlap within the fields in biochemistry, the study of the chemical processes that take place in biological systems, and in chemical biology, the latter term being used to describe the broader area of the application of chemical techniques and principles to biology-related problems. Because of this overlap, this encyclopedia has many entries that relate to biological sciences as well as to chemistry. Similarly, there is overlap with geology, some areas of physics, and any field related to the environment, among others.

While we can define chemistry, it is more difficult to describe what a chemist actually does. The comic book image of a chemist as someone in a white coat surrounded by test tubes and beakers, if it ever had any basis in reality, is far from accurate now. Nowadays, while the white coat may still be in style, a chemist is more likely to be surrounded by complicated instruments such as spectrometers and chromatographs. The type of work a chemist does falls into four general areas: first, synthetic chemistry is involved with the discovery of new materials or finding improved ways of making existing ones; materials can be organic, for example, pharmaceuticals and polymers, or inorganic such as superconducting materials. Second, analytical chemistry is focused on determining what or how much of a substance is present, or identifying its structure, or developing new ways of making these measurements. Third, physical chemistry is the study of reactions and energetics and finding the physical properties of a material and the relation between these properties and composition and structure. Finally, computational or theoretical chemistry involves the use of theoretical methods to calculate expected properties and so guide those doing experimental work. The work of any chemist usually involves several of these aspects, even though one may be predominant. Chemists may also call themselves organic chemists if they work primarily with compounds of carbon; inorganic chemists, if they work mostly with other elements; biochemists if they work with biological materials or systems; geochemists if they are concerned with geological materials; astrochemists if they study the chemistry of stars and other planets; and so on. There are many other combinations.

Chemical engineers, on the other hand, usually have different training than chemists. Both disciplines require knowledge of chemistry, but the chemical engineer is more concerned with practical applications, and there are differences in novelty and scale. A chemist is more likely to be developing new compounds and materials; a chemical engineer is more likely to be working with existing substances. A chemist may make a few grams of a new compound, while a chemical engineer will scale up the process to make it by the ton, and at a profit. The chemical engineer will be more concerned with heating and cooling large reaction vessels, pumps and piping to transfer materials, plant design and operation, and process optimization, while a chemist will be more concerned with establishing the details of the reactions before the plant is designed. These differences are generalizations; there is often much overlap.

The variety of fields in which a chemist can work is extensive. Because chemistry is such a broad science, chemists can work on the interface with many other sciences, and even move into other fields. The primary area, of course, is the chemical industry, pharmaceuticals, polymers and plastics, semiconductor and other solid-state materials, and related fields. Examples of activities include research, quality control and property testing, and customer service. In other areas, modern medicine depends heavily on chemistry and involves many chemists in drug development and testing. Forensic science has a very large chemistry component, and many forensic scientists are in fact chemists. These are just a few of the fields in which chemistry plays a role.

—**Ronald A. Bailey, Ph.D.**, is professor and associate chair
of the department of chemistry and chemical biology
at Rensselaer Polytechnic Institute in Troy, New York

INTRODUCTION

Facts On File's *Encyclopedia of Chemistry* is a reference to understanding the basic concepts in chemistry and its peripheral disciplines—crystallography; analytical, surface, physical, polymer, inorganic, and organic chemistry; bio-, geo-, and electrochemistry; and others. Arranged in alphabetical order, the entries include biographies of individuals who have made major contributions as well as numerous illustrations and photographs to help in visualizing technical concepts.

Chemistry is the study of matter—in its many forms—and the way these forms react with each other. It deals with the smallest of ions that are used in the human body to process energy, with the inner workings of the Earth's core, and even with the faraway study of the chemical composition of rocks on Mars. Chemistry is a pervasive science, or as an anonymous writer once wrote, "What in the world *isn't* chemistry?"

To understand basic chemistry is to have a healthy understanding of the complex world around us. It allows us to be amused by knowing that two of the most dangerous chemicals, sodium and chlorine, when combined, make our food taste good (salt, sodium chloride). It also helps us realize that combining chemicals released to the atmosphere by human activities can have serious health effects on us all, or that the wings of a butterfly may hold the key to a cure for cancer. Chemists today work hard to solve many of the leading problems in industry, agriculture, science, and health.

Throughout the pages of this encyclopedia, you will learn the definitions of chemical terms, literally from A to Z, and will read fascinating biographies of some of the leading chemists of the past and present. The volume also includes a set of essays by today's chemists on the role chemistry plays in our daily lives, ranging from how chemistry helps solve crimes to how it provides dyes for our latest fashions.

While this reference book is designed for high school– and college-level readers, it can also be used by anyone interested in chemistry and the various subdisciplines or by those who simply want to increase their scientific vocabulary. The encyclopedia also includes a set of helpful appendixes with information about Internet Web sites and chemistry-related software.

The most important element in using the *Encyclopedia of Chemistry* can be summarized by a line by novelist Ray Bradbury:

The best scientist is open to experience and begins with romance—the idea that anything is possible.

— Don Rittner
Schenectady, New York

ABO blood groups Blood group antibodies (A, B, AB, O) that can destroy red blood cells bearing the antigen to which they are directed; also called "agglutinins." These red-cell antigens are the phenotypic expression of inherited genes, and the frequency of the four main groups varies in populations throughout the world. The antigens of the ABO system are an integral part of the red-cell membrane as well as all cells throughout the body, and they are the most important in transfusion practice.

abscisic acid (ABA) A plant hormone ($C_{15}H_{20}O_4$) and weak acid that generally acts to inhibit growth, induces dormancy, and helps the plant tolerate stressful conditions by closing stomata. Abscisic acid was named based on a belief that the hormone caused the abscission (shedding) of leaves from deciduous trees during the fall.

At times when a plant needs to slow down growth and assume a resting stage (dormant), abscisic acid is produced in the terminal bud, which slows down growth and directs the leaf primordia to develop scales that protect the dormant bud during winter. Since the hormone also inhibits cell division in the vascular cambium, both primary and secondary growth is put on hold during winter.

This hormone also acts as a stress agent, helping a plant deal with adverse conditions. For example, ABA accumulates on leaves and causes stomata to close, reducing the loss of water when a plant begins to wilt.

In 1963 abscisic acid was first identified and characterized by Frederick Addicott and colleagues. In 1965 the chemical structure of ABA was defined, and in 1967 it was formally called abscisic acid.

absolute configuration The spatial arrangement of the atoms of a CHIRAL molecular entity (or group) and its STEREOCHEMICAL description.

absolute entropy (of a substance) The increase in the entropy of a substance going from a perfectly ordered crystalline form at 0 K (entropy is zero) to the temperature in question.

absolute zero Zero point on the absolute temperature scale, which is –273.15°C or 0 K; theoretically, the temperature at which molecular motion ceases.

absorption spectrum In biology, different pigments absorb light of different wavelengths. For example, chlorophyll effectively absorbs blue and red. The absorption spectrum of a pigment is produced by examining, through the pigment and an instrument called a spectroscope, a continuous spectrum of radiation. The energies removed from the continuous spectrum by the absorbing pigment show up as black lines or bands and can be graphed.

abstraction A CHEMICAL REACTION or TRANSFORMATION, the main feature of which is the bimolecular removal of an atom (neutral or charged) from a MOLECULAR ENTITY. For example: proton abstraction from acetone

$$CH_3COCH_3 + (i\text{-}C_3H_7)_2N^- \rightarrow (CH_3COCH_2)^- + (i\text{-}C_3H_7)_2NH$$

hydrogen abstraction from methane

$$CH_4 + Cl\cdot \rightarrow H_3C\cdot + HCl$$

See also DETACHMENT.

abyssal zone The portion of the ocean floor below 3,281–6,561 ft (1,000–2,000 m), where light does not penetrate and where temperatures are cold and pressures are intense. It lies seaward of the continental slope and covers approximately 75 percent of the ocean floor. The temperature does not rise above 4°C. Since oxygen is present, a diverse community of invertebrates and fishes do exist, and some have adapted to harsh environments such as hydrothermal vents of volcanic creation. Food-producing organisms at this depth are chemoautotrophic prokaryotes and not photosynthetic producers.

See also OCEANIC ZONE.

abzyme An antibody that catalyzes a chemical reaction similar to an enzymatic reaction. It promotes a chemical reaction by lowering the activation energy of a chemical reaction, yet remains unaltered at the end of the reaction.

See also CATALYTIC ANTIBODY.

acceptor number (AN) A quantitative measure of LEWIS ACIDITY.

accessory pigment Plant pigment other than chlorophyll that extends the range of light wavelengths useful in photosynthesis.

acclimatization The progressive physiological adjustment or adaptation by an organism to a change in an environmental factor, such as temperature, or conditions that would reduce the amount of oxygen to its cells. This adjustment can take place immediately or over a period of days or weeks. For example, the human body produces more erythrocytes (red blood cells) in response to low partial pressures of oxygen at high altitudes. Short-term responses include shivering or sweating in warm-blooded animals.

acetylcholine (ACh) One of the most common neurotransmitters of the vertebrate nervous system; a chemical ($CH_3COOCH_2CH_2N{+}(CH_3)_3$) that transmits impulses between the ends of two adjacent nerves or neuromuscular junctions. It is confined largely to the parasympathetic nervous system and is released by nerve stimulation (exciting or inhibiting), where it diffuses across the gap of the synapse and stimulates the adjacent nerve or muscle fiber. It rapidly becomes inactive by the enzyme cholinesterase, allowing further impulses to occur.

acetyl CoA A compound formed in the mitochondria when the thiol group (-SH) of coenzyme A combines with an acetyl group (CH_3CO-). It is important in the KREBS CYCLE in cellular respiration and plays a role in the synthesis and oxidation of fatty acids.

Fritz Albert Lipmann (1899–1986), a biochemist, is responsible for discovering coenzyme A and "cofactor A" (CoA, where A stands for "acetylation") in 1947. He shared the 1953 Nobel Prize for physiology or medicine with Hans Krebs.

achiral *See* CHIRALITY.

acid A chemical capable of donating a hydron (proton, H^+) or capable of forming a covalent bond with an electron pair. An acid increases the hydrogen ion concentration in a solution, and it can react with certain metals, such as zinc, to form hydrogen gas. A strong acid is a relatively good conductor of electricity. Examples of strong acids include hydrochloric (muriatic), nitric, and sulfuric, while examples of mild acids include sulfurous and acetic (vinegar). The strength of an acidic solution is usually measured in terms of its pH (a logarithmic function of the H^+ ion concentration). Strong acid solutions have low pHs (typically around 0–3), while weak acid solutions have pHs in the range 3–6.

See also BASE; BRONSTED ACID; HARD ACID; LEWIS ACID; PH SCALE; SOFT ACID.

acid anhydride The oxide of a nonmetal that reacts with water to form an acid.

acidic salt A salt containing an ionizable hydrogen atom that does not necessarily produce an acidic solution.

acidity

1. Of a compound:
   For BRONSTED ACIDs, acidity refers to the tendency of a compound to act as a HYDRON donor. It can be quantitatively expressed by the acid dissociation constant of the compound in water or some other specified medium. For LEWIS ACIDs it relates to the association constants of LEWIS ADDUCTs and π-ADDUCTs.
2. Of a medium:
   The use of the term is mainly restricted to a medium containing BRONSTED ACIDs, where it means the tendency of the medium to hydronate a specific reference base. It is quantitatively expressed by the appropriate ACIDITY FUNCTION.

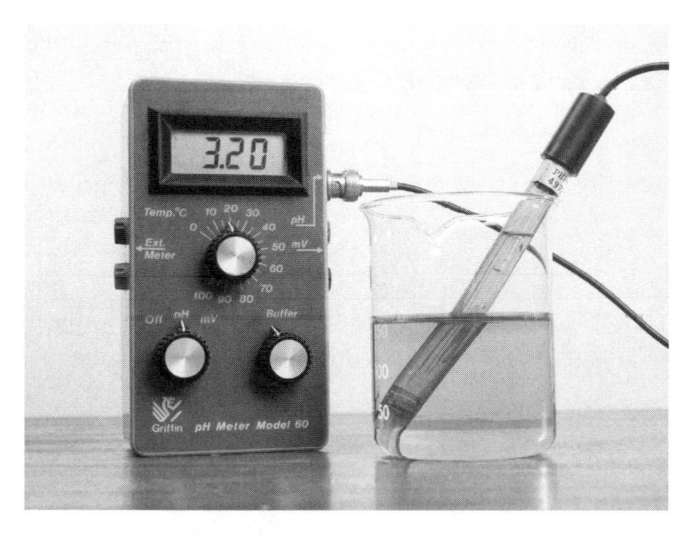

Measuring acidity using a pH meter (left), showing a pH of 3.20 for the solution in the beaker. Solution pH is the measure of the amount of hydrogen ions in a solution and is used to measure its alkalinity or acidity. The probe of the pH meter (right), dipped in the solution, contains electrodes used to measure the electrical potential of the hydrogen ions. This is directly related to the pH. The temperature dial is adjusted to the room temperature, allowing the pH meter to compensate for the temperature. This pH is acidic. The solution contains aluminum ions and bromocresol green, an indicator that displays a yellow color in acidic solutions. *(Courtesy of Andrew Lambert Photography/Science Photo Library)*

acidity function Any function that measures the thermodynamic HYDRON-donating or -accepting ability of a solvent system, or a closely related thermodynamic property, such as the tendency of the LYATE ION of the solvent system to form LEWIS ADDUCTs. (The term *basicity function* is not in common use in connection with basic solutions.) Acidity functions are not unique properties of the solvent system alone but depend on the solute (or family of closely related solutes) with respect to which the thermodynamic tendency is measured. Commonly used acidity functions refer to concentrated acidic or basic solutions. Acidity functions are usually established over a range of composition of such a system by UV/VIS spectrophotometric or NMR measurements of the degree of hydronation (protonation or Lewis adduct formation) for the members of a series of structurally similar indicator bases (or acids) of different strength. The best known of these functions is the HAMMETT ACIDITY FUNCTION, H_0 (for uncharged indicator bases that are primary aromatic amines).

See also NUCLEAR MAGNETIC SPECTROSCOPY; ULTRAVIOLET SPECTRUM.

acidity constant The equilibrium constant for splitting off a HYDRON from a BRONSTED ACID.

acid-labile sulfide Refers to sulfide LIGANDs, for example, the BRIDGING LIGANDs in IRON-SULFUR PROTEINs, which are released as H_2S at acid pH.

See also FERREDOXIN.

acid precipitation Because pure precipitation is slightly acidic (due to the reaction between water droplets and carbon dioxide, creating carbonic acid), with a potential hydrogen (pH) of 5.6, acid precipitation refers to precipitation with a pH less than 5.6. Acid precipitation includes rain, fog, snow, and dry deposition. Anthropogenic (human-made) pollutants (carbon dioxide, carbon monoxide, ozone, nitrogen and sulfur oxides, and hydrocarbons) react with water vapor to produce acid precipitation. These pollutants come primarily from burning coal and other fossil fuels. Sulfur dioxide, which reacts readily with water vapor and droplets (i.e., has a short residence time in the atmosphere as a gas), has been linked to the weathering (eating away) of marble structures and the acidification of freshwater lakes (consequently killing fish). Natural interactions within the biosphere can also lead to acid precipitation.

aconitase A name for citrate (isocitrate) hydro-LYASE (aconitate hydratase), which catalyzes the interconversion of citrate, *cis*-aconitate ([*Z*]-prop-1-ene-1,2,3-tricarboxylate), and isocitrate. The active ENZYME contains a catalytic [4FE-4S] CLUSTER.

actinides Elements 90 to 103 (after actinium).

action potential A localized rapid change in voltage that occurs across the membrane of a muscle or nerve cell when a nerve impulse is initiated. It is caused by a physicochemical change in the membrane during the interaction of the flow and exchange of sodium and potassium ions.

See also DEPOLARIZATION.

activated complex An activated complex, often characterized by the superscript ‡, is defined as that assembly of atoms corresponding to an arbitrary, infinitesimally small region at or near the col (saddle point) of a POTENTIAL ENERGY surface.

See also TRANSITION STATE.

activation energy *See* ENERGY OF ACTIVATION.

active center The location in an ENZYME where the specific reaction takes place.

active metal A metal with low ionization energy that loses electrons readily to form cations.

active site *See* ACTIVE CENTER.

active transport The movement of a substance across a biological membrane, such as living cells, against a concentration (diffusion) gradient with the help of metabolic energy usually provided by ATP (adenosine triphosphate). Active transport serves to maintain the normal balance of ions in cells and, in particular, those of sodium and potassium ions, which play a vital role in nerve and muscle cells. Since a molecule is "pumped" across the membrane against its gradient, requiring the use of metabolic energy, it is referred to as "active" transport.

The sodium-potassium "pump," which exchanges sodium (Na^+) for potassium (K^+) across the plasma membrane of animal cells, is an example of the active transport mechanism.

The carriage of a solute across a biological membrane from low to high concentration requires the expenditure of metabolic energy.

activity series A listing of metals (and hydrogen) in order of decreasing activity.

actual yield Amount of a specified pure product obtained from a given reaction.

acyl group Compound derived from a carbonic acid, usually by replacing the –OH group with a halogen (X).

addend *See* ADDITION REACTION.

addition

1. Refers to ADDITION REACTION or addition TRANSFORMATION
2. Loosely defined, the formation of an ADDUCT, e.g., a LEWIS ACID
3. Loosely defined, any ASSOCIATION or ATTACHMENT.

addition reaction A chemical reaction of two or more reacting MOLECULAR ENTITIES, resulting in a single reaction product containing all atoms of all components, with formation of two chemical bonds and a net reduction in bond multiplicity in at least one of the reactants. The reverse process is called an ELIMINATION reaction. The addition may occur at only one site (α-ADDITION, 1/1 addition), at two adjacent sites (1/2 addition), or at two nonadjacent sites (1/3 or 1/4 addition, etc.). For example:

(a) $H^+ + Br^- + (CH_3)_2C{=}CH_2 \rightarrow (CH_3)_2CBr{-}CH_3$ (1/2 addition)
(b) $Br_2 + CH_2{=}CH{-}CH{=}CH_2 \rightarrow$ $BrCH_2{-}CH{=}CH{-}CH_2Br$ (1/4 addition) and $BrCH_2{-}CH(Br){-}CH{=}CH_2$ (1/2 addition)

If the reagent or the source of the addends of an addition are not specified, then it is called an addition TRANSFORMATION.

See also ADDITION; CHELETROPIC REACTION; CYCLOADDITION.

additivity principle The hypothesis that each of several structural features of a MOLECULAR ENTITY makes a separate and additive contribution to a property of the substance concerned. More specifically, it is the hypothesis that each of the several SUBSTITUENT groups in a parent molecule makes a separate and additive contribution to the standard Gibbs energy change (or GIBBS ENERGY OF ACTIVATION) corresponding to a particular equilibrium (or RATE OF REACTION).

See also TRANSFERABILITY.

address–message concept Refers to compounds in which part of the molecule is required for binding (address) and part for the biological action (message).

adduct A new CHEMICAL SPECIES, AB, each MOLECULAR ENTITY of which is formed by direct combination of two separate molecular entities A and B in such a way that there is change in CONNECTIVITY, but no loss, of atoms within the MOIETIES A and B. Stoichiometries other than 1:1 are also possible, e.g., a bis-adduct (2:1). An "INTRAMOLECULAR adduct" can be formed when A and B are GROUPs contained within the same molecular entity.

This is a general term that, whenever appropriate, should be used in preference to the less explicit term *COMPLEX*. It is also used specifically for products of an ADDITION REACTION. For examples, see π-ADDUCT, LEWIS ADDUCT, and MEISENHEIMER ADDUCT.

adenosine 5′-triphosphate (ATP) Key NUCLEOTIDE in energy-dependent cellular reactions, in combination with Mg(II). The reaction of ATP + water → ADP + phosphate is used to supply the necessary energy.

See also ATP.

adenyl cyclase An enzyme, embedded in the plasma membrane, that converts ATP to cyclic adenosine monophosphate (cyclic AMP [cAMP]) in response to a chemical signal. It is activated when a signal molecule binds to a membrane receptor. Cyclic AMP acts as a second messenger relaying the signal from the membrane to the metabolic machinery of the cytoplasm.

adhesive forces Those forces of attraction that occur between a liquid and another surface.

ADME Abbreviation for absorption, distribution, METABOLISM, excretion.

See also PHARMACOKINETICS; DRUG DISPOSITION.

adrenodoxin A [2FE-2S] FERREDOXIN involved in electron transfer from $NADPH^+$ (the reduced form of NADP [nicotinamide adenine dinucleotide phosphate, a coenzyme]), via a REDUCTASE, to CYTOCHROME P-450 in the adrenal gland.

Adrian, Edgar Douglas (1889–1977) British *Physiologist* Edgar Douglas Adrian was born on November 30, 1889, in London to Alfred Douglas Adrian, a legal adviser to the British local government board. He went to school at Westminster School, London, and in 1908 he entered Trinity College, Cambridge. At Cambridge University, he studied physiology, receiving a bachelor's degree in 1911.

In 1913 he entered Trinity College, studied medicine, did his clinical work at St. Bartholomew's Hospital, London, and received his M.D. in 1915.

In 1929 he was elected Foulerton professor of the Royal Society, and in 1937 he became professor of physiology at the University of Cambridge until 1951, when he was elected master of Trinity College, Cambridge. He was chancellor of the university from 1968 until two years before his death.

He spent most of his research studying the physiology of the human nervous system, particularly the brain, and how neurons send messages. In 1932 he shared the Nobel Prize in physiology or medicine for his work on the function of the neuron. He is considered one of the founders of modern neurophysiology.

He wrote three books, *The Basis of Sensation* (1927), *The Mechanism of Nervous Action* (1932), and *The Physical Basis of Perception* (1947), and was knighted Baron of Cambridge in 1955. He died on August 4, 1977, and is buried at Trinity College.

adsorption Adhesion of a material onto the surfaces of particles.

aerobe An organism that needs dioxygen for respiration and thus for growth.

aerobic Any organism, environmental condition, or cellular process that requires atmospheric oxygen. Aerobic microorganisms, called aerobes, require the presence of oxygen for growth. An aerobe is capable of using oxygen as a terminal electron acceptor and can tolerate oxygen levels higher than that present in the air (21 percent oxygen). They have a respiratory type of metabolism, and some aerobes may also be capable of growing anaerobically with electron accepters other than oxygen.

See also ANAEROBIC.

***A*-factor** *See* ENERGY OF ACTIVATION.

affinity The tendency of a molecule to associate with another. The affinity of a DRUG is its ability to bind to its biological target (RECEPTOR, ENZYME, transport system, etc.). For pharmacological receptors, it can be thought of as the frequency with which the drug, when brought into the proximity of a receptor by diffusion, will reside at a position of minimum free energy within the force field of that receptor.

For an AGONIST (or for an ANTAGONIST), the numerical representation of affinity is the reciprocal of the equilibrium dissociation constant of the ligand-receptor complex, denoted K_A, calculated as the rate constant for offset (k_{-1}) divided by the rate constant for onset (k_1).

agonist An endogenous substance or a DRUG that can interact with a RECEPTOR and initiate a physiological or a pharmacological response characteristic of that receptor (contraction, relaxation, secretion, ENZYME activation, etc.).

agostic The term designates structures in which a hydrogen atom is bonded to both a carbon atom and a metal atom. The term is also used to characterize the interaction between a CH bond and an unsaturated metal center, and to describe similar bonding of a transition metal with Si-H compounds. The expression "μ-hydrido-bridged" is also used to describe the bridging hydrogen.

AIDS (acquired immunodeficiency syndrome) AIDS is the name given to the late stages of HIV (human immunodeficiency virus) infection, a disease discovered and discussed in 1981 in Los Angeles, California. By 1983 the retrovirus responsible for HIV was first described, and since then millions of adults and children worldwide have died from contracting the disease. It is thought to have originated in central Africa from monkeys or developed from contaminated vaccines used in the world's first mass immunization for polio.

AIDS is acquired mostly by unprotected sexual contact—either through homosexual or heterosexual practice—via vaginal and anal intercourse. The routes of infection include infected blood, semen, and vaginal fluid. The virus can also be transmitted by blood by-products through maternofetal infection (where the virus is transmitted by an infected mother to the unborn child in the uterus), by maternal blood during parturition, or by breast milk consumption upon birth. Intravenous drug abuse is also a cause.

The virus destroys a subgroup of lymphocytes that are essential for combating infections, known as helper T cells or CD4 lymphocytes, and suppresses the body's immune system, leaving it prone to infection.

Infection by the virus produces antibodies, but not all of those exposed develop chronic infection. For those that do, AIDS or AIDS-related complex (ARC) bring on a variety of ailments involving the lymph nodes, intermittent fever, loss of weight, diarrhea, fatigue, pneumonia, and tumors. A person infected, known as HIV-positive, can remain disease-free for up to 10 years, as the virus can remain dormant before full-blown AIDS develops.

While HIV has been isolated in substances ranging from semen to breast milk, the virus does not survive outside the body, and it is considered highly unlikely that ordinary social contact can spread the disease. However, the medical profession has developed high standards to deal with handling blood, blood products, and body fluids from HIV-infected people.

In the early-discovery stage of the disease, AIDS was almost certainly fatal, but the development of antiviral drugs such as zidovudine (AZT), didanosine (ddl), zalcitabine (ddc), lamivudine (3TC), stavudine (DAT), and protease inhibitors, used alone or in combination, has showed promise in slowing or eradicating the disease. Initial problems with finding a cure have to do with the fact that glycoproteins encasing the virus display a great deal of variability in their amino-acid sequences, making it difficult to prepare a specific AIDS vaccine.

During the 1980s and 1990s, an AIDS epidemic brought considerable media coverage to the disease as well-known celebrities—such as Rock Hudson, Anthony Perkins, Liberace, and others—died from it. Hudson was the first to admit having the disease in 1985. The gay community became the first population active in lobbying for funds to study the disease, especially as the media naively nicknamed AIDS a "gay" disease. ACT UP, acronym for the AIDS Coalition to Unleash Power, began as a grassroots AIDS organization associated with nonviolent civil disobedience in 1987. ACT UP became the standard-bearer for protest against governmental and societal indifference to the AIDS epidemic. The public attitude changed when heterosexuals became infected, and education on the causes of the disease became more widespread, initiated by celebrities such as Elizabeth Taylor and the American Foundation for AIDS Research, whose fund-raising activities made national news coverage.

There have been significant advances in the treatment for HIV/AIDS by attacking the virus itself, strengthening the immune system, and controlling AIDS-related cancers and opportunistic infections. At present, there is still no cure or vaccine.

albumin A type of protein, especially a protein of blood PLASMA, that transports various substances, including metal ions, drugs, and XENOBIOTICS.

alcohol A hydrocarbon derivative containing an –OH group attached to a carbon atom not in an aromatic ring. See figure on page 8.

alcoholysis *See* SOLVOLYSIS.

aldehydes Aldehydes are organic chemicals that contain the –CHO (aldehyde) group, a carbonyl group (C=O) that has the carbon and hydrogen atom bound. They are the result of the oxidation of alcohols and, when further oxidized, form carboxylic acids. Methanal (formaldehyde) and ethanal (acetaldehyde) are common examples.

alkali metals (Group 1 elements) A group of soft reactive metals, each representing the start of a new period in the periodic table and having an electronic configuration consisting of a rare gas structure plus one outer electron. The metals in this group are cesium (CS), lithium (Li), sodium (Na), potassium (K), rubidium (Rb), and francium (Fr).

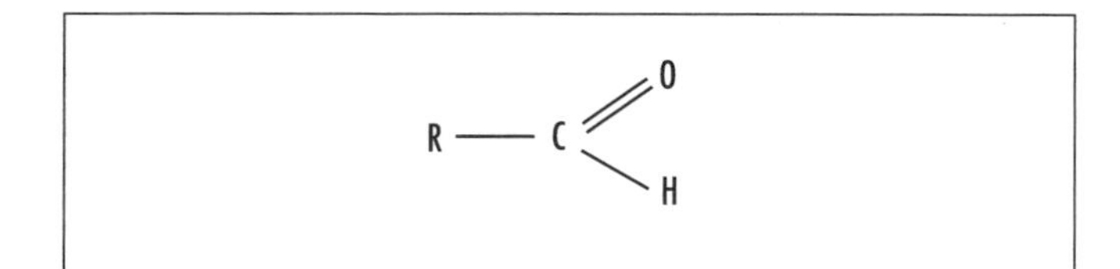

Aldehydes are formed by oxidizing a primary alcohol. Formaldehyde (HCHO) is an example.

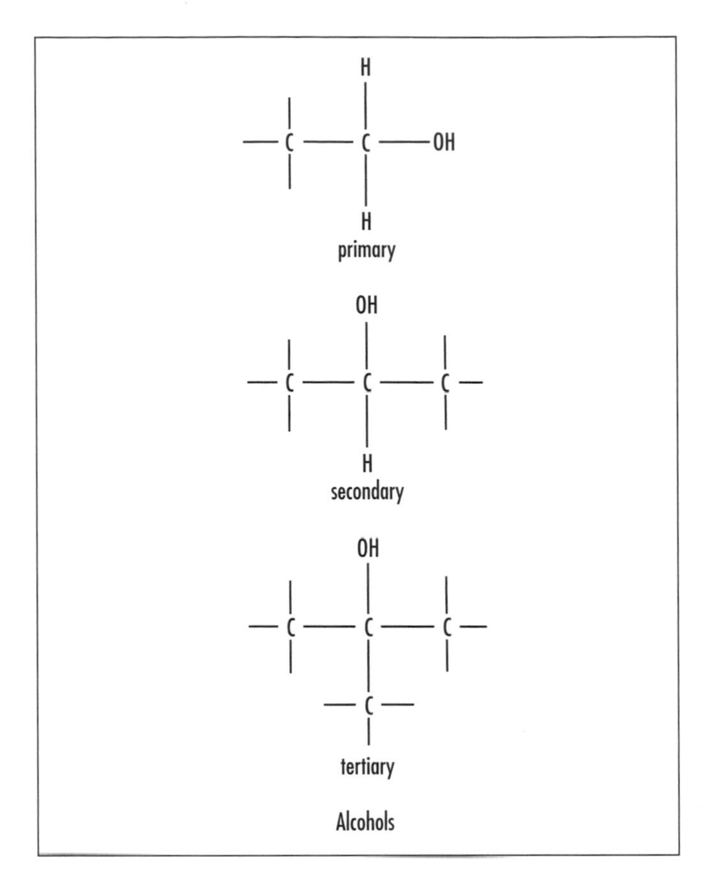

Alcohols

alkaline battery A dry cell.

alkaline earth metals The Group IIA metals.

alkenes (olefins) Unsaturated hydrocarbons containing one or more carbon-carbon double bonds.

alkylbenzene Compound containing an alkyl group bonded to a benzene ring.

alkyl group A group of atoms derived from an alkane by the removal of one hydrogen atom.

alkynes Unsaturated hydrocarbons containing one or more carbon-carbon triple bonds.

allosteric binding sites Contained in many ENZYMES and RECEPTORS. As a consequence of the binding to allosteric binding sites, the interaction with the normal ligand (ligands are molecules that bind to proteins) can be either enhanced or reduced. Ligand binding can change the shape of a protein.

allosteric effector Specific small molecules that bind to a protein at a site other than a catalytic site and modulate (by activation or INHIBITION) the biological activity.

allosteric enzyme An ENZYME that contains a region, separate from the region that binds the SUBSTRATE for catalysis, where a small regulatory molecule binds and affects that catalytic activity. This effector molecule may be structurally unrelated to the substrate or may be a second molecule of substrate. If the catalytic activity is enhanced by binding, the effector is called an activator; if it is diminished, the effector is called an INHIBITOR.

See also REGULATION.

allosteric regulation The regulation of the activity of allosteric ENZYMES.

See also ALLOSTERIC BINDING SITES; ALLOSTERIC ENZYME.

allosteric site A specific receptor site on an enzyme molecule, not to be confused with the active site (the site on the surface of an enzyme molecule that binds the substrate molecule). Molecules bind to the allosteric site and change the shape of the active site, either enabling the substrate to bind to the active site or preventing the binding of the substrate.

The molecule that binds to the allosteric site is an inhibitor because it causes a change in the three-dimensional structure of the enzyme that prevents the substrate from binding to the active site.

allotrope (allotropic modification) A different form of the same element in the same physical state.

alloy A mix of metal with other substances (usually other metals) that modify its properties.

allozyme An enzyme form, a variant of the same enzyme (protein) that is coded for by different alleles at a single locus.

See also ENZYME.

allylic substitution reaction A SUBSTITUTION REACTION at position 1/ of an allylic system, the double bond being between positions 2/ and 3/. The incoming group may be attached to the same atom 1/ as the LEAVING GROUP, or the incoming group becomes attached at the relative position 3/, with movement of the double bond from 2/3 to 1/2. For example:

$$CH_3CH{=}CHCH_2Br \rightarrow CH_3CH{=}CHCH_2OAc$$

or

$$CH_3CH{=}CHCH_2Br \rightarrow CH_3CH(OAc)CH{=}CH_2$$

(written as a TRANSFORMATION).

alpha (α) addition A CHEMICAL REACTION resulting in a single reaction product from two or three reacting chemical species, with formation of two new chemical BONDs to the same atom in one of the reactant MOLECULAR ENTITIES.

The synonymous term 1/1/addition is also used. For example:

$$Cl_2C{:} + CH_3OH \rightarrow Cl_2CHOCH_3$$

(This particular example can also be viewed as an INSERTION REACTION.) In inorganic chemistry, such α-addition reactions, generally to a metallic central atom, are known as "oxidative additions."

α-Addition is the reverse of α-ELIMINATION or 1/1/elimination.

See also ADDITION; ELIMINATION.

alpha (α) effect A positive deviation of a nucleophile (a nucleophile bearing an unshared pair of electrons on an atom adjacent to the nucleophilic site) from a Bronsted-type plot of lg k_{nuc} vs. pKa constructed for a series of related normal nucleophiles. More generally, it is the influence of the atom bearing a lone pair of electrons on the reactivity at the adjacent site.

The use of the term has been extended to include the effect of any substituent on an adjacent reactive center, for example in the case of the "α-silicon effect."

See also BRONSTED RELATION.

alpha (α) elimination A TRANSFORMATION of the general type

$$RR'ZXY \rightarrow RR'Z + XY \text{ (or } X + Y\text{, or } X^+ + Y^-\text{)}$$

where the central atom Z is commonly carbon. The reverse reaction is called α-ADDITION.

alpha (α) helix Most proteins contain one or more stretches of amino acids that take on a particular shape in three-dimensional space. The most common forms are alpha helix and beta sheet.

An alpha helix is spiral shaped, constituting one form of the secondary structure of proteins, arising from a specific hydrogen-bonding structure; the carbonyl group (–C=O) of each peptide bond extends parallel to the axis of the helix and points directly at the –N–H group of the peptide bond four amino acids below it in the helix. A hydrogen bond forms between them and plays a role in stabilizing the helix conformation. The alpha helix is right-handed and twists clockwise, like a corkscrew, and makes a complete turn every 3.6 amino acids. The distance between two turns is 0.54 nm. However, an alpha helix can also be left-handed. Most enzymes contains sections of alpha helix.

The alpha helix was discovered by LINUS PAULING in 1948.

See also HELIX.

alpha (α) particle A helium nucleus or ion with 2+ charge.

alternant A CONJUGATED SYSTEM of pi electrons is termed *alternant* if its atoms can be divided into two sets so that no atom of one set is directly linked to any other atom of the same set.

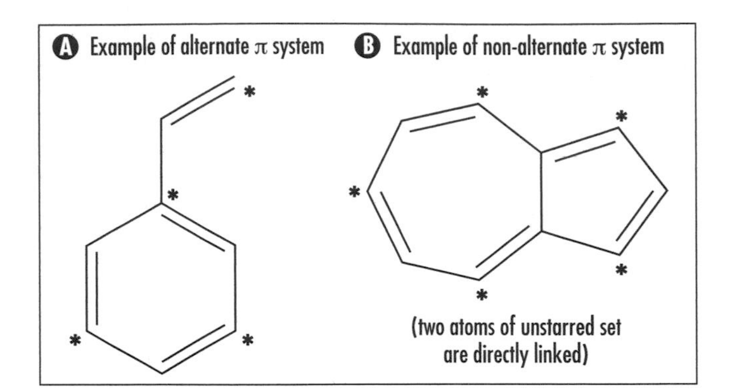

alums Hydrated sulfates of the general formula $M+M_3+(SO_4)_2.12H_2O$.

ambident A description applied to a CHEMICAL SPECIES whose MOLECULAR ENTITIES each possess two alternative and strongly interacting distinguishable reactive centers, to either of which a BOND may be made in a reaction: the centers must be connected in such a way that reaction at either site stops or greatly retards subsequent attack at the second site. The term is most commonly applied to conjugated NUCLEOPHILEs, for example the enolate ion (that may react

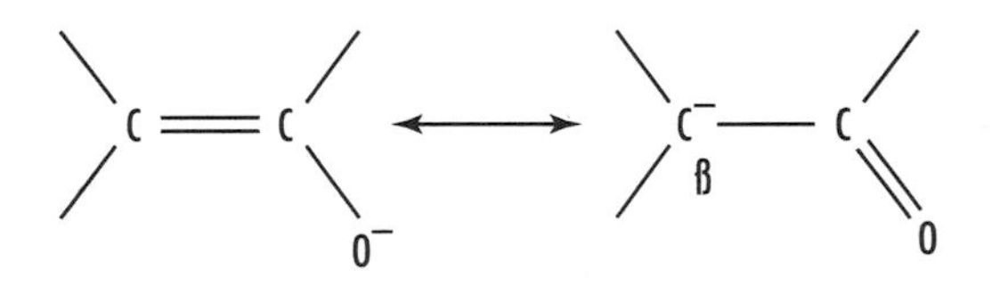

with ELECROPHILEs either at the β-carbon atom or at oxygen) or γ-pyridones, and also to the vicinally ambident cyanide ion, cyanate ion, thiocyanate ion, sulfinate ion, nitrite ion, and unsymmetrical hydrazines. Ambident electrophiles are exemplified by carboxylic esters $RC(O)OCR_3$, which react with nucleophiles either at the carbonyl carbon or the alkoxy carbon. Molecular entities such as dianions of dicarboxylic acids, containing two noninteracting (or feebly interacting) reactive centers, are not generally considered to be ambident and are better described as "bifunctional."

The Latin root of the word implies two reactive centers, but the term has in the past also incorrectly been applied to chemical species with more than two reactive centers. For such species the existing term *polydent* (or, better, *multident*) is more appropriate.

See also CHELATION.

ambidentate LIGANDs, such as $(NCS)^-$, that can bond to a CENTRAL ATOM through either of two or more donor atoms are termed *ambidentate.*

amicyanin An ELECTRON-TRANSFER PROTEIN containing a TYPE 1 COPPER site, isolated from certain bacteria.

amide Compound containing the O–C–N group; a derivative of ammonia in which one or more hydrogens are replaced by alkyl or aryl groups.

amine A compound containing a nitrogen atom bound to hydrogen atoms or hydrocarbon groups. It is

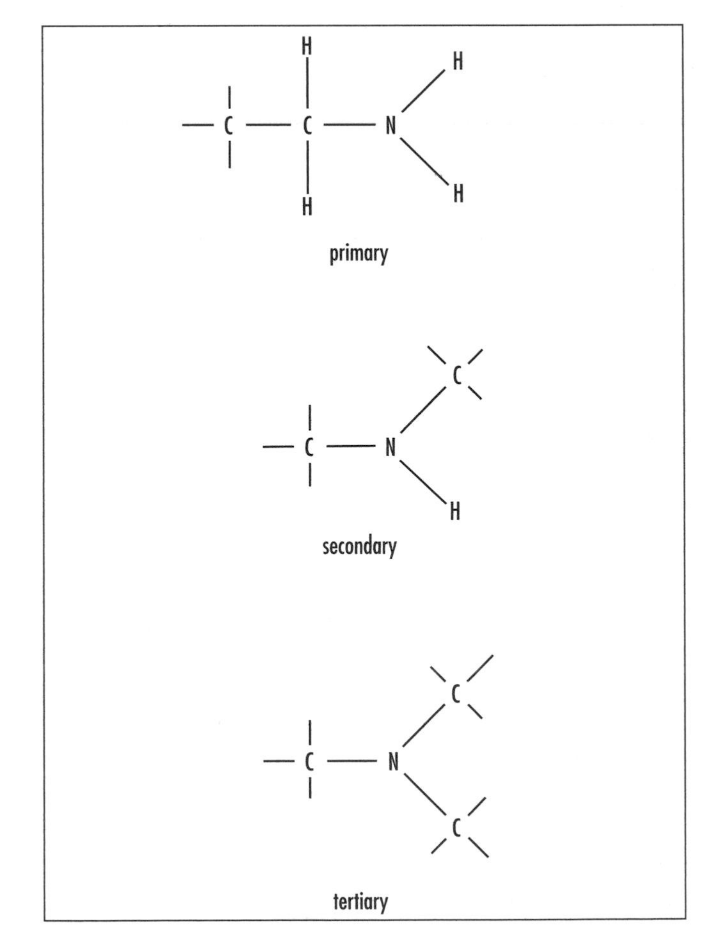

Amines. Compounds with a nitrogen atom bound to hydrogen atoms or hydrocarbon groups

a derivative of ammonia in which one or more hydrogen atoms have been replaced by organic groups. It is classified according to the number of organic groups bonded to the nitrogen atom: one, primary; two, secondary; three, tertiary.

amine complex Complex species containing ammonia molecules bonded to metal ions.

amino acid An organic molecule possessing both acidic carboxylic acid (–COOH) and basic amino (–NH_2) groups attached to the same tetrahedral carbon atom.

Amino acids are the principal building blocks of proteins and enzymes. They are incorporated into proteins by transfer RNA according to the genetic code while messenger RNA is being decoded by ribosomes. The amino acid content dictates the spatial and biochemical properties of the protein or enzyme during and after the final assembly of a protein. Amino acids have an average molecular weight of about 135 daltons. While more than 50 have been discovered, 20 are essential for making proteins, long chains of bonded amino acids.

Some naturally occurring amino acids are alanine, arginine, asparagine, aspartic acid, cysteine, glutamine, glutamic acid, glycine, histidine, isoleucine, leucine, lysine, methionine, phenylalanine, proline, serine, threonine, tryptophan, tyrosine, and valine.

The two classes of amino acids that exist are based on whether the R-group is hydrophobic or hydrophilic. Hydrophobic or nonpolar amino acids tend to repel the aqueous environment and are located mostly in the interior of proteins. They do not ionize or participate in the formation of hydrogen bonds. On the other hand, the hydrophilic or polar amino acids tend to interact with the aqueous environment, are usually involved in the formation of hydrogen bonds, and are usually found on the exterior surfaces of proteins or in their reactive centers. It is for this reason that certain amino acid R-groups allow enzyme reactions to occur.

The hydrophilic amino acids can be further subdivided into polar with no charge, polar with negatively charged side chains (acidic), and polar with positively charged side chains (basic).

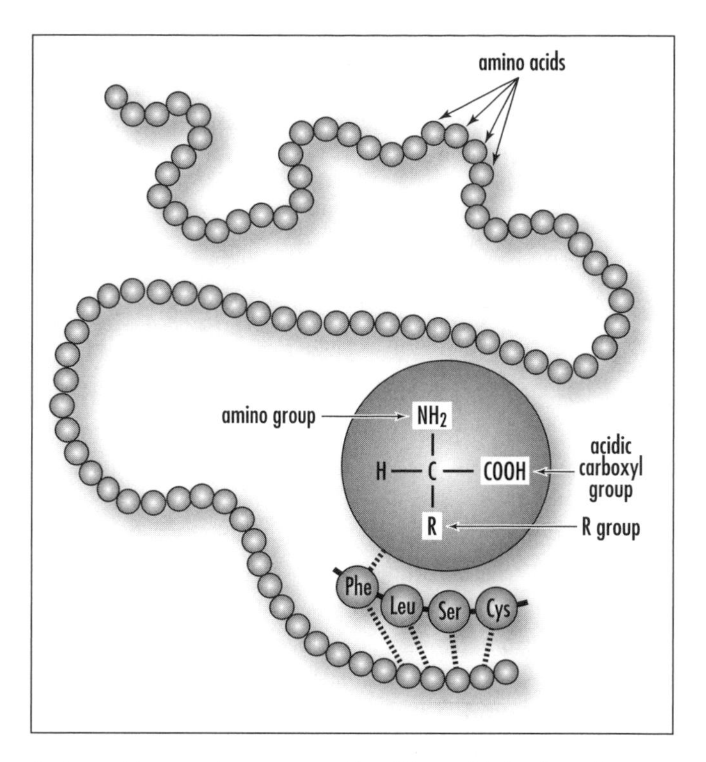

Amino acids comprise a group of 20 different kinds of small molecules that link together in long chains to form proteins. Often referred to as the "building blocks" of proteins. ***(Courtesy of Darryl Leja, NHGRI, National Institutes of Health)***

While all amino acids share some structural similarities, it is the side groups, or "R"-groups as they are called, that make the various amino acids chemically and physically different from each other so that they react differently with the environment. These groupings, found among the 20 naturally occurring amino acids, are ionic (aspartic acid, arginine, glutamic acid, lysine, and histidine), polar (asparagine, serine, threonine, cysteine, tyrosine, and glutamine), and nonpolar amino acids (alanine, glycine, valine, leucine, isoleucine, methionine, phenylalanine, tryptophan, and proline).

Amino acids are also referred to as amphoteric, meaning they can react with both acids and alkali, which makes them effective buffers in biological systems. A buffer is a solution where the pH usually stays constant when an acid or base is added.

In 1986 scientists found a 21st amino acid, selenocysteine. In 2002 two teams of researchers from Ohio State University identified the 22nd genetically encoded amino acid, called pyrrolysine, a discovery that is the biological equivalent of physicists finding a new fundamental particle or chemists discovering a new element.

Amino acid supplements are widely used in exercise and dietary programs.

See also PROTEIN.

amino acid residue (in a polypeptide) When two or more amino acids combine to form a peptide, the elements of water are removed, and what remains of each amino acid is called amino acid residue. Amino acid residues are therefore structures that lack a hydrogen atom of the amino group (–NH–CHR–COOH), or the hydroxy moiety of the carboxy group (NH_2–CHR–CO–), or both (–NH–CHR–CO–); all units of a peptide chain are therefore amino acid residues. (Residues of amino acids that contain two amino groups or two carboxy groups may be joined by isopeptide bonds, and so may not have the formulas shown.) The residue in a peptide that has an amino group that is free, or at least not acylated by another amino acid residue (it may, for example, be acylated or formylated), is called N-terminal; it is the N-terminus. The residue that has a free carboxy group, or at least does not acylate another amino acid residue (it may, for example, acylate ammonia to give –NH–CHR–CO–NH_2), is called C-terminal.

The following is a list of symbols for amino acids (use of the one-letter symbols should be restricted to the comparison of long sequences):

A	Ala	Alanine
B	Asx	Asparagine or aspartic acid
C	Cys	Cysteine
D	Asp	Aspartic acid
E	Glu	Glutamic acid
F	Phe	Phenylalanine
G	Gly	Glycine
H	His	Histidine
I	Ile	Isoleucine
K	Lys	Lysine
L	Leu	Leucine
M	Met	Methionine
N	Asn	Asparagine
P	Pro	Proline
Q	Gln	Glutamine
R	Arg	Arginine
S	Ser	Serine
T	Thr	Threonine
V	Val	Valine
W	Trp	Tryptophan
Y	Tyr	Tyrosine
Z	Glx	Glutamine or glutamic acid

amino group (–NH_2) A functional group (group of atoms within a molecule that is responsible for certain properties of the molecule and reactions in which it takes part), common to all amino acids, that consists of a nitrogen atom bonded covalently to two hydrogen atoms, leaving a lone valence electron on the nitrogen atom capable of bonding to another atom. It can act as a base in solution by accepting a hydrogen ion and carrying a charge of +1. Any organic compound that has an amino group is called an amine and is a derivative of the inorganic compound ammonia, NH_3. A primary amine has one hydrogen atom replaced, such as in the amino group. A secondary amine has two hydrogens replaced. A tertiary amine has all three hydrogens replaced. Amines are created by decomposing organic matter.

amorphous solid Noncrystalline solid having no well-defined ordered structure.

ampere Unit of electrical current; one ampere (A) equals 1 coulomb per second.

amphipathic molecule A molecule that has both a hydrophilic (water soluble, polar) region and a hydrophobic (water hating, nonpolar) region. The hydrophilic part is called the head, while the hydrophobic part is called the tail. Lipids (phospholipids, cholesterol and other sterols, glycolipids [lipids with sugars attached], and sphingolipids) are examples of amphipathic molecules.

Amphipathic molecules act as surfactants, materials that can reduce the surface tension of a liquid at low concentrations. They are used in wetting agents, demisters, foaming agents, and demulsifiers.

amphiphilic A compound containing a large organic cation or anion that possesses a long unbranched hydrocarbon chain, for example

$H_3C(CH_2)_nCO_2^-M^+H_3C(CH_2)_nSO_3^-M^+H_3C(CH_2)_nN(CH_3)_3^+X^-(n > 7)$.

The existence of distinct polar (hydrophilic) and nonpolar (hydrophobic) regions in the molecule promotes the formation MICELLEs in dilute aqueous solution.

amphiprotic solvent Self-ionizing solvent possessing both characteristics of BRONSTED ACIDs and BRONSTED BASEs, for example H_2O and CH_3OH, in contrast to an APROTIC SOLVENT.

amphoteric A CHEMICAL SPECIES that behaves both as an acid and as a base is called *amphoteric.* This property depends upon the medium in which the species is investigated: H_2SO_4 is an acid when studied in water, but becomes amphoteric in SUPERACIDs.

amphoterism The ability to react with both acids and bases.

anabolism The processes of METABOLISM that result in the synthesis of cellular components from precursors of low molecular weight.

anaerobe An organism that does not need dioxygen for growth. Many anaerobes are even sensitive to dioxygen. Obligate (strict) anaerobes grow only in the absence of dioxygen. Facultative anaerobes can grow either in the presence or in the absence of dioxygen.

anaerobic Any organism or any environmental or cellular process that does not require the use of free oxygen. Certain bacteria such as the *Actinomyces israeli, Bacteroides fragilis, Prevotella melaninogenica, Clostridium difficile,* and *Peptostreptococcus* are ANAEROBES.

The term refers to an organism that does not need dioxygen for growth. Many anaerobes are even sensitive to dioxygen. Obligate (strict) anaerobes grow only in the absence of dioxygen. Facultative anaerobes can grow either in the presence or in the absence of dioxygen.

See also AEROBIC.

analog A DRUG whose structure is related to that of another drug but whose chemical and biological properties may be quite different.

See also CONGENER.

anation Replacement of the LIGAND water by an anion in a COORDINATION entity.

anchimeric assistance *See* NEIGHBORING-GROUP PARTICIPATION.

anemia Condition in which there is a reduction in the number of red blood cells or amount of HEMOGLOBIN per unit volume of blood below the reference interval for a similar individual of the species under consideration, often causing pallor and fatigue.

anion An atom or molecule that has a negative charge; a negatively charged ion.

See also ION.

anionotropic rearrangement (anionotropy) A rearrangement in which the migrating group moves with its electron pair from one atom to another.

anion radical *See* RADICAL ION.

anisotropy The property of molecules and materials to exhibit variations in physical properties along different molecular axes of the substance.

annelation Alternative, but less desirable, term for ANNULATION. The term is widely used in the German and French languages.

annulation A TRANSFORMATION; involving fusion of a new ring to a molecule via two new bonds. Some authors use the term *annelation* for the fusion of an additional ring to an already existing one, and *annulation* for the formation of a ring from one or several acyclic precursors, but this distinction is not generally made.

See also CYCLIZATION.

annulene Mancude (i.e., having formally the maximum number of noncumulative double bonds) monocyclic hydrocarbon without side chains of the general formula C_nH_n (n is an even number) or C_nH_n+1 (n is an odd number). Note that in systematic nomenclature, an annulene with seven or more carbon atoms may be named [n]annulene, where n is the number of carbon atoms, e.g., [9]annulene for cyclonona-1,3,5,7-tetraene.

See also AROMATIC HYDROCARBONS; HUCKEL (4N+2) RULE.

anode In a cathode ray tube, the positive electrode. Electrode at which oxidation occurs.

antagonist A DRUG or a compound that opposes the physiological effects of another. At the RECEPTOR level, it is a chemical entity that opposes the receptor-associated responses normally induced by another bioactive agent.

antarafacial, suprafacial When a part of a molecule (molecular fragment) undergoes two changes in bonding (BOND making or bond breaking), either to a common

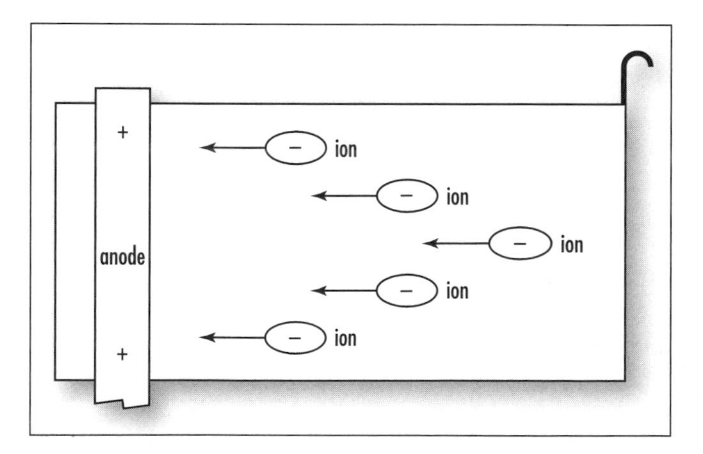

Anodes carry the positive charge.

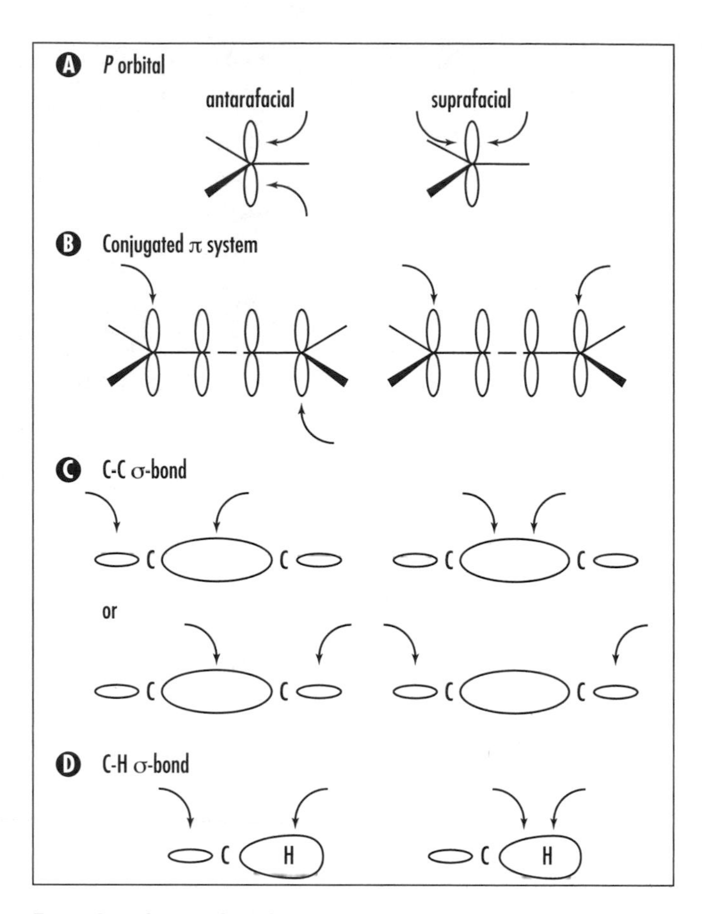

Examples of antarafacial, suprafacial

center or to two related centers, external to itself, these bonding changes may be related in one of two spatially different ways. These are designated as "antarafacial" if opposite faces of the molecular fragment are involved, and "suprafacial" if both changes occur at the same face. The concept of "face" is clear from the diagrams in the cases of planar (or approximately planar) frameworks with isolated or interacting pi orbitals.

The terms *antarafacial* and *suprafacial* are, however, also employed in cases in which the essential part of the molecular fragment undergoing changes in bonding comprises two atoms linked only by a sigma bond. In these cases it is customary to refer to the phases of the local sigma-bonding orbital: occurrence of the two bonding changes at sites of like orbital phase is regarded as suprafacial, whereas that at two sites of opposite phase is antarafacial. The possibilities are shown for C–C and C–H sigma bonds in Figs. c and d. There may be two distinct and alternative stereochemical outcomes of a suprafacial process involving a sigma bond between saturated carbon atoms, i.e., either retention or inversion at both centers. The antarafacial process results in inversion at one center and retention at the second.

For examples of the use of these terms see CYCLOADDITION and SIGMATROPIC REARRANGEMENT.

See also ANTI; PI (π) ORBITAL; SIGMA (σ) ORBITAL.

anti In the representation of stereochemical relationships, "anti" means "on opposite sides" of a reference plane, in contrast to "syn," which means "on the same side," as in the following examples

a. Two substituents attached to atoms joined by a single BOND are anti if the torsion angle (dihedral angle) between the bonds to the substituents is greater than 90 degrees, or syn if it is less than 90 degrees. (A further distinction is made between antiperiplanar, synperiplanar, anticlinal, and synclinal.)
b. In the older literature, the terms *anti* and *syn* were used to designate stereoisomers of oximes and related compounds. That usage was superseded by the terms *trans* and *cis* or *E* and *Z*, respectively.
c. When the terms are used in the context of CHEMICAL REACTIONS or TRANSFORMATIONS, they designate the relative orientation of substituents in the substrate or product:
   1. Addition to a carbon-carbon double bond
   2. Alkene-forming elimination

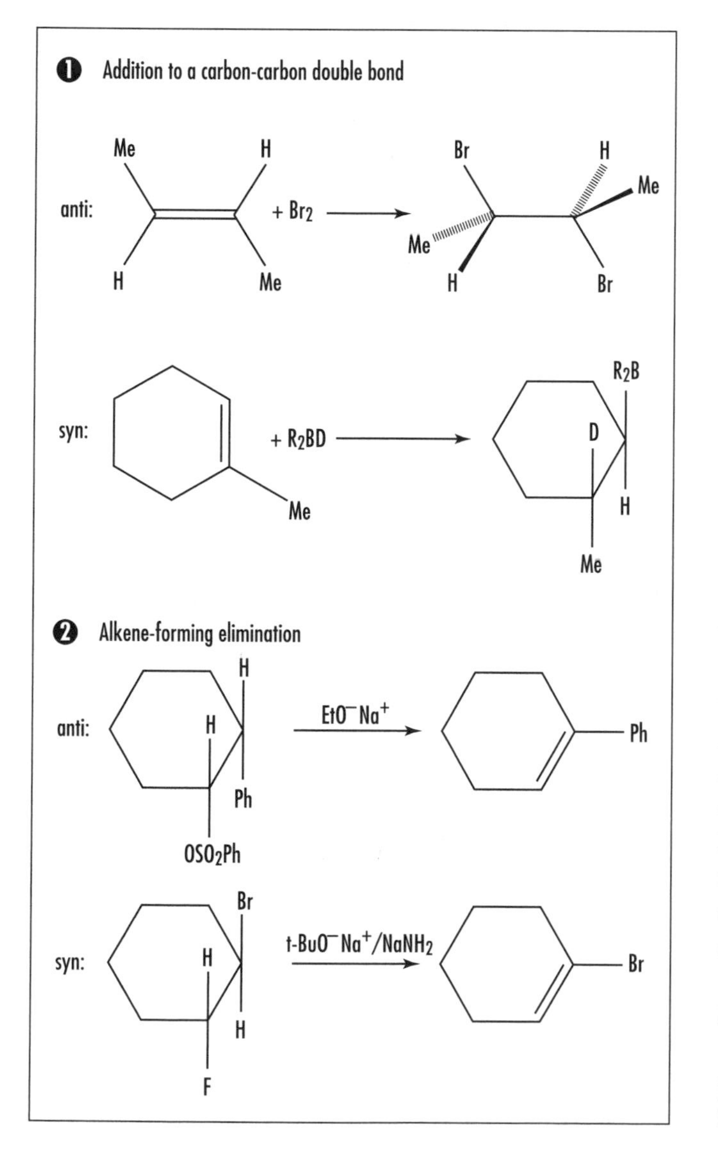

In the above two examples, anti processes are always ANTARAFACIAL, and syn processes are SUPRAFACIAL.

antiaromatic *See* AROMATIC.

antibody A soluble immunoglobulin blood protein produced by the B cells, which are white blood cells that develop in the bone marrow (also known as B lymphocytes, plasma cells) in response to an antigen (a foreign substance). Antibodies are produced in response to disease and help the body fight against a particular disease by binding to the antigen and killing it, or by making it more vulnerable to action by white blood cells. They help the body develop an immunity to diseases.

Each antibody has two light (L) and two heavy (H) immunoglobulin polypeptide chains, linked together by disulfide bonds, with two antigen-binding sites. There are more than 1,000 possible variations, yet each antibody recognizes only one specific antigen. Antibodies are normally bound to a B cell, but when an antibody encounters an antigen, the B cell produces copies of the antibody with the assistance of helper T cells (a lymphocyte that undergoes a developmental stage in the thymus). The released antibodies then go after and bind the antigens, either killing them or marking them for destruction by phagocytes.

There are five immunoglobulins: IgA, IgM, IgG, IgE, and IgD.

IgA, or immunoglobulin A, comprises about 10–15 percent of the body's total immunoglobulins and is found in external secretions such as saliva, tears, breast milk, and mucous, both intestinal and bronchial. They are secreted on the surface of the body as a first defense against bacteria and viral antigens in an attempt to prevent them from entering the body.

IgM, or immunoglobulin M, antibodies are produced in response to new or repeat infections and stay in the body for a short time after infection. They make up from 5 to 10 percent of the total immunoglobulins and are the first to show up in the serum after an antigen enters the system. IgM is produced during the primary immune response. It is the IgMs that capture and bind antigens to form large insoluble complexes that are cleared from the blood.

IgG, or immunoglobulin G (gamma globulin), antibodies remain in the body for long periods of time after infection and are the most common type, comprising about 80 percent of the body's total immunoglobulins. They are in the serum and are produced in substantial quantities during the secondary immune response, and along with IgM, they activate the complement system, which results in the destruction of pathogen membranes. The IgGs act by agglutinating, by opsonising, and by activating complement-mediated reactions against cellular pathogens and by neutralizing toxins.

IgE, or immunoglobulin E, is associated with mast cells, which are basophils, a type of granular white blood cell that has left the bloodstream and entered a tissue. Mast cells release histamine and heparin, chemicals that mediate allergic reactions. Not surprisingly, IgE is responsible for immediate hypersensitivity (allergic) reactions and immune defense against parasites.

IgD, or immunoglobulin D, is a specialized immunoglobulin, but its function is currently unknown. It is found in small amounts in the serum.

antibonding orbital A molecular orbital higher in energy than any of the atomic orbitals from which it is derived; creates instability to a molecule or ion when populated with electrons.

antiferromagnetic *See* FERROMAGNETIC.

antigen A foreign substance, a macromolecule, that is not indigenous to the host organism and therefore elicits an immune response.

anti-Markownikoff addition antigen A foreign substance, a macromolecule, that is not indigenous to the host organism and therefore elicits an immune response.

See also MARKOWNIKOFF RULE.

antimetabolite A structural ANALOG of an intermediate (substrate or COENZYME) in a physiologically occurring metabolic pathway that acts by replacing the natural substrate, thus blocking or diverting the biosynthesis of physiologically important substances.

antisense molecule An OLIGONUCLEOTIDE or ANALOG thereof that is complementary to a segment of RNA (ribonucleic acid) or DNA (deoxyribonucleic acid) and that binds to it and inhibits its normal function.

aphotic zone The deeper part of the ocean beneath the photic zone, where light does not penetrate sufficiently for photosynthesis to occur.

See also OCEANIC ZONE.

apoprotein A protein without its characteristic PROSTHETIC GROUP or metal.

aprotic solvent NonPROTOGENIC (in a given situation). With extremely strong BRONSTED ACIDs or BRONSTED BASEs, solvents that are normally aprotic may accept or lose a proton. For example, acetonitrile is in most instances an aprotic solvent, but it is PROTOPHILIC in the presence of concentrated sulfuric acid and protogenic in the presence of potassium *tert*-butoxide. Similar considerations apply to benzene, trichloromethane, etc.

See also DIPOLAR APROTIC SOLVENT.

aquation The incorporation of one or more integral molecules of water into another chemical species with or without displacement of one or more other atoms or groups.

See also HYDRATION.

aqueous solution A solution in which water is the solvent or dissolving medium, such as salt water, rain, or soda.

Archaea A group of prokaryotes that can be subdivided into three groups (methanogenic, halophilic, thermoacidophilic) and are characterized by special constituents, such as ether-bonded lipids and special COENZYMES. The Archaea are members of a separate kingdom that falls in between eubacterial and eukaryotic organisms.

See also EUKARYOTE; METHANOGEN.

aromatic, aromaticity

1. In the traditional sense, "having a chemistry typified by benzene."
2. A cyclically conjugated MOLECULAR ENTITY with a stability (due to DELOCALIZATION) significantly greater than that of a hypothetical localized structure (e.g., KEKULÉ STRUCTURE) is said to possess aromatic character. If the structure is of higher energy (less stable) than such a hypothetical classical structure, the molecular entity is "antiaromatic."

   The most widely used method for determining aromaticity is the observation of diatropicity in the ^{1}H NMR spectrum.
3. The terms *aromatic* and *antiaromatic* have been extended to describe the stabilization or destabilization of TRANSITION STATES of PERICYCLIC REACTIONS. The hypothetical reference structure is here less clearly defined, and use of the term is based on application of the HUCKEL (4N+2) RULE and on consideration of the topology of orbital overlap in the transition state. Reactions of molecules in the GROUND STATE involving antiaromatic transition states proceed, if at all, much less easily than those involving aromatic transition states.

See also CONJUGATED SYSTEM.

aromatic hydrocarbons Benzene and its derivatives.

Arrhenius, Svante August (1859–1927) Swedish *Chemist, physicist* Svante August Arrhenius was born in Vik (or Wijk), near Uppsala, Sweden, on February 19, 1859. He was the second son of Svante Gustav Arrhenius and Carolina Christina (née Thunberg).

Nobel Prize–winning Swedish chemist, at work in his laboratory. Svante Arrhenius was an infant prodigy and taught himself to read at age three. He studied physics and chemistry at the University of Uppsala before transferring to Stockholm to begin research on aqueous solutions of acids, bases, and salts. He discovered that such solutions conduct electricity because the solute divides into charged ions in the water—a finding that was at first scoffed at but which later won him the 1903 Nobel Prize in chemistry. Subsequent achievements include his elucidation of the effect of temperature on reaction rates (Arrhenius equation) and discovery of the greenhouse effect. ***(Courtesy of Science Photo Library)***

Svante's father was a surveyor and an administrator of his family's estate at Vik. In 1860, a year after Arrhenius was born, his family moved to Uppsala, where his father became a supervisor at the university. He was reading by the age of three.

Arrhenius received his early education at the cathedral school in Uppsala, excelling in biology, physics, and mathematics. In 1876 he entered the University of Uppsala and studied physics, chemistry, and mathematics, receiving his B.S. two years later. While he continued graduate classes for three years in physics at Uppsala, his studies were not completed there. Instead, Arrhenius transferred to the Swedish Academy of Sciences in Stockholm in 1881 to work under Erick Edlund to conduct research in the field of electrical theory.

Arrhenius studied electrical conductivity of dilute solutions by passing electric current through a variety of solutions. His research determined that molecules in some of the substances split apart or dissociated from each other into two or more ions when they were dissolved in a liquid. He found that while each intact molecule was electrically balanced, the split particles carried a small positive or negative electrical charge when dissolved in water. The charged atoms permitted the passage of electricity, and the electrical current directed the active components toward the electrodes. His thesis on the theory of ionic dissociation was barely accepted by the University of Uppsala in 1884, since the faculty believed that oppositely charged particles could not coexist in solution. He received a grade that prohibited him from being able to teach.

Arrhenius published his theories ("Investigations on the Galvanic Conductivity of Electrolytes") and sent copies of his thesis to a number of leading European scientists. Russian-German chemist Wilhelm Ostwald, one of the leading European scientists of the day and one of the principal founders of physical chemistry, was impressed and visited him in Uppsala, offering him a teaching position, which he declined. However, Ostwald's support was enough for Uppsala to give him a lecturing position, which he kept for two years.

The Stockholm Academy of Sciences awarded Arrhenius a traveling scholarship in 1886. As a result, he worked with Ostwald in Riga as well as with physicists Friedrich Kohlrausch at the University of Wurzburg and LUDWIG BOLTZMANN at the University of Graz, and with chemist Jacobus van't Hoff at the University of Amsterdam. In 1889 he formulated his rate equation, which is used for many chemical transformations and processes, in which the rate is exponentially related to temperature. This formulation is known as the "Arrhenius equation."

He returned to Stockholm in 1891 and became a lecturer in physics at Stockholm's Hogskola (high school) and was appointed physics professor in 1895 and rector in 1897. Arrhenius married Sofia Rudbeck in 1894 and had one son. The marriage lasted a short two years. Arrhenius continued his work on electrolytic dissociation and added the study of osmotic pressure.

In 1896 he made the first quantitative link between changes in carbon dioxide concentration and climate. He calculated the absorption coefficients of carbon dioxide and water based on the emission spectrum of the moon, and he also calculated the amount of total heat absorption and corresponding temperature change in the atmosphere for various concentrations of carbon dioxide. His prediction of a doubling of carbon dioxide from a temperate rise of 5–6°C is close to modern predictions. He predicted that increasing reliance on fossil fuel combustion to drive the world's increasing industrialization would, in the end, lead to increases in the concentration of CO_2 in the atmosphere, thereby giving rise to a warming of the Earth.

In 1900 he published his *Textbook of Theoretical Electrochemistry.* In 1901 he and others confirmed the Scottish physicist James Clerk Maxwell's hypothesis that cosmic radiation exerts pressure on particles. Arrhenius went on to use this phenomenon in an effort to explain the aurora borealis and solar corona. He supported the explanation of the origin of auroras proposed by the Norwegian physicist Kristian Birkeland in 1896. He also suggested that radiation pressure could carry spores and other living seeds through space, and he believed that life on Earth was brought here under those conditions. He likewise believed that spores might have populated many other planets, resulting in life throughout the universe.

In 1902 he received the Davy Medal of the Royal Society and proposed a theory of immunology. The following year he was awarded the Nobel Prize in chemistry for his work that originally was believed as improbable by his Uppsala professors. He also published his *Textbook of Cosmic Physics.*

Arrhenius became director of the Nobel Institute of Physical Chemistry in Stockholm in 1905, a post he held until a few months before his death. He married Maria Johansson, his second wife, and had one son and two daughters. The following year he also had time to publish three books: *Theories of Chemistry, Immunochemistry,* and *Worlds in the Making.*

He was elected a foreign member of the Royal Society in 1911, the same year he received the Willard Gibbs Medal of the American Chemical Society. Three years later he was awarded the Faraday Medal of the British Chemical Society. He was also a member of the Swedish Academy of Sciences and the German Chemical Society.

During the latter part of his life, his interests included the chemistry of living matter and astrophysics, especially the origins and fate of stars and planets. He continued to write books such as *Smallpox and Its Combating* (1913), *Destiny of the Stars* (1915), *Quantitative Laws in Biological Chemistry* (1915), and *Chemistry and Modern Life* (1919). He also received honorary degrees from the universities of Birmingham, Edinburgh, Heidelberg, and Leipzig and from Oxford and Cambridge universities. He died in Stockholm on October 2, 1927, after a brief illness, and is buried at Uppsala.

Arrhenius equation *See* ENERGY OF ACTIVATION.

artificial transmutation Artificially induced nuclear reaction made by bombarding a nucleus with subatomic particles or small nuclei.

aryl group Group of atoms remaining after a hydrogen atom is removed from the aromatic system.

aryne A hydrocarbon derived from an arene by abstraction of two hydrogen atoms from adjacent carbon atoms, e.g., 1,2-didehydroarene. Arynes are commonly represented with a formal triple bond. The analogous heterocyclic compounds are called heteroarynes or hetarynes. For example:

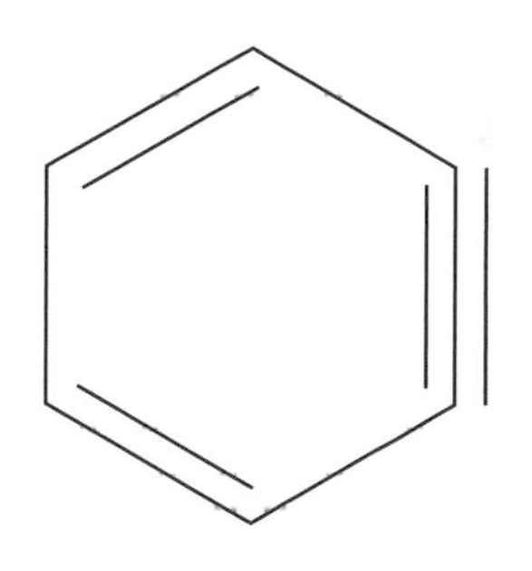

Benzyne

assimilation To transform food and other nutrients into a part of the living organism.

assimilative *See* ASSIMILATION.

assimilator *See* ASSIMILATION.

associated ions Short-lived ions formed by the collision of dissolved ions of opposite charges.

association The assembling of separate MOLECULAR ENTITIES into any aggregate, especially of oppositely charged free ions into ION PAIRs or larger, not necessarily well-defined, clusters of ions held together by electrostatic attraction. The term signifies the reverse of DISSOCIATION, but it is not commonly used for the formation of definite ADDUCTs by COLLIGATION or COORDINATION.

asymmetric carbon A carbon atom covalently bonded to four different atoms or groups of atoms.

asymmetric induction The traditional term describing the preferential formation in a CHEMICAL REACTION of one ENANTIOMER or DIASTEREOISOMER over the other as a result of the influence of a chiral feature in the substrate, reagent, CATALYST, or environment. The term also refers to the formation of a new chiral feature preferentially in one CONFIGURATION under such influence.

asymmetric synthesis A traditional term for stereoselective synthesis. A chemical reaction or reaction sequence in which one or more new elements of CHIRALITY are formed in a SUBSTRATE molecule to produce the STEREOISOMERIC (ENANTIOMERic or DIASTEREOISOMERic) products in unequal amounts.

asymmetry parameter In nuclear quadrupole resonance spectroscopy, the parameter, η, is used to describe nonsymmetric fields. It is defined as $\eta = (q_{xx} - q_{yy})/q_{zz}$, in which q_{xx}, q_{yy}, and q_{zz} are the components of the field gradient q (which is the second derivative of the time-averaged electric potential) along the x-, y-, and z-axes. By convention, q_{zz} refers to the largest field gradient, q_{yy} to the next largest, and q_{xx} to the smallest when all three values are different.

atmosphere A unit of pressure; the pressure that will support a column of mercury 760 mm high at 0°C.

atom The smallest particle of an element.

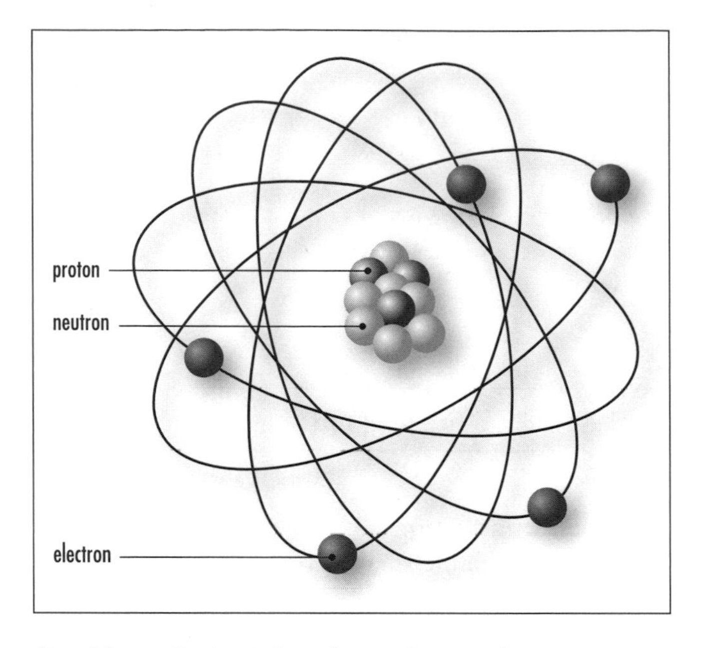

Atom. The smallest part of an element that contains protons, neutrons, and electrons

atomic mass unit (amu) One-twelfth of a mass of an atom of the carbon-12 (^{12}C) isotope; a unit used for stating atomic and formula weights; also called a dalton.

atomic number The atomic number is equal to the number of positively charged protons in an atom's nucleus and determines which element an atom is. The atomic number is unique for each element and is designated by a subscript to the left of the elemental symbol. The atomic number for hydrogen is 1; it has one proton. Elements are substances made up of atoms with the same atomic number. Most of the elements are metals (75 percent), and the others are nonmetals.

atomic orbital A one-electron wave function describing an electron in the effective field provided by a nucleus and the other electrons present.

See also MOLECULAR ORBITAL.

atomic radius Radius of an atom.

atomic weight or mass The total atomic mass (the weighted average of the naturally occurring isotopes), which is the mass in grams of one mole of the atom. The atomic weight is calculated by adding the number of protons and neutrons together. The atomic weight of hydrogen is 1.0079 grams per mole.

ATP (adenosine triphosphate) An adenine (purine base), ribose, and three phosphate units containing nucleoside triphosphate that releases free energy when its phosphate bonds are hydrolyzed, and also produces adenosine diphosphate (ADP) and inorganic phosphorous. This energy is used to drive endergonic reactions in cells (chemical reactions that require energy input to begin). ATP is produced in the cristae of mitochondria and chloroplasts in plants and is the driving force in muscle contraction and protein synthesis in animals. It is the major energy source within cells.

ATP synthase (proton translocating ATPase) A protein complex (a chemiosmotic enzyme), that synthesizes ATP from ADP and from phosphate coupling with electrochemical ion gradient across the membrane. It is found in cellular membranes and the inner membrane of mitochondria, the thylakoid membrane of chloroplasts, and the plasma membrane of prokaryotes. The protein consists of two portions: a soluble fraction that contains three catalytic sites and a membrane-bound portion that contains anion channels. It functions in chemiosmosis (the use of ion gradients across membranes) with adjacent electron transport chains, and it uses the energy stored across the photosynthetic membrane (a hydrogen-ion concentration gradient) to add inorganic phosphate to ADP, thereby creating ATP. This allows hydrogen ions (H^+) to diffuse into the mitochondrion.

attachment A TRANSFORMATION by which one MOLECULAR ENTITY (the SUBSTRATE) is converted into another by the formation of one (and only one) two-center BOND between the substrate and another molecular entity and which involves no other changes in CONNECTIVITY in the substrate. For example, consider the formation of an acyl cation by attachment of carbon monoxide to a CARBENIUM ION (R^+):

$$R^+ + CO \rightarrow (RCO)^+$$

The product of an attachment may also be the ADDUCT of the two reactants, but not all adducts can be represented as the products of an attachment.

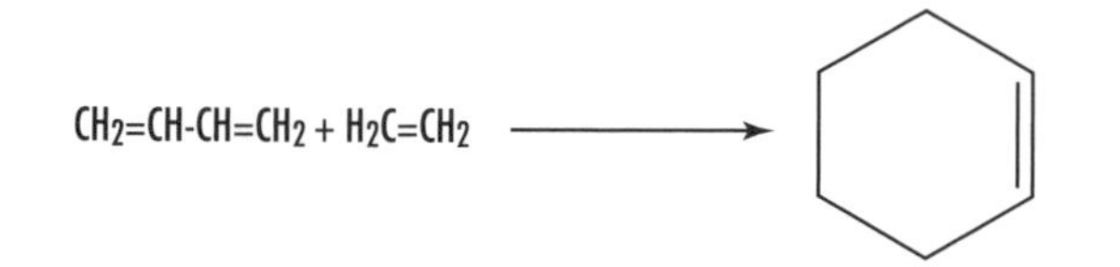

(For example, the Diels-Alder CYCLOADDITION results in an adduct of buta-1,3-diene and ethene, but the reaction cannot be described as an attachment, since bonds are formed between more than two centers.)

See also COLLIGATION; ELECTRON ATTACHMENT.

aufbau ("building up") principle Describes the order in which electrons fill orbitals in atoms.

auranofin *See* GOLD DRUGS.

autacoid A biological substance secreted by various cells whose physiological activity is restricted to the vicinity of its release; it is often referred to as local HORMONE.

autocatalytic reaction A CHEMICAL REACTION in which a product (or a reaction INTERMEDIATE) also functions as a CATALYST. In such a reaction, the observed RATE OF REACTION is often found to increase with time from its initial value.

See also ORDER OF REACTION.

autoionization Ionization reaction between identical molecules.

automerization *See* DEGENERATE REARRANGEMENT.

autophytic The process whereby an organism uses photosynthesis to make complex foods from inorganic substances.

autoprotolysis A PROTON (hydron) TRANSFER REACTION between two identical molecules (usually a solvent), one acting as a BRONSTED ACID and the other as a BRONSTED BASE. For example:

$$2\ H_2O \rightarrow H_3O^+ + OH^-$$

See also HYDRON.

autoprotolysis constant The product of the activities (or, more approximately, concentrations) of the species produced as the result of AUTOPROTOLYSIS. For solvents in which no other ionization processes are significant, the term is synonymous with *ionic product.* The autoprotolysis constant for water, K_w, is equal to the product of activities $a(H_3O^+)a(OH^-) = 1.0 \times 10^{-14}$ at 25°C.

autoreceptor An autoreceptor, present at a nerve ending, is a RECEPTOR that regulates, via positive or negative feedback processes, the synthesis or release of its own physiological ligand.

See also HETERORECEPTOR.

autotrophic organism An organism that is capable of using carbon dioxide as the sole carbon source for growth and product formation. Organisms that use light as a source of energy are said to be PHOTOAUTOTROPHS; those that use the energy from chemical reactions are CHEMOAUTOTROPHS.

auxins A group of plant hormones that produce a number of effects, including plant growth, phototropic response through the stimulation of cell elongation (photopropism), stimulation of secondary growth, apical dominance, and the development of leaf traces and fruit. An important plant auxin is indole-3-acetic acid. (IAA and synthetic auxins such as 2,4-D and 2,4,5-T are used as common weed killers.)

auxotroph An organism that requires a particular organic compound for growth.

***A* value** The conformational preference of an equatorial compared with an axial substituent in a monosubstituted cyclohexane. This steric substituent parameter equals $\Delta_r G^o$ in kcal/mol for the equatorial-to-axial equilibration on cyclohexane. The values are also known as "Winstein–Holness" *A* values.

Avogadro's law At the same temperature and pressure, equal volumes of all gases contain the same number of molecules.

Avogadro's number The number (6.022×10^{23}) of atoms, molecules, or particles found in exactly 1 mole of substance.

azacarbene *See* NITRENE.

azene *See* NITRENE.

azurin An ELECTRON-TRANSFER PROTEIN containing a TYPE 1 COPPER site, isolated from certain bacteria.

azylene *See* NITRENE.

B

background radiation Exposure to various forms of low-level natural radiation from sources such as cosmic rays, radioactive substances from the earth, certain geographic locations and buildings (radon), and trace amounts present in the human body. Considered nonthreatening, with annual exposure in the 1–2 millisievert (mSv) range.

bacteria One of two prokaryotic (no nucleus) domains, the other being the ARCHAEA. Bacteria are microscopic, simple, single-cell organisms. Some bacteria are harmless or, often, are beneficial to human beings. Others are pathogenic, causing disease and even death. All play a major role in the cycling of nutrients in ecosystems via AEROBIC and ANAEROBIC decomposition (saprophytic). Some species form symbiotic relationships with other organisms, such as legumes, and help them survive in the environment by fixing atmospheric nitrogen. Many different species exist as single cells or colonies, and they fall into four shapes based on the shape of their rigid cell wall: coccal (spherical), bacillary (rod-shaped), spirochetal (spiral/helical or corkscrew), and vibro (comma-shaped). Bacteria are also classified on the basis of oxygen requirement (aerobic vs. anaerobic).

In the laboratory, bacteria are classified as Gram-positive (blue) or Gram-negative (pink), following a laboratory procedure called a Gram stain. Gram-negative bacteria, such as those that cause the plague, cholera, typhoid fever, and salmonella, have two outer membranes, which make them more resistant to conventional treatment. They can also easily mutate and transfer these genetic changes to other strains, making them more resistant to antibiotics. Gram-positive bacteria, such as those that cause anthrax and listeriosis, are more rare and treatable with penicillin, but they can cause severe damage either by releasing toxic chemicals (e.g., clostridium botulinum) or by penetrating deep into tissue (e.g., streptococci). Bacteria are often called germs.

bacteriochlorophyll *See* CHLOROPHYLL.

Baldwin's rules A set of empirical rules for certain formations of three- to seven-membered rings. The predicted pathways are those in which the length and nature of the linking chain enables the terminal atoms to achieve the proper geometries for reaction. The disfavored cases are subject to severe distortions of bond angles and bond distances.

Balmer series That portion of the emission spectrum of hydrogen in the visible portion representing electron transitions from energy levels $n > 2$ to $n = 2$.

band theory of metals Accounts for the bonding and properties of metallic solids. Any given metal atom has only a limited number of valence electrons with which to bond to all of its nearest neighbors so that extensive sharing of electrons among individual atoms is required. This is accomplished by the sharing of electrons through the overlap of atomic orbitals of equivalent energy on the metal atoms that are immediately adjacent to one another, leading to molecular orbitals that are delocalized over the entire structure. These orbitals are closely spaced in energy, giving rise to the term *energy bands*.

Banting, Frederick Grant (1891–1941) Canadian *Physician* Frederick Grant Banting was born on November 14, 1891, at Alliston, Ontario, Canada, to William Thompson Banting and Margaret Grant. He went to secondary school at Alliston and then to the University of Toronto to study divinity, changing to the study of medicine. In 1916 he took his M.B. degree and joined the Canadian Army Medical Corps and served in France during World War I. In 1918 he was wounded at the battle of Cambrai, and the following year he was awarded the Military Cross for heroism under fire.

In 1922 he was awarded his M.D. degree and was appointed senior demonstrator in medicine at the University of Toronto. In 1923 he was elected to the Banting and Best Chair of Medical Research, which had been endowed by the legislature of the Province of Ontario.

Also in 1922, while working at the University of Toronto in the laboratory of the Scottish physiologist John James Rickard Macleod, and with the assistance of the Canadian physiologist Charles Best, Banting discovered insulin after extracting it from the pancreas. The following year he received the Nobel Prize in medicine along with Macleod. Angered that Macleod, rather than Best, had received the Nobel Prize, Banting divided his share of the award equally with Best. Banting was the first Canadian recipient of the Nobel Prize. He was knighted in 1934. The word *banting* was associated with dieting for many years.

On February 21, 1941, he was killed in an air disaster in Newfoundland.

bar (unit of pressure) A unit of pressure equal to 1 million dynes per square centimeter, equivalent to 10 newtons per square centimeter.

barometer An instrument that measures atmospheric pressure and was discovered by Evangelista Torricelli in 1643 in Florence.

basalt A hard, black volcanic rock with less than about 52 weight percent silica (SiO_2).

base A substance that reduces the hydrogen ion concentration in a solution. A base has fewer free hydrogen ions (H^+) than hydroxyl ions (OH^-) and gives an aqueous solution with a pH of more than 7 on a scale from 0 to 14. They have a slippery feel in water and a bitter taste. A base turns litmus paper to blue, while acids turn litmus paper to red. The types of bases include Arrhenius (any chemical that increases the number of free hydroxide ions [OH^-] when added to a water-based solution); Bronsted or Bronsted-Lowry bases (any chemical that acts as a proton acceptor in a chemical reaction); and Lewis bases (any chemical that donates two electrons to form a covalent bond during a chemical reaction). Bases are also known as alkali or alkaline substances, and when added to acids, they form salts. Some common examples of bases are soap, ammonia, and lye.

See also ACID; BRONSTED BASE; HARD BASE; LEWIS BASE.

base pairing The specific association between two complementary strands of nucleic acid that results from the formation of hydrogen bonds between the base components of the NUCLEOTIDES of each strand: A(denine)=T(hymine) and G(uanine)≡C(ytosine) in DNA, A=U(racil) and G≡C (and in some cases G≡U) in RNA (the lines indicate the number of hydrogen bonds). Single-stranded nucleic acid molecules can adopt a partially double-stranded structure through intrastrand base pairing.

See also NUCLEOSIDE.

base-pair substitution There are two main types of mutations within a gene: base-pair substitutions and base-pair insertions or deletions. A base-pair substitution is a point mutation. It is the replacement of one nucleotide and its partner from the complementary

DNA strand with another pair of nucleotides. The bases are one of five compounds—adenine, guanine, cytosine, thymine, and uracil—that form the genetic code in DNA and RNA.

basic anhydride A metal oxide that reacts with water and forms a base.

basicity constant *See* ACIDITY CONSTANT.

basicity function *See* ACIDITY FUNCTION.

basic oxide Any ionic oxide that dissolves in water to produce a basic solution.

basic salt A salt that keeps part of the base from which it is formed. The retained portion can be oxygen or a hydroxl group.

basic solution A solution in which the hydroxide ion concentration is higher than the hydronium ion (hydrated proton) concentration.

bathochromic shift (effect) Shift of a spectral band to lower frequencies (longer wavelengths) owing to the influence of substitution or a change in environment. It is informally referred to as a red shift and is opposite to HYPSOCHROMIC SHIFT (blue shift).

bathyal zone A region of the seafloor from the shelf edge (200 m) to the start of the abyssal zone (2,000 m); zone between the edge of the continental shelf and the depth at which the water temperature is 4°C.

bauxite An alumina ore that, when smelted, gives up aluminum. Bauxite usually contains at least 45 percent aluminum oxide (alumina), and the best grades have a low silica content. About 4 lb of bauxite are required to produce 1 lb of aluminum.

Beer-Lambert law The quantitative relationship between the absorbance of radiant energy, concentration of the sample solution, and length of the path through the sample. Named after two scientists, Johann Heinrich Lambert (1728–77) and August Beer (1825–63), the law also serves as the basis of spectroscopic instruments that are being used increasingly in the science curriculum. Research is still in progress to understand and find appropriate models for the light absorbance.

Bell-Evans-Polanyi principle The linear relation between ENERGY OF ACTIVATION (E_a) and enthalpy of reaction (ΔH_r) sometimes observed within a series of closely related reactions.

$$E_a = A + B \, \Delta H_r$$

benthic zone The lowermost region of a freshwater or marine body, pertaining to the bottom terrain of water bodies. It is below the pelagic zone and above the abyssal zone by about 9,000 m. Organisms that live on or in the sediment in these environments are called benthos.

See also OCEANIC ZONE.

benzyne 1,2-Didehydrobenzene (the ARYNE derived from benzene) and its derivatives formed by substitution. The terms *m*- and *p*-benzyne are occasionally used for 1,3- and 1,4-didehydrobenzene, respectively.

Berg, Paul (1926–) American *Chemist* Paul Berg was born in Brooklyn, New York, on June 30, 1926, to Sarah Brodsky and Harry Berg. He attended public school (Abraham Lincoln) in New York and graduated early in 1943. Berg studied biochemistry at Pennsylvania State College from 1943 until 1948, receiving a B.S. in biochemistry. He then attended graduate studies in biochemistry at Western Reserve University, where he received his Ph.D. in 1952.

Berg was an assistant professor of microbiology at Washington University School of Medicine from 1955 to 1959; an associate professor and professor of biochemistry at Stanford University School of Medicine, Stanford, California, from 1959 to the present; and was named Willson Professor of Biochemistry at Stanford University in 1970, when he became chairman of the Department of Biochemistry.

He is considered the father of the controversial branch of biochemistry known as genetic engineering. He was the first person to manufacture a human hormone from a virus combined with genes from a bacterial chromosome. In 1980 he was awarded the Nobel Prize in chemistry "for his fundamental studies of the biochemistry of nucleic acids, with particular regard to recombinant-DNA."

He is a nonresident fellow of Salk Institute (1973 to present) and has been elected to the U.S. National Academy of Sciences and American Academy of Sciences (1966); has received the Distinguished Alumnus Award from Pennsylvania State University (1972); has served as president of the American Society of Biological Chemists (1975); has received an honorary D.Sc. from Yale University and the University of Rochester (1978); and has been named a foreign member of the Japan Biochemical Society (1978). Other awards include the Eli Lilly Prize in Biochemistry (1959); California Scientist of the Year (1963); V.D. Mattia Award of the Roche Institute for Molecular Biology (1974); Sarasota Medical Award (1979); Gairdner Foundation Annual Award (1980); Albert Lasker Medical Research Award (1980); and the New York Academy of Sciences Award (1980). He served in the U.S. Navy (1944–46).

Berg has continued to conduct research in the Department of Biochemistry at Stanford, where his focus is the mechanism of repairing DNA damage. He continues to influence federal policy regarding stem cell research, biotechnology, and human cloning.

beta (β) decay A process in which unstable atoms can become more stable. There are two types of beta decay: beta-minus and beta-plus. During beta-minus decay, a neutron in an atom's nucleus turns into a proton and emits an electron and an antineutrino. During beta-plus decay, a proton in an atom's nucleus turns into a neutron and emits a positron and a neutrino. Both changes add or lose a proton and cause one element to turn into another.

beta (β) particle A charged particle that is emitted from the nucleus of an atom when a neutron decays. The particle has a mass equal to 1/1,837 that of a proton. If the beta particle is negatively charged, it is identical to an electron. If it is positively charged, it is called a positron.

beta (β) sheet Preferentially called a beta pleated sheet; a regular structure in an extended polypeptide chain, stabilized in the form of a sheet by hydrogen bonds between CO and NH groups of adjacent (parallel or antiparallel) chains.

beta (β) strand Element of a BETA SHEET. One of the strands that is hydrogen bonded to a parallel or antiparallel strand to form a beta sheet.

beta (β) turn A hairpin structure in a polypeptide chain reversing its direction by forming a hydrogen bond between the CO group of AMINO-ACID RESIDUE n with the NH group of residue (n+3).

See also HELIX.

bifunctional catalysis Catalysis (usually for hydrogen ion [HYDRON] transfer) by a bifunctional CHEMICAL SPECIES involving a mechanism in which both FUNCTIONAL GROUPs are implicated in the RATE-CONTROLLING STEP so that the corresponding CATALYTIC COEFFICIENT is larger than that expected for catalysis by chemical species containing only one of these functional groups. The term should not be used to describe the concerted action of two different catalysts ("concerted catalysis").

See also CONCERTED PROCESS.

bifunctional ligand A LIGAND that is capable of simultaneous use of two of its donor atoms to bind to one or more CENTRAL ATOMs.

See also AMBIDENTATE.

bimolecular reaction The collision and combination of two reactants involved in the rate-limiting step.

See also RATE-CONTROLLING STEP.

binary acid A BINARY COMPOUND that can act as an acid but contains only one element other than hydrogen.

binary compound A compound that consists of two elements, either of which may be ionic or covalent.

binary fission A type of asexual reproduction in prokaryotes (cells or organisms lacking a membrane-bound, structurally discrete nucleus and other subcellular compartments) in which a cell divides or "splits" into two "daughter" cells, each containing a complete copy of the genetic material of the parent. Examples of organisms that reproduce this way are bacteria, paramecia, and *Schizosaccharomyces pombe* (an ascomycetous species of yeast). Also known as transverse fission.

binding constant *See* STABILITY CONSTANT.

binding energy (nuclear binding energy) Energy that is released as particles (protons and neutrons) are combined to form a nucleus; the amount of energy by which the nucleus is more stable than the separated nucleons.

binding site A specific region (or atom) in a molecular entity that is capable of entering into a stabilizing interaction with another molecular entity. An example of such an interaction is that of an ACTIVE SITE in an enzyme with its SUBSTRATE. Typical forms of interaction are by hydrogen bonding, COORDINATION, and ion pair formation. Two binding sites in different molecular entities are said to be complementary if their interaction is stabilizing.

binomial (binomial name) Each organism is named using a Latin-based code consisting of a combination of two names, the first being a generic (genus) name and the second a specific trivial name which, together, constitute the scientific name of a species. *Lupinus perennis,* or wild blue lupine, is an example. Both names are italicized, and both names used together constitute the species name. This is an example of the binomial nomenclature, critical to the system of classification of plants and animals. Linnaeus, a Swedish naturalist, developed the system in the 18th century. The hierarchy lists the smallest group to largest group: species, genus, family, order, class, division, and kingdom. The first person to formally describe a species is often included, sometimes as an abbreviation, when the species is first mentioned in a research article (e.g., *Lupinus perennis* L., where L. = Linnaeus, who first produced this binomial name and provided an original description of this plant).

binuclear Less frequently used term for the IUPAC recommended term: *dinuclear.*

See also NUCLEARITY.

bioaccumulation A process whereby a toxic chemical enters the food chain, starting at a lower trophic level and, as it moves up the food chain, becoming concentrated in an organism higher in the food chain.

bioassay A procedure for determining the concentration, purity, and/or biological activity of a substance (e.g., vitamin, hormone, plant growth factor, antibiotic, enzyme) by measuring its effect on an ORGANISM, tissue, CELL, ENZYME, or receptor preparation compared with a standard preparation.

bioavailability The availability of a food component or a XENOBIOTIC to an organ or organism.

biocatalyst A CATALYST of biological origin, typically an ENZYME.

bioconjugate A molecular species produced by living systems of biological origin when it is composed of two parts of different origins, e.g., a conjugate of a xenobiotic with some groups, such as glutathione, sulfate, or glucuronic acid, to make it soluble in water or compartmentalized within the cell.

bioconversion The conversion of one substance to another by biological means. The fermentation of sugars to alcohols, catalyzed by yeasts, is an example of bioconversion.

See also BIOTRANSFORMATION.

biodegradability The ability of a material to be broken down or decomposed into simpler substances by the action of microorganisms.

biodiversity (**biological diversity**) The totality of genes, species, and ecosystems in a particular environment, region, or the entire world. Usually refers to the variety and variability of living organisms and the ecological relationships in which they occur. It can be the number of different species and their relative frequencies in a particular area, and it can be organized on several levels from specific species complexes to entire ecosystems or even molecular-level heredity studies.

bioenergetics The study of the energy transfers in and between organisms and their environments and the regulation of those pathways. The term is also used for a form of psychotherapy that works through the body to engage the emotions and is based on the work of Wilhelm Reich and psychiatrist Alexander Lowen in the 1950s.

biogeochemical cycles Both energy and inorganic nutrients flow through ecosystems. However, energy is a one-way process that drives the movement of nutrients and is then lost, whereas nutrients are cycled back into the system between organisms and their environments by way of molecules, ions, or elements. These various nutrient circuits, which involve both biotic and abiotic components of ecosystems, are called biogeochemical cycles. Major biogeochemical cycles include the water cycle, carbon cycle, oxygen cycle, nitrogen cycle, phosphorus cycle, sulfur cycle, and calcium cycle. Biogeochemical cycles can take place on a cellular level (absorption of carbon dioxide by a cell) all the way to global levels (atmosphere and ocean interactions). These cycles take place through the biosphere, lithosphere, hydrosphere, and atmosphere.

biogeography The study of the past and present distribution of life.

bioisostere A compound resulting from the exchange of an atom or of a group of atoms with another broadly similar atom or group of atoms. The objective of a bioisosteric replacement is to create a new compound with similar biological properties to the parent compound. The bioisosteric replacement may be physicochemically or topologically based.

See also ISOSTERES.

bioleaching Extraction of metals from ores or soil by biological processes, mostly by microorganisms.

biological half-life The time at which the amount of a chemical species in a living organism has been reduced by one-half. (The term is used for inorganic materials also, e.g., radionuclides.)

See also HALF-LIFE.

biological magnification The increase in the concentration of heavy metals (i.e., mercury) or organic contaminants (i.e., chlorinated hydrocarbons [CHCs]) in organisms as a result of their consumption within a food chain/web. Another term for this is BIOACCUMULATION. An excellent example is the process by which contaminants such as PCBs accumulate or magnify as they move up the food chain. For example, PCBs concentrate in tissue and internal organs, and as big fish eat little fish, they accumulate all the PCBs that have been eaten by everyone below them in the food chain.

biological oxygen demand (**BOD**) The amount of oxygen used to carry out decomposition by organisms in a particular body of water.

bioluminescence The process of producing light by a chemical reaction by a living organism, e.g., glowworms, fireflies, and jellyfish. Usually produced in organs called photopores or light organs. Can be used for luring prey or as a courting behavior.

biomass The dry weight of organic matter in unit area or volume, usually expressed as mass or weight, comprising a group of organisms in a particular habitat. Also refers to organic matter that is available on a renewable basis, such as forests, agricultural crops, wood and wood wastes, animals, and plants, for example.

biomembrane Organized sheetlike assemblies, consisting mainly of proteins and lipids (bilayers), acting as highly selective permeability barriers and containing specific molecular pumps and gates, receptors, and enzymes.

biomimetic Refers to a laboratory procedure designed to imitate a natural chemical process. Also refers to a compound that mimics a biological material in its structure or function.

biomineralization The synthesis of inorganic crystalline or amorphous mineral-like materials by living organisms. Among the minerals synthesized biologically in various forms of life are: fluoroapatite, ($Ca_5(PO_4)_3F$), hydroxyapatite, magnetite (Fe_3O_4), and calcium carbonate ($CaCO_3$).

biopolymers Macromolecules, including proteins, nucleic acids, and polysaccharides, formed by living organisms.

bioprecursor prodrug A PRODRUG that does not imply the linkage to a carrier group, but results from a molecular modification of the active principle itself. This modification generates a new compound that can be transformed metabolically or chemically, the resulting compound being the active principle.

biosensor A device that uses specific biochemical reactions mediated by isolated enzymes, immunosystems, tissues, organelles, or whole cells to detect chemical compounds, usually by electrical, thermal, or optical signals.

biosphere The entire portion of Earth between the outer portion of the geosphere (the physical elements of the Earth's surface crust and interior) and the inner portion of the atmosphere that is inhabited by life; it is the sum of all the planet's communities and ECOSYSTEMS.

biotechnology The industrial or commercial manipulation and use of living organisms or their components to improve human health and food production either on the molecular level (genetics, gene splicing, or use of recombinant DNA) or in more visible areas such as cattle breeding.

biotic Pertaining to the living organisms in the environment, including entire populations and ECOSYSTEMS.

biotransformation A chemical TRANSFORMATION mediated by living organisms or ENZYME preparations. The chemical conversion of substances by living organisms or enzyme preparations.

See also BIOCONVERSION.

biradical Per the figure below, an even-electron MOLECULAR ENTITY with two (possibly delocalized) radical centers that act nearly independently of each other:

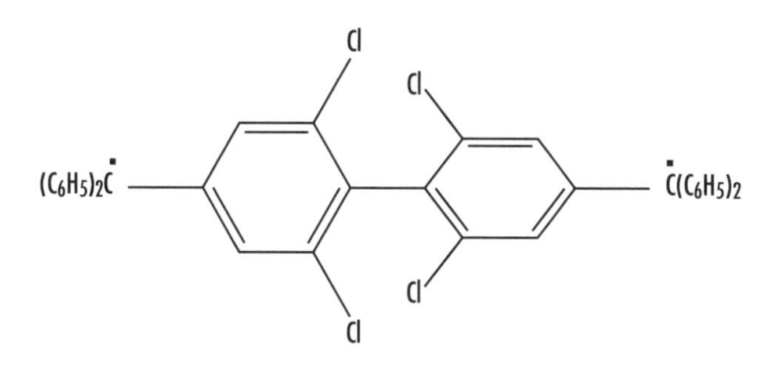

Species in which the two radical centers interact significantly are often referred to as "biradicaloids." If the two radical centers are located on the same atom, the species are more properly referred to by their generic names: CARBENES, NITRENES, etc.

The lowest-energy triplet state of a biradical lies below or at most only a little above its lowest singlet state (usually judged relative to $k_B T$, the product of the Boltzmann constant k_B, and the absolute temperature T). The states of those biradicals whose radical centers interact particularly weakly are most easily understood in terms of a pair of local doublets.

Theoretical descriptions of low-energy states of biradicals display the presence of two unsaturated valences (biradicals contain one fewer bond than permitted by the rules of valence): the dominant valence bond structures have two dots; the low-energy molecular orbital CONFIGURATIONs have only two electrons in two approximately nonbonding molecular orbitals; two of the natural orbitals have occupancies close to one; etc.

The term is synonymous with "diradical."

black smoker Chimneylike accumulations of mineral deposits found at hydrothermal vents.

bleomycin (BLM) A glycopeptide molecule that can serve as a metal chelating ligand. The Fe(III) complex of bleomycin is an antitumor agent, and its activity is associated with DNA cleavage.

BLM *See* BLEOMYCIN.

blood An animal fluid that transports oxygen from the lungs to body tissues and returns carbon dioxide from body tissues to the lungs through a network of vessels such as veins, arteries, and capillaries. It transports nourishment from digestion, hormones from glands, disease-fighting substances to tissues, as well as wastes to the kidneys. Blood contains red and white blood cells and platelets that are responsible for a variety of functions from transporting substances to fighting invasion from foreign substances. Some 55 percent of blood is a clear liquid called plasma. The average adult has about five liters of blood.

blotting A technique used for transferring DNA, RNA, or protein from gels to a suitable binding matrix, such as nitrocellulose or nylon paper, while maintaining the same physical separation.

blue copper protein An ELECTRON-TRANSFER PROTEIN containing a TYPE 1 COPPER site. Characterized by a strong absorption in the visible region and an EPR signal with an unusually small hyperfine coupling to the copper nucleus. Both characteristics are attributed to COORDINATION of the copper by a cysteine sulfur.

See also ELECTRON PARAMAGNETIC RESONANCE SPECTROSCOPY.

Bodenstein approximation *See* STEADY STATE.

body-centered unit cell A structure in which every atom is surrounded by eight adjacent atoms, regardless of whether the atom is located at a corner or at the center of a unit cell.

Computer artwork of the lattice of a body-centered cubic crystal over water. This structure is adopted by the metals lithium, sodium, potassium, and iron below 906°C. The lattice has a single atom at the center and another eight on the corners of a cube, and the pattern is repeated again and again to form a crystal. It is a relatively open structure, utilizing about 68 percent of the available space. Metals that form this structure cannot use a lattice with less space because the thermal vibrations of the atoms are able to overcome the relatively weak binding forces between them. ***(Courtesy of Laguna Design/Science Photo Library)***

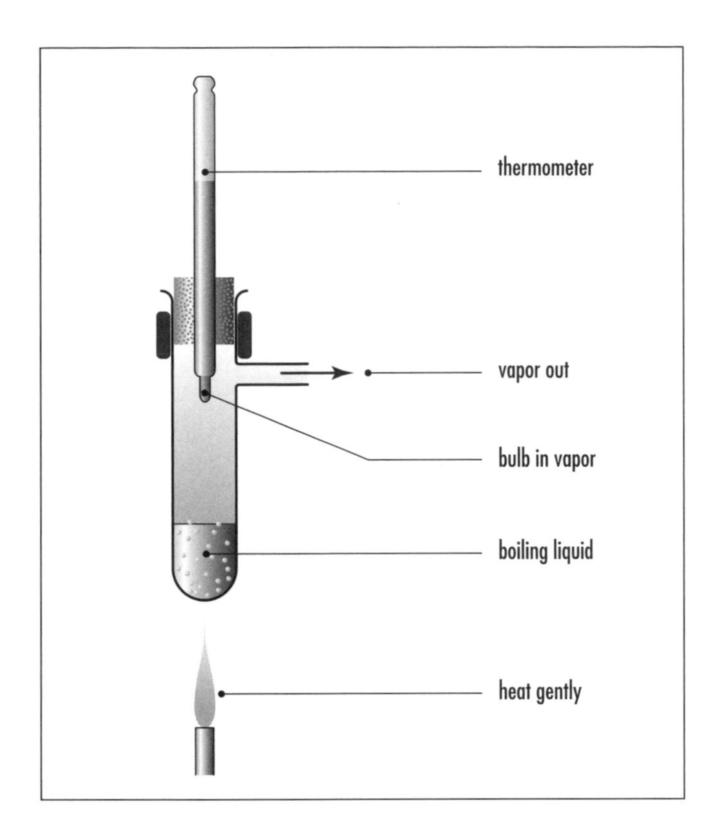

Boiling point. When a substance changes from liquid to gas at a fixed temperature

Bohr model A model of the atom proposed by Niels Bohr in 1913 that showed electrons in fixed orbits around the nucleus but acting in some ways like waves.

boiling point The temperature at which the vapor pressure of a liquid is equal to the external atmospheric pressure. A normal boiling point is considered to be the boiling point at normal atmospheric pressure (101.325 kPa).

boiling-point elevation The increase in the boiling point of a liquid due to the presence of a solute. The extent of the increase is based on concentration and molecular weight.

Boltzmann distribution A formula for calculating the populations of different energy states at a given temperature. Higher energy quantum levels have a lower probability of being occupied.

bomb calorimeter A rigid device used to measure the heat transfer (released or absorbed) during a chemical reaction under conditions of constant volume.

bond There is a chemical bond between two atoms or groups of atoms when the forces acting between them are such as to lead to the formation of an aggregate with sufficient stability to make it convenient for the chemist to consider it as an independent "molecular species." In the context of this encyclopedia, the term refers usually to the COVALENT BOND.

See also AGOSTIC; COORDINATION; HYDROGEN BOND; MULTICENTER BOND.

bond dissociation *See* HETEROLYSIS; HOMOLYSIS.

bond-dissociation energy The energy required to break a given BOND in a compound. For example: $CH_4 \rightarrow H_3C + H$, symbolized as $D(CH_3\text{–}H)$.

See also BOND ENERGY; HETEROLYTIC BOND-DISSOCIATION ENERGY.

bond energy Atoms in a molecule are held together by covalent bonds, and to break these bonds, atoms need bond energy. The source of energy to break the

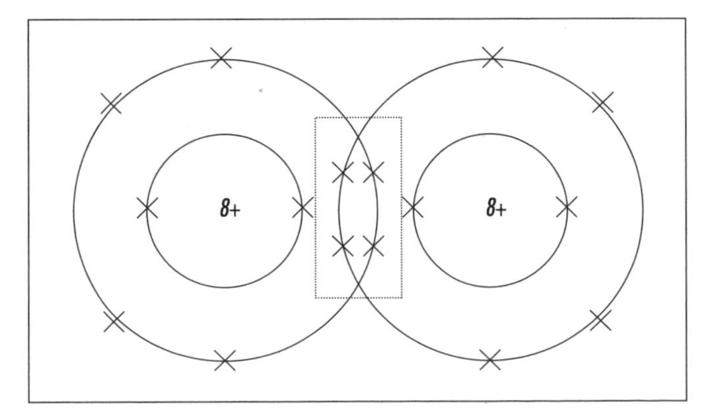

Bond + double covalent bond in oxygen. A bond is what holds atoms together in a molecule.

bonds may be in the form of heat, electricity, or mechanical means. Bond energy is the quantity of energy that must be absorbed to break a particular kind of chemical bond. It is equal to the quantity of energy the bond releases when it forms. It can also be defined as the amount of energy necessary to break one mole of bonds of a given kind (in gas phase).

See also BOND-DISSOCIATION ENERGY.

bond energy (mean bond energy) The average value of the gas-phase BOND-DISSOCIATION ENERGIES (usually at a temperature of 298 K) for all BONDs of the same type within the same CHEMICAL SPECIES. The mean bond energy for methane, for example, is one-fourth the enthalpy of reaction for

$$CH_4(g) \rightarrow C(g) + 4H(g)$$

Tabulated bond energies are generally values of bond energies averaged over a number of selected typical chemical species containing that type of bond.

bonding orbital A molecular orbital that is located between two atomic nuclei, the orbitals of which overlap and reinforce each other. Electrons in a bonding orbital tend to stabilize a molecule. The term also refers to a molecular orbital, the energy of which is lower than that of the atomic orbitals that are used in its construction.

bonding pair A pair of electrons used to form a covalent bond.

bond migration *See* MIGRATION.

bond number The number of electron-pair BONDS between two nuclei in any given LEWIS FORMULA. For example, in ethene, the bond number between the carbon atoms is two and between the carbon and hydrogen atoms is one.

bond order A theoretical index of the degree of bonding between two atoms relative to that of a normal single bond, i.e., the bond provided by one localized electron pair. In the VALENCE BOND THEORY, it is a weighted average of the bond numbers between the respective atoms in the CONTRIBUTING STRUCTURES. In MOLECULAR-ORBITAL THEORY, it is calculated from the weights of the atomic orbitals in each of the occupied molecular orbitals. For example, in valence-bond theory (neglecting other than KEKULÉ STRUCTURES), the bond order between adjacent carbon atoms in benzene is 1.5; in Hückel molecular-orbital theory, it is 1.67. Other variants of molecular-orbital theory provide other values for bond orders.

bone imaging The construction of bone tissue images from the radiation emitted by RADIONUCLIDES that have been absorbed by the bone. Radionuclides such as 18F, 85Sr, and ^{99m}Tc are introduced as complexes with specific LIGANDs (very often phosphonate ligands) and are absorbed in the bones by metabolic activity.

See also IMAGING.

borderline mechanism A mechanism intermediate between two extremes, e.g., a nucleophilic substitution intermediate between S_N1 and S_N2 or an intermediate between electron transfer and S_N2.

Born-Haber cycle A series of thermochemical reactions or cycles used for calculating the lattice energies of ionic crystalline solids.

boron hydrides Binary compounds of boron and hydrogen.

Bovet, Daniele (1907–1992) Swiss *Physiologist* Daniele Bovet was born in Neuchâtel, Switzerland, on March 23, 1907, to Pierre Bovet, professor of pedagogy at the University of Geneva, and Amy Babut. He graduated from the University of Geneva in 1927 and then worked on a doctorate in zoology and comparative anatomy, which he received in 1929.

During the years 1929 until 1947, he worked at the Pasteur Institute in Paris, starting as an assistant and later as chief of the Laboratory of Therapeutic Chemistry. Here he discovered the first synthetic antihistamine, pyrilamine (Mepyramine). In 1947 he went to Rome to organize a laboratory of therapeutic chemistry and became an Italian citizen. He became the laboratory's chief at the Istituto Superiore di Sanità, Rome. Seeking a substitute for curare, a muscle relaxant, for anesthesia, he discovered gallamine (trade name Flaxedil), a neuromuscular blocking agent used today as a muscle relaxant in the administration of anesthesia.

He and his wife Filomena Nitti published two important books: *Structure chimique et activité pharmacodynamique des médicaments du système nerveux végétatif* (The chemical structure and pharmacodynamic activity of drugs of the vegetative nervous system) in 1948 and, with G. B. Marini-Bettòlo, *Curare and Curare-like Agents* in 1959. In 1957 he was awarded the Nobel Prize in physiology or medicine for his discovery relating to synthetic compounds that blocked the effects of certain substances occurring in the body, especially in its blood vessels and skeletal muscles.

Bovet published more than 300 papers and received numerous awards. He served as the head of the psychobiology and psychopharmacology laboratory of the National Research Council (Rome) from 1969 until 1971, when he became professor of psychobiology at the University of Rome (1971–82). He died on April 8, 1992, in Rome.

Boyle's law The volume of a given mass of gas held at constant temperature is inversely proportional to the pressure under which it is measured. Articulated as $PV = k$.

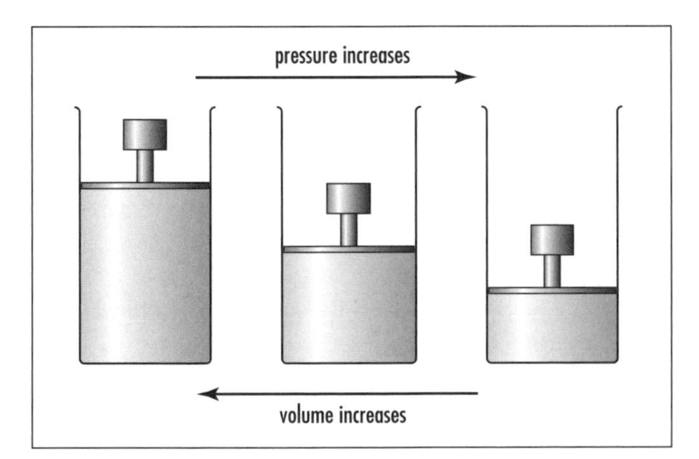

The volume of a given mass of gas varies inversely with its pressure at constant temperature.

Bragg equation An equation ($n\lambda = 2d\sin\theta$) in which:

n = order of diffracted beam
λ = wavelength of X-ray beam (in angstroms)
d = distance between diffracting planes (in angstroms)
θ = angle between incident X rays and the diffracting planes (in degrees)

Discovered by Lawrence Bragg in 1912, the equation deduces the angles at which X rays scatter from a crystal to the spacing between the layers of molecules.

brain imaging In addition to MAGNETIC RESONANCE IMAGING, which is based on the brain's absorption of electromagnetic radiation, brain images can be acquired by scintillation counting (scintigraphy) of radiation emitted from radioactive nuclei that have crossed the blood–brain barrier. The introduction of radionuclides into brain tissue is accomplished with the use of specific 99mTc(V) COMPLEXes with lipophilic ligands.

See also IMAGING.

Bredt's rule A double bond cannot be placed with one terminus at the bridgehead of a bridged ring system unless the rings are large enough to accommodate the double bond without excessive STRAIN. For example, while bicyclo[2.2.1]hept-1-ene is only capable of existence as a TRANSIENT SPECIES, its higher homologues having a double bond at the bridgehead position have been isolated. For example:

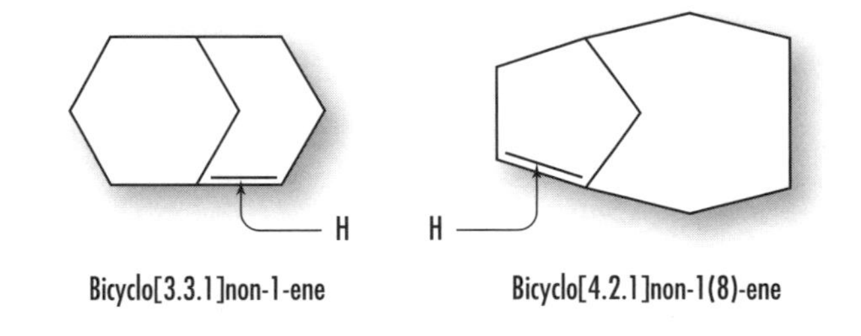

breeder reactor A nuclear reactor that produces and consumes fissionable fuel but creates more fissionable nuclear fuel than it consumes. The fission chain reaction is sustained by thermal neutrons.

bremsstrahlung German for "breaking radiation." Electromagnetic radiation is emitted when a charged particle changes its velocity or as it changes direction due to near collisions with other particles.

bridged carbocation A CARBOCATION (real or hypothetical) in which there are two (or more) carbon atoms that could, in alternative LEWIS FORMULAe, be designated as CARBENIUM CENTERs but which are instead represented by a structure in which a GROUP (a hydrogen atom or a hydrocarbon residue, possibly with substituents in noninvolved positions) bridges these potential carbenium centers. One can distinguish "electron-sufficient bridged carbocations" from "electron-deficient bridged carbocations." Examples of the former are phenyl-bridged ions (for which the trivial name "phenonium ion" has been used), such as depicted in Figure (A). These ions are straightforwardly classified as CARBENIUM IONs. The latter type of ion necessarily involves three-center bonding.

The hydrogen-bridged carbocation (B) contains a two-coordinate hydrogen atom. Hypercoordination—which includes two-coordination for hydrogen and at least five-coordination for carbon—is generally observed in bridged carbocations. Structures (C) and (D) contain five-coordinate carbon atoms.

See also MULTICENTER BOND; NEIGHBORING-GROUP PARTICIPATION.

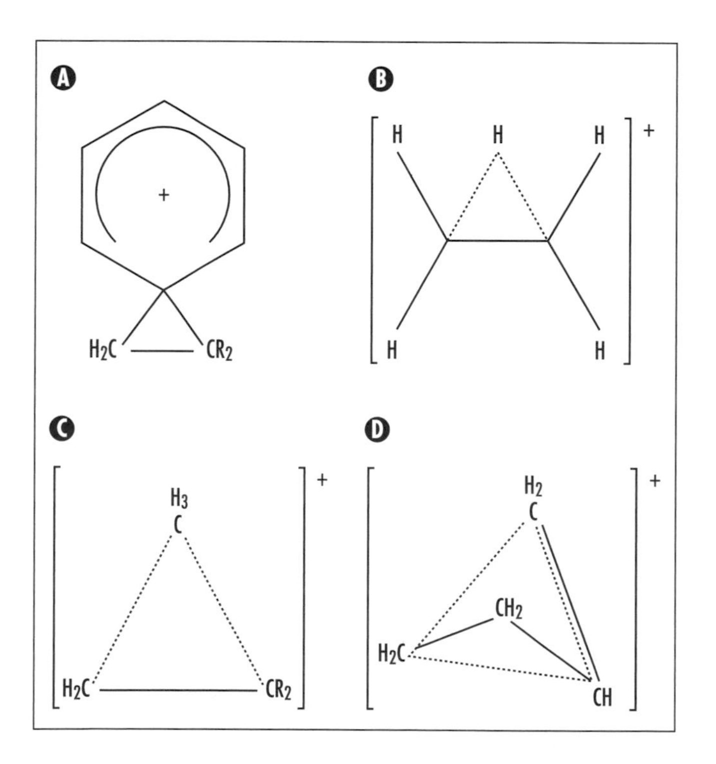

bridging ligand A bridging ligand binds to two or more CENTRAL ATOMs, usually metals, thereby linking them together to produce polynuclear COORDINATION entities. Bridging is indicated by the Greek letter μ appearing before the ligand name and separated by a hyphen.

See also FEMO COFACTOR.

Bronsted acid (Bronsted-Lowry acid) A molecular entity capable of donating a HYDRON to a base (i.e., a "hydron donor") or the corresponding chemical species. For example: H_2O, H_3O^+, CH_3CO_2H, H_2SO_4, HSO_4^-, HCl, CH_3OH, and NH_3.

See also CONJUGATE ACID–BASE PAIR.

Bronsted base (Bronsted-Lowry base) A molecular entity capable of accepting a HYDRON from an acid (i.e., a "hydron acceptor") or the corresponding chemical species. For example: OH^-, H_2O, $CH_3CO_2^-$, HSO_4^-, SO_4^{2-}, and Cl^-.

See also CONJUGATE ACID–BASE PAIR.

Bronsted relation The term applies to either of the following equations:

$$k_{HA}/p = G(K_{HA}q/p)^{\alpha}$$
$$k_A/q = G(K_{HA}q/p)^{-\beta}$$

(or their logarithmic forms) where α, β, and G are constants for a given reaction series (α and β are called "Bronsted exponents"), and k_{HA} and k_A are catalytic coefficients (or rate coefficients) of reactions whose rates depend on the concentrations of HA and/or of A^-. K_{HA} is the acid dissociation constant of the acid HA; p is the number of equivalent acidic protons in the acid HA; and q is the number of equivalent basic sites in its conjugate base A^-. The chosen values of p and q should always be specified. (The charge designations of HA and A^- are only illustrative.) The Bronsted relation is often termed the Bronsted CATALYSIS LAW. Although justifiable on historical grounds, this name is not recommended, since Bronsted relations are known to apply to many uncatalyzed and pseudo-catalyzed reactions (such as simple proton [HYDRON] transfer reactions). The term *pseudo-Bronsted relation* is sometimes used for reactions that involve NUCLEOPHILIC CATALYSIS instead of acid-base catalysis. Various types of Bronsted parameters have been proposed, such as β_{lg}, β_{nuc}, and β_{eq} for leaving group, nucleophile, and equilibrium constants, respectively.

See also LINEAR FREE-ENERGY RELATION.

Brownian movement The rapid but random motion of particles colliding with molecules of a gas or liquid in which they are suspended.

buckminsterfullerene (fullerene or buckyball) An ALLOTROPE of carbon containing clusters of 60 carbon atoms that vary in size, bound in a symmetric polyhedral structure. Robert Curl, Harold Kroto, and Richard Smalley discovered buckminsterfullerene, C_{60}, the third allotrope of carbon, in 1985. Using laser evaporation of graphite, they found clusters of which the most common were found to be C_{60} and C_{70}. For this discovery they were awarded the 1996 Nobel Prize in chemistry. The actual molecule was named after American architect Richard Buckminster Fuller (1895–1983) because its structure resembled his geodesic dome.

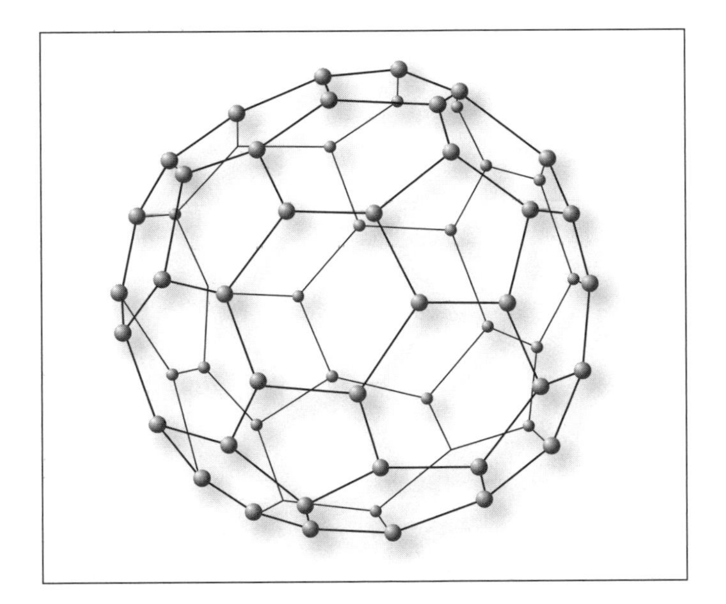

A carbon allotrope that contains clusters of 60 carbon atoms bound in a highly symmetrical polyhedral structure

buffer A molecule or chemical used to control the pH of a solution. It consists of acid and base forms and minimizes changes in pH when extraneous acids or bases are added to the solution. It prevents large changes in pH by either combining with H^+ or by releasing H^+ into solution.

See also PH SCALE.

buffer capacity The ability of a buffer solution to absorb added alkali or acid while maintaining the solution's pH.

bulk flow (pressure flow) Movement of water (or any other fluid) due to a difference in pressure between two locations. The movement of solutes in plant phloem tissue is an example.

Bunnett-Olsen equations The equations for the relation between $\lg([SH^+]/[S]) + H_o$ and $H_o + \lg[H^+]$ for base S in aqueous mineral acid solution, where H_o is Hammett's acidity function and $H_o + \lg[H^+]$ represents the activity function $\lg(\gamma_S \gamma_{H^+})/\gamma_{SH^+}$ for the nitroaniline reference bases to build H_o.

$$\lg([SH^+]/[S]) - \lg[H^+] = (\Phi - 1)(H_o + \lg[H^+]) + pK_{SH^+}$$
$$\lg([SH^+]/[S]) + H_o = \Phi\,(H_o + \lg[H^+]) + pK_{SH^+}$$

See also COX-YATES EQUATION.

buret A length of volumetric glass, usually graduated in 0.1-ml intervals, that is used to deliver solutions in a quantitative manner.

C

CADD *See* COMPUTER-ASSISTED DRUG DESIGN.

cage Aggregate of molecules, generally in the condensed phase, that surrounds the fragments formed, for example, by thermal or photochemical dissociation. Because the cage hinders the separation of the fragments by diffusion, they may preferentially react with one another ("cage effect") but not necessarily to re-form the precursor species. For example

$$R–N{=}N–R, \text{heat} \rightarrow [R^{\cdot} + N_2 + R^{\cdot}]\text{cage} \rightarrow R–R + N_2$$

See also GEMINATE RECOMBINATION.

cage compound A polycyclic compound having the shape of a cage. The term is also used for INCLUSION COMPOUNDS.

calmodulin A Ca^{2+} binding protein involved in METABOLIC REGULATION.

See also EF-HAND; HELIX.

calorie An energy measurement unit; the amount of energy required to raise the temperature of 1 gram of water by 1°C, equal to 4.1868 joules. The term *Calorie* (capitalized) is used in food science to represent a kilocalorie (1,000 calories) to describe the energy content of food products.

calorimeter An instrument used to measure quantities of heat.

calpain A calcium-activated neutral protease.

Calvin, Melvin (1911–1997) American *Chemist* Melvin Calvin was born on April 8, 1911, in St. Paul, Minnesota. He received a B.S. degree in chemistry in 1931 at the Michigan College of Mining and Technology and a Ph.D. degree in chemistry from the University of Minnesota in 1935. He conducted postdoctoral studies in England and then began his academic career as an instructor at the University of California at Berkeley in 1937 and as a full professor from 1947 until his death in 1997. He served as director of the big-organic chemistry group in the Lawrence Radiation Laboratory beginning in 1946. This group became the Laboratory of Chemical Biodynamics in 1960.

Calvin is the man known to have unlocked the secrets of photosynthesis. Calvin received the 1961 Nobel Prize in chemistry for identifying the path of carbon in photosynthesis. Shortly thereafter he established the Chemical Biodynamics Division (now Structural Biology Division), which he directed for 20 years.

Using carbon-14 isotope as a tracer, Calvin and his research team mapped the complete route that carbon travels through a plant during photosynthesis, beginning with its absorption as atmospheric carbon dioxide to its conversion into carbohydrates and other organic compounds. They showed that sunlight acts on the chlorophyll in a plant to fuel the manufacturing of organic compounds, rather than on carbon dioxide as was previously believed.

He married Genevieve Jemtegaard in 1951, and they had two daughters and a son.

In his final years of active research, he studied the use of oil-producing plants as renewable sources of energy and spent years testing the chemical evolution of life.

Throughout his distinguished career, Calvin received many awards and honors, including the National Medal of Science, which he received from President Bush in 1989; the Priestley Medal from the American Chemical Society; the Davy Medal from the Royal Society of London; and the Gold Medal from the American Institute of Chemists.

Calvin died on January 9, 1997, at Alta Bates Hospital in Berkeley, California, after a long illness.

Calvin cycle Discovered by chemist Melvin Calvin (1911–97), it is the second major stage in photosynthesis after light reactions whereby carbon molecules from

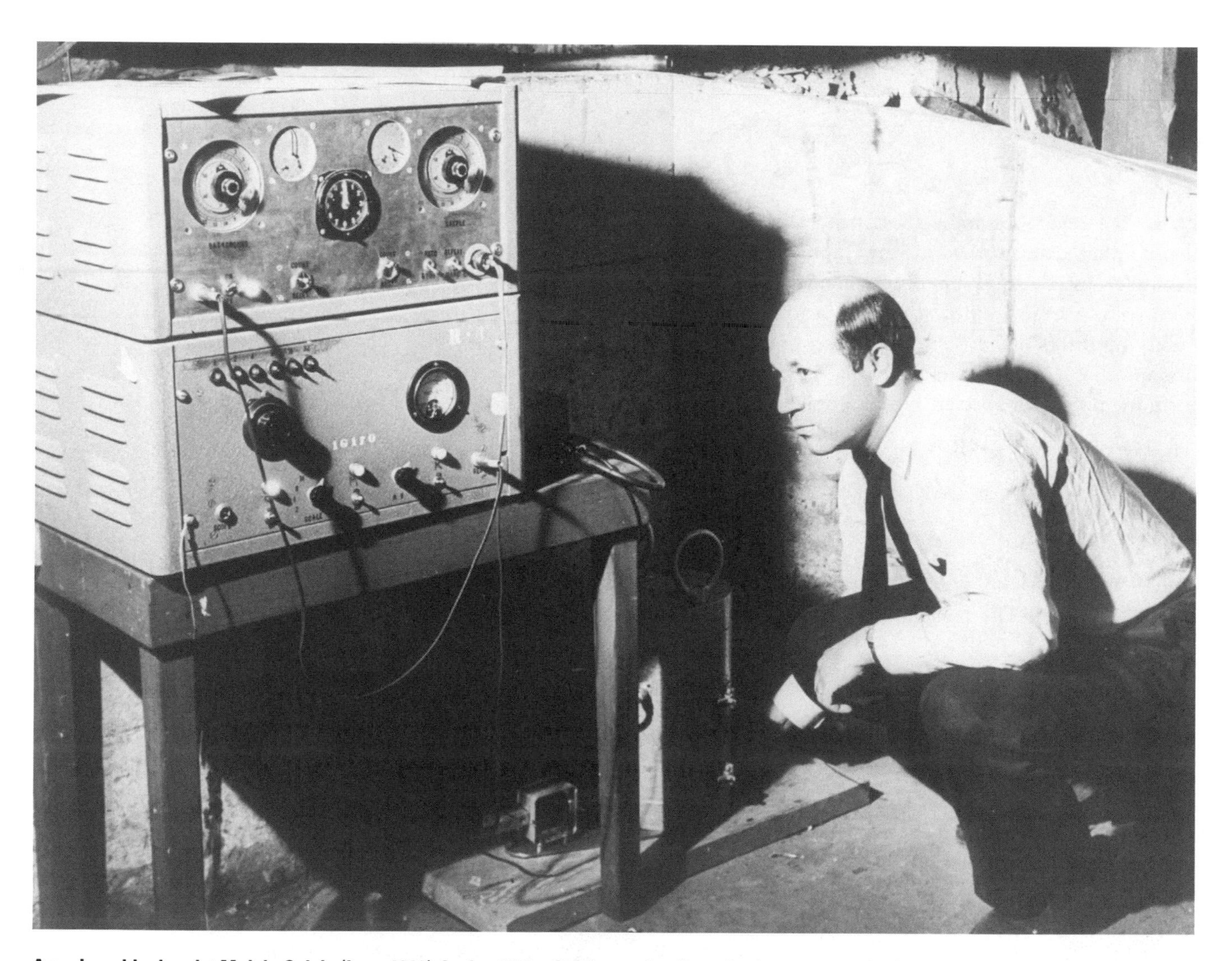

American biochemist Melvin Calvin (born 1911). In the 1950s, Calvin used radioactive isotopes to elucidate the chemical details of the process of photosynthesis. He won the Nobel Prize in chemistry in 1961. The photograph was taken at the University of California at Berkeley, where Calvin directed the chemical biodynamics laboratory in the Lawrence Radiation Laboratory (later the Lawrence Berkeley Laboratory). ***(Courtesy of Lawrence Berkeley Laboratory/Science Photo Library)***

CO_2 are fixed into sugar (glucose) and mediated by the enzyme rubisco (ribulose-1-5-biphosphate carboxylase). It occurs in the stroma of chloroplasts. The Calvin cycle is also known as the dark reaction, as opposed to the first-stage light reactions.

canonical form *See* CONTRIBUTING STRUCTURE.

capillary A tube with a very small inside diameter.

capillary action The rising of a liquid up the inside of a small-bore tube or vertical space when adhesive forces (the attractive forces between the capillary material and the liquid) exceed cohesive forces (the attractive forces between the molecules of the liquid itself).

captodative effect Effect on the stability of a carbon-centered RADICAL determined by the combined action of a captor (electron withdrawing) and a dative (electron releasing) substituent, both attached to the radical center. The term is also used for certain unsaturated compounds.

carbanion Generic name for anions containing an even number of electrons and having an unshared pair of electrons on a tervalent carbon atom (e.g., Cl_3C^- or $HC{\equiv}C^-$) or—if the ion is mesomeric (see MESOMERISM)—having at least one significant CONTRIBUTING STRUCTURE with an unshared pair of electrons on a tervalent carbon atom, for example,

$$CH_3C(-O^-){=}CH{-}C({=}O)CH_3 \leftrightarrow CH_3C({=}O){-}CH^-{-}C({=}O)CH_3$$

See also RADICAL ION.

carbene Generic name for the species H_2C: and substitution derivatives thereof, containing an electrically neutral bivalent carbon atom with two nonbonding electrons. The nonbonding electrons may have antiparallel spins (singlet state) or parallel spins (triplet state). Use of the alternative name *methylene* as a generic term is not recommended.

See also BIRADICAL.

carbenium center The three-coordinate carbon atom in a CARBENIUM ION, of which the excess positive charge of the ion (other than that located on heteroatoms) can be formally considered to be largely attributed, i.e., which has one vacant *P* ORBITAL. (It is not always possible to uniquely identify such an atom.) This formal attribution of charge often does not express the real charge distribution.

carbenium ion A generic name for CARBOCATION, real or hypothetical, that has at least one important CONTRIBUTING STRUCTURE containing a tervalent carbon atom with a vacant *P* ORBITAL. (The name implies a protonated carbene or a substitution derivative thereof.)

The term was proposed (and rejected) as a replacement for the traditional usage of the name CARBONIUM ION.

To avoid ambiguity, the name should not be used as the root for the systematic nomenclature of carbocations. The corresponding difficulty confused carbonium ion nomenclature for many years. For example, the term *ethylcarbonium ion* has at times been used to refer either to $CH_3CH_2^+$ (ethyl cation) or (correctly) to $CH_3CH_2CH_2^+$ (propyl cation).

carbenoid A CARBENE-like CHEMICAL SPECIES but with properties and REACTIVITY differing from the free carbene itself, e.g., $R^1R^2C(Cl)M$ (M = metal).

carbocation A cation containing an even number of electrons with a significant portion of the excess positive charge located on one or more carbon atoms. This is a general term embracing CARBENIUM IONs, all types of carbenium ions, vinyl cations, etc. Carbocations can be named by adding the word *cation* to the name of the corresponding RADICAL. Such names do not imply structure (e.g., whether three-coordinated or five-coordinated carbon atoms are present).

See also BRIDGED CARBOCATION; RADICAL ION.

carbohydrate A large class of compounds that contain carbon, hydrogen, and oxygen in a general formula of $C_n(H_2O)_n$. Classified from simple to complex, they form mono-, di-, tri-, poly-, and heterosaccharides.

Examples include sugars (monosaccharide, di- and polysaccharides), starches, and cellulose. Carbohydrates are used as an energy source by organisms, and most are formed by green plants and are obtained by animals via food intake.

carbonates Chemical compounds derived from carbonic acid or carbon dioxide.

carbon cycle All parts (reservoirs) and fluxes of carbon. The cycle is usually thought of as four main reservoirs of carbon interconnected by pathways of exchange. The reservoirs are the atmosphere, terrestrial biosphere (usually includes freshwater systems), oceans, and sediments (includes fossil fuels). The annual movements of carbon, the carbon exchanges between reservoirs, occur because of various chemical, physical, geological, and biological processes. The ocean contains the largest pool of carbon near the surface of the Earth, but most of that pool is not involved with rapid exchange with the atmosphere.

carbon dioxide (CO_2) A colorless, odorless gas that makes up the fourth most-abundant gas in the atmosphere, used by plants in carbon fixation. Atmospheric CO_2 has increased about 25 percent since the early 1800s due to burning of fossil fuels and deforestation. Increased amounts of CO_2 in the atmosphere enhance the greenhouse effect by blocking heat from escaping into space, thus contributing to the warming of Earth's lower atmosphere and having an effect on the world's biota. This is a major issue currently being debated by scientists around the world.

See also GREENHOUSE EFFECT.

carbon fixation The process by which carbon atoms from CO_2 gas are incorporated into sugars. Carbon fixation occurs in the chloroplasts of green plants or any photosynthetic or chemoautotrophic organism.

See also PHOTOSYNTHESIS.

carbon-14 dating Estimating the age of once-living material by measuring the amount of radioactive isotope of carbon present in the material tested.

Molecular models of assorted greenhouse gases, including carbon dioxide (right, with double bonds), methane (top, center), water (top, left), and several unidentified chlorofluorocarbons (CFCs). The buildup of these gases in the Earth's atmosphere traps an increased amount of solar radiation and leads to a gradual warming of the whole planet. The gases are generated by industry, the burning of fossil fuels, modern agricultural methods, and deforestation. Global warming is expected to cause massive changes in weather patterns and a rise in sea level due to polar ice melting that could flood coastal regions. ***(Courtesy of Adam Hart-Davis/Science Photo Library)***

carbonic anhydrase A zinc-containing ENZYME (carbonate hydrolyase, carbonate dehydratase) that catalyzes the reversible decomposition of carbonic acid to carbon dioxide and water.

carbonium ion The term should be used with great care, since several incompatible meanings are currently in use. It is not acceptable as the root for systematic nomenclature for CARBOCATIONs.

1. In most of the existing literature the term is used in its traditional sense for what is here defined as CARBENIUM ION.
2. A carbocation, real or hypothetical, that contains at least one five-coordinate carbon atom.
3. A carbocation, real or hypothetical, whose structure cannot adequately be described by two-electron two-center BONDs only. (The structure may involve carbon atoms with a COORDINATION NUMBER greater than five.)

carbon monoxide (CO) A colorless, odorless gas that is toxic.

carbon monoxide dehydrogenases ENZYMEs that catalyze the oxidation of carbon monoxide to carbon dioxide. They contain IRON-SULFUR CLUSTERs and either nickel and zinc or MOLYBDOPTERIN. Some nickel-containing enzymes are also involved in the synthesis of acetyl coenzyme A from CO_2 and H_2.

carbon sequestration The uptake and storage of carbon. Trees and plants, for example, absorb CARBON DIOXIDE, release the oxygen, and store the carbon. Fossil fuels were at one time biomass and continue to store the carbon until burned. Sometimes used to refer to any way in which carbon can be removed from active participation in the carbon cycle, such as by injecting carbon dioxide into depleted oil wells or the deep ocean as a means of controlling greenhouse gas emissions.

See also GREENHOUSE EFFECT.

carbonyl group A functional group with an oxygen atom double bonded to a carbon atom, e.g., aldehydes (joined to at least one hydrogen atom) and ketones (carbonyl group is joined to ALKYL GROUPs or ARYL GROUPs).

carboplatin A second-generation platinum drug effective in cancer chemotherapy named *cis*-diammine (cyclobutane-1,1-dicarboxylato)platinum(II). Carboplatin is less toxic than the first-generation antitumor drug, CISPLATIN.

carboxyl group A functional group that consists of a carbon atom joined to an oxygen atom by a double bond and to a HYDROXYL GROUP; present in all CARBOXYLIC ACIDs.

carboxylic acid Organic molecules with a CARBONYL GROUP in which the carbon is bonded to a HYDROXYL GROUP.

carbyne Generic name for the species HC: and substitution derivatives thereof, such as EtO_2C–C: containing an electrically neutral univalent carbon atom with three nonbonding electrons. Use of the alternative name *methylidyne* as a generic term is not recommended.

carcinogen Any substance known that may produce cancer.

cardiotech A species radiolabeled with ^{99m}Tc with the formula $[Tc(CNR)_6]^+$(R=*tert*-butyl) known for IMAGING the heart after a heart attack.

Carnot cycle Composed of four reversible processes—two isothermal and two adiabatic—and can be executed either in a closed or a steady-flow system. First proposed in 1824 by French engineer Sadi Carnot (1796–1832).

carotenoids A large family of natural phytochemicals, accessory pigments, found in plants (in chloroplasts) and animals. They are composed of two small six-carbon rings connected by a carbon chain that must be attached to cell membranes. Their variety of colors absorb wavelengths that are not available to chlorophyll and so serve to transfer their captured energy from the Sun to help in photosynthesis. Carotenoids color fruits and vegetables and give them their characteristic red, orange, and yellow colors and serve as antioxidants in human nutrition. Over 600 carotenoids are known.

carrier-linked prodrug (**carrier prodrug**) A PRODRUG that contains a temporary linkage of a given active substance with a transient carrier group that produces improved physicochemical or pharmacokinetic properties and that can be easily removed *in vivo*, usually by a hydrolytic cleavage.

cascade prodrug A PRODRUG for which the cleavage of the carrier group becomes effective only after unmasking an activating group.

catabolic pathway The process for taking large complex organic molecules and breaking them down into smaller ones, which release energy that can be used for metabolic processes.

catabolism Reactions involving the breaking down of organic SUBSTRATES, typically by oxidative breakdown, to provide chemically available energy (e.g., ATP) or to generate metabolic intermediates used in subsequent anabolic reactions.

See also ANABOLISM; METABOLISM.

catabolite A naturally occurring METABOLITE.

catabolite activator protein (CAP) A protein that binds cyclic adenosine monophosphate (cAMP), a regulatory molecule, to DNA in organisms. When this interaction takes place, the gene promoter is made accessible to the enzyme RNA polymerase, and transcription of the gene can begin.

catalase A HEME protein that catalyzes the DISPROPORTIONATION of dihydrogen peroxide to O_2 and water. It also catalyzes the oxidation of other compounds, such as ethanol, by dihydrogen peroxide. A nonheme protein containing a dinuclear manganese CLUSTER with catalase activity is often called *pseudocatalase*.

catalysis The action of a CATALYST.

catalysis law *See* BRONSTED RELATION.

catalyst A substance that participates in a particular CHEMICAL REACTION and thereby increases its rate, but without a net change in the amount of that substance in the system. At the molecular level, the catalyst is used and regenerated during each set of MICROSCOPIC CHEMICAL EVENTs leading from a MOLECULAR ENTITY of reactant to a molecular entity of product.

See also AUTOCATALYTIC REACTION; BIFUNCTIONAL CATALYSIS; CATALYTIC COEFFICIENT; ELECTRON-TRANSFER CATALYSIS; GENERAL ACID CATALYSIS; GENERAL BASE CATALYSIS; HOMOGENEOUS CATALYST; HETEROGENEOUS CATALYST; INTRAMOLECULAR CATALYSIS; MICELLAR CATALYSIS; MICHAELIS-MENTEN KINETICS; PHASE-TRANSFER CATALYSIS; PSEUDOCATALYSIS; RATE OF REACTION; SPECIFIC CATALYSIS.

catalytic antibody (abzyme) An ANTIBODY that catalyzes a chemical reaction analogous to an enzymatic reaction, such as an ester hydrolysis. It is obtained by using a hapten that mimics the transition state of the reaction.

See also ENZYME.

catalytic coefficient If the RATE OF REACTION *(v)* is expressible in the form

$$v = (k_0 + \Sigma k_i[C_i]^{n_i}) [A]^\alpha [B]\beta\ldots$$

where A, B, ... are reactants and C_i represents one of a set of catalysts, then the proportionality factor k_i is the catalytic coefficient of the particular CATALYST C_i. Normally, the partial order of reaction (n_i) with respect to a catalyst will be unity, so that k_i is an $(\alpha + \beta + \ldots + 1)$th-order rate coefficient. The proportionality factor k_0 is the $(\alpha + \beta + \ldots)$th-order rate coefficient of the uncatalyzed component of the total reaction.

catenation The tendency of an element to form bonds to itself into chains or rings.

cathode The negative part of an electric field; electrode where reduction occurs.

cathode-ray tube Closed glass tube containing a gas under low pressure; produces cathode rays (electrons) when high voltage is applied.

cathodic protection Protection of a metal against corrosion by making it the cathodic part of an electrochemical cell in which the anode is a more easily oxidized metal.

cation A positively charged ION.

cation exchange The ability of some natural and human-made substances to attract and exchange cations with the solution with which they are in contact. An important characteristic of soils, where the ability is high for clays and humus and low for sand.

cationotropic rearrangement (**cationotropy**) *See* TAUTOMERISM.

cation radical *See* RADICAL ION.

CBS (**colloidal bismuth subcitrate**) *See* DE-NOL.

CD *See* CIRCULAR DICHROISM.

cell The basic unit of life, capable of growing and multiplying. All living things are either single, independent cells or aggregates of cells. A cell is composed of cytoplasm and a nucleus and is surrounded by a membrane or wall. Cells can be categorized by the presence of specific cell surface markers called *clusters of differentiation.*

See also UNIT CELL.

cellular respiration The process in which ATP is created by metabolizing glucose and oxygen and the release of carbon dioxide. Occurs in the MITOCHONDRIA of EUKARYOTES and in the CYTOPLASM of prokaryotes.

cellulose A polysaccharide, polymer of glucose, that is found in the cell walls of plants. A fiber that is used in many commercial products, notably paper.

Celsius, Anders (1701–1744) Swedish *Astronomer, physicist* Anders Celsius was a Swedish astronomer, physicist, and mathematician who introduced the Celsius temperature scale that is used today by scientists in most countries. He was born in Uppsala, Sweden, a city that has produced six Nobel Prize winners. Celsius was born into a family of scientists all originating from the province of Hälsingland. His father Nils Celsius was a professor of astronomy, as was his grandfather Anders Spole, and his other grandfather, Magnus Celsius, was a professor of mathematics; both grandfathers were at the University in Uppsala. Several of his uncles also were scientists.

Celsius's important contributions include determining the shape and size of the Earth; gauging the magnitude of the stars in the constellation Aries; publishing a catalog of 300 stars and their magnitudes; observing eclipses and other astronomical events; and preparing a study that revealed that the Nordic countries were slowly rising above the sea level of the Baltic. His most famous contribution falls in the area of temperature, and the one he is remembered most for is the creation of the Celsius temperature scale.

In 1742 he presented to the Swedish Academy of Sciences his paper, "Observations on Two Persistent Degrees on a Thermometer," in which he presented his observations that all thermometers should be made on a fixed scale of 100 divisions (centigrade), based on two points: 0 degrees for boiling water, and 100 degrees for freezing water. He presented his argument on the inaccuracies of existing scales and calibration methods and correctly presented the influence of air pressure on the boiling point of water.

After his death, the scale that he designed was reversed, giving rise to the existing 0° for freezing and 100° for boiling water, instead of the reverse. It is not known if the reversal was done by his student Martin Stromer; by botanist Carolus Linnaeus, who in 1745 reportedly showed the senate at Uppsala University a thermometer so calibrated; or by Daniel Ekström, who manufactured most of the thermometers used by both Celsius and Linnaeus. However, Jean Christin from France made a centigrade thermometer with the current calibrations (0° freezing, 100° boiling) a year after Celsius and independent of him, and so he may therefore equally claim credit for the existing "Celsius" thermometers.

For years Celsius thermometers were referred to as "Centigrade" thermometers. However, in 1948, the Ninth General Conference of Weights and Measures ruled that "degrees centigrade" would be referred to as "degrees Celsius" in his honor. The Celsius scale is still used today by most scientists.

Anders Celsius was secretary of the oldest Swedish scientific society, the Royal Society of Sciences in

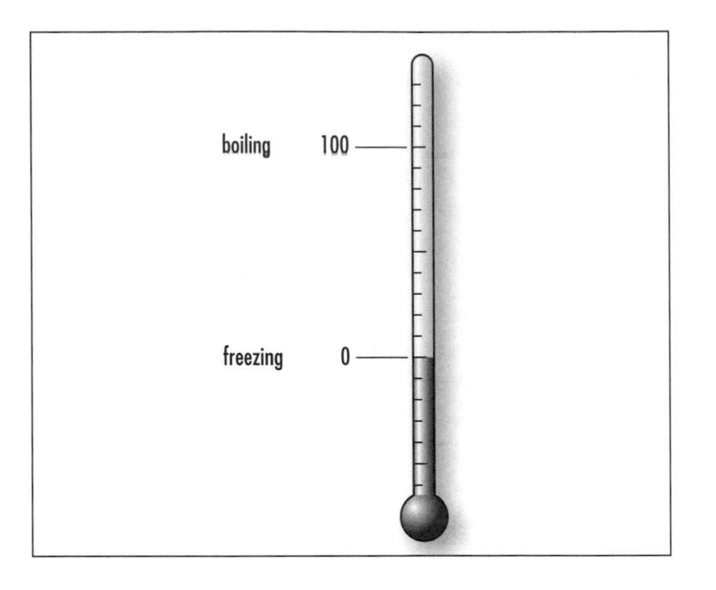

A temperature scale where boiling is 100°C and freezing is 0°C

Uppsala, between 1725 and 1744 and published much of his work through that organization, including a math book for youth in 1741. He died of tuberculosis on April 25, 1744, in Uppsala.

Celsius scale (**centigrade scale**) A temperature scale with the range denoted by °C, as seen in the above figure. The normal freezing point of water is 0°C, and the normal boiling point of water is 100°C. The scale was named after ANDERS CELSIUS, who proposed it in 1742 but designated the freezing point to be 100 and the boiling point to be 0 (reversed after his death).

central atom The atom in a COORDINATION entity that binds other atoms or group of atoms (LIGANDs) to itself, thereby occupying a central position in the coordination entity.

ceramic Formed of clay (aluminosilicates), in whole or in part, and baked. Also used to describe other refractory materials, such as oxides and nitrides, with network structures.

ceruloplasmin A copper protein present in blood plasma, containing type 1, type 2, and type 3 copper centers, where the type 2 and type 3 are close together, forming a triNUCLEAR copper CLUSTER.

See also MULTICOPPER OXIDASES; TYPE 1, 2, 3 COPPER.

Chain, Ernst Boris (1906–1979) German/British *Biochemist* Ernst Boris Chain was born in Berlin on June 19, 1906, to Dr. Michael Chain, a chemist and industrialist. He was educated at the Luisen gymnasium, Berlin, with an interest in chemistry. He attended the Friedrich-Wilhelm University, Berlin, and graduated in chemistry in 1930. After graduation he worked for three years at the Charité Hospital, Berlin, on enzyme research. In 1933, after the rise of the Nazi regime in Germany, he left for England.

In 1935 he was invited to Oxford University, and in 1936 he became demonstrator and lecturer in chemical pathology. In 1948 he was appointed scientific director of the International Research Center for Chemical Microbiology at the Istituto Superiore di Sanità, Rome. He became professor of biochemistry at Imperial College, University of London, in 1961, until 1973. Later, he became professor emeritus and senior research fellow (1973–76) and fellow (1978–79).

From 1935 to 1939 he worked on snake venoms, tumor metabolism, the mechanism of lysozyme action, and the invention and development of methods for biochemical microanalysis. In 1939 he began a systematic study of antibacterial substances produced by microorganisms and the reinvestigation of penicillin. Later he worked on the isolation and elucidation of the chemical structure of penicillin and other natural antibiotics.

With pathologist Howard Walter Florey (later Baron Florey), he isolated and purified penicillin and performed the first clinical trials of the antibiotic. For their pioneering work on penicillin Chain, Florey, and Fleming shared the 1945 Nobel Prize in physiology or medicine.

Later, his research topics included carbohydrate-amino acid relationship in nervous tissue, a study of the mode of action of insulin, fermentation technology, 6-aminopenicillanic acid and penicillinase-stable penicillins, lysergic acid production in submerged culture, and the isolation of new fungal metabolites.

Chain was the author of many scientific papers and a contributor to important monographs on penicillin and antibiotics, and he was the recipient of many awards, including being knighted in 1969. He died of heart failure in Ireland on August 12, 1979.

chain reaction A reaction in which one or more reactive reaction INTERMEDIATES (frequently RADICALS) are continuously regenerated, usually through a repetitive cycle of elementary steps (the "propagation step"). For example, in the chlorination of methane by a radical MECHANISM, Cl· is continuously regenerated in the chain propagation steps:

$$Cl\cdot + CH_4 \rightarrow HCl + H_3C\cdot$$
$$H_3C\cdot + Cl_2 \rightarrow CH_3Cl + Cl\cdot$$

In chain polymerization reactions, reactive intermediates of the same types, generated in successive steps or cycles of steps, differ in relative molecular mass, as in

$$RCH_2C\cdot HPh + H_2C{=}CHPh \rightarrow RCH_2CHPhCH_2C\cdot HPh$$

See also CHAIN TRANSFER; INITIATION; TERMINATION.

chain transfer The abstraction, by the RADICAL end of a growing chain polymer, of an atom from another molecule. The growth of the polymer chain is thereby terminated, but a new radical, capable of chain propagation and polymerization, is simultaneously created. For the example of alkene polymerization cited for a CHAIN REACTION, the reaction

$$RCH_2C\cdot HPh + CCl_4 \rightarrow RCH_2CHClPh + Cl_3C\cdot$$

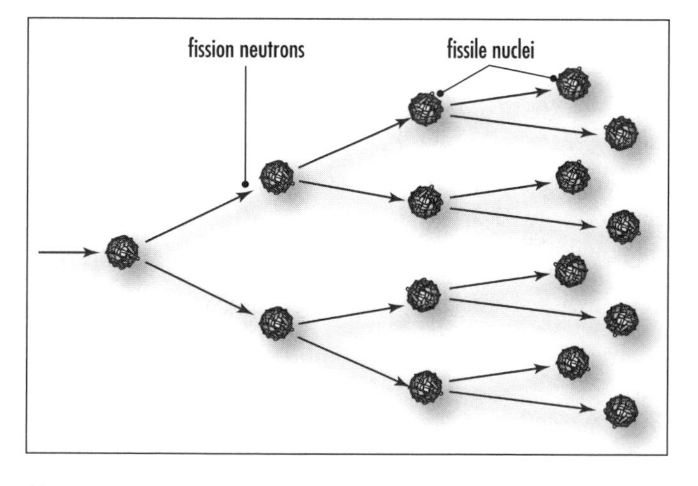

Chain reaction. Any reaction where one reaction leads to another that leads to another, and so forth

represents a chain transfer, with the radical $Cl_3C\cdot$ inducing further polymerization

$$H_2C{=}CHPh + Cl_3C\cdot \rightarrow Cl_3CCH_2C\cdot HPh$$
$$Cl_3CCH_2C\cdot HPh + H_2C{=}CHPh \rightarrow$$
$$Cl_3CCH_2CHPhCH_2C\cdot HPh$$

The phenomenon occurs also in other chain reactions such as cationic polymerization.

See also TELOMERIZATION.

chalcogen One of the elements in the same column of the periodic table as oxygen.

channels Transport proteins that act as gates to control the movement of sodium and potassium ions across the plasma membrane of a nerve cell.

See also ACTIVE TRANSPORT.

chaperonin Member of the set of molecular chaperones, located in different organelles of the cell and involved either in transport of proteins through BIOMEMBRANES by unfolding and refolding the proteins or in assembling newly formed polypeptides.

charge density *See* ELECTRON DENSITY.

charge population The net electric charge on a specified atom in a MOLECULAR ENTITY, as determined by some prescribed definition.

See also ELECTRON DENSITY.

charge-transfer complex An aggregate of two or more molecules in which charge is transferred from a donor to an acceptor.

charge-transfer transition An electronic transition in which a large fraction of an electronic charge is transferred from one region of a molecular entity, called the electron donor, to another, called the electron acceptor (intramolecular charge-transfer), or from one molecular entity to another (intermolecular charge-transfer).

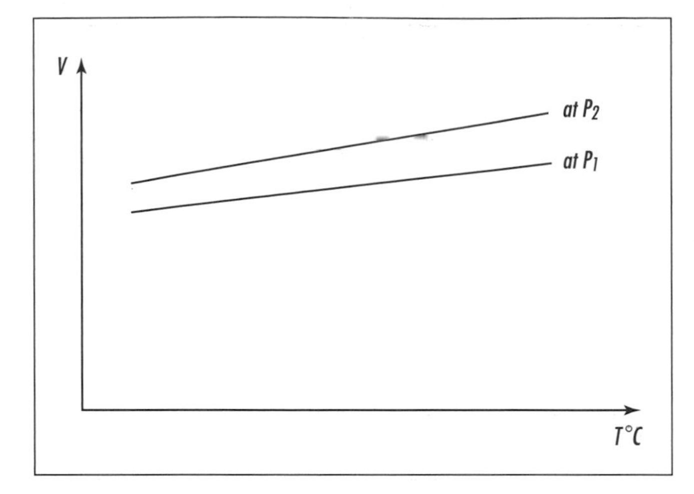

Charles' law. The volume *(V)* of a fixed mass of gas at constant pressure *(P)* is dependent on its temperature *(T)*.

Charles' law At constant pressure, the volume occupied by a definite mass of gas is directly proportional to its absolute temperature.

chelation The formation or presence of BONDs (or other attractive interactions) between two or more separate BINDING SITEs within the same LIGAND and a single central atom. A MOLECULAR ENTITY in which there is chelation (and the corresponding CHEMICAL SPECIES) is called a "chelate." The terms *bidentate* (or *didentate*), *tridentate, tetradentate, ..., multidentate* are used to indicate the number of potential binding sites of the ligand, at least two of which must be used by the ligand in forming a "chelate." For example, the bidentate ethylenediamine forms a chelate with CuI in which both nitrogen atoms of ethylenediamine are bonded to copper. (The use of the term is often restricted to metallic central atoms.)

The phrase "separate binding sites" is intended to exclude cases such as $[PtCl_3(CH_2{=}CH_2)]^-$, ferrocene, and (benzene)tricarbonylchromium, in which ethene, the cyclopentadienyl group, and benzene, respectively, are considered to present single binding sites to the respective metal atom, and which are not normally thought of as chelates (see HAPTO).

See also CRYPTAND.

chelation therapy The judicious use of chelating (metal binding) agents for the removal of toxic amounts of metal ions from living organisms. The metal ions are sequestered by the chelating agents and are rendered harmless or excreted. Chelating agents such as 2,3-dimercaptopropan-1-ol, ethylenediaminetetraacetic acid, DESFERRIOXAMINE, and D-penicillamine have been used effectively in chelation therapy for arsenic, lead, iron, and copper, respectively.

See also CHELATION.

cheletropic reaction A form of CYCLOADDITION across the terminal atoms of a fully CONJUGATED SYSTEM, with formation of two new SIGMA BONDs to a single atom of the ("monocentric") reagent. There is formal loss of one PI BOND in the substrate and an increase in COORDINATION NUMBER of the relevant atom of the reagent. An example is the ADDITION of sulfur dioxide to butadiene:

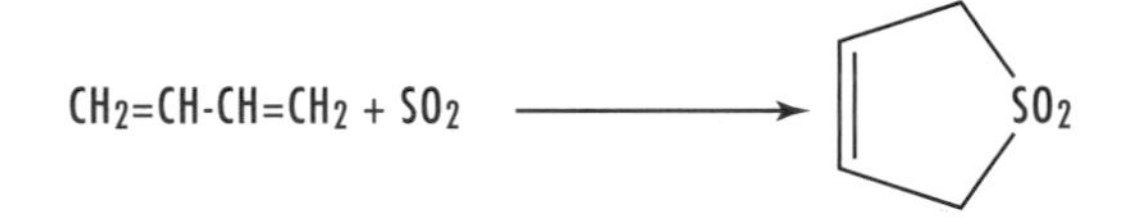

The reverse of this type of reaction is designated "cheletropic elimination."

See also CHELOTROPIC REACTION.

chelotropic reaction Alternative (and etymologically more correct) name for CHELETROPIC REACTION.

chemical bond The attractive force that binds atoms together in elements or compounds.

chemical change Occurs when atoms in a substance are rearranged so that a new substance with a new chemical identity is formed.

chemical equation A graphic representation of a chemical reaction.

chemical equilibrium The condition when the forward and reverse reaction rates are equal and the con-

Metal displacement reaction. Experiment demonstrating the displacement of silver from solution by copper. ***(Courtesy Jerry Mason/Photo Researchers, Inc.)***

centrations of the products remain constant. Also called the law of chemical equilibrium.

chemical flux A concept related to RATE OF REACTION, particularly applicable to the progress in one direction only of component reaction steps in a complex system or to the progress in one direction of reactions in a system at dynamic equilibrium (in which there are no observable concentration changes with time).

See also ORDER OF REACTION; RATE-LIMITING STEP; STEADY STATE.

chemical formula A scientific notation in which the composition of a compound is illustrated. It consists of atomic symbols for the various elements of the compound together with numerical subscripts indicating the ratio in which the atoms combine.

chemical kinetics The study of rates and mechanisms of CHEMICAL REACTIONS.

chemical periodicity Properties of the elements are periodic functions of atomic number. As you move from left to right in a row of the periodic table, the properties of the elements change gradually, but at the end of each row a drastic shift in chemical properties occurs. The next element in order of atomic number is more chemically similar to the first element in the row above it; thus a new row begins on the table. Therefore, chemical periodicity is the variations in properties of elements with their position in the periodic table.

chemical property The characteristics of a substance that describe how it undergoes or resists change to form a new substance.

chemical reaction A process that results in the interconversion of CHEMICAL SPECIES. Chemical reactions may

Female scientist in a laboratory writes a chemical formula on glass while observed by a male scientist. Both scientists are wearing protective white coats and safety glasses. H_3O^+, written on the glass, is a water molecule with an extra hydrogen atom. At ordinary temperatures, two molecules of water are in equilibrium with H_3O^+ and OH^-. ***(Courtesy of Tek Image/Science Photo Library)***

be ELEMENTARY REACTIONS or STEPWISE REACTIONS. (It should be noted that this definition includes experimentally observable interconversions of conformers.)

Detectable chemical reactions normally involve sets of MOLECULAR ENTITIES, as indicated by this definition, but it is often conceptually convenient to use the term also for changes involving single molecular entities (i.e., microscopic chemical events).

See also IDENTITY REACTION.

chemical relaxation If the equilibrium mixture of a CHEMICAL REACTION is disturbed by a sudden change, especially of some external parameter (such as temperature, pressure, or electrical field strength), the system will readjust itself to a new position of the chemical equilibrium or return to the original position, if the perturbation is temporary. The readjustment is known as chemical relaxation.

In many cases, and in particular when the displacement from equilibrium is slight, the progress of the system toward equilibrium can be expressed as a first-order law

$$[C_t-(C_{eq})_2] = [(C_{eq})_1-(C_{eq})_2]e^{-t/\tau}$$

where $(C_{eq})_1$ and $(C_{eq})_2$ are the equilibrium concentrations of one of the chemical species involved in the reaction before and after the change in the external parameter, respectively, and C_t is its concentration at time t. The time parameter t, named relaxation time, is related to the RATE CONSTANTs of the chemical reaction involved.

Measurements of the relaxation times by relaxation methods (involving a temperature jump [T-jump], pressure jump, electric field jump, or a periodic disturbance of an external parameter, as in ultrasonic techniques) are commonly used to follow the kinetics of very fast reactions.

See also RELAXATION.

chemical shift **(NMR),** δ (SI unit: 1) The variation of the resonance frequency of a nucleus in NUCLEAR MAGNETIC RESONANCE SPECTROSCOPY (NMR) in consequence of its magnetic environment. The chemical shift of a nucleus, δ, is expressed in ppm by its frequency, ν_{cpd}, relative to a standard, ν_{ref}, and is defined as

$$\delta = 10^6(\nu_{cpd}-\nu_{ref})/\nu_o$$

where ν_o is the operating frequency of the spectrometer. For ^1H and ^{13}C NMR, the reference signal is usually that of tetramethylsilane ($SiMe_4$). Other references are used in the older literature and in other solvents, such as D_2O.

If a resonance signal occurs at lower frequency or higher applied field than an arbitrarily selected reference signal, it is said to be upfield, and if resonance occurs at higher frequency or lower applied field, the signal is downfield. Resonance lines upfield from $SiMe_4$ have positive δ-values, and resonance lines downfield from $SiMe_4$ have negative δ-values.

chemical species An ensemble of chemically identical MOLECULAR ENTITIES that can explore the same set of molecular energy levels on the time scale of the experiment. The term is applied equally to a set of chemically identical atomic or molecular structural units in a solid array.

For example, two conformational ISOMERs may be interconverted sufficiently slowly to be detectable by separate NMR spectra and hence to be considered as separate chemical species on a time scale governed by the radiofrequency of the spectrometer used. On the other hand, in a slow chemical reaction, the same mixture of conformers may behave as a single chemical species, i.e., there is virtually complete equilibrium population of the total set of molecular energy levels belonging to the two conformers.

Except where the context requires otherwise, the term is taken to refer to a set of molecular entities containing isotopes in their natural abundance.

The wording of the definition given in the first paragraph is intended to embrace cases such as graphite, sodium chloride, or a surface oxide, where the basic structural units may not be capable of isolated existence, as well as those cases where they are.

In common chemical usage, and in this encyclopedia, generic and specific chemical names (such as RADICAL or hydroxide ion) or chemical formulae refer either to a chemical species or to a molecular entity.

chemical weight The weight of a molar sample as determined by the weight of the molecules (the molecular weight); calculated from the weights of the atoms in the molecule.

Assorted chemical weights. The different amount of each chemical represents a measurement known as the mole. One mole of any sample contains the same number of molecules or atoms. The weight (shown in grams) of each molar sample is determined by the weight of the molecules (the molecular weight) and is calculated from the weights of the atoms in the molecule. Three of the chemicals are hydrated with water (H_2O). The chemical formulae are, clockwise from lower left: NaCl, $FeCl_3 \cdot 6(H_2O)$, $CuSO_4 \cdot 5(H_2O)$, KI, $Co(NO_3)_2 \cdot 6(H_2O)$, and $KMnO_4$. The heaviest atom here is iodine (I), eight times heavier than oxygen (O). The transition-metal compounds are colored. ***(Courtesy of Andrew Lambert Photography/Science Photo Library)***

chemiluminescence Spontaneous light emission created by the chemical (enzymatic) conversion of a non-light-emitting substrate.

chemiosmosis A method of making ATP that uses the ELECTRON-TRANSPORT CHAIN and a PROTON PUMP to transfer hydrogen protons across certain membranes and then utilize the energy created to add a phosphate group (phosphorylate) to adenosine diphosphate (ADP), creating ATP as the end product.

chemistry The science that studies matter and all its transformations.

chemoautotroph (chemolithotroph) An organism that uses carbon dioxide as its carbon source and obtains energy by oxidizing inorganic substances.

chemoheterotroph Any organism that derives its energy by oxidizing organic substances for both carbon source and energy.

chemoreceptor A sense organ, cell, or structure that detects and responds to chemicals in the air or in solution.

chemoselectivity, chemoselective Chemoselectivity is the preferential reaction of a chemical reagent with one of two or more different functional groups. A reagent has a high chemoselectivity if reaction occurs with only a limited number of different functional groups. For example, sodium tetrahydroborate is a more chemoselective reducing agent than is lithium tetrahydroaluminate. The concept has not been defined in more quantitative terms. The term is also applied to reacting molecules or intermediates that exhibit selectivity toward chemically different reagents.

Some authors use the term *chemospecificity* for 100 percent chemoselectivity. However, this usage is discouraged.

See also REGIOSELECTIVITY; STEREOSELECTIVITY; STEREOSPECIFICITY.

chemospecificity *See* CHEMOSELECTIVITY, CHEMOSELECTIVE.

chemotherapy The treatment of killing cancer cells by using chemicals.

chirality A term describing the geometric property of a rigid object (or spatial arrangement of points or atoms) that is nonsuperimposable on its mirror image; such an object has no symmetry elements of the second kind (a mirror plane, a center of inversion, a rotation reflection axis). If the object is superimposable on its mirror image, the object is described as being achiral.

chi-square test An enumeration-statistic exercise that compares the frequencies of various kinds or categories of items in a random sample to the frequencies that are expected if the population frequencies are as hypothesized by the researcher.

chitin The long-chained structural polysaccharide found in the exoskeleton of invertebrates such as crustaceans, insects, and spiders and in some cell walls of fungi. A beta-1,4-linked homopolymer of N-acetyl-D-glucosamine.

chloralkali cell Consists of two inert electrodes in a salt solution.

chlorin 2,3-Dihydroporphyrin. An unsubstituted, reduced PORPHYRIN with two nonfused saturated carbon atoms (C-2, C-3) in one of the pyrrole rings.

chlorofluorocarbon Gases formed of chlorine, fluorine, and carbon whose molecules normally do not react with other substances. Formerly used as spray-can propellants, they are known to destroy the Earth's protective ozone layer.

chlorophyll Part of the photosynthetic systems in green plants. Generally speaking, it can be considered as a magnesium complex of a PORPHYRIN in which a double bond in one of the pyrrole rings (17–18) has been reduced. A fused cyclopentanone ring is also present (positions 13–14–15). In the case of chlorophyll *a*, the substituted porphyrin ligand further contains four methyl groups in positions 2, 7, 12, and 18, a vinyl group in position 3, an ethyl group in position 8, and a $-(CH_2)_2CO_2R$ group (R = phytyl, (2*E*)-(7*R*, 11*R*)-3,7,11,15-tetramethylhexadec-2-en-1-yl) in position 17. In chlorophyll *b*, the group in position 7 is a –CHO group. In bacteriochlorophyll *a* the porphyrin ring is further reduced (7–8), and the group in position 3 is now a $-COCH_3$ group. In addition, in bacteriochlorophyll *b*, the group in position 8 is a $=CHCH_3$ group.

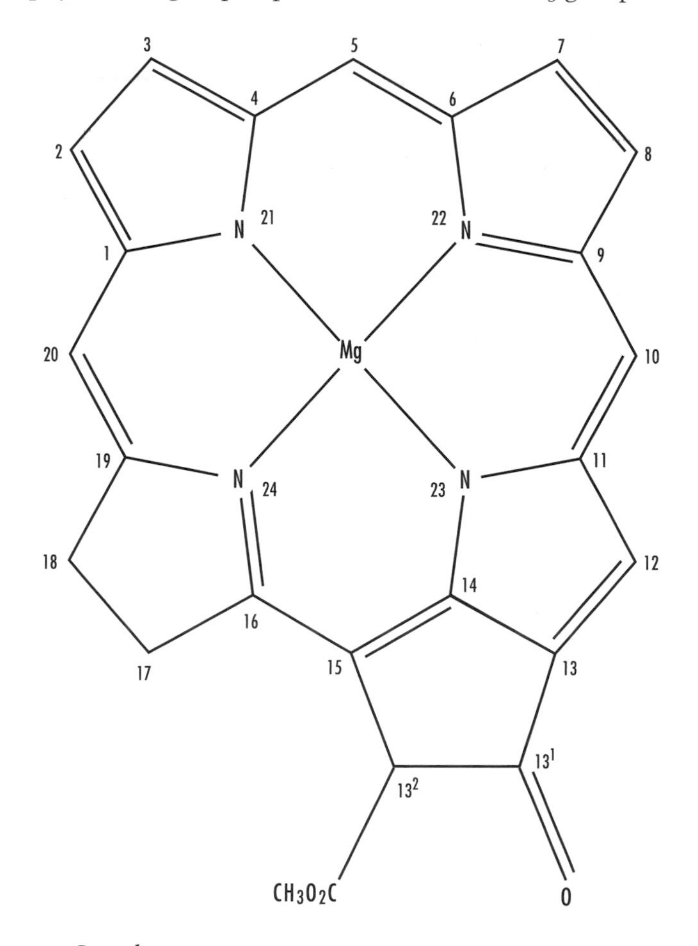

See also PHOTOSYNTHESIS.

chloroplast The double-membrane organelle of eukaryotic photosynthesis; contains enzymes and pigments that perform photosynthesis.

See also EUKARYOTE.

cholesterol A soft, waxy, fat-soluble steroid formed by the liver and a natural component of fats in the bloodstream (as LIPOPROTEINs); the most common steroid in the human body and used by all cells in permeability of their membranes. It is used in the formation of many products such as bile acids, vitamin D, progesterone, estrogens, and androgens. In relation to human health, there is "good," high-density cholesterol (HDL), which protects the heart, and "bad," low-density cholesterol (LDL), which causes heart disease and other problems.

chromatin The combination of DNA and proteins that make up the chromosomes of EUKARYOTEs. Exists as long, thin fibers when cells are not dividing; not visible until cell division takes place.

chromatography A method of chemical analysis where compounds are separated by passing a mixture in a suitable carrier over an absorbent material. Compounds with different absorption coefficients move at different rates and are separated.

See also GAS CHROMATOGRAPHY; HIGH-PERFORMANCE LIQUID CHROMATOGRAPHY.

chromophore That part of a molecular entity consisting of an atom or group of atoms in which the electronic transition responsible for a given spectral band is approximately localized.

chromosome The self-replicating GENE-carrying member found in the CELL nucleus and composed of a DNA molecule and proteins (chromatin). Prokaryote organisms contain only one chromosome (circular DNA), while EUKARYOTES contain numerous chromosomes that comprise a genome. Chromosomes are divided into functional units called genes, each of which contains the genetic code (instructions) for making a specific protein.

See also NUCLEUS.

Series of chromatograms showing the separation of black ink. Chromatography is an analytical process, which separates a compound into its constituent chemicals. Chromatography paper is dipped vertically in a solvent with the ink painted on it (left). Capillary action draws the solvent up through the paper (center) and dissolves the ink. As the solvent travels up the paper, it takes the various chemicals in the ink with it, separating them into a series of colored bands. *(Courtesy of Andrew Lambert Photography/ Science Photo Library)*

CIDNP (chemically induced dynamic nuclear polarization) Non-Boltzmann nuclear spin state distribution produced in thermal or photochemical reactions, usually from COLLIGATION and DIFFUSION, or DISPROPORTIONATION of RADICAL PAIRs, and detected by NUCLEAR MAGNETIC RESONANCE SPECTROSCOPY by enhanced absorption or emission signals.

cine-substitution A SUBSTITUTION REACTION (generally AROMATIC) in which the entering group takes up a position adjacent to that occupied by the LEAVING GROUP. For example,

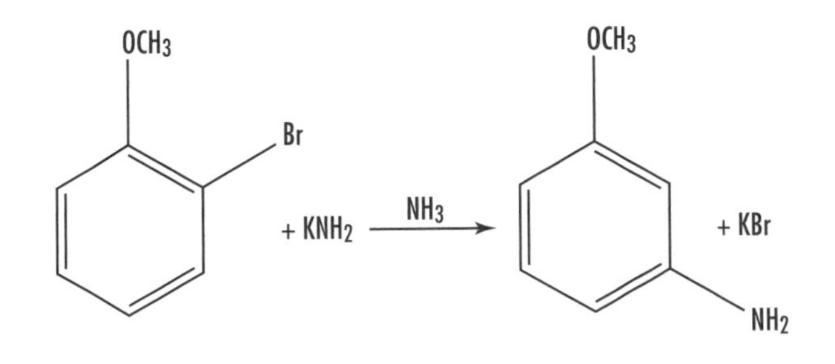

See also TELE-SUBSTITUTION.

circular dichroism (CD) A spectroscopic method that measures the difference in absorbance of left- and right-handed circularly polarized light by a material as a function of the wavelength. Most biological molecules, including PROTEINS and NUCLEIC ACIDS, are chiral and show circular dichroism in their ultraviolet absorption bands, which can be used as an indication of SECONDARY STRUCTURE. Metal centers that are bound to such molecules, even if they have no inherent CHIRALITY, usually exhibit CD in absorption bands associated with LIGAND-based or ligand-metal CHARGE- TRANSFER TRANSITIONS. CD is frequently used in combination with absorption and MAGNETIC CIRCULAR DICHROISM (MCD) studies to assign electronic transitions.

cis In inorganic nomenclature, a structural prefix designating two groups occupying adjacent positions (not generally recommended for precise nomenclature purposes of complicated systems).

See also TRANS-.

cisplatin *cis*-Diamminedichloroplatinum(II). An antitumor drug highly effective in the chemotherapy of many forms of cancer. Of major importance in the antitumor activity of this drug is its interaction with the NUCLEIC ACID bases of DNA.

***cis-trans* isomerism** Compounds with double bonds, or alicyclic rings, may exhibit isomerism (a molecule possessing the same molecular formula but with the atoms arranged in a different way) due to the attached groups lying above or below the plane of the double bond or ring. The *cis* compound has the groups on the same side of the bond, while the *trans* has the groups on the opposite sides. The different ISOMERS have different physical and chemical properties.

class (a) metal ion A metal ion that combines preferentially with LIGANDS containing ligating atoms that are the lightest of their periodic group.

See also CLASS (B) METAL ION; HARD ACID.

class (b) metal ion A metal ion that combines preferentially with LIGANDS containing ligating atoms other than the lightest of their periodic group.

See also CLASS (A) METAL ION; HARD ACID.

clathrate *See* HOST; INCLUSION COMPOUND.

Clausius-Clapeyron equation The differential equation relating pressure of a substance to temperature in a system in which two phases of the substance are in equilibrium. Also referred to as Clapeyron equation, or Clapeyron-Clausius equation.

clay A very fine-grained soil that is plastic when wet but hard when fired. Typical clays consists of silicate and aluminosilicate minerals that are the products of weathering reactions of other minerals, but the term is also used to refer to any mineral of very small particle size.

clone A population of organisms, cells, viruses, or DNA molecules that is derived from the replication of a single genetic progenitor. In the case of B cells, each B cell has a typical IMMUNOGLOBULIN (Ig), and so all the cells that descend from one B cell (the clone) have the same Ig. Typically, a cancer is a clone of cells. Sometimes the term *clone* is also used for a number of recombinant DNA molecules that are all carrying the same inserted SEQUENCE.

close packing The structure of compounds based on the stacking of spheres in arrangements, where the occupied volume of the structure is maximized.

cloud chamber An instrument or chamber filled with a supersaturated vapor and designed for observing the paths of speeding ionizing particles, which appear as a trail of condensed liquid droplets as they pass through. A related device is a bubble chamber, which uses a liquid close to its boiling point and leaves a trail of bubbles to reveal the path of the ionizing particle.

cluster A number of metal centers grouped close together that can have direct metal-bonding interactions or interactions through a BRIDGING LIGAND, but that are not necessarily held together by these interactions. Examples can be found under the entries [2FE-2S],

[4FE-4S], HIPIP, IRON-SULFUR CLUSTER, FEMO COFACTOR, FERREDOXIN, FERRITIN, METALLOTHIONEIN, NITROGENASE, and RIESKE IRON-SULFUR PROTEIN.

coagulation The clotting of blood.

coal A solid that is composed of tightly bound three-dimensional networks of hydrocarbon chains and rings, formed from the decomposition of living things and compressed under high pressure. Used for fuel and heating.

cobalamin Vitamin B12. A vitamin synthesized by microorganisms and conserved in animals in the liver. Deficiency or collective uptake of vitamin B12 leads to pernicious ANEMIA. Cobalamin is a substituted CORRIN–Co(III) complex in which the cobalt atom is bound to the four nitrogen atoms of the corrin ring, an axial group R, and 5,6-dimethylbenzimidazole.The latter is linked to the cobalt by the N-3 nitrogen atom and is bound to the C-1 carbon of a ribose molecule by the N-1 nitrogen atom. Various forms of the vitamin are known with different R groups such as R=CN, cyanocobalamin; R=OH, hydroxocobalamin; R=CH_3, methylcobalamin; R=adenosyl, COENZYME B12.

codon A sequence of three consecutive NUCLEOTIDES that occurs in MRNA and directs the incorporation of a specific amino acid into a protein, or represents the starting or termination signal of protein synthesis.

coefficient of variation A measure of dispersion around the mean (average).

coenzyme A low-molecular-weight, nonprotein organic compound (often a NUCLEOTIDE) participating in enzymatic reactions as a dissociable acceptor or donor of chemical groups or electrons.

See also ENZYME.

cofactor An organic molecule or ion (usually a metal ion) that is required by an ENZYME for its activity. It can be attached either loosely (COENZYME) or tightly (PROSTHETIC GROUP).

cohesion The force of attraction between molecules of the same substance that allows them to bind.

cohesive force Force that enables something to stick to itself.

coke A hard, dry substance containing carbon that is produced by heating bituminous coal to a very high temperature in the absence of air. Used as a fuel for iron and steel foundries.

collagen The most abundant fibrous protein in the human body (about 30 percent) and the animal kingdom; shapes the structure of tendons, bones, and connective tissues. There are several types (I, II, III, IV) that are found in bone, skin, tendons, cartilage, embryonic tissues, and basement membranes.

colligation The formation of a covalent bond by the combination or recombination of two RADICALS (the reverse of unimolecular HOMOLYSIS). For example:

$$HO\cdot + H_3C\cdot \rightarrow CH_3OH$$

See also MOLECULARITY.

colligative properties Physical properties of solutions that depend on the concentration of solute molecules or ions in a solution, but not on the kind or identity of solute particles present in the solution.

collision theory A theory of reaction rates that assumes that molecules must collide in order to react.

colloid A heterogeneous mixture in which very small particles of a substance are dispersed in another medium. Although sometimes referred to as colloidal solutions, the dispersed particles are typically much larger than molecular scale. However, they do not settle out.

colloidal bismuth subcitrate (CBS) *See* DE-NOL.

combination reaction A reaction in which two substances, either elements or compounds, combine with each other to form one substance or compound.

combinatorial library A set of compounds prepared by COMBINATORIAL SYNTHESIS.

combinatorial synthesis A process for preparing large sets of organic compounds by combining sets of building blocks in all possible combinations.

combustible Refers to any material that will burn.

combustion An exothermic oxidation-reduction reaction between molecular oxygen and a nonmetallic substance.

CoMFA *See* COMPARATIVE MOLECULAR FIELD ANALYSIS.

common-ion effect (on rates) A reduction in the RATE OF REACTION of a SUBSTRATE RX in solution (by a path that involves a PRE-EQUILIBRIUM with formation of R^+ [or R^-] ions as reaction intermediates) caused by the addition to the reaction mixture of an electrolyte solute containing the "common ion" X^- (or X^+). For example, the rate of solvolysis of diphenylmethyl chloride in acetone-water is reduced by the addition of salts of the common ion Cl^-, which causes a decrease in the quasi-equilibrium concentration of the diphenylmethyl cation in the scheme

$$Ph_2CHCl \rightleftharpoons Ph_2CH^+ + Cl^- \text{ (free ions, not ion pairs)}$$

$$Ph_2CH^+ + OH_2 \xrightarrow{2} Ph_2CHOH + H^+\text{(solvated)}$$

This phenomenon is a direct consequence of the MASS-LAW EFFECT on ionization equilibria in electrolytic solution.

More generally, the common-ion effect is the influence of the "common ion" on the reactivity due to the shift of the dissociation equilibrium. It may also lead to an enhancement of the rates of reaction.

comparative molecular field analysis (CoMFA) A THREE-DIMENSIONAL QUANTITATIVE STRUCTURE-ACTIVITY RELATIONSHIP (3D-QSAR) method that uses statistical correlation techniques for the analysis of (a) the quantitative relationship between the biological activity of a set of compounds with a specified alignment and (b) their three-dimensional electronic and steric properties. Other properties such as hydrophobicity and hydrogen bonding can also be incorporated into the analysis.

compensation effect In a considerable number of cases, plots of $T\Delta^{\ddagger}S$ vs. $\Delta^{\ddagger}H$, for a series of reactions, e.g., for a reaction in a range of different solvents, are straight lines of approximately unit slope. Therefore, the terms $\Delta^{\ddagger}H$ and $T\Delta^{\ddagger}S$ in the expression partially compensate, and $\Delta^{\ddagger}G = \Delta^{\ddagger}H - T\Delta^{\ddagger}S$ often is a much simpler function of solvent (or other) variation than $\Delta^{\ddagger}H$ or $T\Delta^{\ddagger}S$ separately.

See also ISOKINETIC RELATIONSHIP.

competitive exclusion principle The condition where one species is driven out of a community by extinction due to interspecific competition; one species will dominate the use of resources and have a reproductive advantage, forcing the others to disappear. Also called Gauss's law.

competitive inhibitor A substance that resembles the substrate for an enzyme, both in shape and size, and competes with the substrate for the substrate-binding site on the enzyme, thereby reducing the rate of reaction by reducing the number of enzyme molecules that are successful.

complementary binding site *See* BINDING SITE.

complementary DNA (cDNA) A laboratory-produced DNA section that is created by extracting a single-stranded RNA from an organism as a template and transcribing it back into a double-stranded DNA using the enzyme reverse transcriptase. However, the cDNA does not include introns, those portions of the DNA that were spliced out while still in the cell. Complementary DNA is used for research purposes and can be cloned into plasmids for storage.

complex A MOLECULAR ENTITY formed by loose ASSOCIATION involving two or more component molecular entities (ionic or uncharged), or the corresponding CHEMICAL SPECIES. The bonding between the components is normally weaker than in a covalent BOND.

The term has also been used with a variety of shades of meaning in different contexts; it is therefore best avoided when a more explicit alternative is applicable. In inorganic chemistry the term *coordination entity* is recommended instead of *complex*.

See also ACTIVATED COMPLEX; ADDUCT; CHARGE-TRANSFER COMPLEX; ELECTRON-DONOR-ACCEPTOR COMPLEX; ENCOUNTER COMPLEX; INCLUSION COMPOUND; PI (π) ADDUCT; SIGMA (σ) ADDUCT, TRANSITION STATE.

complex ions Ions composed of two or more ions or radicals, both of which can exist alone.

composite reaction A CHEMICAL REACTION for which the expression for the rate of disappearance of a reactant (or rate of appearance of a product) involves rate constants of more than a single ELEMENTARY REACTION. Examples are opposing reactions (where rate constants of two opposed chemical reactions are involved), parallel reactions (for which the rate of disappearance of any reactant is governed by the rate constants relating to several simultaneous reactions to form different respective products from a single set of reactants), and STEPWISE REACTIONS.

composition stoichiometry Descibes the quantitative (mass) relationships among elements in compounds. Stoichiometry is the calculation of the quantities of reactants and products involved in a chemical reaction.

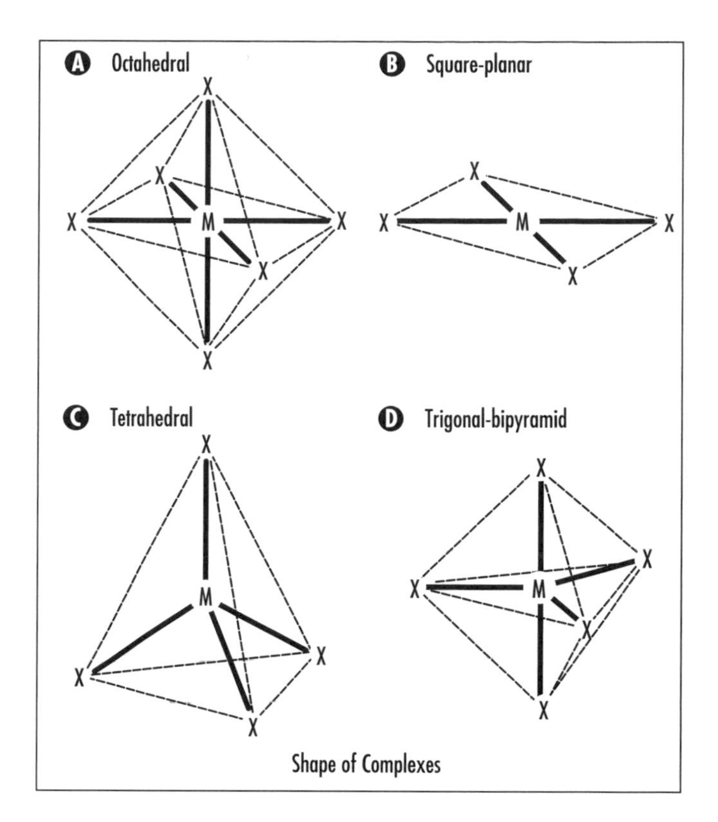

Shape of Complexes

compost Fertilizer formed by decaying organic matter.

compound The combination of two or more different elements, held together by chemical bonds. The elements in a given compound are always combined in the same proportion by mass (law of definite proportion).

compressed gas A gas or mixture of gases having, in a container, an absolute pressure exceeding 40 psi at 21.1°C (70°F).

comproportionation Refers to a mixture of species in different oxidation states that react and produce a product that is in a different but more stable intermediate oxidation state. A type of redox reaction, e.g., when iodide and iodate ions react to form elemental iodine. The reverse of DISPROPORTIONATION.

computational chemistry A discipline using mathematical methods to calculate molecular properties or to simulate molecular behavior.

computer-assisted drug design (CADD) Involves all computer-assisted techniques used to discover, design, and optimize biologically active compounds with a putative use as DRUGs.

concanavalin A A protein from jack beans, containing calcium and manganese, that agglutinates red blood cells and stimulates T lymphocytes to undergo mitosis.

concentration A quantitative measure of the amount of a solute in a solution. Can be an amount of solute per unit volume or mass of solvent or of solution.

concerted process Two or more PRIMITIVE CHANGES are said to be concerted (or to constitute a concerted process) if they occur within the same ELEMENTARY REACTION. Such changes will normally (though perhaps not inevitably) be "energetically coupled." (In the present context, the term *energetically coupled* means that the simultaneous progress of the primitive changes involves a TRANSITION STATE of lower energy than that for their successive occurrence.) In a concerted process, the primitive changes may be SYNCHRONOUS or asynchronous.

See also BIFUNCTIONAL CATALYSIS; POTENTIAL-ENERGY (REACTION) SURFACE.

condensation The transformation of gas to a liquid.

condensation polymer A polymer made by condensation polymerization; formed by the combination of MONOMERs and the release of a small molecules at the point where monomers are joined.

condensation reaction A (usually stepwise) reaction in which two or more reactants (or remote reactive sites within the same MOLECULAR ENTITY) yield a single main product, with accompanying formation of water or of some other small molecule, e.g., ammonia, ethanol, acetic acid, hydrogen sulfide.

The *mechanism* of many condensation reactions has been shown to comprise consecutive ADDITION and ELIMINATION reactions, as in the base-catalyzed formation of *(E)*-but-2-enal (crotonaldehyde) from acetaldehyde, via 3-hydroxybutanal (aldol). The overall reaction in this example is known as the aldol condensation.

The term is sometimes also applied to cases where the formation of water or another simple molecule does not occur. Also called dehydration reaction.

condensed phases The liquid and solid phases, not gases; phases in which particles interact strongly.

conduction band A partially filled or empty energy level in which electrons are free to move. It allows the material to conduct an electrical current when an electric field is applied by means of an applied voltage.

confidence limits The results of a statistical analysis. The lower and upper boundaries or values of a confidence interval. A range of values that is estimated from a study group that is highly likely to include the true, although unknown, value.

configuration In the context of stereochemistry, the term is restricted to the arrangements of atoms of a molecular entity in space that distinguishes STEREOISOMERs, the isomerism between which is not due to CONFORMATIONal differences.

configuration (electronic) A distribution of the electrons of an atom or a MOLECULAR ENTITY over a set of one-electron wave functions called ORBITALs, according to the Pauli principle. From one configuration, several states with different multiplicities may result. For

The Power of Chemistry: Natural versus Synthetic Compounds

by Theresa Beaty, Ph.D.

Indigo is the dye that makes the "blue" in blue jeans. It has been used for thousands of years to color textiles. Some civilizations have also used indigo in paint and cosmetics. Until the late 1800s, indigo was produced from plant sources on tropical plantations, mainly in India. In fact, the name *indigo* is derived from India. After harvesting the *Indigofera* plants, workers on the indigo plantations spent many days preparing just one batch of the dye. The plants were soaked in water to extract the colorless compound, indecan. This mixture was fermented for up to 15 hours, during which the indecan was converted into indoxyl. This yellow liquid was agitated while the color changed to green and then blue. Dark flakes were formed, and the mixture was boiled to remove impurities. The flakes were filtered and pressed to remove moisture, then cut into cubes and air-dried. This dried *indigo* was ready for market. Because of the lengthy processing required to produce the dye, indigo was very expensive.

During the last half of the 19th century, organic chemists discovered how to synthesize different dyes in the laboratory. Although some of the chemical steps involved in the synthesis reactions were complicated, these synthetic dyes were still much cheaper than the natural dyes isolated from plants or shellfish. The chemical synthesis of indigo was first published in 1882. This chemical reaction started with *o*-nitrobenzaldehyde, a component of coal tar. Acetone was added under basic conditions (dilute NaOH), and the resulting compound formed a dimer, indigo.

This initial synthesis reaction was modified in the late 1890s for large-scale commercial production. It required petroleum-based starting reagents and generated toxic by-products. However, synthetic indigo was easier to produce than natural indigo, and therefore cheaper. The synthetic dye became increasingly popular. By World War I, nearly all of the indigo sold on international markets was synthesized in laboratories.

The effects of synthetic indigo on society were more widespread than one might think. Synthetic indigo was not merely the cheaper source of dye. It also contributed to the eventual independence of India from the British Empire. Because the process of generating natural indigo was so labor-intensive, thousands of workers were affected when the plantations became too expensive to efficiently operate during the early 1900s. The plantation workers had to work longer hours in order to try and produce more natural indigo. Since most of the plantations were in India, this contributed to the social and political unrest in that country. Mahatma Gandhi was one of the leaders who used the terrible conditions on the plantations as a means to organize the Indian population to protest British rule. Thus, the economic turmoil caused by the chemical synthesis of indigo contributed to the efforts of the Indian people to become independent of Great Britain.

While synthetic indigo has enjoyed a virtual monopoly for nearly a century, another method for generating an environmentally friendly indigo is under development. At the end of the 20th century, the enzymes required for cellular indigo synthesis were cloned into bacteria. When these genetically modified bacteria are fed tryptophan, they synthesize indigo and secrete it into the growth medium. This bioindigo is not yet economical because the bacteria produce it so slowly. However, scientists continue to enhance the growth conditions of these biological indigo factories. Perhaps one day, the "blue" in blue jeans will be primarily produced by genetically engineered bacteria.

Besides indigo, many other natural products have been produced in the laboratory. One of these is quinine, a compound used for hundreds of years to treat malaria. Quinine is naturally derived from the bark of the tropical *Cinchona* trees found in Amazonia. Chemists had been trying to synthesize this important drug since the mid-1800s, and finally succeeded in the 1940s. The availability of synthetic quinine helped the Allied troops combat malaria in the Pacific during World War II, and thus it may indirectly be partly responsible for the outcome of the war.

Another compound derived from bark, this time from the Pacific yew tree, is paclitaxol, better known as Taxol. This drug has potent antitumor activity. However, each yew tree makes so little of the compound that the bark of several old trees is required to treat just one cancer patient. Since harvesting the bark kills the tree, there is a tremendous drive to generate a high-yield synthesis reaction for Taxol. While the compound was first synthesized in the laboratory in the 1990s, the yields are too small to be practical. Chemists are continuing to work on a better synthesis pathway for this drug. In addition, derivatives of Taxol are being created that may have even more potent anticancer activity.

Other natural products that have synthetic or semisynthetic versions include antibiotics, antifungals, and anesthetics.

Creating synthetic versions of useful natural products certainly benefits society by producing cheaper compounds. However, synthetic chemistry can also be environmentally friendly. Improved synthetic pathways can reduce the amount of toxic by-products formed during some chemical reactions. In addition, the availability of synthetic compounds eliminates the need to continually harvest large quantities of rare plants or other organisms in order to isolate the natural product.

— **Theresa Beaty, Ph.D.**, is an associate professor
in the department of chemistry & physics at
Le Moyne College in Syracuse, New York.

example, the ground electronic configuration of the oxygen molecule (O_2) is

$$1\,\sigma_g^2, 1\,\sigma_u^2, 2\,\sigma_g^2, 2\,\sigma_u^2, 1\,\pi_u^4, 3\,\sigma_g^2, 1\,\pi_g^2$$

resulting in the

$$^3\Sigma_g,\ ^1\Delta_g, \text{ and } ^3\Sigma_g^+ \text{ multiplets}$$

configuration (molecular) In the context of stereochemistry, the term is restricted to the arrangements of atoms of a MOLECULAR ENTITY in space that distinguishes stereoisomers, the isomerism of which is not due to conformational differences.

conformation The spatial arrangements of atoms affording distinction between STEREOISOMERS that can be interconverted by rotations about formally single bonds. One of the possible spatial orientations of a single molecule.

congener A substance literally *con-* (with) *generated* or synthesized by essentially the same synthetic chemical reactions and the same procedures. ANALOGs are substances that are analogous in some respect to the prototype agent in chemical structure.

Clearly congeners may be analogs or vice versa, but not necessarily. The term *congener,* while most often a synonym for HOMOLOGUE, has become somewhat more diffuse in meaning so that the terms *congener* and *analog* are frequently used interchangeably in the literature.

conjugate acid-base pair The BRONSTED ACID BH^+, formed on protonation of a base B, is called the conjugate acid of B, and B is the conjugate base of BH^+. (The conjugate acid always carries one unit of positive charge more than the base, but the absolute charges of the species are immaterial to the definition.) For example, the Bronsted acid HCl and its conjugate base Cl^- constitute a conjugate acid-base pair.

conjugated double bonds Double bonds that are separated from each other by one single bond such as $CH_2{=}CH{-}CH{=}CH_2$.

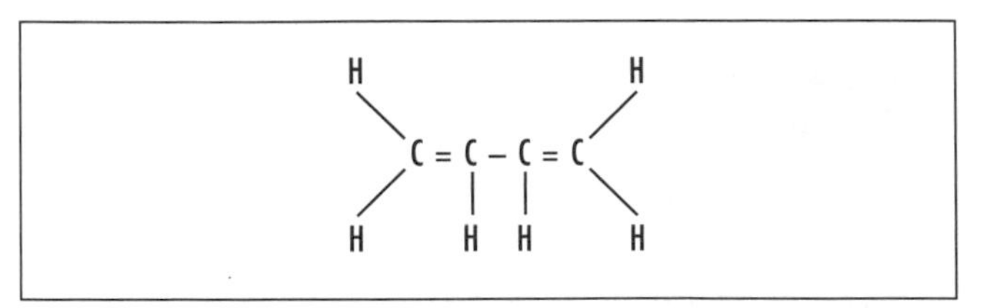

A conjugated structure has alternate single and double bonds between carbon atoms in an organic compound.

conjugated system (conjugation) In the original meaning, a conjugated system is a molecular entity whose structure can be represented as a system of alternating single and multiple bonds, e.g.,

$$CH_2{=}CH{=}CH{=}CH_2 \quad CH_2{=}CH{-}C{\equiv}N$$

In such systems, conjugation is the interaction of one *p*-orbital with another across an intervening sigma bond in such structures. (In appropriate molecular entities, *d*-orbitals may be involved.) The term is also extended to the analogous interaction involving a *p*-orbital containing an unshared electron pair, e.g.,

$$:Cl{-}CH{=}CH_2$$

See also DELOCALIZATION; HOMOCONJUGATION; RESONANCE.

conjugative mechanism *See* ELECTRONIC EFFECT.

connectivity In a chemical context, the information content of a line formula, but omitting any indication of BOND multiplicity.

conrotatory *See* ELECTROCYCLIC REACTION.

consensus sequence A SEQUENCE of DNA, RNA, protein, or carbohydrate derived from a number of similar molecules that comprises the essential features for a particular function.

constitution The description of the identity and connectivity (and corresponding bond multiplicities) of the atoms in a MOLECULAR ENTITY (omitting any distinction from their spatial arrangement).

consumer Any organism that ingests matter and energy of other organisms.

contact ion pair *See* ION PAIR.

contact process A method in which sulfur trioxide and sulfuric acid are produced from sulfur dioxide.

continuous spectrum A spectrum in which there are no absorption or emission lines. It contains all wavelengths in a specified region of the electromagnetic spectrum.

contrast agent A PARAMAGNETIC (or FERROMAGNETIC) metal complex or particle causing a decrease in the RELAXATION times (increase in relaxivity) of nuclei detected in an IMAGE, usually of water. Contrast agents, sometimes referred to as "dyes," are used in medical imaging techniques such as computed tomography (CT) scans to highlight specific areas such as blood vessels, organs, or tissues to make them more visible.

contributing structure The definition is based on the valence-bond formulation of the quantum mechanical idea of the wave function of a molecule as composed of a linear combination of wave functions, each representative of a formula containing bonds that are only single, double, or triple, with a particular pairing of electron spins. Each such formula represents a contributing structure, also called "resonance structure" to the total wave function, and the degree to which each contributes is indicated by the square of its coefficient in the linear combination. The contributing structures, also called "canonical forms," themselves thus have a purely formal significance: they are the components from which wave functions can be built. Structures can be covalent (or nonpolar) or ionic (or polar). The representation is frequently kept qualitative so that we speak of important or major contributing structures and minor contributing structures. For example, two major nonequivalent contributing structures for the conjugate base of acetone are

$$CH_2{=}C(CH_3){-}O^- \leftrightarrow H_2C^-{-}C(CH_3){=}O$$

See also DELOCALIZATION; KEKULÉ STRUCTURE; RESONANCE.

control rod A rod containing neutron-absorbing materials (boron or cadmium). Control rods are used to move in and out of the core of the reactor to control the rate of the nuclear reaction.

control test A test to increase the conclusiveness of an experiment.

convection Fluid or air circulation driven by temperature gradients; the rising of warm air and the sinking of cool air. The transfer of heat by circulation or movement of heated liquid or gas.

cooperativity The phenomenon that binding of an effector molecule to a biological system either enhances or diminishes the binding of a successive molecule, of the same or different kind, to the same system. The system may be an ENZYME or a protein that specifically binds another molecule such as oxygen or DNA. The effector molecule may be an enzyme SUBSTRATE or an ALLOSTERIC EFFECTOR. The enzyme or protein exists in different CONFORMATIONs, with different catalytic rates or binding affinities, and the binding of the effector molecule changes the proportion of these conformations. Enhanced binding is named positive cooperativity; diminished binding is named negative cooperativity. A well-known example of positive cooperativity is in HEMOGLOBIN. In BIOCATALYSIS, it was originally proposed that only multi-SUBUNIT enzymes could respond in this way. However, single-subunit enzymes may give such a response (so-called mnemonic enzymes).

coordinate covalence (coordinate link) *See* COORDINATION.

coordinate covalent bond A bond between two atoms in which the shared electrons are contributed by only one of the atoms.

coordination A coordination entity is composed of a CENTRAL ATOM, usually that of a metal, to which is attached a surrounding array of other atoms or group of atoms, each of which is called a LIGAND. A coordination entity may be a neutral molecule, a cation, or an anion. The ligands can be viewed as neutral or ionic entities that are bonded to an appropriately charged central atom. It is standard practice to think of the ligand atoms that are directly attached to the central atom as defining a coordination polyhedron (tetrahedron, square plane, octahedron, etc.) about the central atom. The coordination number is defined as being equal to the number of sigma bonds between ligands and the central atom; this definition is not necessarily appropriate in all areas of (coordination) chemistry. In a coordination formula, the central atom is listed first. The formally anionic ligands appear next, and they are listed in alphabetical order according to the first symbols of their formulas. The neutral ligands follow, also in alphabetical order, according to the same principle. The formula for the entire coordination entity, whether charged or not, is enclosed in square brackets. In a coordination name, the ligands are listed in alphabetical order, without regard to charge, before the name of the central atom. Numerical prefixes indicating the number of ligands are not considered in determining that order. All anionic coordination entities take the ending *-ate,* whereas no distinguishing termination is used for cationic or neutral coordination entities.

See also DIPOLAR BOND; PI (π) ADDUCT.

coordination compound **(coordination complex)** A compound containing coordinate covalent bonds.

coordination isomers One of two or more coordination compounds or complexes having the same chemical composition but differing in which ligands are attached to the metal ion(s).

coordination number The coordination number of a specified atom in a CHEMICAL SPECIES is the number of other atoms directly linked to that specified atom. For example, the coordination number of carbon in methane is four, and it is five in protonated methane, CH_5^+. (The term is used in a different sense in the crystallographic description of ionic crystals.)

coordination sphere The metal ion and its coordinating ligands.

coordinatively saturated A transition-metal complex that has formally 18 outer-shell electrons at the central metal atom.

coordinatively unsaturated A transition-metal complex that possesses fewer ligands than exist in the coordinatively saturated complex. These complexes usually have fewer than 18 outer-shell electrons at the central metal atom.

copolymer A material created by polymerizing a mixture of two (or more) starting compounds (MONOMERS).

Cori, Carl Ferdinand (1896–1984) Austrian *Biochemist* Carl Ferdinand Cori was born in Prague on December 5, 1896, to Carl I. Cori, director of the Marine Biological Station in Trieste. He studied at the gymnasium in Trieste and graduated in 1914, when he entered the German University of Prague to study medicine. During World War I, he served as a lieutenant in the sanitary corps of the Austrian army on the Italian front; he returned to the university to graduate as a doctor of medicine in 1920. He spent a year at the University of Vienna and a year as assistant in pharmacology at the University of Graz until, in 1922, he accepted a position as biochemist at the State Institute for the Study of Malignant Diseases in Buffalo, New York. In 1931 he was appointed professor of pharmacology at the Washington University Medical School in St. Louis, where he later became professor of biochemistry.

He married Gerty Theresa Cori (née Radnitz) in 1920, and they worked together in Buffalo. When he moved to St. Louis, she joined him as a research associate. Gerty Cori was made professor of biochemistry in 1947.

Jointly, they researched the biochemical pathway by which glycogen, the storage form of sugar in liver and muscle, is broken down into glucose. They also determined the molecular defects underlying a number of genetically determined glycogen storage diseases. For these discoveries, the Coris received the 1947 Nobel Prize for physiology or medicine.

They became naturalized Americans in 1928. He died on October 20, 1984, in Cambridge, Mass. His wife died earlier, in 1957.

Cori, Gerty Theresa (née **Radnitz**) (1896–1957) Austrian *Biochemist* Gerty Theresa Cori (née Radnitz) was born in Prague on August 15, 1896, and received her primary education at home before entering a lyceum for girls in 1906. She entered the medical school of the German University of Prague and received the doctorate in medicine in 1920. She then spent two years at the Carolinian Children's Hospital before emigrating to America with her husband, Carl, whom she married in 1920. They worked together in Buffalo, New York, and when he moved to St. Louis, she joined him as a research associate. She was made professor of biochemistry in 1947.

Jointly, they researched the biochemical pathway by which glycogen, the storage form of sugar in liver and muscle, is broken down into glucose. They also determined the molecular defects underlying a number of genetically determined glycogen storage diseases. For these discoveries, the Coris received the 1947 Nobel Prize for physiology or medicine. She died on October 26, 1957.

Carl Cori (b. 1896) graduated in medicine at Prague in 1920 and married his classmate Gerty Radnitz in the same year. They moved to America in 1922 and formed a team until Gerty's death in 1957. Their best-known joint research was in determining the precise biochemical process involved in the conversion of glucose to glycogen in the body. They also described the reverse reaction, where glycogen (an energy stored in the liver and muscles) is converted back to glucose. In 1947 they shared the Nobel Prize for physiology or medicine. ***(Courtesy of Science Photo Library)***

coronate *See* CROWN.

corphin The F-430 cofactor found in methyl-coenzyme M REDUCTASE, a nickel-containing ENZYME that catalyzes one step in the conversion of CO_2 to methane in METHANOGENic bacteria. The Ni ion in F-430 is coordinated by the tetrahydrocorphin LIGAND. This ligand combines the structural elements of both PORPHYRINs and CORRINs.

See also COORDINATION.

correlation analysis The use of empirical correlations relating one body of experimental data to another, with the objective of finding quantitative estimates of the factors underlying the phenomena involved. Correlation analysis in organic chemistry often uses LINEAR FREE-ENERGY RELATIONs for rates or equilibria of reactions, but the term also embraces similar analysis of physical (most commonly spectroscopic) properties and of biological activity.

See also QUANTITATIVE STRUCTURE-ACTIVITY RELATIONSHIPS.

corrin A ring-contracted PORPHYRIN derivative that is missing a carbon from one of the mesopositions (C-20).

It constitutes the skeleton $C_{19}H_{22}N_4$ upon which various B_{12} vitamins, COFACTORS, and derivatives are based.

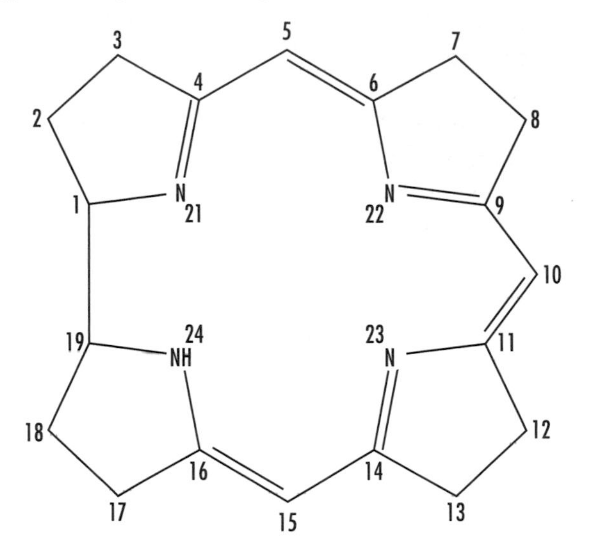

corrosion Oxidation of a metal in the presence of air and moisture.

cosphere *See* CYBOTACTIC REGION.

cotransport A simultaneous transporting of two solutes across a MEMBRANE by a transporter going one way (symport) or in opposite directions (antiport).

coulomb Unit of electrical charge. A quantity of electricity when a current of one ampere flows for 1 second; 1 coulomb is abbreviated as 1 C.

Coulomb's law The force of attraction or repulsion of two electric charges is proportional to the product of the charges and inversely proportional to the square of the distance between them. $F = k(Q_4Q_b / d^2)$, where F represents the electrostatic force, k represents a constant of proportionality, Q_4 and Q_b represent quantities of electrostatic charge, and d represents the distance between the charges.

countercurrent exchange The effect caused when two fluids move past each other in opposite directions and facilitate the efficient exchange of heat, gas, or substance; for example, passage of heat from one blood vessel to another; *rete mirabile,* the countercurrent exchange structure of capillaries that allows gas uptake in a fish swim bladder; the kidney nephron loop, a tubular section of nephron between the proximal and distal convoluted tubules where water is conserved and urine concentrates by a countercurrent exchange system; and the upper airway where, on expiration, heat and moisture are retained and then given up to the relatively cooler and drier inspired gases.

coupled system Two or more processes that affect one another.

coupling constant **(spin-spin coupling constant),** J (SI unit: Hz) A quantitative measure for nuclear spin-spin, nuclear-electron (hyperfine coupling), and electron–electron (fine coupling in EPR) coupling in magnetic resonance spectroscopy. The "indirect" or scalar NMR coupling constants are in a first approximation independent of the external magnetic field and are expressed in units of hertz (Hz).

See also ELECTRONIC PARAMAGNETIC RESONANCE SPECTROSCOPY; NUCLEAR MAGNETIC RESONANCE SPECTROSCOPY.

covalent bond A region of relatively high electron density between nuclei that arises at least partly from sharing of electrons and gives rise to an attractive force and characteristic internuclear distance.

See also AGOSTIC; COORDINATION; HYDROGEN BOND; MULTICENTER BOND.

covalent compounds Compounds held together by covalent bonds.

Cox-Yates equation A modification of the BUNNETT-OLSEN EQUATION of the form

$$\lg ([SH^+])/([S])–\lg[H^+] = m^*X + pK_{SH^+}$$

where X is the activity function $\lg(\gamma_S\ \gamma_{H^+}/\gamma_{SH^+})$ for an arbitrary reference base. The function X is called the excess acidity because it gives a measure of the difference between the acidity of a solution and that of an ideal solution of the same concentration. In practice $X = -(Ho + \lg[H^+])$ and $m^* = 1 - \Phi$.

cracking The process in which large molecules are broken down into smaller molecules. Used especially in the petroleum refining industry.

critical mass The amount of fissionable material necessary for a CHAIN REACTION to sustain itself.

critical micelle concentration (cmc) There is a relatively small range of concentrations separating the limit below which virtually no MICELLEs are detected and the limit above which virtually all additional surfactant molecules form micelles. Many properties of surfactant solutions, if plotted against the concentration, appear to change at a different rate above and below this range. By extrapolating the loci of such a property above and below this range until they intersect, a value may be obtained known as the critical micellization concentration (critical micelle concentration), symbol *c*M, abbreviated as cmc (or c.m.c.). As values obtained using different properties are not quite identical, the method by which the cmc is determined should be clearly stated.

See also INVERTED MICELLE.

critical point The combination of critical temperature and critical pressure. The temperature and pressure at which two phases of a substance in equilibrium become identical and form a single phase.

critical pressure The pressure required to liquefy a gas at its critical temperature.

critical temperature The temperature above which a gas cannot be liquefied, regardless of the amount of pressure applied.

cross conjugation In a system XC_6H_4GY, cross conjugation involves the substituent X, the benzene ring, and the side-chain connective-plus-reaction site GY, i.e., either X is a +R group and GY is a –R group, or X is a –R group and GY is a +R group. In Hammett correlations this situation can lead to the need to apply exalted substituent constants σ^+ or σ^-, respectively, as in electrophilic or nucleophilic aromatic substitution. The term *through resonance* is synonymous. Cross conjugation has also been used to describe the interactions occurring in 2-phenylallyl and and similar systems.

cross reactivity The ability of an IMMUNOGLOBULIN, specific for one ANTIGEN, to react with a second antigen. A measure of relatedness between two different antigenic substances.

crown A MOLECULAR ENTITY comprising a monocyclic LIGAND assembly that contains three or more BINDING SITEs held together by covalent bonds and capable of binding a GUEST in a central (or nearly central) position. The ADDUCTs formed are sometimes known as "coronates." The best known members of this group are macrocyclic polyethers, such as the below "18-crown-6," containing several repeating units $-CR_2-CR_2O-$ (where R is most commonly H), and known as crown ethers.

See also HOST.

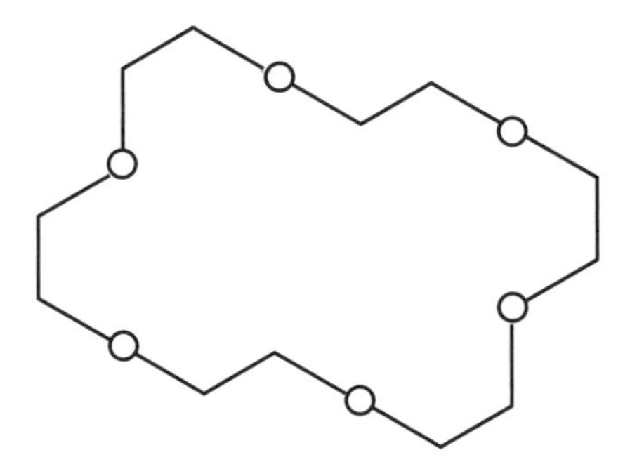

cryoscopy The depression of the freezing point for a material in solution; the study or practice of determining the freezing point of liquids.

cryptand A MOLECULAR ENTITY comprising a cyclic or polycyclic assembly of BINDING SITEs that contains three or more binding sites held together by COVALENT BONDs, and which defines a molecular cavity in such a way as to bind (and thus "hide" in the cavity) another molecular entity, the GUEST (a cation, an anion, or a neutral species), more strongly than do the separate parts of the assembly (at the same total concentration of binding sites). The ADDUCT thus formed is called a "cryptate." The term is usually restricted to bicyclic or oligocyclic molecular entities, as seen here.

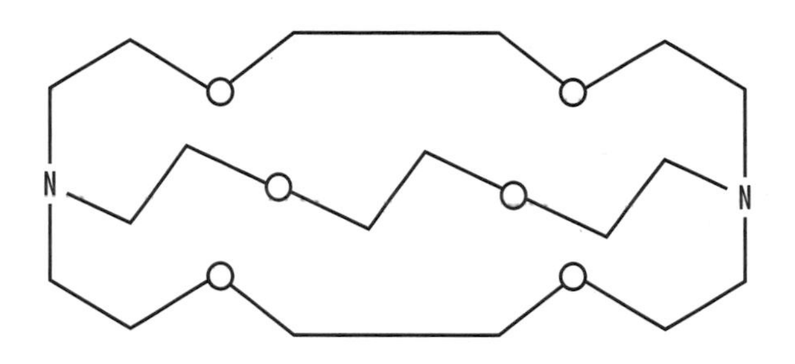

Corresponding monocyclic ligand assemblies (CROWNs) are sometimes included in this group, if they can be considered to define a cavity in which a guest can hide. The terms *podand* and *spherand* are used for certain specific LIGAND assemblies. Coplanar cyclic polydentate ligands, such as PORPHYRINs, are not normally regarded as cryptands.

See also HOST.

crystal A three-dimensional solid formed by regular repetition of the packing of atoms, ions, or molecules.

crystal classes Three-dimensional point groups. The 32 possible unique combinations of rotation, reflection, inversion, and rotoinversion symmetry elements in three dimensions. The morphology of every crystal must be based on one of these collections of symmetry.

crystal field Crystal field theory is the theory that interprets the properties of COORDINATION entities on the basis that the interaction of the LIGANDs and the CENTRAL ATOM is a strictly ionic or ion-dipole interaction resulting from electrostatic attractions between the central atom and the ligands. The ligands are regarded as point negative (or partially negative) charges surrounding a central atom; covalent bonding is completely neglected. The splitting or separation of energy levels of the five degenerate *d* orbitals in a transition metal, when the metal is surrounded by ligands arranged in a particular geometry with respect to the metal center, is called the crystal field splitting.

crystal lattice A pattern of particle arrangements in a crystal. There are 14 possible lattice patterns.

crystal lattice energy The amount of energy that holds a crystal together.

crystalline solid A solid characterized by a regular, ordered arrangement of particles; atoms, ions, or molecules assume ordered positions.

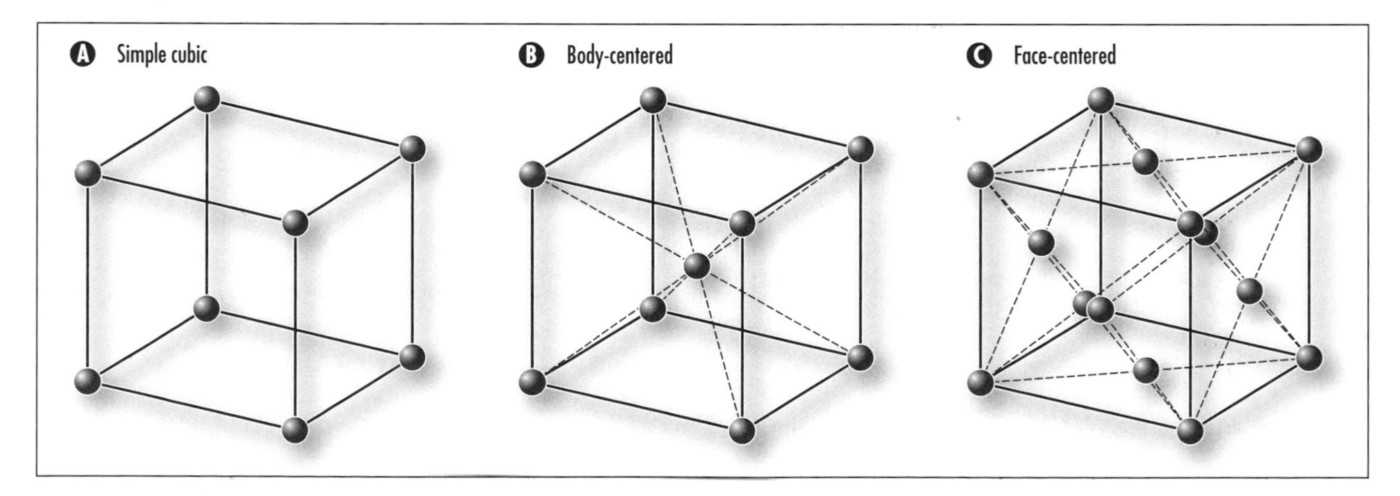

Examples of structures of cubic crystal

crystallization The process of forming crystals from the melt or solution.

crystal systems The six (or seven) groupings of the 32 crystal classes based on the presence of common symmetry elements: triclinic, monoclinic, orthorhombic, tetragonal, hexagonal (rhombohedral), and isometric.

C-terminal amino acid residue *See* AMINO ACID RESIDUE.

cubic close packing A way of packing atoms to minimize unoccupied space that results in a crystal lattice with cubic symmetry. There are three kinds of cubic lattice: simple cubic, body-centered cubic, and face-centered cubic, which is the close-packed type. Many metals feature a cubic close-packed structure, including gold (Au), silver (Ag), copper (Cu), aluminium (Al), lead (Pb), nickel (Ni), and platinum (Pt).

curie (Ci) The basic unit used to describe the intensity of radioactivity in a sample of material. One Ci = 3.700 × 1010 disintegrations per second. The unit used in modern terminology is the Becquerel, which is 1 dps.

Curie, Marie (1867–1934) Polish *Chemist* Marie Curie (née Sklodowska) was born in Warsaw on November 7, 1867, to a secondary school teacher. Educated in local schools, and by her father, she went to Paris in 1891 to continue studying at the Sorbonne and received licenciateships in physics and the mathematical sciences.

She met her future husband Pierre Curie, a professor in the school of physics, in 1894, and they were married the following year. She succeeded her husband as head of the physics laboratory at the Sorbonne, where she gained her doctor of science degree in 1903. Following the death of her husband in 1906, she took his place as professor of general physics in the faculty of sciences, the first time a woman had held this position. She was appointed director of the Curie Laboratory in the Radium Institute of the University of Paris, founded in 1914.

Their early researches led to the isolation of polonium, named after the country of Marie's birth, and radium, and she later developed methods for separating radium from radioactive residues in sufficient quantities to allow for the study of its properties, and especially its therapeutic properties.

Throughout her life she actively promoted the use of radium to alleviate suffering. She promoted this during World War I, assisted by her daughter Iréne (who won the Nobel Prize in chemistry in 1935).

Together with Pierre, she was awarded half of the Nobel Prize for physics in 1903 for their study into the spontaneous radiation discovered by Becquerel, who was awarded the other half of the prize. In 1911 Marie received a second Nobel Prize, in chemistry, in recognition of her work in radioactivity.

She received, jointly with her husband, the Davy Medal of the Royal Society in 1903, and in 1921, President Harding of the United States, on behalf of the women of America, presented her with one gram of radium in recognition of her service to science. Marie Curie died in Savoy, France, on July 4, 1934.

Curie relation *See* MAGNETIC SUSCEPTIBILITY.

Curtin-Hammett principle In a CHEMICAL REACTION that yields one product (X) from one conformational ISOMER (A′) and a different product (Y) from another conformational isomer (A″) (and provided that these two isomers are rapidly interconvertible relative to the rate of product formation, whereas the products do not undergo interconversion), the product composition is not in direct proportion to the relative concentrations of the conformational isomers in the SUBSTRATE; it is controlled only by the difference in standard free energies ($d\Delta^{\ddagger} G$) of the respective TRANSITION STATES.

It is also true that the product composition is formally related to the relative concentrations of the conformational isomers A′ and A″ (i.e., the conformational equilibrium constant) and the respective rate constants of their reactions. These parameters are generally, though not invariably, unknown.

The diagram on page 66 represents the energetic situation for transformation of interconverting isomers A and A′ into products X and Y.

See also CONFORMATION.

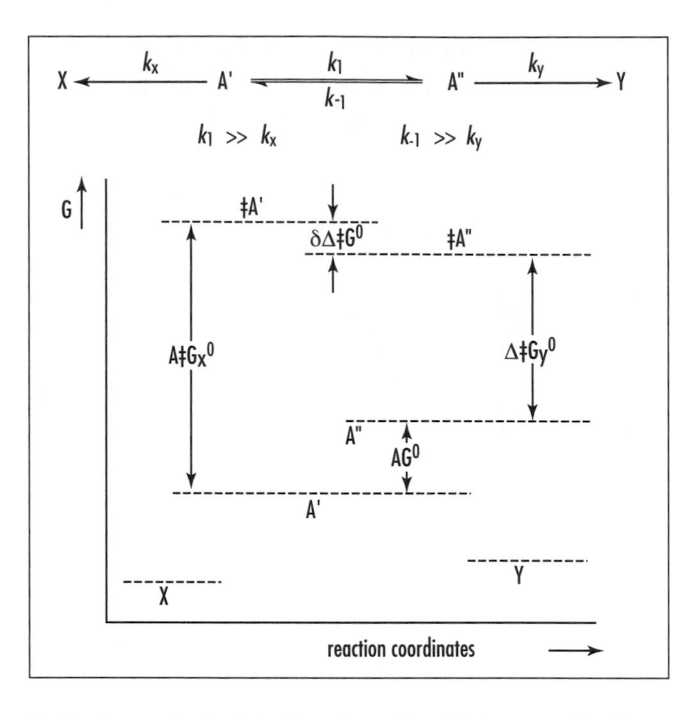

Curtin-Hammett principle. Transformation of interconverting isomers A′ and A″ into products X and Y.

cyanobacteria Bacteria, formerly known as blue-green algae; aquatic and photosynthetic organisms that live in water and manufacture their own food. Their fossils go back more than 3.5 billion years, making them the oldest known species, and they are the contributors to the origin of plants.

cybernetics The science that studies the methods to control behavior and communication in animals (and machines).

cybotactic region That part of a solution in the vicinity of a solute molecule in which the ordering of the solvent molecules is modified by the presence of the solute molecule. The term *solvent "cosphere" of the solute* has also been used

See also SOLVATION.

cyclic AMP (**cAMP or 3′,5′-AMP**) Cyclic adenosine monophosphate. A compound synthesized from ATP (by the enzyme adenylyl cyclase) in living cells that acts as an intercellular and extracellular second messenger mediating peptide and amine hormones.

cyclic electron flow Two photosystems are present in the thylakoid membrane of chloroplasts: photosystem I and photosystem II. The two photosystems work together during the light reactions of PHOTOSYNTHESIS. The light-induced flow of electrons beginning with and returning to photosystem I to produce ATP without production of NADPH (nicotine adenine dinucleotide phosphate with hydrogen) is cyclic electron flow. The generation of ATP by this process is called noncyclic photophosphorylation.

cyclin A protein found in dividing cells that activates protein kinases (cyclin-dependent protein kinases), an enzyme that adds or removes a phosphate group from a target protein and controls the progression of one phase of the cell cycle to the next. The concentration of the cyclin increases and decreases during the cell cycle.

cyclin-dependent kinase A protein kinase, an enzyme involved in regulating cell growth and division, that must be attached to cyclin to occur.

cyclization Formation of a ring compound from a chain by formation of a new bond.

See also ANNULATION.

cycloaddition A reaction in which two or more unsaturated molecules (or parts of the same molecule) combine with the formation of a cyclic ADDUCT, in which there is a net reduction of the bond multiplicity.

The following two systems of notation have been used for the more detailed specification of cycloadditions, of which the second, more recent system (described under (2)) is preferred:

(1) An $(i+j+ \ldots)$ cycloaddition is a reaction in which two or more molecules (or parts of the same molecule) provide units of $i, j, \ldots$ linearly connected atoms. These units become joined at their respective termini by new sigma bonds so as to form a cycle containing $(i+j+ \ldots)$ atoms. In this notation, (a) a Diels-

Alder reaction is a (4+2) cycloaddition; (b) the initial reaction of ozone with an alkene is a (3+2) cycloaddition; and (c) the reaction shown below is a (2+2+2) cycloaddition. (Parentheses (...) are used in the description based on numbers of atoms.)

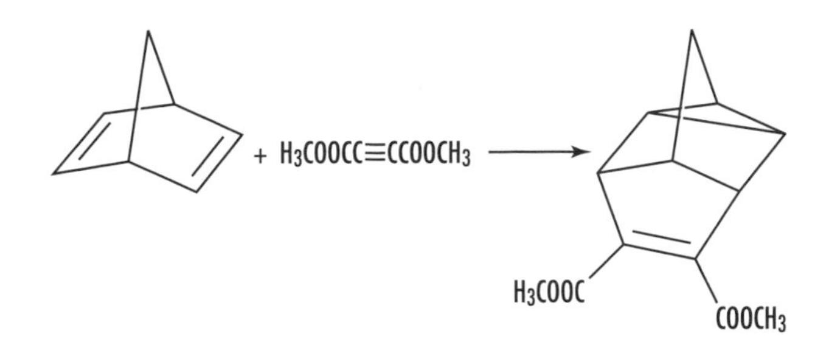

(2) The symbolism [*i*+*j*+...] for a cycloaddition identifies the numbers *i*, *j*, ... of electrons in the interacting units that participate in the transformation of reactants to products. In this notation, the reactions (a) and (b) of the preceding paragraph would both be described as [2+4] cycloadditions, and (c) as a [2+2+2] cycloaddition. The symbol *a* or *s* (*a* = ANTARAFACIAL, *s* = suprafacial) is often added (usually as a subscript after the number to designate the stereochemistry of addition to each fragment). A subscript specifying the orbitals, viz., σ, π (sigma, pi), with their usual significance, or *n* (for an orbital associated with a single atom only) may be added as a subscript before the number. Thus the normal Diels-Alder reaction is a [4*s*+2*s*] or [π4*s* + π2*s*] cycloaddition, while the reaction

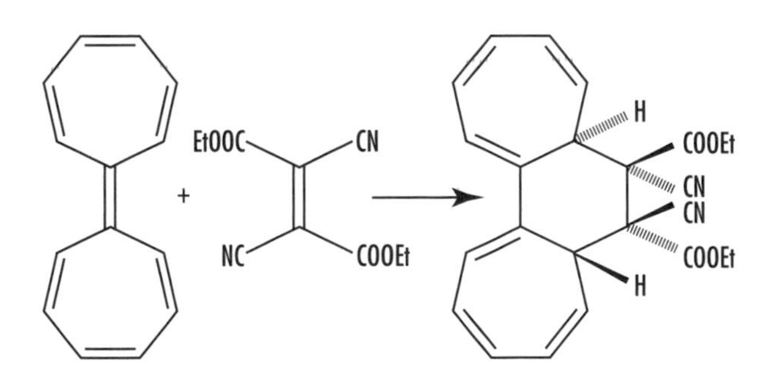

would be a [14*a*+2*s*] or [π14*a* + π2*s*] cycloaddition. (Square brackets [...] are used in the descriptions based on numbers of electrons.)

Cycloadditions may be PERICYCLIC REACTIONS or (nonconcerted) STEPWISE REACTIONS. The term *dipolar cycloaddition* is used for cycloadditions of 1,3-dipolar compounds.

See also CHELETROPIC REACTIONS; CONCERTED PROCESS.

cycloalkane An alkane that contains a ring of carbon atoms.

cycloelimination The reverse of CYCLOADDITION. The term is preferred to the synonyms *cycloreversion, retroaddition,* and *retrocycloaddition.*

cycloreversion *See* CYCLOELIMINATION.

cyclotron A machine for accelerating charged particles along a spiral path to high energies. The first cyclotron was built by Ernest Orlando Lawrence and his graduate student, M. Stanley Livingston, at the University of California, Berkeley, in the early 1930s.

cytochrome A HEME protein that transfers electrons and exhibits intense absorption bands (the α-band and β-bands, with the α-band having the longer wavelength) between 510 and 615 nm in the reduced form. Cytochromes are designated types *a, b, c,* or *d,* depending on the position of the α-band, which depends on the type of heme. The iron undergoes oxidation-reduction between oxidation states Fe(II) and Fe(III). Most cytochromes are hemochromes, in which the fifth and sixth COORDINATION sites in the iron are occupied by strong field LIGANDs, regardless of the oxidation state of iron. Cytochromes may be distinguished by the wavelength of the α-band, such as cytochrome *c*-550. Certain specific cytochromes with particular functions are designated with suffixes, such as cytochrome a_1,b_2, etc., but this practice is discouraged.

cytochrome-c oxidase An ENZYME, ferrocytochrome-c: dioxygen OXIDOREDUCTASE, CYTOCHROME aa_3. The major respiratory protein of animal and plant MITOCHONDRIA, it catalyzes the oxidation of Fe(II)-cytochrome c and the reduction of dioxygen to water. Contains two HEMEs and three copper atoms, arranged in three centers. Heme a_3 and copper-B form a center that reacts with dioxygen; the second heme is cytochrome *a*; the third site, copper-A, is a DINUCLEAR center.

cytochrome P-450 General term for a group of HEME-containing MONOOXYGENASEs. Named from the prominent absorption band of the Fe(II)-carbonyl complex. The heme comprises PROTOPORPHYRIN IX, and the proximal LIGAND to iron is a cysteine sulfur. Cytochromes P-450 of microsomes in tissues such as liver are responsible for METABOLISM of many XENOBIOTICs, including drugs. Others, such as the MITOCHONDRIAl ENZYMEs from adrenal glands, are involved in biosynthetic pathways such as those of steroids. The reaction with dioxygen appears to involve higher oxidation states of iron, such as Fe(IV)=O.

cytokine Cytokines are soluble GLYCOPROTEINs released by cells of the immune system (secreted primarily from leukocytes), which act nonenzymatically through specific RECEPTORs to regulate IMMUNE RESPONSEs. Cytokines resemble HORMONEs in that they act at low concentrations bound with high affinity to a specific receptor.

cytoplasm The part of PROTOPLASM in a CELL outside of and surrounding the NUCLEUS. The contents of a cell other than the nucleus. Cytoplasm consists of a fluid containing numerous structures, known as organelles, that carry out essential cell functions.

D

Dale, Henry Hallett (1875–1968) British *Physiology* Sir Henry Hallett Dale was born in London on June 9, 1875, to Charles James Dale, a businessman, and Frances Ann Hallett. He attended Tollington Park College in London, Leys School, Cambridge, and in 1894 he entered Trinity College with a scholarship. He graduated through the Natural Sciences Tripos, specializing in physiology and zoology, in 1898.

In 1900 he gained a scholarship and entered St. Bartholomew's Hospital, London, for the clinical part of the medical course. He received a B.Ch. at Cambridge in 1903 and became an M.D. in 1909.

He took an appointment as pharmacologist at the Wellcome Physiological Research Laboratories in 1904 and became director of these laboratories in 1906, working for some six years. In 1914 he was appointed director of the department of biochemistry and pharmacology at the National Institute for Medical Research in London. In 1928 he became the director of this institute, serving until his retirement in 1942, when he became professor of chemistry and a director of the Davy–Faraday Laboratory at the Royal Institution, London.

In 1911 he was the first to identify the compound histamine in animal tissues, and he studied its physiological effects, concluding that it was responsible for some allergic and anaphylactic reactions. After successfully isolating ACETYLCHOLINE in 1914, he established that it was found in animal tissue, and in the 1930s he showed that it is released at nerve endings in the parasympathetic nervous system, establishing acetylcholine's role as a chemical transmitter of nerve impulses.

In 1936 he shared the Nobel Prize for physiology or medicine with his friend German pharmacologist OTTO LOEWI for their discoveries in the chemical transmission of nerve impulses.

He was knighted in 1932 and appointed to the Order of Merit in 1944. In addition to numerous articles in medical and scientific journals that record his work, he was the author of *Adventures in Physiology* (1953) and *An Autumn Gleaning* (1954).

Dale was president of the Royal Society (1940–45) and others and received many awards. He married his first cousin Ellen Harriet Hallett in 1904. He died on July 23, 1968, in Cambridge.

dalton A unit of measurement of molecular weight based on the mass of 1/12th the mass of C_{12}, i.e., 1.656 $\times 10^{-24}$. A dalton is also called an atomic mass unit, or amu, and is used to measure atomic mass. Protein molecules are expressed in kilodaltons (kDa). It was named in honor of John Dalton (1766–1844), an English chemist and physicist.

Dalton's law The total pressure of a mixture of gases is equal to the sum of the partial pressures of each constituent gas. Partial pressure is the pressure each gas would exert if it occupied the volume of the mixture alone.

Dam, Henrik (1895–1976) Danish *Biochemist* Carl Peter Henrik Dam was born in Copenhagen on February 21, 1895, to druggist Emil Dam and his wife Emilie (née Peterson), a teacher. He attended and graduated in chemistry from the Polytechnic Institute, Copenhagen, in 1920, and the same year was appointed assistant instructor in chemistry at the School of Agriculture and Veterinary Medicine, advancing to full instructor in biochemistry at the physiological laboratory of the University of Copenhagen in 1923.

In 1925 Dam became assistant professor at the Institute of Biochemistry, University of Copenhagen, and three years later was promoted to associate professor, where he served until 1941. Upon submitting a thesis "Nogle Undersøgelser over Sterinernes Biologiske Betydning" (Some investigations on the biological significance of the sterines) to the University of Copenhagen in 1934, he received a Ph.D. in biochemistry.

He discovered vitamin K and its anticoagulant effects while studying the sterol metabolism of chicks in Copenhagen and was awarded the Nobel Prize in physiology or medicine in 1943 for this work.

Dam conducted research at Woods Hole Marine Biological Laboratories during the summer and autumn of 1941 and at the University of Rochester, N.Y., where he served as a senior research associate between 1942 and 1945. He was at the Rockefeller Institute for Medical Research in 1945 as an associate member.

Dam was appointed professor of biochemistry at the Polytechnic Institute, Copenhagen, in 1941, though the designation of his chair at the Polytechnic Institute was changed to professor of biochemistry and nutrition in 1950.

After his return to Denmark in 1946, he concentrated his research on vitamin K, vitamin E, fats, cholesterol, and nutritional studies in relation to gallstone formation.

He published over 300 articles in biochemistry and was a member of numerous scientific organizations. Dam died on April 17, 1976.

Daniel cell (gravity cell) A special copper/zinc battery, where a spontaneous chemical reaction between zinc metal and aqueous copper (II) sulfate is conducted in which the chemical energy released is in the form of electrical energy. Invented in 1836 by Englishman John F. Daniel.

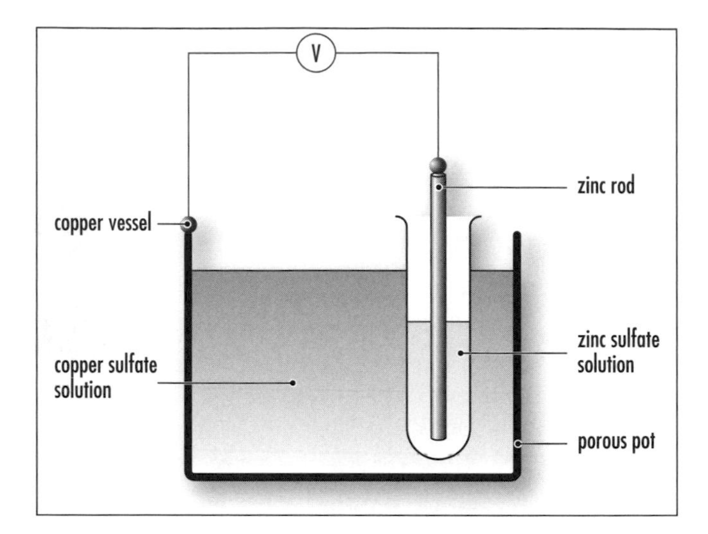

Daniel cell. An electrical cell that has zinc as the negative electrode

data A collection of facts, concepts, or instructions in a formalized manner suitable for communication or processing by human beings or by computer.

dative bond *See* COORDINATION.

daughter nuclide Nuclide that is produced in a nuclear decay of another NUCLIDE and can decay further or become stable.

debye The unit used to express dipole moments. Equal to the moment that exists between a unit of positive charge and a unit of negative charge separated by a distance of 1 cm; 1 debye = 10^{-18} statcoulomb·cm.

Debye, Petrus (Peter) Josephys Wilhelmus (1884–1966) Dutch *Physical chemist* Peter Debye was born on March 24, 1884, at Maastricht, in the Netherlands. He received elementary and secondary school education in his home town. Thereafter, he attended the Aachen Institute of Technology (Technische Hochschule) and gained a degree in electrical technology in 1905, serving as assistant in technical mechanics at the Aachen Technological Institute for two years. In 1906 Debye obtained a similar position in theoretical physics at Munich University and qualified as a university lecturer in 1910 after having obtained a Ph.D. in physics in 1908 there.

In 1911 he went on to become professor of theoretical physics at Zurich University for two years and then returned to the Netherlands in 1912 as appointed professor of theoretical physics at Utrecht University. In 1913 he married Mathilde Alberer, with whom he had a son and a daughter. His son (Peter P. Debye) became a physicist and collaborated with Debye on some of his research. The following year, he moved to the University of Göttingen, heading the theoretical department of the Physical Institute. In 1915 he became editor of *Physikalische Zeitschrift* (until 1940). He became director of the entire Physical Institute and lectured on experimental physics until 1920.

Debye returned to Zurich in 1920 as professor of physics and principal of the Eidgenössische Technische Hochschule, and seven years later he held the same post at Leipzig. From 1934 to 1939 he was director of the Max Planck Institute of the Kaiser Wilhelm Institute for Physics in Berlin-Dahlem and professor of physics at the University of Berlin. During this period he was awarded the 1936 Nobel Prize in chemistry "for his contributions to our knowledge of molecular structure through his investigations on dipole moments and on the diffraction of X-rays and electrons in gases."

In 1940 he became professor of chemistry and principal of the chemistry department of Cornell University, in Ithaca, New York, and he became an American citizen in 1946.

In 1952 he resigned his post as head of the chemistry department at Cornell University and became emeritus professor of chemistry there.

Much of his later work at Cornell dealt with light-scattering techniques (derived from his X-ray scattering work of years earlier) to determine the size and molecular weight of polymer molecules. This interest was a carryover from his work during World War II on synthetic rubber and extended to proteins and other macromolecules.

During his career Debye was awarded the Rumford Medal of the Royal Society, London, the Franklin and Faraday Medals, the Lorentz Medal of the Royal Netherlands Academy, the Max Planck Medal (1950), the Willard Gibbs Medal (1949), the Nichols Medal (1961), the Kendall Award (Miami, 1957), and the Priestley Medal of the American Chemical Society (1963). He was appointed Kommandeur des Ordens by Leopold II in 1956 and died of a heart attack on November 2, 1966.

decant To draw off the upper layer of liquid after the heaviest material, which can be a solid or another liquid, has settled to the bottom.

decay constant (disintegration constant) A constant that expresses the probability that an atom or molecule of a chemical will decay in a given time interval.

decomposers A trophic level or group of organisms such as fungi, bacteria, insects, and others that, as a group, digest or break down organic matter, such as dead animals, plants, or other organic waste, by ingesting the matter, secreting enzymes or other chemicals, and turning it into simpler inorganic molecules or compounds that are released back into the environment.

decomposition The breakdown of matter by bacteria and fungi. It changes the chemical makeup and physical appearance of materials.

decomposition, chemical The breakdown of a single entity (normal molecule, reaction INTERMEDIATE, etc.) into two or more fragments.

de-electronation *See* OXIDATION.

degree A unit of angular measure, often represented by the symbol °. The circumference of a circle contains 360 degrees. When applied to the roughly spherical shape of the Earth for geographic and cartographic purposes, degrees are each divided into 60 min. Also refers to temperature, as in degrees Celsius or degrees Fahrenheit, usually represented by the symbol °.

degenerate Describing different quantum states that have the same energy.

degenerate chemical reaction *See* IDENTITY REACTION.

degenerate rearrangement A MOLECULAR REARRANGEMENT in which the principal product is indistinguishable (in the absence of isotopic labeling) from the principal reactant. The term includes both "degenerate INTRAMOLECULAR rearrangements" and reactions that involve INTERMOLECULAR transfer of atoms or groups ("degenerate intermolecular rearrangements"): both are degenerate ISOMERIZATIONs. The occurrence of degenerate rearrangements may be detectable by isotopic labeling or by dynamic NUCLEAR MAGNETIC RESONANCE (NMR) techniques. For example, consider the [3,3]SIGMATROPIC REARRANGEMENT of hexa-1,5-diene (Cope rearrangement),

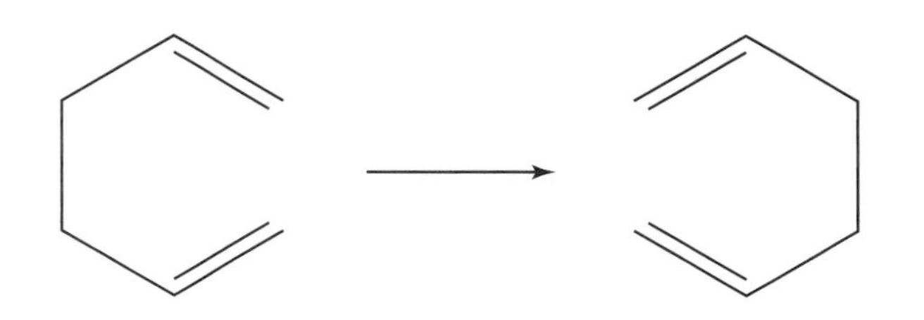

Synonymous but less preferable terms are *automerization, permutational isomerism, isodynamic transformation, topomerization.*

See also FLUXIONAL; MOLECULAR REARRANGEMENT; VALENCE ISOMER.

dehydration reaction **(condensation reaction)** A chemical reaction in which two organic molecules become linked to each other via COVALENT BONDs with the removal of a molecule of water; common in synthesis reactions of organic chemicals.

dehydrogenase An OXIDOREDUCTASE that catalyzes the removal of hydrogen atoms from a SUBSTRATE.

delocalization A quantum mechanical concept usually applied in organic chemistry to describe the pi bonding in a CONJUGATED SYSTEM. This bonding is not localized between two atoms: instead, each link has a "fractional double-bond character" or BOND ORDER. There is a corresponding "delocalization energy," identifiable with the stabilization of the system compared with a hypothetical alternative in which formal (localized) single and double BONDs are present. Some degree of delocalization is always present and can be estimated by quantum mechanical calculations. The effects are particularly evident in AROMATIC systems and in symmetrical MOLECULAR ENTITIES in which a lone pair of electrons or a vacant *p*-orbital is conjugated with a double bond (e.g., carboxylate ions, nitro compounds, enamines, the allyl cation). Delocalization in such species can be represented by partial bonds or as RESONANCE (here symbolized by a two-headed arrow) between CONTRIBUTING STRUCTUREs.

The examples on the next page also illustrate the concomitant delocalization of charge in ionic conjugated systems. Analogously, delocalization of the spin of an unpaired electron occurs in conjugated RADICALs.

See also MESOMERISM.

demodulation The process of retrieving information (data) from a modulated carrier wave, the reverse of modulation.

denaturation In DNA denaturation, two strands of DNA are separated as a result of the disruption of the hydrogen bonds following exposure to extreme conditions such as high temperature, chemical treatment, pH changes, salt concentration, and others. Denaturation in proteins by heat, acids, bases, or other means results in the change in the three-dimensional structure of the protein so that it cannot perform its function and becomes biologically inactive.

denatured A commercial term used to describe ethanol that has been rendered unfit for human consumption by the addition of harmful ingredients.

denitrification The reduction of nitrates to nitrites, including nitrogen monoxide (nitric oxide), dinitrogen oxide (nitrous oxide), and ultimately dinitrogen catalyzed by microorganisms, e.g., facultative AEROBIC soil bacteria under ANAEROBIC conditions.

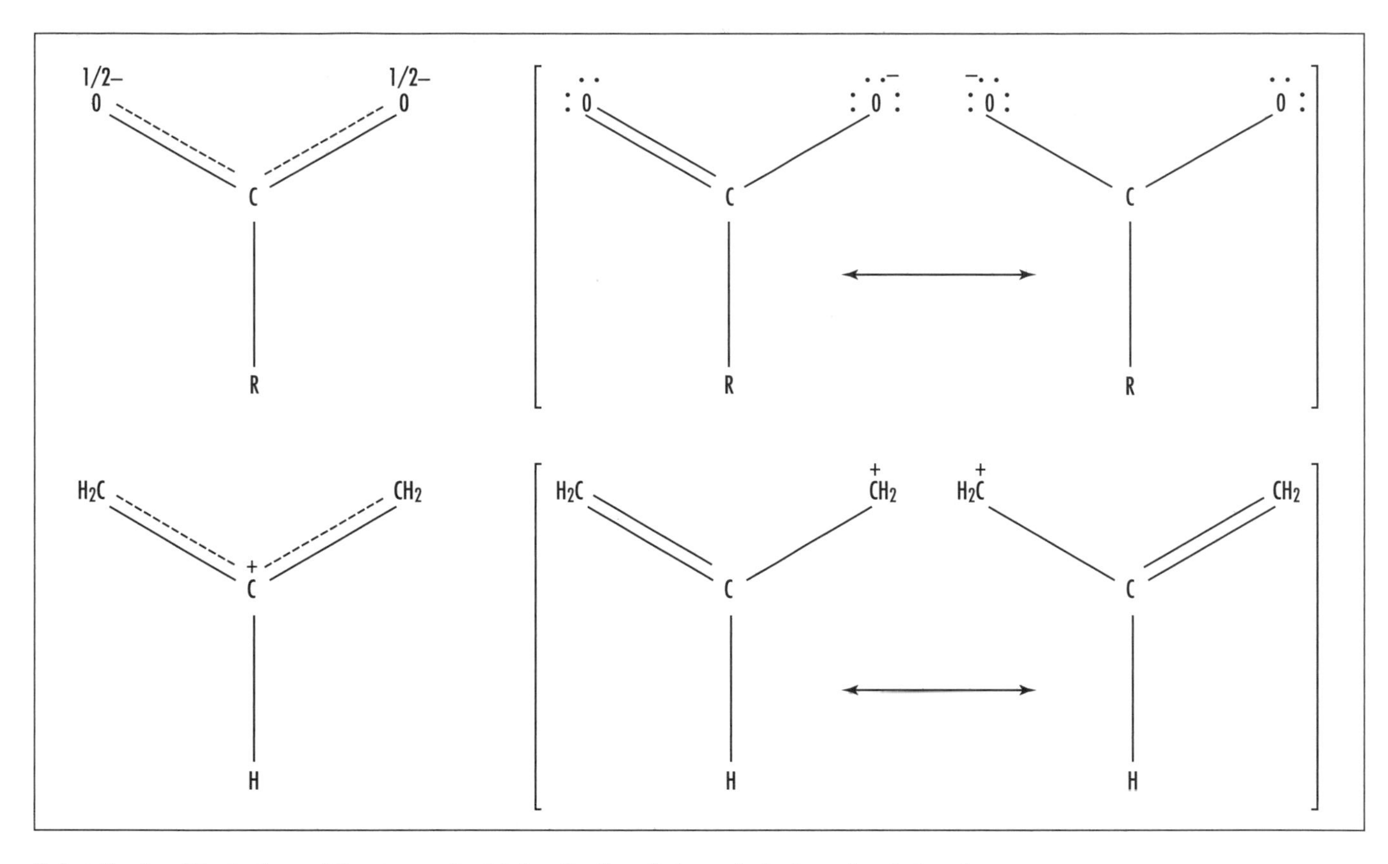

Delocalization. Illustrations of the concomitant delocalization of charge in ionic conjugated systems

De-nol Trade name for the potassium salt or mixed ammonium potassium salt of a bismuth citrate complex, used in the treatment of ulcers.

de novo design The design of bioactive compounds by incremental construction of a ligand model within a model of the RECEPTOR or ENZYME-active site, the structure of which is known from X-ray or NUCLEAR MAGNETIC RESONANCE data.

density An object's mass divided by its volume. The higher the density, the higher its mass per volume. Mass per unit volume: d = m÷v.

denticity The number of donor groups from a given LIGAND attached to the same CENTRAL ATOM.

deoxyribonucleic acid *See* DNA.

deoxyribose A five-carbon sugar ($C_5H_{10}O_4$) component of DNA. Joins with a phosphate group and base to form a deoxyribose nucleotide, the subunit of nucleic acids.

depolarization A process where an electrical charge of a neuron becomes less negative as the MEMBRANE potential moves from resting potential (70 mV) toward 0 mV; a decrease in voltage. The loss of membrane polarity is caused by the inside of the cell membrane becoming less negative in comparison with the outside. Depolarization is caused by an influx of Na^+ ions through voltage-gated Na^+ channels in axons.

Depolarization is a reduction in potential that usually ends with a more-positive and a less-negative charge. HYPERPOLARIZATION is the opposite, with an increase in potential becoming more negatively and less positively charged. Repolarization is when the state returns to its resting potential. ACTION POTENTIALS are caused by depolarization in nerve cells. An action potential is a one-way, self-renewing wave of membrane depolarization that propagates at rapid speed (up to 120 m/sec) along the length of a nerve axon. Julius Bernstein first proposed the concept of depolarization in 1868. The term is also used in reference to electrochemical cells to identify a process that reduces concentration gradients at the electrodes, thus reducing the output potential. The term is also used in optics to refer to depolarization of polarized light.

deposition Solidification of vapors by cooling; the opposite of SUBLIMATION. Process by which water changes phase directly from vapor into a solid without first becoming a liquid. More generally, a process that leads to formation of a solid material on a surface.

derivative A compound that resembles or arises from the original compound, except that some modifications in atomic structure are evident, such as the replacement of one atom with another or with a group. The term also refers to the product of certain mathematical operations.

dermal toxicity Adverse health effects resulting from skin exposure to a toxic substance.

desferal *See* DESFERRIOXAMINE.

desferrioxamine (DFO) CHELATING agent used worldwide in the treatment of iron overload conditions, such as HEMOCHROMATOSIS and THALASSEMIA. The iron-free form of ferrioxamine.

deshielding *See* SHIELDING.

detachment The reverse of an ATTACHMENT.

See also ELECTRON ATTACHMENT.

detailed balancing, principle of When equilibrium is reached in a reaction system (containing an arbitrary number of components and reaction paths), as many atoms, in their respective MOLECULAR ENTITIES, will pass forward as will pass backward along each individual path in a given finite time interval. Accordingly, the reaction path in the reverse direction must in every detail be the reverse of the reaction path in the forward direction (provided always that the system is at equilibrium).

The principle of detailed balancing is a consequence for macroscopic systems of the principle of MICROSCOPIC REVERSIBILITY.

detector A device in a radiometer that senses the presence and intensity of radiation. The incoming radiation is usually modified by filters or other optical components that restrict the radiation to a specific spectral band. The information can either be transmitted immediately or recorded for transmittal at a later time.

More generally, the term refers to any device used to sense or measure the amount or kind of radiation or substance in an analytical system. The term is also used to refer to any device that can respond to a particular signal or substance and provide qualitative or quantitative information about it, e.g., electrochemical detectors in HPLC, thermal conductivity detectors in GC, etc.

See also GAS CHROMATOGRAPHY; HIGH-PERFORMANCE LIQUID CHROMATOGRAPHY.

detergent An organic compound or compounds composed of molecules containing both hydrophilic (polar) and hydrophobic (nonpolar) portions. A petroleum-based surfactant or emulsifier other than soap.

detonate To cause an explosion.

deuterium A stable isotope of hydrogen whose atoms are twice as massive as ordinary hydrogen. Discovered in 1932 by Harold C. Urey. An atom of deuterium consists of one proton, one neutron, and one electron. About 0.015 percent of natural hydrogen is composed of deuterium.

dextrorotatory Refers to an optically active substance or compound that rotates plane-polarized light clockwise (to the right) when viewed in the direction of the light source.

DFO *See* DESFERRIOXAMINE.

diagonal similarities Chemical similarities in the periodic table. Refers to elements of period 2 and elements of period 3 (one group to the right), and is especially evident toward the left of the periodic table.

diamagnetic Substances having a negative MAGNETIC SUSCEPTIBILITY are diamagnetic. They are repelled by a magnetic field.

See also PARAMAGNETIC.

diamagnetism Weak repulsion by a magnetic field; a property exhibited by materials to oppose applied magnetic fields.

diastereoisomerism Stereoisomerism other than ENANTIOMERism. DIASTEREOISOMERS (or diastereomers) are stereoisomers not related as mirror images. Diastereoisomers are characterized by differences in physical properties and by some differences in chemical behavior toward ACHIRAL as well as chiral reagents.

diastereoisomers (diastereomers) STEREOISOMERS not related as mirror images.

See also DIASTEREOISOMERISM.

diatom The Bacillariophyceae or diatoms are unicellular algae that are found in single, colonial, or filamentous states. Under the microscope, they are often beautifully symmetrical, as their cell walls, or frustules, are composed of silica and are bivalved, one of which overlaps the other, and the frustule is often punctuated and ornamented. The two orders, Centrales and Pennales, occupy two different environments. The centric diatoms (Centrales) are circular in shape with radial symmetry and live mostly in marine environments. The pennate diatoms (Pennales) are elliptical in shape, have bilateral symmetry, and are found in freshwater environments.

Deposits of fossil diatoms known as diatomaceous earth have been mined and used for years in paints, abrasives, and other products such as chalk. The famous White Cliffs of Dover in England (rising to 300 feet) are composed of massive amounts of diatoms—coccoliths—that were laid down some 790 million years ago when Great Britain was submerged by a shallow sea.

diatomic Containing two atoms in the molecule.

dielectric constant A measure for the effect of a medium on the potential energy of interaction between two charges. It is measured by comparing the capacity of a capacitor with and without the sample present.

dienophile The olefin component of a Diels-Alder reaction.

See also CYCLOADDITION.

differential scanning calorimetry (DSC) A process that scans temperature and measures heat capacity of a specimen. Records the energy required to keep a zero temperature difference between a sample cell and a reference cell that are either heated or cooled at a controlled rate.

differential thermal analysis (DTA) Thermal analysis comprises a group of methods based on the determination of changes in chemical or physical properties of a material as a function of temperature in a controlled atmosphere. DTA is a technique for recording differences in temperature between a substance and some reference material against time or temperature while the two specimens are subjected to identical temperature controls in an environment that is heated or cooled at a controlled rate.

differential thermometer A thermometer used for accurate measurement of very small changes in temperature; usually consists of a U-shaped tube terminating in two air bulbs and a colored liquid used for indicating the difference between the temperatures to which the two bulbs are exposed.

diffraction The scattering of light from a regular array of points or lines that can produce a constructive or destructive interference. X-ray diffraction is a technique used for seeing the structures of crystalline solids. X rays of a single wavelength are directed at a crystal to obtain a diffraction pattern in which interatomic spaces are then determined.

diffusion The random dispersion or spreading out of molecules from a region of high concentration to one of low concentration, stopping when the concentration is equally dispersed.

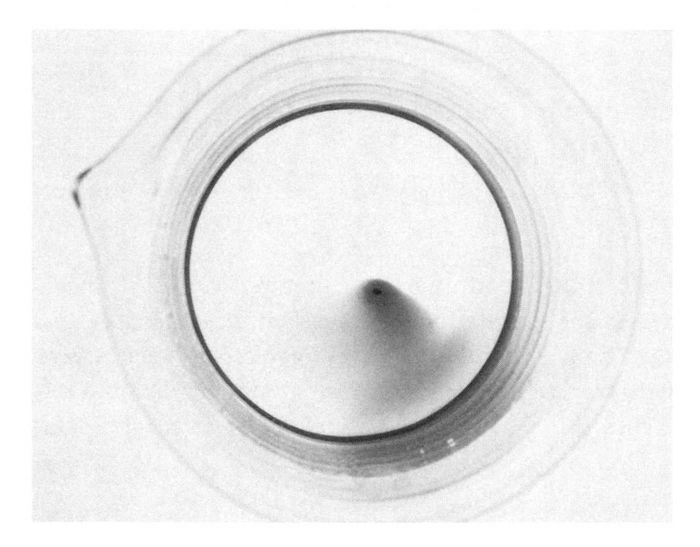

Purple cloud of potassium permanganate diffusing outward as a solid piece dissolves in a beaker of water, seen from above. The solid piece (black) takes up much less space than the solution of potassium permanganate ions. This demonstrates the principle of diffusion, where a liquid or gas will spread out to fill the available space. This in turn demonstrates the fundamental differences between a solid and a gas or liquid. A solid is a rigid, constrained structure, which contrasts with the fluid, chaotic nature of a liquid or gas. The molecules of a solid are bonded, but solvation or melting breaks the bonds between the molecules. ***(Courtesy of Andrew Lambert Photography/Science Photo Library)***

diffusion-controlled rate *See* ENCOUNTER-CONTROLLED RATE; MICROSCOPIC DIFFUSION CONTROL; MIXING CONTROL.

digestion The process by which living organisms break down ingested food in the alimentary tract into more easily absorbed and assimilated products using enzymes and other chemicals. Digestion can occur in aerobic conditions, where waste is decomposed by microbial action in the presence of oxygen, or anaerobic conditions, where waste is decomposed under microbial action in the absence of oxygen. In anaerobic conditions such as in a large-animal facility (i.e., dairy farm), the by-product, a biogas, a low-energy gas that is made with the combination of methane and carbon dioxide, can be used as an energy source. In analytical chemistry, the term *digestion* is used to describe the process of decomposing materials and bringing them into solution by heating with a liquid reagent, usually an acid.

digital In signal processing, this refers to the representation of quantities in discrete units. The information is contained and manipulated as a series of discrete numbers, as opposed to an analog representation, where the information is represented as a continuous signal. In practice, even analog signals are usually processed digitally, in that the analog signal is sampled to create a digital signal that can be processed by inherently digital computers.

dihydrofolate An oxidation product of TETRAHYDROFOLATE that appears during DNA synthesis and other reactions. It must be reduced to tetrahydrofolate to be of further use.

See also FOLATE COENZYMES.

dilution The process of reducing the concentration of a solute in solution.

dimer A molecule or compound formed by the combination of two smaller identical molecules.

dimerization The TRANSFORMATION of a MOLECULAR ENTITY A to give a molecular entity A_2. For example:

$$CH_3^{\cdot} + H_3C^{\cdot} \rightarrow CH_3CH_3$$
$$2\ CH_3COCH_3 \rightarrow (CH_3)_2C(OH)CH_2COCH_3$$
$$2\ RCOOH \rightarrow (RCOOH)_2$$

See also ASSOCIATION.

Dimroth-Reichardt E_T parameter A measure of the IONIZING POWER (loosely POLARITY) of a solvent, based on the maximum wave number of the longest wavelength electronic absorption band of

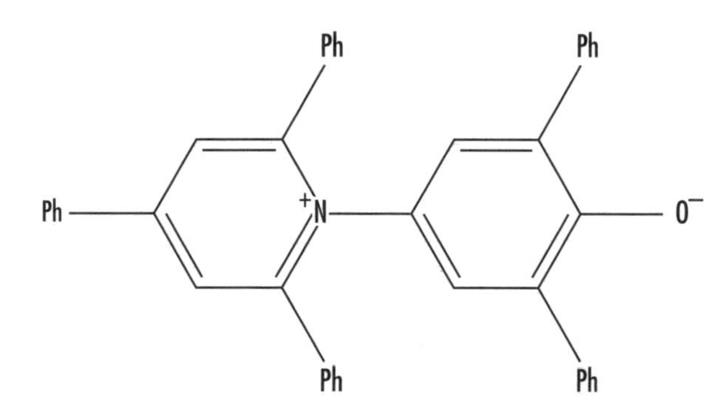

in a given solvent. E_T, called $E_T(30)$ by its originators, is given by

$$E_T = 2.859 \times 10^{-3}\nu = 2.859 \times 10^{4}\lambda^{-1}$$

where E_T is in kcal/mol, ν is in cm^{-1}, and λ is in nm.

The so-called normalized E_T^N scale is defined as

$$E_T^N = [E_T(\text{solvent}) - E_T(\text{SiMe}_4)]/[E_T(\text{water}) - E_T(\text{SiMe}_4)] = [E_T(\text{solvent}) - 30.7]/32.4$$

See also GRUNWALD-WINSTEIN EQUATION; Z-VALUE.

dinuclear *See* NUCLEARITY.

dioxygenase An ENZYME that catalyzes the INSERTION of two oxygen atoms into a SUBSTRATE, both oxygens being derived from O_2.

dipolar aprotic solvent A solvent with a comparatively high relative permittivity (or DIELECTRIC CONSTANT), greater than ca. 15, and a sizable permanent dipole moment that cannot donate suitably labile hydrogen atoms to form strong hydrogen bonds, e.g., dimethyl sulfoxide. The term (and its alternative, *polar aprotic solvent*) is a misnomer and is therefore discouraged. Such solvents are usually not APROTIC, but PROTOPHILIC (and at most weakly PROTOGENIC). In describing a solvent, it is better to be explicit about its essential properties, e.g., dipolar and nonprotogenic.

dipolar bond A BOND formed (actually or conceptually) by COORDINATION of two neutral moieties, the combination of which results in charge-separated structures, e.g.,

$$R_3N: + O \rightarrow R_3N^{+}\text{–}O^{-}$$

The term is preferred to the obsolescent synonyms *coordinate link, coordinate covalence, dative bond,* and *semipolar bond.*

dipolar cycloaddition *See* CYCLOADDITION.

dipole The separation of charge between two covalently bonded atoms; a pair of separated opposite electric charges.

dipole–dipole interaction Intermolecular or intramolecular interaction between molecules or groups having a permanent electric DIPOLE MOMENT. The strength of the interaction depends on the distance and relative orientation of the dipoles. The term applies also to intramolecular interactions between bonds having permanent dipole moments.

See also VAN DER WAALS FORCES.

dipole-induced dipole forces *See* VAN DER WAALS FORCES.

dipole moment The product of the positive charge and the distance between the charges; a measure of the POLARITY of a bond or molecule.

diprotic acid An acid that can furnish two H^+ per molecule; an acid having two dissociable protons. Examples are sulfuric acid (H_2SO_4) and carbonic acid (H_2CO_3).

diradical *See* BIRADICAL.

direct effect *See* FIELD EFFECT.

disaccharide A class of sugar, a carbohydrate, created by linking a pair of MONOSACCHARIDES, which are simple sugars. Examples of disaccharides are sucrose, which is glucose joined to fructose. Other examples include lactose, which is glucose joined with galactose, and maltose, which is two glucoses joined together. While disaccharides can be decomposed into monosaccharides, monosaccharides cannot be degraded by HYDROLYSIS. However, disaccharides can be degraded by hydrolysis into monosaccharides.

dismutase An ENZYME that catalyzes a DISPROPORTIONATION reaction.

dismutation *See* DISPROPORTIONATION.

dispersed phase The solute-like substance in a colloid.

dispersing medium A substance in which another substance is colloidally dispersed.

dispersion forces *See* LONDON FORCES.

displacement reaction A chemical reaction in which one element displaces another from a compound.

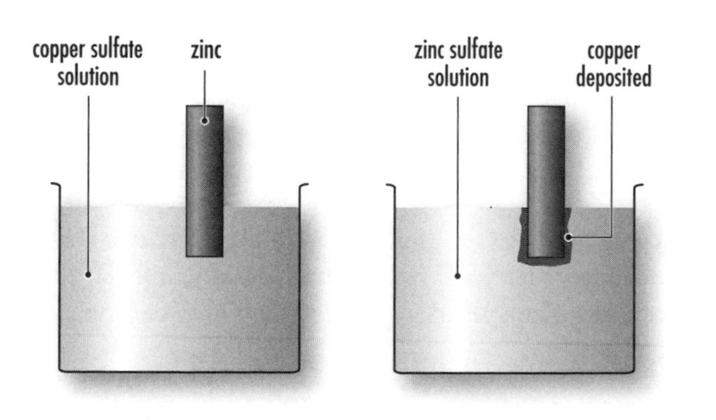

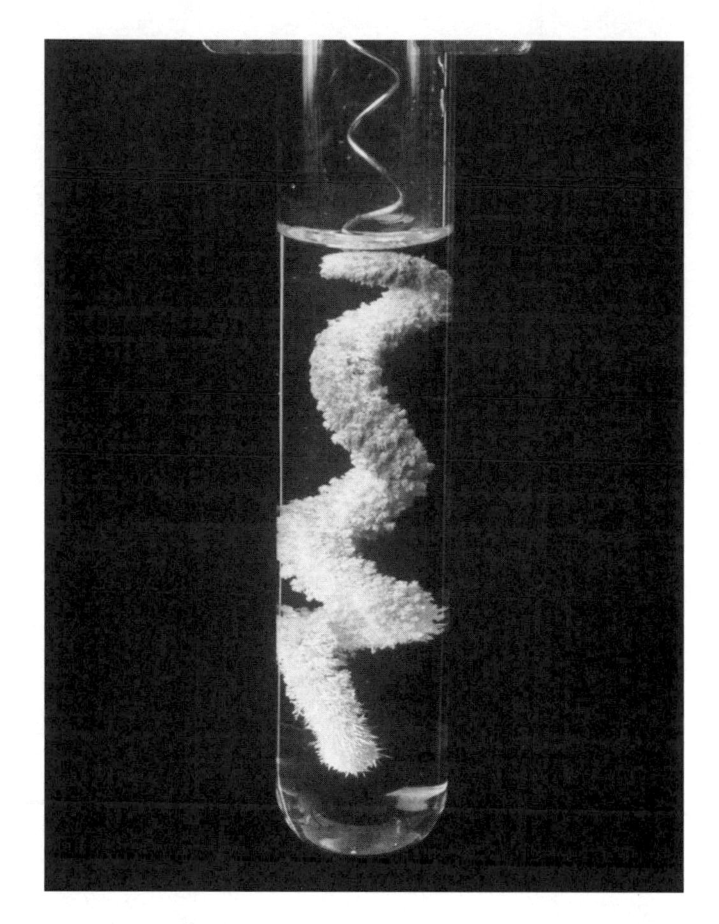

Experiment demonstrating the displacement of silver from solution by copper. A copper coil is placed in a solution of silver nitrate (colorless). After some time, the solution turns blue, and dendritic crystals form on the coil. The copper from the coil displaces silver ions in solution. The copper becomes ionized as copper(II), seen as a characteristic blue color, and the silver is deposited as a metal crystal. ***(Courtesy of Jerry Mason/Science Photo Library)***

disposition *See* DRUG DISPOSITION.

disproportionation Any chemical reaction of the type A + A → A′ + A″, where A, A′, and A″ are different chemical species. For example:

$$2\,ArH^{+} \rightarrow ArH + ArH^{2+}$$

The reverse of disproportionation (or dismutation) is called COMPROPORTIONATION. A special case of disproportionation (or dismutation) is "radical disproportionation," exemplified by

$$\cdot CH_2CH_3 + \cdot CH_2CH_3 \rightarrow CH_2{=}CH_2 + CH_3CH_3$$

Reactions of the more general type

$RC{\cdot}HCH_3 + R'C{\cdot}HCH_3 \rightarrow RCH{=}CH_2 + R'CH_2CH_3$

are also loosely described as radical disproportionations.

disrotatory *See* ELECTROCYCLIC REACTION.

dissimilatory Related to the conversion of food or other nutrients into products plus energy-containing compounds.

dissociation (1) The separation of a MOLECULAR ENTITY into two or more molecular entities (or any similar separation within a polyatomic molecular entity). Examples include unimolecular HETEROLYSIS and HOMOLYSIS, and the separation of the constituents of an ION PAIR into free ions. (2) The separation of the constituents of any aggregate of molecular entities.

In both senses dissociation is the reverse of ASSOCIATION.

See also MOLECULARITY.

dissociation constant *See* STABILITY CONSTANT.

dissociation energy The energy needed to separate a pair of atoms.

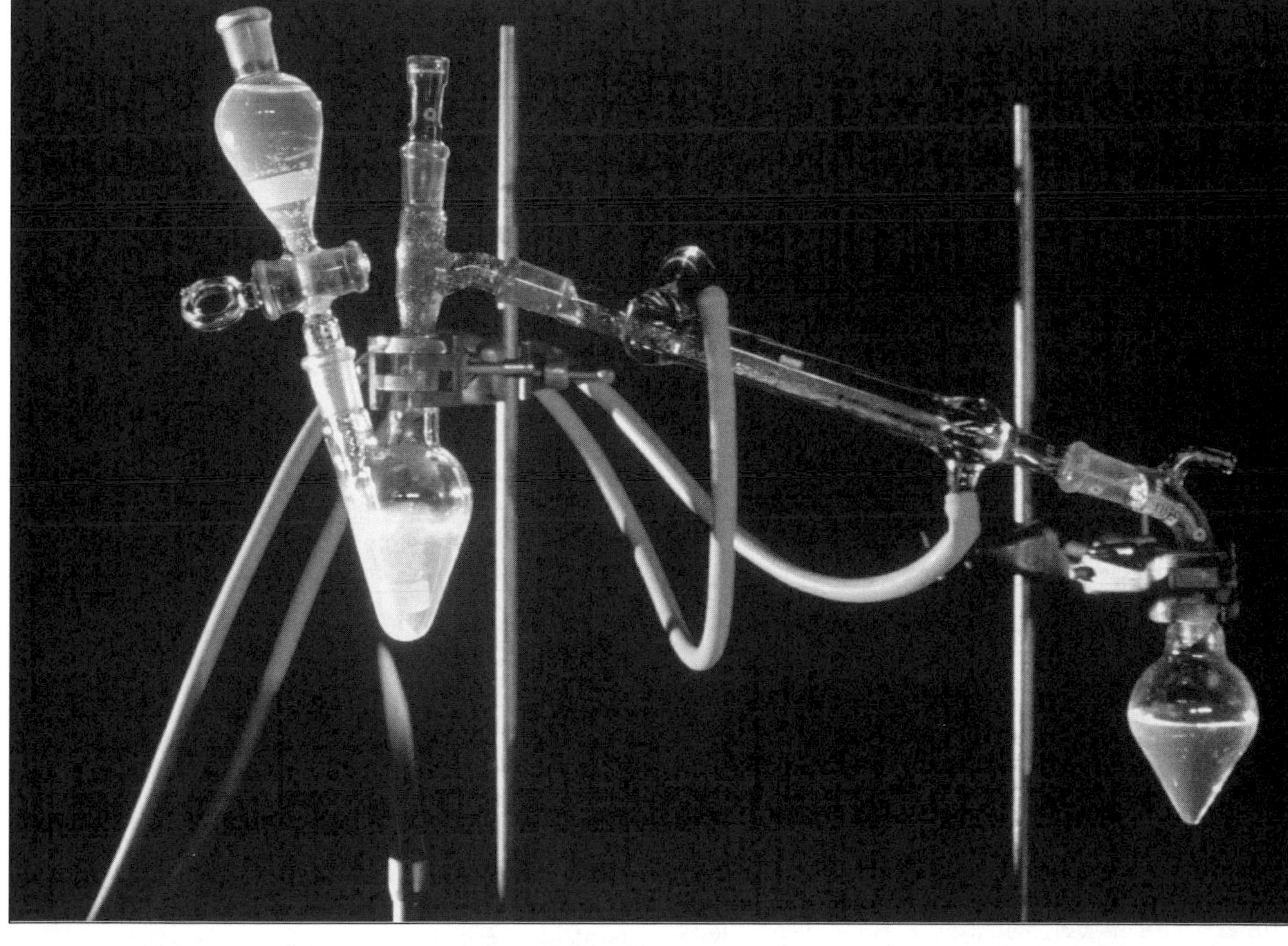

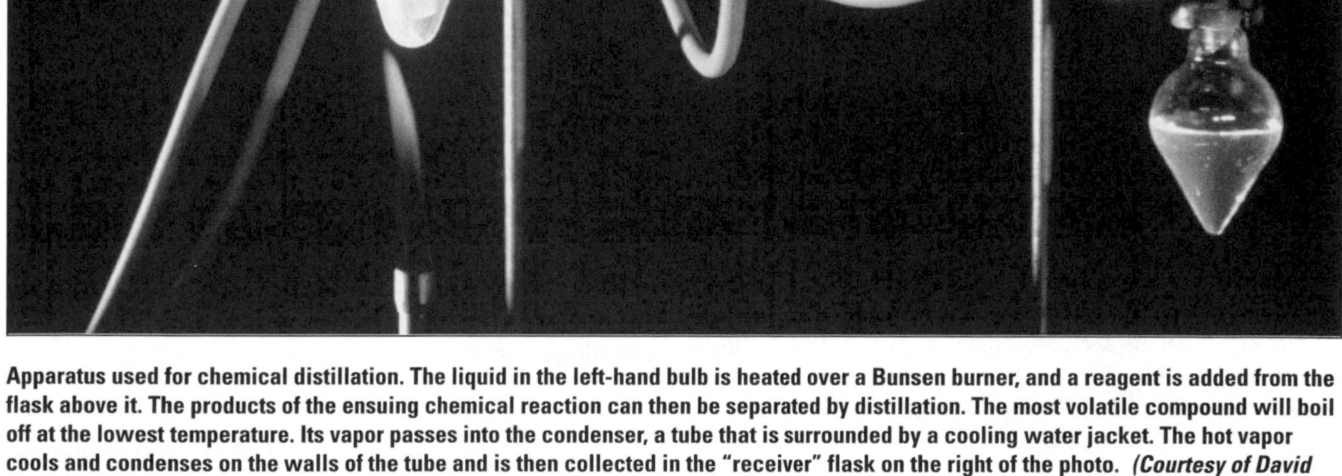

Apparatus used for chemical distillation. The liquid in the left-hand bulb is heated over a Bunsen burner, and a reagent is added from the flask above it. The products of the ensuing chemical reaction can then be separated by distillation. The most volatile compound will boil off at the lowest temperature. Its vapor passes into the condenser, a tube that is surrounded by a cooling water jacket. The hot vapor cools and condenses on the walls of the tube and is then collected in the "receiver" flask on the right of the photo. ***(Courtesy of David Taylor/Science Photo Library)***

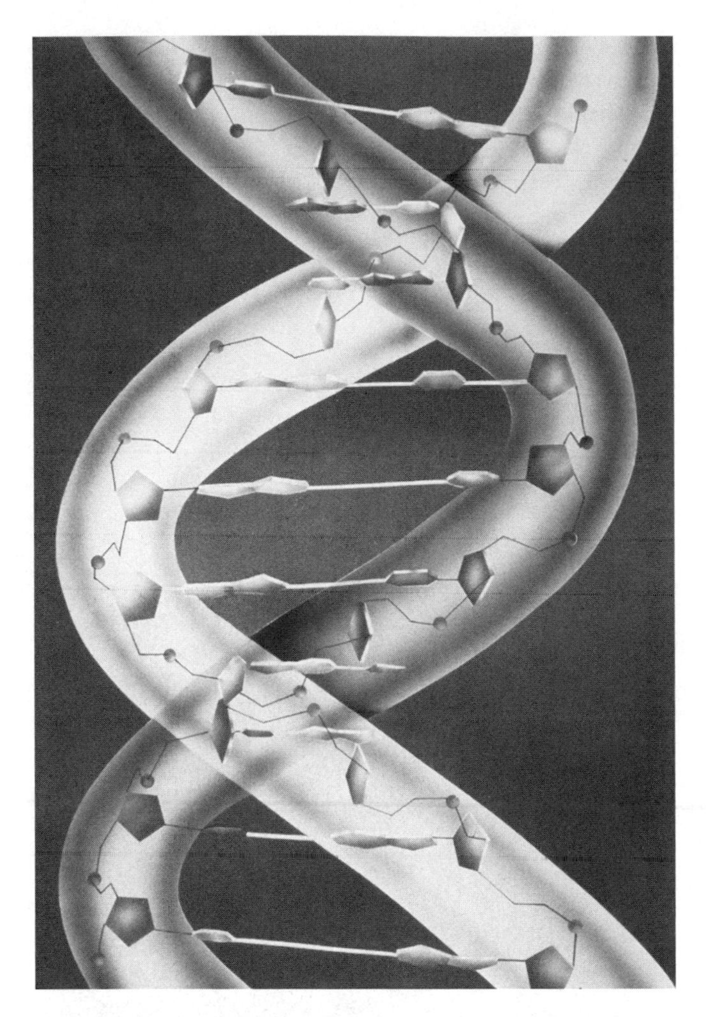

Artwork that represents the structure of deoxyribonucleic acid (DNA). This is the structure that contains the inherited coded instructions responsible for the development of an organism. A DNA molecule is composed of two interwound strands twisted into a helical shape that gives it great stability. Each strand alternates a sugar group (pentagon) with a phosphate group (circle). The two strands are linked by pairs of substances called bases, of which there are four types, known as adenine, guanine, thymine, and cytosine. The sequence of bases along a strand provides the code for the activities of a cell. ***(Courtesy of Kairos, Latin Stock/Science Photo Library)***

distillation The creation of a purified liquid by condensing from a vapor produced during distilling (boiling).

distomer The ENANTIOMER of a chiral compound that is the less potent for a particular action. This definition does not exclude the possibility of other effects or side effects of the distomer.

See also CHIRALITY; EUTOMER.

distonic radical cation A radical cation in which charge and radical sites are separated.

DNA (deoxyribonucleic acid) A high-molecular-mass linear polymer, composed of NUCLEOTIDES containing 2-deoxyribose and linked between positions 3′ and 5′ by phosphodiester groups. DNA contains the GENETIC information of organisms. The double-stranded form consists of a DOUBLE HELIX of two complementary chains that run in opposite directions and are held together by hydrogen bonds between pairs of the complementary nucleotides. The way the helices are constructed may differ and is usually designated as A, B, Z, etc. Occasionally, alternative structures are found, such as those with Hoogsteen BASE PAIRING.

DNA ligase A linking enzyme involved in replicating and repairing DNA molecules. It seals "nicks" in the backbone of a single strand of a double-stranded DNA molecule; connects Okazaki fragments, short, single-stranded DNA fragments on the lagging strand during DNA replication, producing a complementary strand of DNA; and links two DNA molecules together by catalyzing the formation of a (phosphodiester) bond between the 5′ and 3′ ends of the nicked DNA backbone.

DNA methylation A biochemical event that adds a methyl group ($-CH_3$) to DNA, usually at the base cytosine or adenosine, and may be a signal for a gene or part of a chromosome to turn off gene expression and become inactive.

DNA polymerase An enzyme that catalyzes the synthesis of new complementary DNA molecules from single-stranded DNA templates and primers. Different DNA polymerases are responsible for replication and repair of DNA, and they extend the chain by adding nucleotides to the 3′ end of the growing DNA; catalyzes the formation of covalent bonds between the 3′ end of a new DNA fragment and the 5′ end of the growing strand. Used in DNA fingerprinting for genealogical studies and forensics.

DNA probe A single strand of DNA that is labeled or tagged with a fluorescent or radioactive substance

What a Crime Lab Does, It Does with Chemistry

by Harry K. Garber

Much as a car runs on gasoline, a crime lab runs on physical evidence. The O. J. Simpson murder trial in 1995 put a nationwide, if not a worldwide, spotlight on many aspects of what a crime lab does. Physical evidence encompasses any and all objects that can either establish that a crime has been committed or provide a link between a crime and its victim or perpetrator. Forensic scientists in a crime lab analyze this physical evidence; however, much of the analysis is in fact chemical analysis.

Let's briefly examine physical and biological analysis first. The hit-and-run death of a pedestrian can involve both. Fragments of broken headlight glass are collected from the scene. Samples of the victim's blood are collected from the autopsy. Later, a suspect car is located. Evidence collected from the car includes the broken headlight, more fragments of broken headlight glass from the headlight well, and dried bloodstains from the undercarriage. The stains are collected by rubbing with cotton swabs moistened with purified water; the swabs are dried and submitted to the crime lab. Having first kept careful track of which fragments of glass came from the car (the "known" evidence) and which came from the crime scene (the "questioned" evidence), forensic scientists attempt to reassemble part of the headlight in jigsaw-puzzle fashion. Any jigsaw fit that includes both questioned and known fragments of glass shows that the two were originally part of the same headlight and therefore links the car to the crime scene. This is an example of physical analysis. And if DNA profiles of the victim's blood (the "known" evidence) and the blood on one or more swabs (the "questioned" evidence) match, then the car is linked to the victim. This is an example of biological analysis.

A crime lab is typically divided into several sections. Due to the degree of expertise required to perform the necessary analysis, a forensic scientist is typically hired to work in just one section. However, cross training sometimes occurs, depending upon a lab's needs. In many labs, a single forensic scientist analyzes a given case in a given section. In some labs, scientists team up. In either instance, peer review by a scientist not involved precedes issue of the final report in many labs for quality assurance. And what is considered one "case" by an agency bringing evidence to the lab may require analysis by more than one section.

A scientist in the trace-evidence section might have performed the physical analysis in the hit-and-run case mentioned earlier (whereas one in the bioscience section might have performed the biological analysis). If a jigsaw fit cannot be obtained, chemical analysis can be performed to determine whether the questioned and known glass could have had a common source. Properties compared may include ultraviolet fluorescence, refractive index, density, and elemental analysis. The more properties that agree, the more likely it is that the two at least had a common source. Conversely, a mismatch in any property rules out a common source. Trace-evidence scientists also examine fire debris for the presence of accelerants, paint chips (from auto crashes or burglaries) for common source (can compare paint vehicle itself, pigments, elemental analysis), tear gas samples (to identify the active agent), explosive residues (to identify either traces of parent explosive or products of the explosion), and headlight filaments to determine whether a broken headlight was on or off at the instant of a crash. This last analysis is actually physical (simply examination under an ordinary microscope) but is in part a result of chemistry. An intact headlight is filled with an inert gas, since air would oxidize the white-hot filament. A headlight that was off at impact has a shiny, silver-colored filament. A headlight that was on at impact has a dark blue- to black-colored filament; when the glass breaks, the admitted air forms a layer of metal oxide(s). Another feature is often evident on such a filament—microscopic globules of glass sticking to its surface. These globules arise from small glass fragments melting on contact and then freezing into position as the filament cools off.

A primary function of the questioned-documents section is to determine whether the same person executed both questioned or known handwriting. For example, a murderer might attempt to conceal the crime by writing a suicide note. Comparison of handwriting in the note to known samples of both the victim's and the suspect's handwriting would expose the ruse. This is not chemical analysis. However, if the note was typed or computer printed, both physical and chemical analysis could compare properties, such as letter size and shape or chemical composition of the paper and ink, to see if the note matches either a typewriter, a computer printer, or paper accessible to either the victim or the suspect. If a document is suspected of being altered, modern instrumentation allows it to be examined in a matter of minutes for inks with different light absorption or emission properties. Sometimes, determining whether a document was executed on an alleged date is desired. For example, a document alleged to be executed in 1980 could not be authentic if analysis reveals that it contains either paper or ink that did not exist until 2000.

(continues)

What a Crime Lab Does, It Does with Chemistry
(continued)

A primary function of the firearms section is to determine whether a questioned bullet (for example, removed from the body of a shooting victim) was fired from a known (suspected) weapon. Rifling in a weapon's barrel cuts impressions into a bullet's surface, but fine striations unique to a given barrel are superimposed on these impressions. Analysts fire the suspected weapon two or more times into a water tank (for handguns) or a cotton waste recovery box (for rifles), which stops the bullets without damaging them. If the impressions and striations around the entire circumference of these recovered bullets match those on the questioned bullet (as viewed under a comparison microscope), then the same weapon fired all the bullets. This analysis is physical. But chemistry also enters into the firearms section; one example is serial-number restoration. Often weapons (or other metal parts such as motor-vehicle engine blocks) arrive at the lab with their serial numbers obliterated. Serial numbers are usually die stamped; this process cold-works the metal in the area immediately surrounding and a short distance below the penetration of the die. The cold-worked metal is less resistant to chemical attack than the base metal. Polishing the obliterated area, then etching the area with a chemical suitable for the particular metal (sometimes with an applied voltage to speed up the attack), can make the numbers visible again. Serial-number restoration also works with plastic parts; cold-worked plastic is less heat-resistant than the base plastic, so one can replace etching with a high-intensity lamp.

The drug-chemistry section analyzes evidence believed to contain illegal or controlled substances. Evidence is most often tablets, capsules, powder, crystals, or vegetation. Scientists isolate the active ingredient(s) using solvent extraction and then perform several chemical tests. Final identification is accomplished by one or more techniques considered to produce a "chemical fingerprint" for a substance: infrared (IR) spectroscopy, nuclear magnetic resonance (NMR) spectroscopy, or mass spectrometry (MS). Rarely is the mass spectrometer used alone; it is linked instead to either a gas chromatograph or a liquid chromatograph, hence the names GC-MS and LC-MS, respectively. The hyphenated techniques are especially useful if solvent extraction yields a mixture rather than one pure compound. The chromatograph separates the mixture; one by one, each compound is identified as it passes into the mass spectrometer.

The author is assigned to the toxicology section, which will provide the final examples of chemical analysis. Toxicologists analyze body fluids (such as blood and urine) and body tissues (such as liver, brain, and stomach contents, although the last item is not strictly a tissue) for the presence of (and often the amount of) drugs and poisons. These items are submitted in two main types of cases. The first is in arrests of motorists for driving under the influence of alcohol and/or drugs. The second is in unattended deaths, where a drug or poison overdose might be the cause of death. Occasionally, nonbiological specimens will be submitted. One example is suspected alcoholic beverages in cases involving violations of open-container laws (open-container laws prohibit the possession of any open alcoholic beverage container and the consumption of any alcoholic beverage in the passenger area of a motor vehicle). Toxicologists isolate drugs, poisons, and their by-products (metabolites) from blood, urine, and tissue specimens using solvent extraction or solid-phase extraction. A complex mixture normally results; indentification by IR or NMR is impossible, since these techniques require a pure substance. Hence GC-MS and LC-MS are employed.

Toxicology can reveal the unnatural nature of what might at first appear to be a natural death. For example, a murderer might attempt to conceal the crime by setting a fire. But a fire victim normally has a high blood carbon monoxide level. A murder victim, dead before the fire began, couldn't inhale any carbon monoxide. In another example, the author analyzed autopsy specimens from a 43-year-old female found dead in bed. Lethal levels of chloroform (an industrial solvent) were found. When confronted by this evidence, the woman's estranged husband admitted entering the house while she slept and covering her mouth and nose with a chloroform-soaked tissue.

Other times a death really is natural. The author analyzed autopsy specimens from Sergei Grinkov, the Olympic pairs figure skating champion who collapsed and died during a practice session in Lake Placid, New York. Only lidocaine and atropine (given during resuscitation attempts) were found.

Occasionally the toxicology section analyzes unusual items. The author examined both fighters' boxing gloves for foreign substances after the 1996 Madison Square Garden heavyweight bout in which Evander Holyfield defeated Bobby Czyz—and found none. Czyz's camp argued that he could not continue because his eyes were burned from a foreign substance on Holyfield's gloves.

— **Harry K. Garber** is a forensic scientist in the toxicology section of the New York State Police Forensic Investigation Center in Albany, New York.

and binds specifically to a complementary DNA sequence. The probe is used to detect its incorporation through hybridization with another DNA sample. DNA probes can provide rapid identification of certain species like mycobacterium.

See also NUCLEIC ACID.

Dobson unit The standard way to express ozone amounts in the atmosphere. One DU is 2.7×10^{16} ozone molecules per square centimeter. One Dobson unit refers to a layer of ozone that would be 0.001 cm thick under conditions of standard temperature (0°C) and pressure (the average pressure at the surface of the Earth). For example, 300 Dobson units of ozone brought down to the surface of the Earth at 0°C would occupy a layer only 0.3 cm thick in a column. Dobson was a researcher at Oxford University who, in the 1920s, built the first instrument (now called the Dobson meter) to measure total ozone from the ground.

docking studies MOLECULAR MODELING studies aiming at finding a proper fit between a LIGAND and its BINDING SITE.

Doisy, Edward Adelbert (1893–1986) American *Biochemist* Edward Adelbert Doisy was born at Hume, Illinois, on November 3, 1893, to Edward Perez and his wife Ada (née Alley). Doisy was educated at the University of Illinois, receiving a B.A. degree in 1914 and a M.S. degree in 1916. He received a Ph.D. in 1920 from Harvard University.

From 1915 to 1917 he was assistant in biochemistry at Harvard Medical School, and during the following two years he served in the war in the Sanitary Corps of the U.S. Army. In 1919 he became an instructor, associate, and associate professor at Washington University School of Medicine, advancing to professor of biochemistry at St. Louis University School of Medicine in 1923. The following year he was appointed director of the department of biochemistry, retiring in 1965 (emeritus 1965–86).

Doisy and his associates isolated the sex hormones estrone (1929), estriol (1930), and estradiol (1935). He also isolated two forms of vitamin K and synthesized it in 1936–39. For his work on vitamin K, Doisy was awarded the Nobel Prize in physiology or medicine for 1943.

Later, Doisy improved the methods used for the isolation and identification of insulin and contributed to the knowledge of antibiotics, blood buffer systems, and bile acid metabolism.

In 1936 he published *Sex Hormones,* and in 1939, in collaboration with Edgar Allen and C. H. Danforth, he published a book entitled *Sex and Internal Secretions.* He died on October 23, 1986, in St. Louis, Missouri.

Domagk, Gerhard Johannes Paul (1895–1964) German *Biochemist* Gerhard Johannes Paul Domagk was born on October 30, 1895, at Lagow, a small town in the Brandenburg Marches. He attended school in Sommerfeld until age 14, where his father was assistant headmaster. His mother, Martha Reimer, came from farming stock in the Marches; she lived in Sommerfeld until 1945, when she was expelled from her home and died from starvation in a refugee camp.

Domagk became a medical student at Kiel and during World War I served in the army. After being wounded in 1914, he worked in the cholera hospitals in Russia. He noticed that medicine of the time had little success and was moved by the helplessness of the medical men of that time with cholera, typhus, diarrhea infections, and other infectious diseases, noting that surgery had little value in the treatment of these diseases. He also noticed that amputations and other radical treatments were often followed by severe bacterial infections.

In 1918 he resumed his medical studies at Kiel and graduated in 1921. In 1923 he moved to Greifswald, and a year later he became a university lecturer in pathological anatomy, holding the same post in the University of Münster from 1925 until he became a professor in 1958.

During the years 1927–1929 he was given a leave of absence from the University of Münster to do research in the laboratories of the I. G. Farbenindustrie, at Wuppertal. In 1932 he tested a red dye, Prontosil rubrum. While the dye itself had no antibacterial properties, when he slightly changed its chemical makeup, it showed a remarkable ability to stop infections in mice caused by streptococcal bacteria. He had discovered the sulfa drugs that have since revolutionized medicine and saved many thousands of lives. He was awarded the 1939 Nobel Prize in physiology or medicine for his discovery. He died on April 24, 1964.

domain An independently folded unit within a PROTEIN, often joined by a flexible segment of the POLYPEPTIDE chain. Domain is also the highest taxonomic rank comprising the Eukarya, Bacteria, and Archaea. The Archaea are commonly known as extremophiles, occurring in deep-sea vents and hot sulfur springs, whereas the Eukarya comprise the higher life forms, including humans.

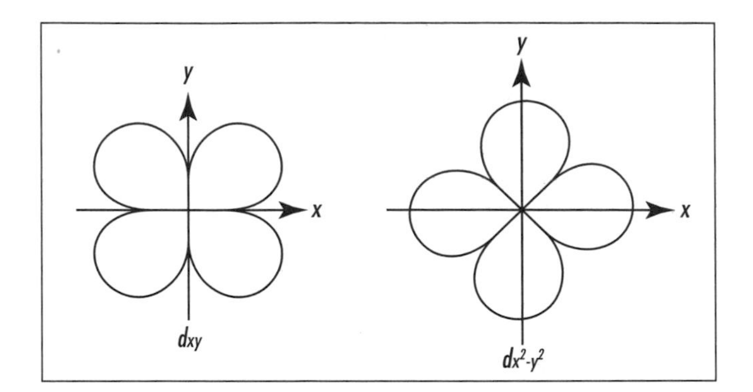

Each *d* orbital can hold two electrons.

donor-atom symbol A polydentate LIGAND possesses more than one donor site, some or all of which may be involved in COORDINATION. To indicate the points of ligation, a system is needed. The general and systematic system for doing this is called the kappa convention: single ligating atom attachments of a polyatomic ligand to a coordination center are indicated by the italic element symbol preceded by a Greek kappa, κ. In earlier practice, the different donors of the ligand were denoted by adding to the end of the name of the ligand the italicized symbol(s) for the atom or atoms through which attachment to the metal occurs.

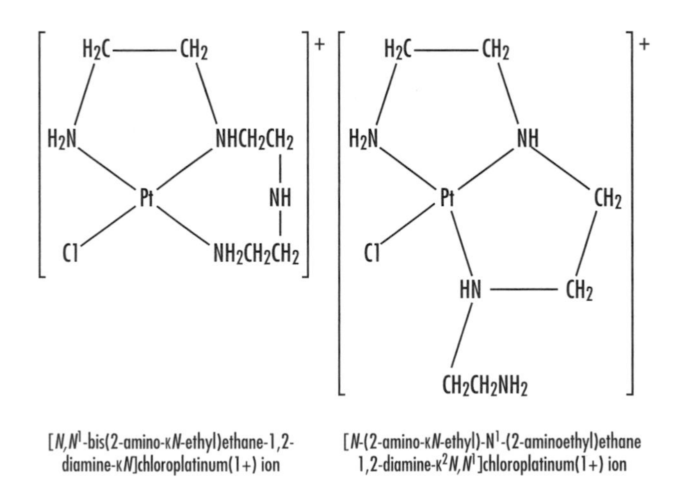

[*N*,*N*'-bis(2-amino-κ*N*-ethyl)ethane-1,2-diamine-κ*N*]chloroplatinum(1+) ion

[*N*-(2-amino-κ*N*-ethyl)-N'-(2-aminoethyl)ethane 1,2-diamine-κ²*N*,*N*']chloroplatinum(1+) ion

donor number (DN) A quantitative measure of Lewis basicity.

See also LEWIS BASE.

Doppler radar The weather radar system that uses the Doppler shift of radio waves to detect air motion that can result in tornadoes and precipitation, as previously developed weather radar systems do. It can also measure the speed and direction of rain and ice, as well as detect the formation of tornadoes sooner than older radars.

***d* orbitals** Atomic-level organization; a set of five degenerate orbitals per energy level, beginning in the third energy level, higher in energy than *s* and *p* orbitals of the same energy level. Orbitals with subshell quantum number $l = 2$.

dosimeter A small, calibrated device designed to detect and measure incident ionizing radiation or chemical exposure. Usually worn as a badge.

double-blind study A clinical study of potential and marketed DRUGS, where neither the investigators nor the subjects know which subjects will be treated with the active principle and which ones will receive a placebo.

double bond Covalent bond resulting from the sharing of two pairs of electrons (four electrons) between two atoms.

double helix Two strands of DNA coiled about a central axis, usually a right-handed HELIX. As seen on page 80, the two sugar phosphate backbones wind around the outside of the bases (A = adenine, G = guanine, T = thymine, C = cytosine) and are exposed to the phosphate solvent. The strands are antiparallel, thus the phosphodiester bonds run in opposite directions. As a result, the structure has major and minor grooves at the surface. Each adenine in one strand of DNA is hydrogen-bonded to a thymine in the second strand; each guanine is hydrogen-bonded to a cytosine.

See also DNA.

double prodrug (pro-prodrug) A biologically inactive molecule that is transformed in vivo in two steps (enzymatically or chemically) to the active species.

double salt A compound of two salts. That is, the ions in solution appear to come from two salts, but a single substance is formed on crystallization. Examples are alum, potassium aluminum sulfate, and zinc ammonium chloride.

downfield *See* CHEMICAL SHIFT.

Downs cell Electrolytic cell used for the commercial electrolysis of molten sodium chloride to produce commercial-grade sodium.

DP number The number of MONOMER units jointed together in a polymer; the degree of polymerization; the average number of monomer units per polymer unit.

driving force (affinity of a reaction), (SI unit: kJ mol^{-1}) The decrease in Gibbs energy on going from the reactants to the products of a chemical reaction ($-\Delta G$).

See also GIBBS FREE ENERGY.

drug Any substance presented for treating, curing, or preventing disease in human beings or in animals. A drug may also be used for making a medical diagnosis, managing pain, or for restoring, correcting, or modifying physiological functions (e.g., the contraceptive pill).

drug disposition Refers to all processes involved in the absorption, distribution, METABOLISM, and excretion of DRUGs in a living organism.

drug latentiation The chemical modification of a biologically active compound to form a new compound that in vivo will liberate the parent compound. Drug latentiation is synonymous with PRODRUG design.

drug targeting A strategy aiming at the delivery of a compound to a particular tissue of the body.

dry cells The term used for ordinary batteries (voltaic cells) that are used in everyday flashlights, etc. A dry-cell battery contains electrolytes that are in the form of paste rather than liquid; the electrolyte is absorbed in a porous medium or is otherwise restrained from flowing. Also known as a Leclanché cell, which is the common commercial type.

dry ice A solid form of the gas carbon dioxide (CO_2), used for keeping items cold or transporting food materials over long distances. At –78.5°C and ambient pressure, it changes directly to a gas as it absorbs heat.

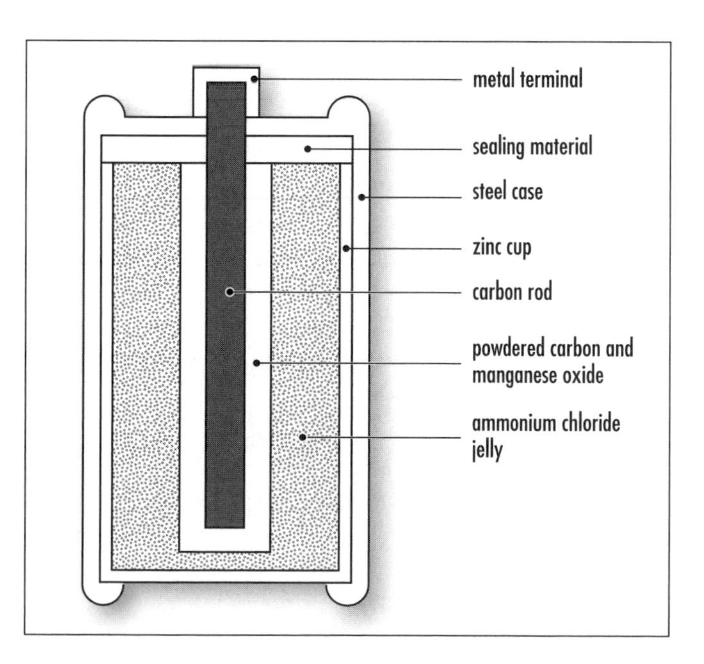

A dry-cell battery contains electrolytes that are in the form of paste rather than liquid.

d-transition elements (metals) Transition elements are elements that can form bonds with the electrons from the outer two shells. They are characterized by having an incompletely filled d subshell. There are three series of transition elements, corresponding to filling of the *3d, 4d,* and *5d* subshells. They are all metals, and most are hard, strong, and lustrous, have high melting and boiling points, and are good conductors of heat and electricity.

Transition elements that are common in minerals, or occur in significant amounts, are titanium, chromium, manganese, iron, cobalt, nickel, copper, molybdenum, silver, tungsten, gold, and platinum.

d-Transition elements are B Group elements except for IIB in the periodic table. In the newer labeling scheme, they are groups 3 through 11.

dual-action drug A compound that combines two desired different pharmacological actions in a similarly efficacious dose.

dual substituent-parameter equation In a general sense, this is any equation that expresses substituent effects in terms of two parameters. However, in practice, the term is used more specifically for an equation for summarizing the effects of meta- or para-substituents (i = m or p) X on chemical reactivity, spectroscopic properties, etc., of a probe site Y in a benzene or other aromatic system.

$$P^i = \rho_I^i \sigma_I + \rho_R^i \sigma_R$$

P is the magnitude of the property Y for substituent X, expressed relative to the property for X=H; σ_I and σ_R are inductive or polar and resonance substituent constants, respectively, there being various scales for σ_R; ρ_I and ρ_R are the corresponding regression coefficients.

See also EXTENDED HAMMETT EQUATION.

ductile The ability of a material to be drawn out or hammered in thin pieces or fashioned into a new form.

Dumas method A way to determine the MOLECULAR WEIGHTs of volatile liquids.

dynamic equilibrium An EQUILIBRIUM in which processes occur continuously or the actions oppose each other so that no net change occurs.

dynein A molecular motor, a complex believed to be made of 12 distinct protein parts, that performs basic transportation tasks critical to the cell. Converts chemical energy stored in an ATP (adenosine triphosphate) molecule into mechanical energy that moves material through the cell along slender filaments called microtubules. One of the most important functions occurs during cell division, when it helps move chromosomes into proper position. It also plays a part in the movement of eukaryotic flagella and cilia.

Molecular motors play a critical role in a host of cell functions such as membrane trafficking and cell movement during interphase, and for cell asymmetry development. During cell division, they are responsible for establishing the mitotic or meiotic spindle, as well as segregating chromosomes and dividing the cell at cytokinesis. It is the last part of the mitotic cycle, during which the two daughter cells separate. Motors either move along actin tracks (members of the myosin superfamily) or microtubules (the dynein and kinesin superfamilies). Based on the Greek *dunamis,* meaning power.

dyotropic rearrangement An uncatalyzed process in which two SIGMA (σ) BONDs simultaneously migrate intramolecularly, for example,

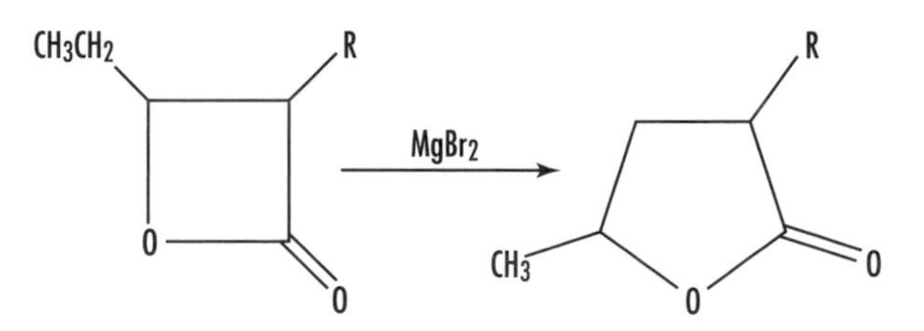

E

EC nomenclature for enzymes A classification of ENZYMEs according to the Enzyme Commission of the International Union of Biochemistry and Molecular Biology. Enzymes are allocated four numbers, the first of which defines the type of reaction catalyzed; the next two define the SUBSTRATEs, and the fourth is a catalogue number. Categories of enzymes are EC 1, OXIDOREDUCTASEs; EC 2, TRANSFERASEs; EC 3, HYDROLASEs; EC 4, LYASEs; EC 5, ISOMERASEs; EC 6, LIGASEs (Synthetases).

ecosystem Any natural system including biotic and abiotic parts that interact as a unit to produce a stable functioning system through cyclical exchange of materials.

EDRF *See* ENDOTHELIUM-DERIVED RELAXING FACTOR.

educt Used mainly in the German literature as a term for starting material (reactant). It should be avoided in English because in that context it means "something that comes out" and not "something that goes in." The German use of the term is in fact also incorrect.

effective atomic number Represents the total number of electrons surrounding the nucleus of a metal atom in a metal complex, and is calculated from the composition and atomic numbers of a compound or mixture. In the organometallic chemistry of transition metals, the metal atom often forms bonds such that all of the valence shell orbitals are filled, so that the total number of electrons in its valence shell is 18, which is effectively the configuration of the next noble gas. Also called the 18-electron rule. The term is used in different contexts to represent an effective average atomic number of a multicomponent system.

effective charge Change in effective charge is a quantity obtained by comparison of the POLAR EFFECT of substituents on the free energies of rate or equilibrium processes with that on a standard ionization equilibrium. Provided that the effective charge on the states in the standard equilibrium is defined, it is possible to measure effective charges for states in the reaction or equilibrium under consideration.

effective collisions A collision between two molecules or reactants that results in a reaction or that produces a product, and where the colliding particles have the necessary amount of energy and orientation to produce the reaction.

effective molarity (effective concentration) The ratio of the first-order rate constant of an INTRAMOLECULAR reaction involving two functional groups within the same MOLECULAR ENTITY to the second-order RATE CONSTANT of an analogous INTERMOLECULAR elementary reaction. This ratio has the dimension of concentration. The term can also apply to an equilibrium constant.

See also INTRAMOLECULAR CATALYSIS.

effective nuclear charge That portion of the nuclear charge that is experienced by the highest-energy-level electrons (outermost electrons) in an atom.

efficacy Describes the relative intensity with which AGONISTs vary in the response they produce, even when they occupy the same number of RECEPTORs with the same AFFINITY. Efficacy is not synonymous with INTRINSIC ACTIVITY.

Efficacy is the property that enables DRUGs to produce responses. It is convenient to differentiate the properties of drugs into two groups: those that cause them to associate with the receptors (affinity) and those that produce stimulus (efficacy). This term is often used to characterize the level of maximal responses induced by agonists. In fact, not all agonists of a receptor are capable of inducing identical levels of maximal responses. Maximal response depends on the efficiency of receptor coupling, i.e., from the cascade of events that, from the binding of the drug to the receptor, leads to the observed biological effect.

effusion The flow of gases through small openings in comparison with the distance between molecules.

EF-hand A common structure to bind Ca^{2+} in CALMODULIN and other Ca^{2+}-binding proteins consisting of a HELIX (E), a loop, and another helix (F).

See also METABOLIC REGULATION.

eighteen-electron rule An electron-counting rule to which an overwhelming majority of stable diamagnetic transition metal complexes adhere. The number of nonbonded electrons at the metal plus the number of electrons in the metal-ligand bonds should be 18. The 18-electron rule in transition metal chemistry is a full analog of the Lewis octet rule, also known as the effective atomic number rule.

Eijkman, Christiaan (1858–1930) Nordic *Physician* Christiaan Eijkman was born on August 11, 1858, at Nijkerk in Gelderland (the Netherlands) to Christiaan Eijkman, the headmaster of a local school, and Johanna Alida Pool. He received his education at his father's school in Zaandam. In 1875 he entered the Military Medical School of the University of Amsterdam and received training as a medical officer for the Netherlands Indies Army. From 1879 to 1881 he wrote his thesis "On Polarization of the Nerves," which gained him his doctor's degree, with honors, on July 13, 1883. On a trip to the Indies he caught malaria and returned to Europe in 1885.

Eijkman was director of the Geneeskundig Laboratorium (medical laboratory) in Batavia from 1888 to 1896, and during that time he made a number of important discoveries in nutritional science. In 1893 he discovered that the cause of beriberi was the deficiency of vitamins, not of bacterial origin as thought by the scientific community. He discovered vitamin B, and this discovery led to the whole concept of vitamins. For this discovery he was given the Nobel Prize in physiology or medicine for 1929.

He wrote two textbooks for his students at the Java Medical School, one on physiology and the other on organic chemistry. In 1898 he became professor in hygiene and forensic medicine at Utrecht, but also engaged in problems of water supply, housing, school hygiene, physical education, and, as a member of the Gezondheidsraad (health council) and the Gezondheids commissie (health commission), he participated in the struggle against alcoholism and tuberculosis. He was the founder of the Vereeniging tot Bestrijding van de Tuberculose (Society for the Struggle against Tuberculosis). Eijkman died in Utrecht on November 5, 1930.

Eijkman's syndrome, a complex of nervous symptoms in animals deprived of vitamin B1, is named for him.

Einthoven, Willem (1860–1927) Nordic *Physiologist* Willem Einthoven was born on May 21, 1860, in Semarang on the island of Java, Indonesia, to Jacob Einthoven, an army medical officer in the Indies, and Louise M.M.C. de Vogel, daughter of the director of finance in the Indies.

At the death of his father, the family moved to Holland and settled in Utrecht, where he attended school. In 1878 he entered the University of Utrecht as a medical student. In 1885, after receiving his medical doctorate, he was appointed successor to A. Heynsius, professor of physiology at the University of Leiden, and stayed there until his death.

He conducted a great deal of research on the heart. To measure the electric currents created by the heart, he invented a string galvanometer (called the Einthoven galvanometer) and was able to measure the changes of electrical potential caused by contractions of the heart muscle and to record them, creating the electrocardiogram (EKG), a word he coined, which is a graphic record of the action of the heart. This work earned him the Nobel Prize in physiology or medicine for 1924. He published many scientific papers in journals of the time. He died on September 29, 1927.

electrical conductivity Having the ability to conduct electricity.

electrical resonance An effect in which the resistance to the flow of an electrical current becomes very small over a narrow frequency range.

electrochemical gradient The relative concentration of charged ions across a membrane. Ions move across the membrane due to the concentration difference on the two sides of the membrane plus the difference in electrical charge across the membrane.

electrochemistry The branch of chemistry that is involved in studying reactions of electrodes and chemical changes resulting from electrical current or the production of electricity by chemical means.

electrocyclic reaction A MOLECULAR REARRANGEMENT that involves the formation of a SIGMA (σ) BOND between the termini of a fully conjugated linear pi-electron system (or a linear fragment of a pi-electron system) and a decrease by one in the number of PI (π) BONDs, or the reverse of that process. For example,

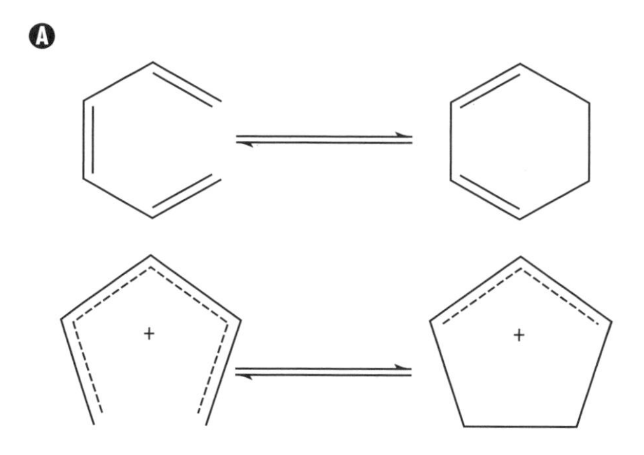

The stereochemistry of such a process is termed "conrotatory" or ANTARAFACIAL if the substituents at the interacting termini of the conjugated system both rotate in the same sense, for example,

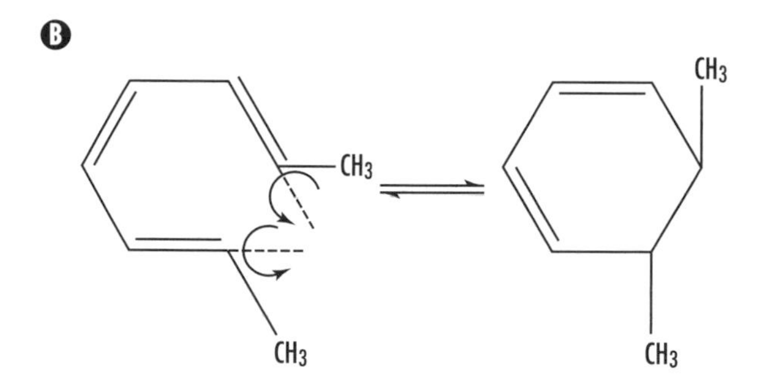

or "disrotatory" or SUPRAFACIAL if one terminus rotates in a clockwise and the other in a counterclockwise sense, for example,

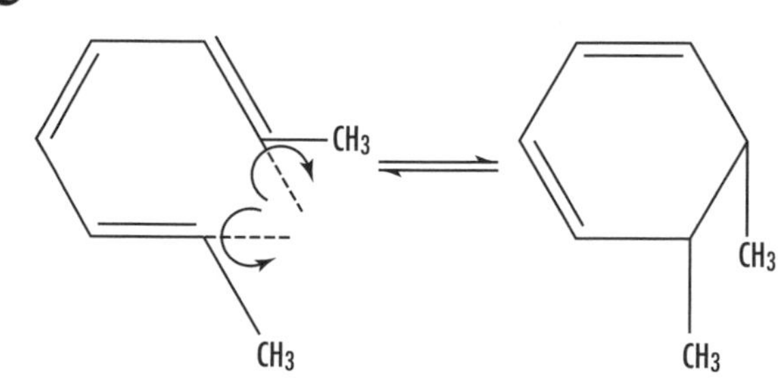

See also CONJUGATED SYSTEM; PERICYCLIC REACTION.

electrode A conductive material that, in an electric field, gives off, captures, or controls electrons (anode) or ions (cathode), such as in a battery. It is a surface where oxidation and reduction occurs. The anode is where OXIDATION takes place; the cathode is where REDUCTION takes place.

electrode potential Electrode potential of an electrode is defined as the electromotive force (emf) of a cell in which the electrode on the left is a standard hydrogen electrode, and the electrode on the right is the electrode in question.

See also REDOX POTENTIAL.

electrofuge A LEAVING GROUP that does not carry away the bonding electron pair. For example, in the nitration of benzene by NO_2^+, H^+ is the electrofuge. The adjective form of *electrofuge* is *electrofugal.*

See also ELECTROPHILE; NUCLEOFUGE.

electrogenic pump Any large, integral membrane protein (pump) that mediates the movement of a substance (ions or molecules) across the plasma membrane against its energy gradient (active transport). The pump can be ATP-dependent or Na^+-dependent; moves net electrical charges across the membrane.

electrolysis The process whereby an electric current passes through a substance to cause a chemical change; produces an oxidation–reduction reaction.

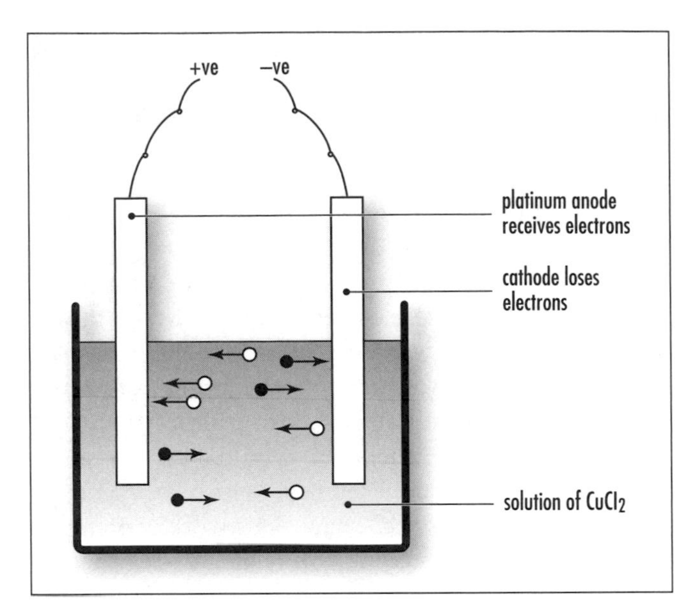

Electrolysis. The process whereby an electric current passes through a substance to cause a chemical change; produces an oxidation–reduction reaction.

electrolyte A substance whose liquid form conducts electricity. It may be a solution or a pure ionic liquid.

electromagnetic radiation Energy propagated as time-varying electric and magnetic fields. These two fields are inextricably linked as a single entity, since time-varying electric fields produce time-varying magnetic fields and vice versa. Light and radar are examples of electromagnetic radiation, differing only in their wavelengths (or frequency). Electric and magnetic fields propagate through space at the speed of light.

electromagnetic spectrum The entire spectrum of radiation arranged according to frequency and wavelength that includes visible light, radio waves, microwaves, infrared, ultraviolet light, X rays, and gamma rays. Ranges in wavelength from less than a nanometer, i.e., X and gamma rays (1 nm is about the length of 10 atoms in a row) to more than a kilometer, i.e., radio waves. Wavelength is directly related to the amount of energy the waves carry. The shorter the radiation's wavelength, the higher is its energy, ranging from high frequency (gamma rays) to low frequency (AM radio). All electromagnetic radiation travels through space at the speed of light, or 186,000 miles (300,000 km) per second.

Electromagnetic Spectrum
(note: the figures are only approximate)

Radiation	Wavelength (m)	Frequency (Hz)
gamma radiation	-10^{-12}	10^{19}–
X rays	10^{-12}–10^{-9}	10^{17}–10^{20}
ultraviolet radiation	10^{-9} –10^{-7}	10^{15}–10^{18}
visible radiation	10^{-7} –10^{-6}	10^{14}–10^{15}
infrared radiation	10^{-6} –10^{-4}	10^{12}–10^{-14}
microwaves	10^{-4} –1	10^{9}–10^{13}
radio waves	1–	-10^{9}

electromagnetic wave Method of travel for radiant energy (all energy is both particles and waves), so called because radiant energy has both magnetic and electrical properties. Electromagnetic waves are produced when electric charges change their motion. Whether the frequency is high or low, all electromagnetic waves travel at 186,000 miles (300,000 km) per second.

electromeric effect A molecular polarizability effect occurring by an INTRAMOLECULAR electron displacement (sometimes called the CONJUGATIVE MECHANISM and, previously, the "tautomeric mechanism") characterized by the substitution of one electron pair for another within the same atomic octet of electrons. It can be indicated by curved arrows symbolizing the displacement of electron pairs, as in

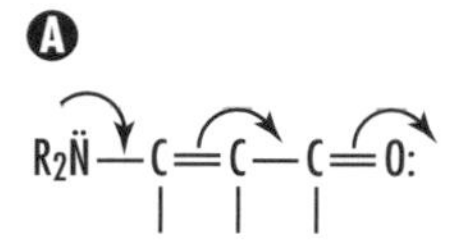

that represents the hypothetical electron shift

B

$R_2N—C=C—C=O$ --------------→ $R_2N^+=C—C=C—O^-$

The term has been deemed obsolescent or even obsolete (*see* MESOMERIC EFFECT, RESONANCE EFFECT). Many have used phrases such as "enhanced substituent resonance effect" that imply the operation of the electromeric effect without using the term, and various modern theoretical treatments parametrize the response of substituents to "electronic demand," which amounts to considering the electromeric effect together with the INDUCTOMERIC EFFECT.

electron A negatively charged subatomic particle of an atom or ion that is outside of the nucleus. A neutral atom contains the same number of electrons as there are protons in the nucleus. A negatively charged BETA PARTICLE is an electron that is emitted from the nucleus as a result of a nuclear decay process.

electron acceptor A substance that receives electrons in an oxidation–reduction reaction.

electron affinity The energy released when an additional electron (without excess energy) attaches itself to a MOLECULAR ENTITY (usually an electrically neutral molecular entity, but also used with single atomic species). The direct measurement of this quantity involves molecular entities in the gas phase.

electronation *See* ELECTRON ATTACHMENT; REDUCTION.

electron attachment The transfer of an electron to a MOLECULAR ENTITY, resulting in a molecular entity of (algebraically) increased negative charge.

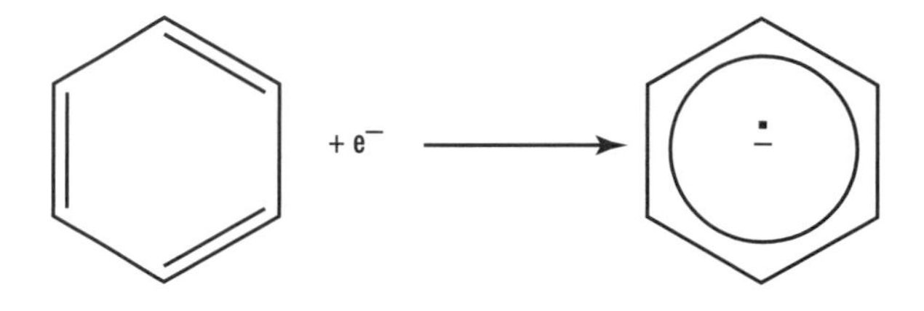

See also OXIDATION (1); REDUCTION.

electron capture When one of the inner-orbital electrons in an atom is captured by the nucleus. In the nucleus, the captured electron reacts with a proton to produce a neutron, changing the element to one of lower atomic number.

electron configuration *See* CONFIGURATION (ELECTRONIC).

electron-deficient bond A single bond between adjacent atoms that is formed by less than two electrons, as in B_2H_6:

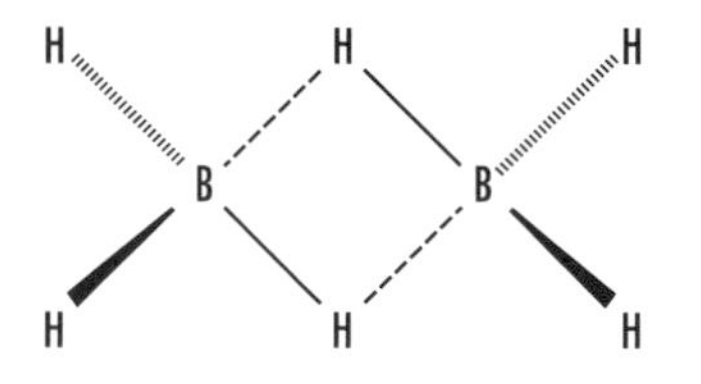

The B–H–B bonds are called "two-electron three-center bonds."

electron density If $P(x,y,z)\, dx\, dy\, dz$ is the probability of finding an electron in the volume element $dx\, dy\, dz$ at the point of a molecular entity with coordinates x,y,z, then $P(x,y,z)$ is the electron density at this point. For many purposes (e.g., X-ray scattering, forces on atoms), the system behaves exactly as if the electrons were spread out into a continuously distributed charge. The term has frequently been wrongly applied to negative CHARGE POPULATION.

See also CHARGE DENSITY.

electron detachment The reverse of an ELECTRON ATTACHMENT.

electron donor (1) A MOLECULAR ENTITY that can transfer an electron to another molecular entity or to the corresponding CHEMICAL SPECIES.

(2) A LEWIS BASE. This use is discouraged.

electron-donor-acceptor complex A term sometimes employed instead of CHARGE-TRANSFER COMPLEX or LEWIS ADDUCT.

See also ADDUCT; COORDINATION.

electronegativity Each kind of atom has a certain attraction for the electrons involved in a chemical bond. This attraction can be listed numerically on a scale of electronegativity. Since the element fluorine has the greatest attraction for electrons in bond-forming, it has the highest value on the scale. Metals usually have a low electronegativity, while nonmetals usually have high electronegativity. When atoms react with one another, the atom with the higher electronegativity value will always pull the electrons away from the atom that has the lower electronegativity value.

electroneutrality principle The principle expresses the fact that all pure substances carry a net charge of zero.

electronic effect of substituents: symbols and signs The INDUCTIVE EFFECT has universally been represented by the symbol I. This is now commonly taken to include both through-bond and through-space transmission, but I is also used specifically for through-bond transmission; through-space transmission is then symbolized as F (for FIELD EFFECT). The symbols for the influence of substituents exerted through electron delocalization have variously been M (MESOMERIC EFFECT), E (ELECTROMERIC EFFECT), T (TAUTOMERIC EFFECT), C (conjugative), K (konjugativ), and R (RESONANCE EFFECT). Since the present fashion is to use the term *resonance effect,* R is the most commonly used symbol, although M is still seen quite often.

Both the possible sign conventions are in use. The Ingold sign convention associates ELECTRONEGATIVITY (relative to the hydrogen atom) with a negative sign, electropositivity with a positive sign. Thus the nitro group is described as electron-withdrawing by virtue of its –I and –M effects; chloro is described as a –I, +M substituent, etc. For CORRELATION ANALYSIS and LINEAR FREE-ENERGY RELATIONships, this convention has been found inconvenient, for it is in contradiction to the sign convention for polar substituent constants (σ-constants). Authors concerned with these fields often avoid this contradiction by adopting the opposite-sign convention originally associated with Robinson for electronic effects. This practice is almost always associated with the use of R for the electron delocalization effect: thus the nitro group is a +I, +R substituent; chloro a +I, –R substituent, etc.

electron magnetic resonance spectroscopy (EMR spectroscopy) *See* ELECTRON PARAMAGNETIC RESONANCE SPECTROSCOPY.

electron microscope (EM) A very large tubular microscope that focuses a highly energetic electron beam instead of light through a specimen, resulting in resolving power thousands of times greater than that of a regular light microscope. A transmission EM (TEM) is used to study the internal structure of thin sections of cells, while a scanning EM (SEM) is used to study the ultrastructure of surfaces. The transmission electron microscope was the first type of electron microscope, developed in 1931 by Max Knoll and Ernst Ruska in Germany, and was patterned exactly on the light transmission microscope except for the focused beam of

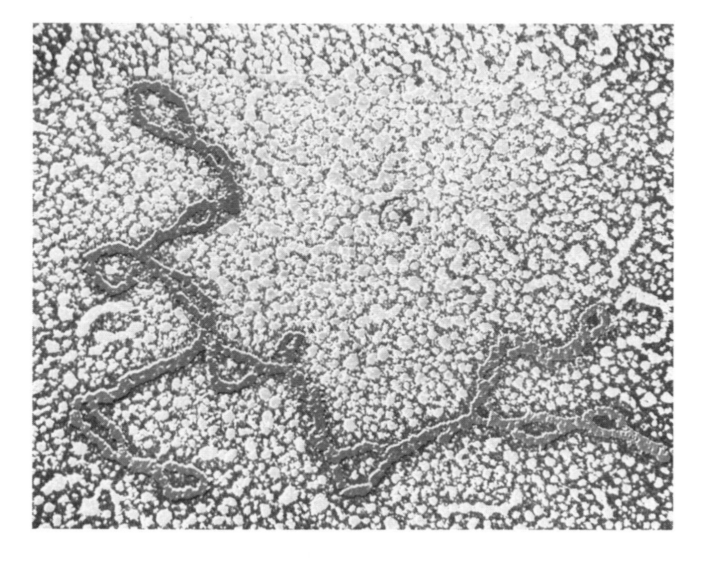

False-color transmission electron micrograph (TEM) of circular DNA from a mitochondrion, the site of synthesis of chemical energy within the cell. The filament seen in image was collected from a fragmented mitochondrion. Such circular DNA molecules are 5–6 μm long, with a molecular weight of around 10 million; each mitochondrion has between three to six. This relatively tiny amount of DNA is estimated to have a coding capacity of about 5,000 amino acids and suggests that the mitochondrion needs to rely on a supplement of external DNA (from the cell nucleus) to synthesize its constituents. Magnification: ×300,000 at 6×7cm size. ***(Courtesy of CNRI/Science Photo Library)***

electrons instead of light to see through the specimen. The first scanning electron microscope was available in 1942, but the first commercial availability was not until 1965.

electron-nuclear double resonance (ENDOR) A magnetic resonance spectroscopic technique for the determination of HYPERFINE interactions between electrons and nuclear spins. There are two principal techniques. In continuous-wave ENDOR, the intensity of an ELECTRON PARAMAGNETIC RESONANCE (EPR) signal, partially saturated with microwave power, is measured as radio frequency is applied. In pulsed ENDOR, the radio frequency is applied as pulses, and the EPR signal is detected as a spin echo. In each case an enhancement of the EPR signal is observed when the radio frequency is in resonance with the coupled nuclei.

electron-pair acceptor A synonym for LEWIS ACID.

electron-pair donor A synonym for LEWIS BASE.

electron paramagnetic resonance spectroscopy (EPR spectroscopy) The form of spectroscopy concerned with microwave-induced transitions between magnetic energy levels of electrons having a net spin and orbital angular momentum. The spectrum is normally obtained by magnetic-field scanning. Also known as electron spin resonance (ESR) spectroscopy or electron magnetic resonance (EMR) spectroscopy. The frequency (ν) of the oscillating magnetic field to induce transitions between the magnetic energy levels of electrons is measured in gigahertz (GHz) or megahertz (MHz). The following band designations are used: L (1.1 GHz), S (3.0 GHz), X (9.5 GHz), K (22.0 GHz), and Q (35.0 GHz). The static magnetic field at which the EPR spectrometer operates is measured by the magnetic flux density *(B)*, and its recommended unit is the tesla (T). In the absence of nuclear hyperfine interactions, B and ν are related by: $h\nu = g\mu_B B$, where h is the Planck constant, μ_B is the Bohr magneton, and the dimensionless scalar g is called the g-factor. When the PARAMAGNETIC species exhibits an ANISOTROPY, the spatial dependency of the g-factor is represented by a 3×3 matrix. The interaction energy between the electron spin and a magnetic nucleus is characterized by the hyperfine-coupling constant A. When the paramagnetic species has anisotropy, the hyperfine coupling is expressed by a 3×3 matrix called a hyperfine-coupling matrix. Hyperfine interaction usually results in the splitting of lines in an EPR spectrum. The nuclear species giving rise to the hyperfine interaction should be explicitly stated, e.g., "the hyperfine splitting due to ^{65}Cu." When additional hyperfine splittings due to other nuclear species are resolved ("superhyperfine"), the nomenclature should include the designation of the nucleus and the isotopic number.

electron spin-echo envelope modulation (ESEEM) *See* ELECTRON SPIN-ECHO SPECTROSCOPY.

electron spin-echo spectroscopy (ESE spectroscopy) A pulsed technique in ELECTRON PARAMAGNETIC RESONANCE, in some ways analogous to pulsed techniques in NMR (NUCLEAR MAGNETIC RESONANCE

SPECTROSCOPY). ESE may be used for measurements of electron spin relaxation times, as they are influenced by neighboring paramagnets or molecular motion. It may also be used to measure anisotropic nuclear HYPERFINE couplings. The effect is known as electron spin-echo envelope modulation (ESEEM). The intensity of the electron spin echo resulting from the application of two or more microwave pulses is measured as a function of the temporal spacing between the pulses. The echo intensity is modulated as a result of interactions with the nuclear spins. The frequency-domain spectrum corresponds to hyperfine transition frequencies.

See also ANISOTROPY; PARAMAGNETIC INTERACTION.

electron spin quantum number The quantum number representing one of two possible values for the electron spin; either –1/2 or 1/2.

electron spin-resonance spectroscopy (ESR spectroscopy) *See* ELECTRON PARAMAGNETIC RESONANCE SPECTROSCOPY.

electron transfer The transfer of an electron from one MOLECULAR ENTITY to another, or between two localized sites in the same molecular entity.

See also INNER-SPHERE ELECTRON TRANSFER; MARCUS EQUATION; OUTER-SPHERE ELECTRON TRANSFER.

electron-transfer catalysis The term indicates a sequence of reactions such as shown in equations (1)–(3), leading from A to B:

$$A + e^- \rightarrow A^{\cdot -} (1)$$
$$A^{\cdot -} \rightarrow B^{\cdot -} (2)$$
$$B^{\cdot -} + A \rightarrow B + A^{\cdot -} (3)$$

An analogous sequence involving radical cations ($A^{\cdot +}$, $B^{\cdot +}$) is also observed.

The most notable example of electron-transfer catalysis is the $S_{RN}1$ (or $T+D_N+A_N$) reaction of aromatic halides.

The term has its origin in a suggested analogy to acid–base catalysis with the electron instead of the proton. However, there is a difference between the two catalytic mechanisms, since the electron is not a true catalyst, but rather behaves as the initiator of a CHAIN REACTION. *Electron-transfer-induced chain reaction* is a more appropriate term for the situation described by equations (1)–(3).

Demonstration of an electron-transfer reaction in a laboratory. At left is a test tube containing copper(II) sulfate solution; below this is a watch glass containing zinc powder. When the zinc is added to the sulfate, the solution loses its blue color (right). This is due to the electron transfer between copper(II) ions (blue) and the zinc to form zinc(II) ions (colorless). The copper is deposited as a metal. ***(Courtesy of Jerry Mason/Science Photo Library)***

electron-transfer protein A protein, often containing a metal ion, that oxidizes and reduces other molecules by means of electron transfer.

electron-transport chain A chain of electron acceptors embedded in the inner membrane of the mitochondrion. These acceptors separate hydrogen protons from their electrons. When electrons enter the transport chain, the electrons lose their energy, and some of it is used to pump protons across the inner membrane of the mitochondria, creating an electrochemical gradient across the inner membrane that provides the energy needed for ATP synthesis. The function of this chain is to permit the controlled release of free energy to drive the synthesis of ATP.

See also PROTON PUMP.

electrophile (electrophilic reagent) An electrophile (or electrophilic reagent) is a reagent that forms a BOND to its reaction partner (the NUCLEOPHILE) by accepting both bonding electrons from that reaction partner.

An electrophilic SUBSTITUTION REACTION is a heterolytic reaction in which the reagent supplying the entering group acts as an electrophile. For example

$C_6H_6 + NO_2^+$ (electrophile) $C_6H_5NO_2 + H^+$
(ELECTROFUGE)

Electrophilic reagents are LEWIS ACIDs. Electrophilic catalysis is catalysis by Lewis acids.

The term *electrophilic* is also used to designate the apparent polar character of certain RADICALS, as inferred from their higher relative reactivities with reaction sites of higher electron density.

See also ELECTROPHILICITY; HETEROLYSIS.

electrophilicity (1) The property of being electrophilic. (*See* ELECTROPHILE.)

(2) The relative reactivity of an electrophilic reagent. (It is also sometimes referred to as "electrophilic power.") Qualitatively, the concept is related to LEWIS ACIDITY. However, whereas Lewis acidity is measured by relative equilibrium constants, ELECTROPHILICITY is measured by relative RATE CONSTANTs for reactions of different electrophilic reagents toward a common SUBSTRATE (usually involving attacking a carbon atom).

See also NUCLEOPHILICITY.

electrophoresis Migrating charged particles (e.g., colloidal particles that acquire a charge through adsorption of ions, charged macromolecules such as proteins) in an electric field by virtue of their charge.

A method of separating charged molecules through their different rates of migration under an electric field. Usually a medium that minimizes diffusion is used, e.g., gel, paper, or capillary electrophoresis.

element A substance that consists of atoms that have the same number of protons in their nuclei. Elements are defined by the number of protons they possess.

elementary process Each step of a reaction mechanism.

elementary reaction A reaction for which no reaction intermediates have been detected or need to be postulated in order to describe the chemical reaction on a molecular scale. An elementary reaction is assumed to occur in a single step and to pass through a single TRANSITION STATE.

See also STEPWISE REACTION.

element effect The ratio of the rate constants of two reactions that differ only in the identity of the element of the atom in the LEAVING GROUP, e.g., k_{Br}/k_{Cl}. As for ISOTOPE EFFECTs, a ratio of unity is regarded as a "null effect."

elimination (1) The process achieving the reduction of the concentration of a XENOBIOTIC compound including its METABOLISM.

(2) The reverse of an ADDITION REACTION or TRANSFORMATION. In an elimination, two groups (called eliminands) are lost most often from two different centers (1/2/elimination or 1/3/elimination, etc.) with concomitant formation of an unsaturation in the molecule (double bond, triple bond) or formation of a new ring.

If the groups are lost from a single center (α-elimination, 1/1/elimination), the resulting product is a carbene or a carbene analog.

See also ALPHA (α) ELIMINATION.

empirical formula The simplest whole-number ratio of atoms in a compound.

EMR (electron magnetic resonance) *See* ELECTRON PARAMAGNETIC RESONANCE SPECTROSCOPY.

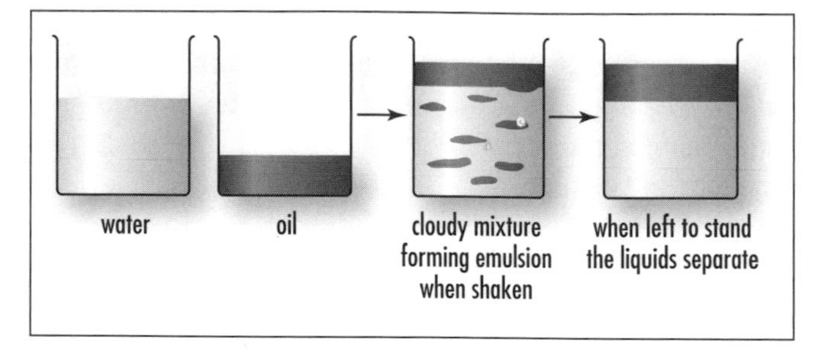

Emulsion. Droplets of a liquid substance dispersed in another immiscible liquid

emulsifying agent Any material having both HYDROPHILIC and HYDROPHOBIC characters acting to stabilize an EMULSION.

emulsion Droplets of a liquid substance dispersed in another immiscible liquid. Milk and salad dressing are emulsions. These are examples of colloidal systems in which both phases are liquid.

See also COLLOID.

enantiomer One of a pair of molecular entities that are mirror images of each other and nonsuperimposable.

enantioselectivity *See* STEREOSELECTIVITY.

encounter complex A COMPLEX of MOLECULAR ENTITIES produced at an ENCOUNTER-CONTROLLED RATE that occurs as an intermediate in a reaction MECHANISM. When the complex is formed from two molecular entities, it is called an encounter pair. A distinction between encounter pairs and (larger) encounter complexes may be relevant in some cases, e.g., for mechanisms involving PRE-ASSOCIATION.

encounter-controlled rate A RATE OF REACTION corresponding to the rate of encounter of the reacting MOLECULAR ENTITIES. This is also known as DIFFUSION-CONTROLLED RATE, since rates of encounter are themselves controlled by diffusion rates (which in turn depend on the viscosity of the MEDIUM and the dimensions of the reactant molecular entities).

For a BIMOLECULAR REACTION between solutes in water at 25°C, an encounter-controlled rate is calculated to have a second-order rate constant of about 10^{10} dm^3 mol^{-1} sec^{-1}.

See also MICROSCOPIC DIFFUSION CONTROL.

endergonic reaction A chemical reaction that consumes energy rather than releasing energy. Endergonic reactions are not spontaneous because they do not release energy.

endogenous Originating internally. In the description of metal ion COORDINATION in metalloproteins, the term *endogenous* refers to internal, or protein-derived, LIGANDS.

ENDOR *See* ELECTRON-NUCLEAR DOUBLE RESONANCE.

endothelium-derived relaxing factor (EDRF) The factor originally described as EDRF is NO·, produced by a specific P-450-type of ENZYME from arginine upon response of a cell to a biological signal (molecule). Different types of cells respond differently to the presence of NO·.

See also CYTOCHROME P-450.

endotherm A warm-blooded animal, one in which the internal temperature does not fluctuate with the temperature of the environment but is maintained by a constant internal temperature regulated by metabolic processes. Examples include birds and mammals.

endothermal reaction A chemical reaction in which heat is absorbed.

endothermic The state of being warm-blooded or producing heat internally. In chemistry, a reaction where heat enters into a system; energy is absorbed by the reactant.

See also EXOTHERMIC.

ene reaction The addition of a compound with a double bond having an allylic hydrogen (the ene) to a compound with a multiple bond (the enophile) with transfer of the allylic hydrogen and a concomitant reorganization of the bonding, as illustrated below for propene (the ene) and ethene (the enophile). The reverse is a "retro-ene" reaction.

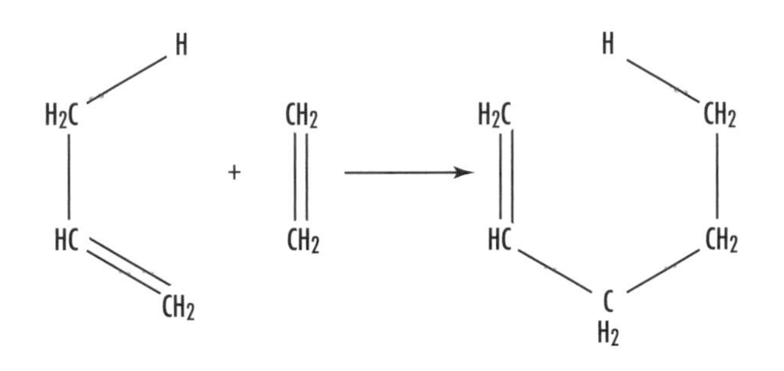

energy Classically defined as the capacity for doing work, energy can occur in many forms such as heat (thermal), light, movement (mechanical), electrical, chemical, sound, or radiation. The first law of thermodynamics is often called the law of conservation of energy and states that energy cannot be created or destroyed, only transformed from one form into another.

energy of activation (Arrhenius energy of activation; activation energy E_a), (SI unit: kJ mol^{-1}) An operationally defined quantity expressing the dependence of a rate constant on temperature according to

$$E_a = RT^2(\ln k/T)_p$$

as derived from the Arrhenius equation, $k = A \exp(-E_a/RT)$, where A (SI unit: as for the corresponding rate constant) is termed the "pre-exponential factor."

See also ENTHALPY OF ACTIVATION.

energy profile *See* GIBBS ENERGY DIAGRAM; POTENTIAL-ENERGY PROFILE.

enforced concerted mechanism Variation of the reaction parameters in a series of reactions proceeding in nonconcerted steps may lead to a situation where the putative intermediate will possess a lifetime shorter than a bond vibration, so that the steps become concerted. The TRANSITION STATE structure will lie on the coordinate of the More O'Ferrall–Jencks diagram leading to that of the putative intermediate.

enhanced greenhouse effect The natural GREENHOUSE EFFECT has been enhanced by anthropogenic emissions of greenhouse gases. Increased concentrations of carbon dioxide, methane, nitrous oxide, CFCs, HFCs, PFCs, SF_6, NF_3, and other photochemically important gases caused by human activities, such as fossil fuel consumption and added waste to landfills, trap more infrared radiation, thereby exerting a warming influence on the climate.

See also GLOBAL WARMING.

enophile *See* ENE REACTION.

entatic state A state of an atom or group that, due to its binding in a protein, has its geometric or electronic condition adapted for function. Derived from *entasis* (Greek), meaning tension.

entering group An atom or GROUP that forms a BOND to what is considered to be the main part of the SUBSTRATE during a reaction, for example, the attacking NUCLEOPHILE in a bimolecular nucleophilic SUBSTITUTION REACTION.

enterobactin A SIDEROPHORE found in enteric bacteria such as *Escherichia coli;* sometimes called enterochelin.

enterochelin *See* ENTEROBACTIN.

enthalpy The property of a system that is equal to *E* + *PV*. *E* is the internal energy of the system, *P* is the pressure, and V is the volume of the system. The change in enthalpy equals the energy flow as heat at a constant pressure. Used to predict the heat flow in chemical reactions.

enthalpy of activation (**standard enthalpy of activation**), $\Delta^{\ddagger}H^{\circ}$ (SI unit: kJ mol^{-1}) The standard enthalpy difference between the TRANSITION STATE and the GROUND STATE of the reactants at the same temperature and pressure. It is related to the temperature coefficient of the rate constant according to the equation:

$$H = RT^2(\partial \ln k/\partial T)_p - RT = E_a - RT$$
$$= -R(\partial \ln(k/T) / \partial(1/T))_p$$

where E_a is the ENERGY OF ACTIVATION, providing that the rate constants for reactions other than first-order reactions are expressed in temperature-independent concentration units (e.g., mol dm^{-3}, measured at a fixed temperature and pressure). If ln$k$ is expressed as

$$\ln k = (a/T) + b + c \ln T + dT$$

then

$$H = -aR + (c-1)RT + dRT^2$$

If enthalpy of activation and ENTROPY OF ACTIVATION are assumed to be temperature independent, then

$$H = -aR$$

If the concentration units are mol dm^{-3}, the true and apparent enthalpies of activation differ by $(n-1)/(\alpha RT^2)$, where n is the order of reaction and α the thermal expansivity.

See also ENTROPY OF ACTIVATION; GIBBS ENERGY OF ACTIVATION.

enthalpy of formation (**ΔH_f**) The change in the enthalpy that occurs during a chemical reaction.

entropy The amount of energy in a closed system that is not available for doing work; disorder and randomness in a system. The higher the entropy, the less energy is available for work. The second law of thermodynamics states that the entropy of the universe will always increase.

entropy of activation (**standard entropy of activation**), $\Delta^{\ddagger}S^{\circ}$ (SI unit: J mol^{-1} K^{-1}) The standard entropy difference between the TRANSITION STATE and the GROUND STATE of the reactants, at the same temperature and pressure.

It is related to the GIBBS ENERGY OF ACTIVATION and ENTHALPY OF ACTIVATION by the equations

$$\Delta^{\ddagger}S = (\Delta^{\ddagger}H - \Delta^{\ddagger}G)/T$$
$$= \Delta^{\ddagger}H/T - R\ln(k_B/h) + R \ln(k/T)$$

or, if ln$k$ is expressed as $\ln k = a/T + b + c \ln T + dT$,

$$\Delta^{\ddagger}S = R\ [b - \ln(k_B/h) + (c-1)(1 + \ln T) + 2\ dT]$$

provided that rate constants for reactions other than first-order reactions are expressed in temperature-independent concentration units (e.g., mol dm^{-3}, measured at a fixed temperature and pressure). The numerical value of $\Delta^{\ddagger}S$ depends on the standard state (and therefore on the concentration units selected). If entropy of activation and ENTHALPY OF ACTIVATION are assumed to be temperature-independent,

$$\Delta^{\ddagger}S = R[b - \ln(k_B/h)]$$

Strictly speaking, the quantity defined is the entropy of activation at constant pressure, from which the entropy of activation at constant volume can be deduced.

The information represented by the entropy of activation can alternatively be conveyed by the pre-exponential factor A (*see* ENERGY OF ACTIVATION).

environment The total living and nonliving conditions of an organism's internal and external surroundings that affect an organism's complete life span.

enzyme A macromolecule that functions as a BIOCATALYST by increasing the reaction rate, frequently containing or requiring one or more metal ions. In general, an enzyme catalyzes only one reaction type (reaction specificity) and operates on only a narrow range of SUBSTRATEs (substrate specificity). Substrate molecules are attacked at the same site (regiospecificity), and only one, or preferentially one of the ENANTIOMERs of CHIRAL substrate or of RACEMIC mixtures, is attacked (enantiospecificity).

See also COENZYME.

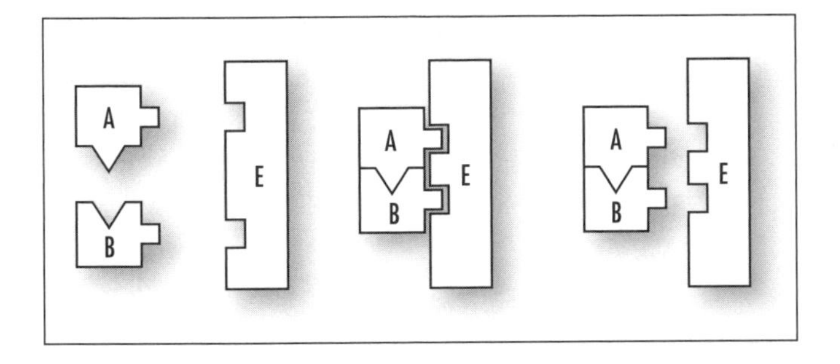

Enzyme action. In general, an enzyme catalyzes only one reaction type (reaction specificity) and operates on only a narrow range of substrates (substrate specificity).

enzyme induction The process whereby an (inducible) ENZYME is synthesized in response to a specific inducer molecule. The inducer molecule (often a substrate that needs the catalytic activity of the inducible enzyme for its METABOLISM) combines with a repressor and thereby prevents the blocking of an operator by the repressor, leading to the translation of the gene for the enzyme. An inducible enzyme is one whose synthesis does not occur unless a specific chemical (inducer) is present, which is often the substrate of that enzyme.

enzyme repression The mode by which the synthesis of an ENZYME is prevented by repressor molecules.

In many cases, the end product of a synthesis chain (e.g., an amino acid) acts as a feedback corepressor by combining with an intracellular aporepressor protein, so that this complex is able to block the function of an operator. As a result, the whole operation is prevented from being transcribed into mRNA, and the expression of all enzymes necessary for the synthesis of the end product enzyme is abolished.

epimer A DIASTEREOISOMER that has the opposite CONFIGURATION at only one of two or more tetrahedral "stereogenic" centers present in the respective MOLECULAR ENTITY.

epimerization Interconversion of EPIMERS by reversal of the configuration at one of the "stereogenic" centers.

epitope An alternative term for an antigenic determinant. These are particular chemical groups on a molecule that are antigenic, i.e., that elicit a specific immune response. Also called antigenic determinant.

epoch A period or date in time, shorter than or part of an era, and used in geological time tables and to mark historical events, for example. Usually refers to an event (mountain building, appearance of a species, etc.). Also called a series.

See also GEOLOGICAL TIME.

EPR *See* ELECTRON PARAMAGNETIC RESONANCE SPECTROSCOPY.

equatorial bonds Those bonds that radiate outwardly from the center of a molecule around its equator.

equilibrium The ideal state in a system where opposing forces or rates are in balance, resulting in the occurrence of no net change.

equilibrium, chemical Reversible processes can be made to proceed in the forward or reverse direction by the (infinitesimal) change of one variable, ultimately reaching a point where the rates in both directions are identical, so that the system gives the appearance of having a static composition at which the Gibbs energy, G, is a minimum. At equilibrium, the sum of the chemical potentials of the reactants equals that of the products, so that

$$G_r = G_r^0 + RT \ln K = 0$$
$$G_r^0 = -RT \ln K$$

The equilibrium constant, K, is given by the MASS-LAW EFFECT.

equilibrium constant *See* ACIDITY CONSTANT; STABILITY CONSTANT.

equilibrium control *See* THERMODYNAMIC CONTROL.

equilibrium isotope effect *See* ISOTOPE EFFECT.

equivalent The mass of acid required to provide one mole of hydrogen ions in a reaction, or the mass of base required to react with a mole of hydrogen ions.

equivalent weight (1) Acid, the mass that provides one mole of H^+. (2) Base, the mass that neutralizes one mole of H^+. (3) Reduction–oxidation reaction, the mass of reactant that either gains or loses one mole of electrons.

Erlanger, Joseph (1874–1965) American *Neuroscientist* Joseph Erlanger was born on January 5, 1874, in San Francisco, California, to Herman and Sarah Erlanger. He received a B.S. in chemistry at the University of California and later attended Johns Hopkins University to study medicine, receiving an M.D. in 1899. He was appointed assistant in the department of physiology at the medical school after spending a year of hospital training at Johns Hopkins Hospital until 1906, moving up successively as instructor, associate, and associate professor. He was then appointed the first professor of physiology in the newly established Medical School of the University of Wisconsin. In 1910 he was appointed professor of physiology in the reorganized Medical School of the Washington University in St. Louis, retiring in 1946 as chairman of the school.

In 1922 in collaboration with his student Herbert Gasser, Erlanger adapted the cathode-ray oscillograph for studying nerve action potentials. They amplified the electrical responses of a single nerve fiber and analyzed them with the use of the oscilloscope. The characteristic wave pattern of an impulse generated in a stimulated nerve fiber could be observed on the screen and the components of the nerve's response studied.

Erlanger and Gasser were given the Nobel Prize for medicine or physiology in 1944 for this work. Erlanger later worked on the metabolism of dogs with shortened intestines, on traumatic shock, and on the mechanism of the production of sound in arteries.

With Gasser he wrote *Electrical Signs of Nervous Activity* (1937). He died on December 5, 1965, in St Louis, Missouri.

ESE (**electron spin echo**) *See* ELECTRON SPIN-ECHO SPECTROSCOPY.

ESEEM (**electron spin-echo envelope modulation**) *See* ELECTRON SPIN-ECHO SPECTROSCOPY.

ESR (**electron spin resonance**) *See* ELECTRON PARAMAGNETIC RESONANCE SPECTROSCOPY.

ester Any organic compound produced through the reaction of a carboxylic acid and alcohol that removes the water from the compound. An example is ethyl acetate.

eta (**η**) **symbol** *See* HAPTO.

ethanol Another word for ethyl alcohol, C_2H_5OH.

ether A compound with an oxygen atom attached to two hydrocarbon groups. Any carbon compound containing the functional group C–O–C, such as diethyl ether.

ethylene (C_2H_4) A reactive chemical made from natural gas or crude oil components (occurs naturally in both petroleum and natural gas) that acts as a plant hormone. The only gaseous hormone, it is used for accelerating fruit ripening (bananas), maturing citrus fruit color, leaf abscission, aging, and increasing the growth rate of seedlings, vegetables, and fruit trees.

E_T-value *See* DIMROTH-REICHARDT E_T PARAMETER; Z-VALUE.

euchromatin Within a nucleus of eukaryotes there are two types of a mixture of nucleic acid and protein called CHROMATIN that make up a CHROMOSOME: euchromatin and heterochromatin. During interphase, the genetically active euchromatin is uncoiled and is available for transcription, while heterochromatin is denser and usually is not transcribed.

eudismic ratio The POTENCY of the EUTOMER relative to that of the DISTOMER.

eukaryote Organism whose cells have their GENETIC material packed in a membrane-surrounded, structurally discrete nucleus, and that have well-developed cell organelles.

eutomer The ENANTIOMER of a CHIRAL compound that is the more potent for a particular action.

See also DISTOMER.

eutrophication The accelerated loading or dumping of nutrients in a lake by natural or human-induced causes. Natural eutrophication will change the character of a lake very gradually, sometimes taking centuries, but human-induced or cultural eutrophication speeds up the aging of a lake, changing its qualities quickly, often in a matter of years.

eutrophic lake Any lake that has an excessive supply of nutrients, usually nitrates and phosphates. Eutrophic lakes are usually not deep, contain abundant algae or rooted plants, and contain limited oxygen in the bottom layer of water.

evaporation When surface molecules of liquids break loose of the intermolecular forces that hold them in the liquid and enter the vapor phase.

evaporative cooling Temperature reduction that occurs when water absorbs latent heat from the surrounding air as it evaporates; cooling of the skin from the evaporation of sweat is evaporative cooling and is a process for the body to lose excess heat.

evapotranspiration The sum of evaporation and plant transpiration. Potential evapotranspiration is the amount of water that could be evaporated or transpired at a given temperature and humidity when there is plenty of water available. Actual evapotranspiration cannot be any greater than precipitation, and will usually be less because some water will run off in rivers and flow to the oceans. If potential evapotranspiration is greater than actual precipitation, then soils are extremely dry during at least a major part of the year.

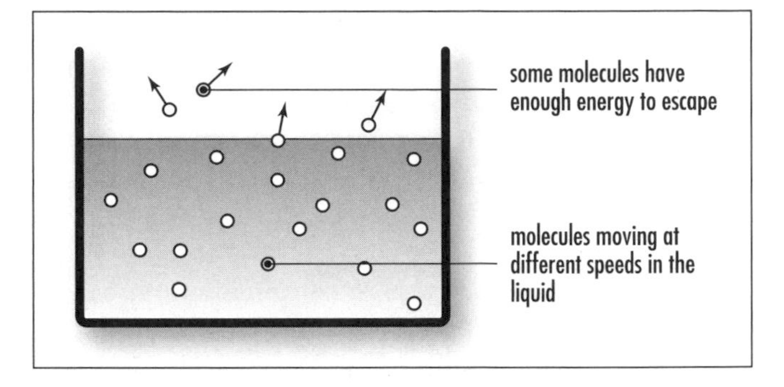

Evaporation. When surface molecules of liquids break loose of the intermolecular forces that hold them in the liquid and enter the vapor phase

Ewens-Bassett number *See* OXIDATION NUMBER.

EXAFS *See* EXTENDED X-RAY ABSORPTION FINE STRUCTURE.

excess acidity *See* BUNNETT-OLSEN EQUATIONS; COX-YATES EQUATION.

excimer An excited dimer, "nonbonding" in the GROUND STATE. For example, a complex formed by the interaction of an excited MOLECULAR ENTITY with a ground-state counterpart of this entity.

See also EXCIPLEX.

exciplex An electronically excited complex of definite stoichiometry, "nonbonding" in the GROUND STATE. For example, a complex formed by the interaction of an excited MOLECULAR ENTITY with a ground-state counterpart of a different structure.

See also EXCIMER.

excited state State of a system with energy higher than that of the GROUND STATE. This term is most commonly used to characterize a molecule in one of its electronically excited states, but can also refer to vibrational and/or rotational excitation in the electronic ground state.

exclusion principle No two electrons existing in an atom can have the same four quantum numbers.

exegetic reaction A spontaneous reaction in which energy flows out of the system; a decrease in free energy; a reaction that liberates heat.

exogenous Originating externally. In the context of metalloprotein LIGANDS, *exogenous* describes ligands added from an external source, such as CO or O_2.

exon A section of DNA that carries the coding SEQUENCE for a protein or part of it. Exons are separated by intervening, noncoding sequences (called INTRONS). In EUKARYOTES, most GENES consist of a number of exons.

exosphere The uppermost layer of the atmosphere, its lower boundary is estimated at 500 km to 1,000 km above the Earth's surface. It is only from the exosphere that atmospheric gases can, to any appreciable extent, escape into outer space.

exothermic A reaction that produces heat and absorbs heat from the surroundings.

See also ENDOTHERMIC.

exotoxin A toxic substance produced by bacteria and then released outside its cell into its environment.

expression The cellular production of the protein encoded by a particular GENE. The process includes TRANSCRIPTION of DNA, processing of the resulting mRNA product and its TRANSLATION into an active protein. A recombinant gene inserted into a host cell by means of a vector is said to be expressed if the synthesis of the encoded polypeptide can be demonstrated. For the expression of metalloproteins, other gene products will usually be required.

extended Hammett equation This term applies in a general way to any multiparametric extension of the HAMMETT EQUATION. It is sometimes used specifically for a form of DUAL SUBSTITUENT-PARAMETER EQUATION in which the actual value of the correlated property P under the influence of the substituent X is used, rather than the value relative to that for $X = H$. An intercept term h corresponding to the value of P for $X = H$ is introduced, for example

$$P = \alpha\sigma_I + \beta\sigma_R + h$$

The equation can be applied to systems for which the inclusion of further terms to represent other effects, e.g., steric, is appropriate.

extended X-ray absorption fine structure (EXAFS) EXAFS effects arise because of electron scattering by atoms surrounding a particular atom of interest as that special atom absorbs X rays and emits electrons. The atom of interest absorbs photons at a characteristic wavelength, and the emitted electrons, undergoing constructive or destructive interference as they are scattered by the surrounding atoms, modulate the

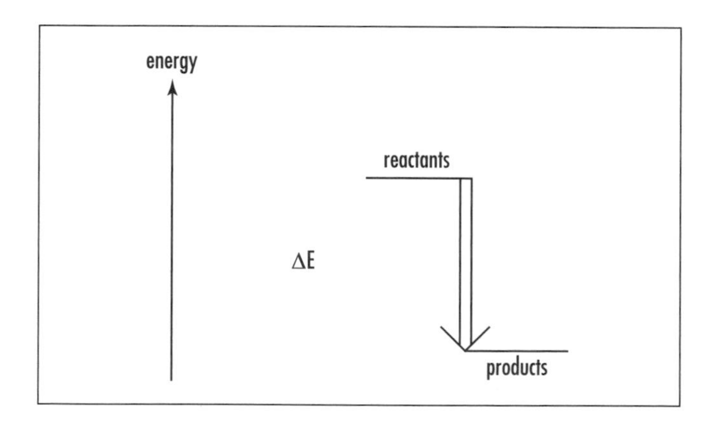

Exothermic. A reaction that produces heat and absorbs heat from the surroundings

absorption spectrum. The modulation frequency corresponds directly to the distance of the surrounding atoms, while the amplitude is related to the type and number of atoms. EXAFS studies are a probe of the local structure. EXAFS can be applied to systems that have local structure, but not necessarily long-range structure, such as noncrystalline materials. In particular, bond lengths and local symmetry (COORDINATION numbers) may be derived. The X-ray absorption spectrum may also show detailed structure below the absorption edge. This X-ray absorption near edge structure (XANES) arises from excitation of core electrons to high-level vacant orbitals.

external return *See* ION-PAIR RETURN.

extinct species A species no longer in existence.

extrusion transformation A TRANSFORMATION in which an atom, or GROUP Y, connected to two other atoms, or groups X and Z, is lost from a molecule, leading to a product in which X is bonded to Z, for example

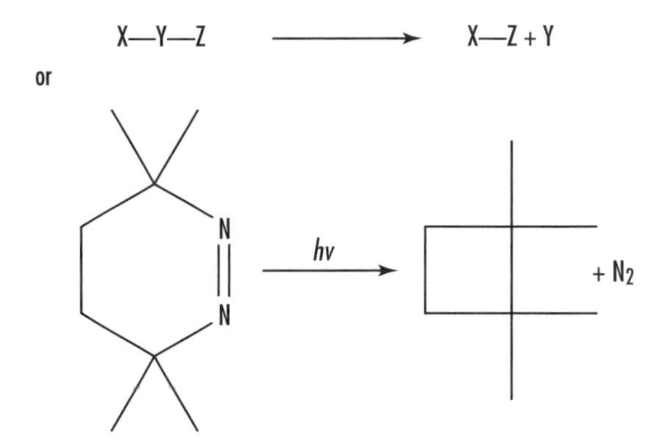

The reverse of an extrusion is called INSERTION. *See also* CHELETROPIC REACTION.

F

F-430 A tetrapyrrole structure containing nickel, a component of the ENZYME methyl-coenzyme M REDUCTASE, that is involved in the formation of methane in METHANOGENIC bacteria. The highly reduced macrocyclic structure, related to PORPHYRINs and CORRINs, is termed a CORPHIN.

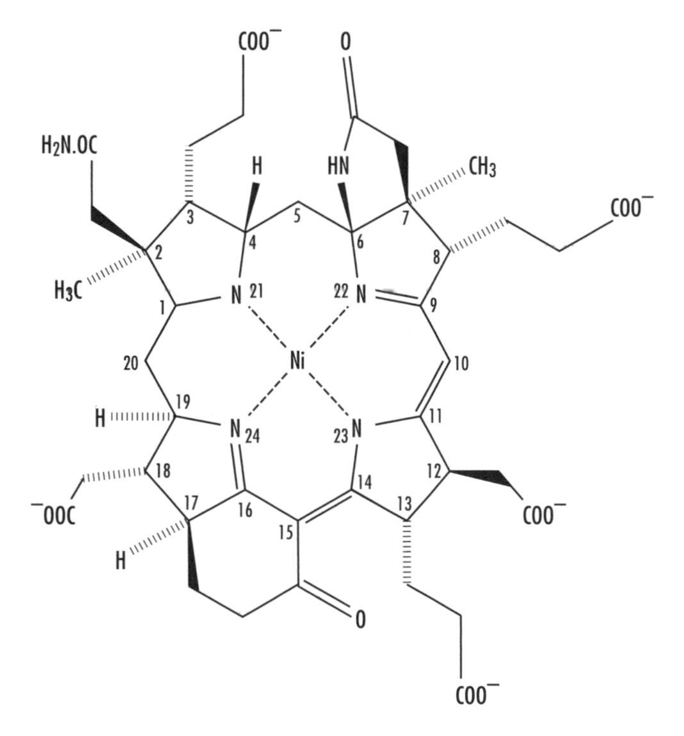

face-centered cubic A cubic unit cell having atoms, molecules, or ions at the corners and in the center of each face; the unit cell produced by cubic close packing.

facilitated diffusion A process by which carrier proteins, also called permeases or transporters, or ion channels in the cell membrane transport substances such as glucose, sodium, and chloride ions into or out of cells down a concentration (electrochemical) gradient; does not require the use of metabolic energy.

See also VOLTAGE-GATED CHANNELS; ACTIVE TRANSPORT.

facultative anaerobe A facultative anaerobe is a microorganism that makes ATP by aerobic respiration if oxygen is present, but if absent switches to fermentation under anaerobic conditions.

facultative organism Any organism that changes a metabolic pathway to another when needed.

Fahrenheit, Daniel Gabriel (1686–1736) German *Instrument maker, physicist* Daniel Gabriel Fahrenheit, a German instrument maker and physicist, was born in Danzig, Germany (now Gdansk, Poland), in 1686, the oldest of five children. Fahrenheit's major contributions lay in the creation of the first accurate thermometers in 1709 and a temperature scale in 1724 that bears his name today.

When Daniel was 15 years of age, his parents died of mushroom poisoning. The city council placed the four younger Fahrenheit children in foster homes and apprenticed Daniel to a merchant who taught him bookkeeping. He was sent to Amsterdam around 1714, where he learned of the Florentine thermometer, invented in Italy 60 years prior in 1654 by the Grand Duke of Tuscany, Ferdinand II (1610–70), a member of the great Medici family. For some unknown reason it sparked his curiosity, and he decided to make thermometers for a living. He abandoned his bookkeeping apprenticeship, whereupon Dutch authorities issued warrants for his arrest. While on the run, he spent several years traveling around Europe and meeting scientists, such as Danish astronomer Olaus Romer. Eventually he came back to Amsterdam in 1717 and remained in the Netherlands for the rest of his life.

What seems so simple today—having a fixed scale and fixed points on a thermometer—did not exist for a long time, since several makers of thermometers used different types of scales and liquids for measuring. In 1694 Carlo Renaldini, a member of the Academia del Cimento and professor of philosophy at the University of Pisa, was the first to suggest taking the boiling and freezing points of water as the fixed points. The academy was founded by Prince Leopoldo de' Medici and the Grand Duke Ferdinand II in 1657 with the purpose of examining the natural philosophy of Aristotle. The academy was active sporadically over 10 years and concluded its work in 1667 with the publication of the *Saggi di Naturali Esperienze.*

Unfortunately, Florentine thermometers, or any thermometers of the time, were not very accurate; no two thermometers gave the same temperature, since there was no universal acceptance of liquid type or agreement on what to use for a scale. Makers of Florentine thermometers marked the lowest scale as the coldest day in Florence that year and the highest scale for the hottest day. Since temperature fluctuations naturally occur over the years, no two thermometers gave the same temperature. For several years Fahrenheit experimented with this problem, finally devising an accurate alcohol thermometer in 1709 and the first mercury or "quicksilver" thermometer in 1714.

Fahrenheit's first thermometers, from about 1709 to 1715, contained a column of alcohol that directly expanded and contracted, based on a design made by Danish astronomer Olaus Romer in 1708, which Fahrenheit personally reviewed. Romer used alcohol (actually wine) as the liquid, but his thermometer had two fixed reference points. He selected 60 degrees for the temperature of boiling water and $7^1/_2$ degrees for melting ice.

Fahrenheit eventually devised a temperature scale for his alcohol thermometers with three points calibrated at 32 degrees for freezing water, 96 degrees for body temperature (based on the thermometer being in a healthy man's mouth or under the armpit), and zero degrees fixed at the freezing point of ice and salt, which was believed at that time to be the coldest possible temperature. The scale was etched in 12 major points, with zero, four, and 12 as the three points and eight graduations between the major points, giving him a total of 96 points for his scale for body temperature on his thermometer.

Since his thermometers showed such consistency between them, mathematician Christian Wolf of Halle, Prussia, devoted a whole paper in an edition of *Acta Eruditorum,* one of the most important scientific journals of the time, to two of Fahrenheit's thermometers that Wolf received in 1714. In 1724 Fahrenheit published a paper entitled "Experimenta circa gradum caloris liquorum nonnullorum ebullientium instituta" (Experiments done on the degree of heat of a few boiling liquids) in the Royal Society's publication *Philosophical Transactions* and was admitted to the Royal Society the same year.

Fahrenheit decided to substitute mercury for the alcohol because its rate of expansion was more constant and could be used over a wider range of temperatures. Fahrenheit, like Isaac Newton before him, realized that it was more accurate to base the thermometer on a substance that changed consistently based on temperature, not simply the hottest or coldest day of the year like the Florentine models. Mercury also had a much wider temperature range than alcohol. This was contrary to the common thought at the time, promoted by Halley as late as 1693, that mercury could not be used for thermometers due to its low coefficient of expansion.

Fahrenheit later adjusted his temperature scale to ignore body temperature as a fixed point, bringing the scale to the freezing and boiling of water alone. When he died, scientists recalibrated his thermometer so that the boiling point of water was the highest point, changing it to 212 degrees, as Fahrenheit had earlier indicated in a publication on the boiling points of various liquids. The freezing point became 32 degrees (body temperature became 98.6 degrees, which we use today). This is the scale that is presently used in today's thermometers in the United States and some English-speaking countries, although most scientists use the Celsius scale.

By 1779 there were some 19 different scales in use on thermometers, but it was Fahrenheit, as well as astronomer ANDERS CELSIUS and Jean Christin, whose scales, when presented in 1742 and 1743, helped finally set the standards for an accurate thermometer that are still in use today. Besides making thermometers, Fahrenheit was the first to show that the boiling point of liquids varies at different atmospheric pressures, and he suggested this as a principle for the construction of barometers. Among his other contributions were a pumping device for draining the Dutch polders and a hygrometer for measuring atmospheric humidity.

Fahrenheit died on September 16, 1736, at The Hague at age 50. There is virtually no one in the English-speaking countries today who does not have a thermometer with his initials on it.

fallout The ground fall of dust and other debris raised to great heights in the atmosphere by a violent explosion. Also applied to radioactive fallout from an atomic or thermonuclear explosion.

faraday The charge carried by one mole of electrons is known as one faraday (symbol *F*). One faraday of electricity corresponds to the charge on 6.022×10^{23} electrons, or 96,487 coulombs.

Faraday's law of electrolysis Relates to the number of electrons required to discharge one mole of an element. One equivalent weight of a substance is produced at each electrode during the passage of 96,487 coulombs of charge through an electrolytic cell.

See also FARADAY.

fast-atom bombardment mass spectroscopy (**FAB mass spectroscopy**) A method in which ions are produced in a mass spectrometer from nonvolatile or thermally fragile organic molecules by bombarding the compound in the condensed phase with energy-rich neutral particles.

fast neutron In a nuclear reaction, a neutron ejected at high kinetic energy.

fat (general) Any substance made up of LIPIDS or FATTY ACIDS that supply calories to the body and can be found in solid or liquid form (e.g., margarine, vegetable oil); three fatty acids linked to a glycerol molecule form fat.

fat (triacylglycerol) Triacylglycerols are storage LIPIDS, consisting of three similar to different FATTY ACIDS attached to a glycerol molecule. They are found mostly stored in adipose (fat) cells and tissues and are highly concentrated regions of metabolic energy. Since there are abundant reduced CH groups available in them for oxidation-required energy production, they are excellent storage containers of energy. Can be found in plants, animals, and animal plasma lipoproteins for lipid transport. Formerly known as triglycerides.

fatty acid Fatty acids are the components of two LIPID types mostly found in cells in the form of large lipids or small amounts in free form: storage fats and structural phospholipids. They consist of long hydrocarbon chains of varying length (from four to 24 carbon atoms), containing a terminal CARBONYL GROUP at one end, and they may be saturated (having only a single carbon-to-carbon bond), or unsaturated (one or more double or triple carbon-to-carbon bonds). The number and location of double bonds also vary. More than 70 different kinds have been found in cells. Saturated fatty acids cause higher levels of blood CHOLESTEROL, since they have a regulating effect on its synthesis, but unsaturated ones do not have that effect and nutritionally are promoted more. Some fatty acids are palmitic acid, palmitoleic acid, alpha-linolenic acid, eleostearic acid, linoleic acid, oleic acid, and elaidic acid. Three fatty acids linked to a glycerol molecule form FAT.

[2Fe-2S] Designation of a two-iron, two-labile-sulfur CLUSTER in a protein, comprising two sulfido-bridged iron atoms. The oxidation levels of the clusters are indicated by adding the charges on the iron and sulfide atoms, i.e., $[2Fe\text{-}2S]^{2+}$; $[2Fe\text{-}2S]^{+}$. The alternative designation, which conforms to inorganic chemical convention, is to include the charges on the LIGANDs; this is more appropriate where the ligands are other than the usual cysteine sulfurs, such as in the RIESKE PROTEINS.

See also FERREDOXIN.

[4Fe-4S] Designation of a four-iron, four-labile-sulfur CLUSTER in a protein. (*See* [2FE-2S].) Possible oxidation levels of the clusters are $[4Fe\text{-}4S]^{3+}$; $[4Fe\text{-}4S]^{2+}$; $[4Fe\text{-}4S]^{+}$.

See also FERREDOXIN; HIPIP.

feedback inhibition A way for the end product of a cell's biosynthetic pathway to stop the activity of the first enzymes in that pathway, thereby controlling the enzymatic activity; it stops the synthesis of the product. Also called end-product inhibition.

FeMo cofactor An inorganic CLUSTER that is found in the FeMo protein of the molybdenum-NITROGENASE and is essential for the catalytic reduction of N_2 to ammonia. This cluster contains Fe, Mo, and S in a 7:1:9 ratio. The structure of the COFACTOR within the FeMo protein can be described in terms of two cuboidal SUBUNITs, Fe_4S_3 and $MoFe_3S_3$ bridged by three S^{2-} ions and anchored to the protein by a histidine bound via an imidazole group to the Mo atom and by a cysteine bound via a deprotonated SH group to an Fe atom of the Fe_4S_3 subunit. The Mo atom at the periphery of the molecule is six-COORDINATE and, in addition to the three sulfido LIGANDs and the histidine imidazole, is also bound to two oxygen atoms from an (R)-homocitrate molecule.

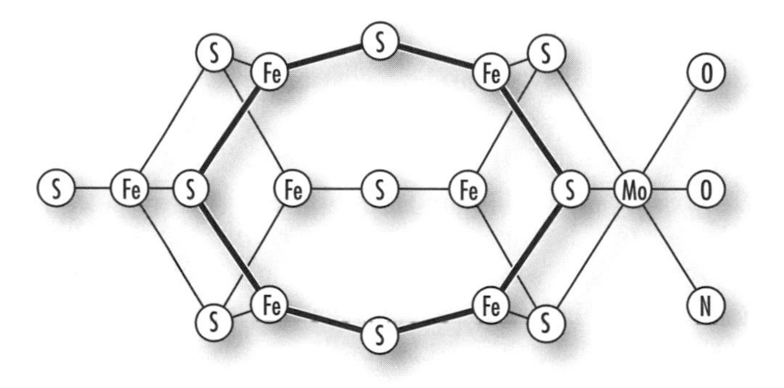

Fenton reaction

$$Fe^{2+} + H_2O_2 \rightarrow Fe^{3+} + OH^{\cdot} + OH^{-}$$

This equation describes the iron-salt-dependent decomposition of dihydrogen peroxide, generating the highly reactive hydroxyl radical, possibly via an oxo-iron(IV) intermediate. Addition of a reducing agent such as ascorbate leads to a cycle that increases the damage to biological molecules.

See also HABER-WEISS REACTION.

fermentation The anaerobic decomposition of complex organic substances by microorganisms such as bacteria, molds, or yeast (called ferments) on a fermentation substrate that produces simpler substances or other desired effects, such as the yielding of ethanol and carbon dioxide from yeast for commercial purposes, the production of ATP and energy production, and the development of antibiotics and enzymes. Fermentation is used by microflora of the large intestine to break down indigestible carbohydrates.

Large fermentors are used to culture microorganisms for the production of some commercially valuable products such as bread, beer, wine, and other beverages.

ferredoxin A protein—containing more than one iron and ACID-LABILE SULFIDE—that displays electron-transfer activity but not classical ENZYME function.

See also HIPIP.

ferriheme An iron(III) PORPHYRIN COORDINATION complex.

ferritin An iron-storage protein consisting of a shell of 24 protein SUBUNITs encapsulating up to 4,500 iron atoms in the form of a hydrated iron(III) oxide.

ferrochelatase An ENZYME that catalyzes the insertion of iron into PROTOPORPHYRIN IX to form HEME. The mammalian enzyme contains an IRON-SULFUR CLUSTER.

ferroheme An iron(II) PORPHYRIN COORDINATION complex.

ferromagnetic If there is coupling between the individual magnetic dipole moments of a PARAMAGNETIC sample, spontaneous ordering of the moments will occur at low temperatures. If this ordering results in an electronic GROUND STATE in which the moments are aligned in the same direction (parallel), the substance is said to be ferromagnetic. If the ordering results in an electronic ground state in which the moments are aligned in opposite directions, the substance is said to be antiferromagnetic.

ferrous metal A metal composed chiefly of iron.

field effect An experimentally observable effect—symbolized by F (on reaction rates, etc.) of INTRAMOLECULAR coulombic interaction between the center of interest and a remote unipole or dipole—by direct action through space rather than through bonds. The magnitude of the field effect (or direct effect) depends on the unipolar charge/dipole moment, orientation of dipole, shortest distance between the center of interest and the remote unipole or dipole, and on the effective dielectric constant.

See also ELECTRONIC EFFECT; INDUCTIVE EFFECT; POLAR EFFECT.

filter A device, instrument, or material that removes something from whatever passes through it.

filtration The passage of a liquid or material through a FILTER, utilizing gravity, pressure, or vacuum (suction).

first law of thermodynamics Simply put, energy can neither be created nor destroyed, only transformed or transferred from one molecule to another; in effect, the total amount of energy in the universe is constant. Also known as the law of conservation of energy. THERMODYNAMICS is the study of the conversion of energy between heat and other forms, e.g., mechanical.

See also SECOND LAW OF THERMODYNAMICS.

Fischer, Hermann Emil (1852–1919) German *Chemist* Hermann Emil Fischer was born on October 9, 1852, in Euskirchen (the Cologne district). He spent three years with a private tutor, then attended local school. He then spent two years in school at Wetzlar and two more at Bonn, passing in 1869 with great distinction. Instead of going into his family's lumber business, he went to the University of Bonn in 1871, originally studying chemistry but later changing to physics and mineralogy.

The following year he attended the new University of Strasbourg with his cousin Otto Fischer and became influenced by Adolf von Baeyer. Studying under von Baeyer, Fischer received his Ph.D. in 1874 for his work on fluoresceine and orcin-phthalein. In the same year he was appointed assistant instructor at Strasbourg University and discovered the first hydrazine base, phenylhydrazine, and demonstrated its relationship to hydrazobenzene and to a sulfonic acid described by Strecker and Römer.

In 1875 von Baeyer became an assistant in organic chemistry, succeeding Liebig at the University of Munich. In 1878 he qualified as a privatdozent at Munich and was appointed associate professor of analytical chemistry in 1879 after refusing the chair of chemistry at Aix-la-Chapelle. In 1881 he was appointed professor of chemistry at the University of Erlangen. In 1888 he became professor of chemistry at the University of Würzburg until 1892, and then succeeded A. W. Hofmann in the chair of chemistry at the University of Berlin, where he stayed until his death in 1919.

He and his cousin Otto continued to work on hydrazines, and both worked out a new theory of the constitution of the dyes derived from triphenylmethane, which they proved through experimentation.

While at Erlangen, Fischer studied the active principles of tea, coffee, and cocoa (caffeine and theobromine) and eventually synthesized them. Between 1882 and 1906, his work on purines and sugars gained him fame. In 1906 he was awarded the Nobel Prize in chemistry for his work in this area.

Fischer was awarded the Prussian Order of Merit and the Maximilian Order for Arts and Sciences.

In 1888 he married Agnes Gerlach, daughter of J. von Gerlach, professor of anatomy at Erlangen, but she died seven years into their marriage after bearing three sons. Fischer died in Berlin on July 15, 1919. The German Chemical Society instituted the Emil Fischer Memorial Medal in 1919.

fission *See* NUCLEAR FISSION.

flammable A liquid having a flash point below 37.8°C (100°F).

flash point The temperature at which liquids yield enough flammable vapor to ignite. It is the lowest temperature at which a liquid can form an ignitable mixture in air near the surface of the liquid. The lower the flash point, the easier it is to ignite.

flash vacuum pyrolysis (FVP) Thermal reaction of a molecule by exposing it to a short thermal shock at high temperature, usually in the gas phase.

flavin A PROSTHETIC GROUP found in flavoproteins and involved in biological oxidation and reduction. Forms the basis of natural yellow pigments like riboflavin.

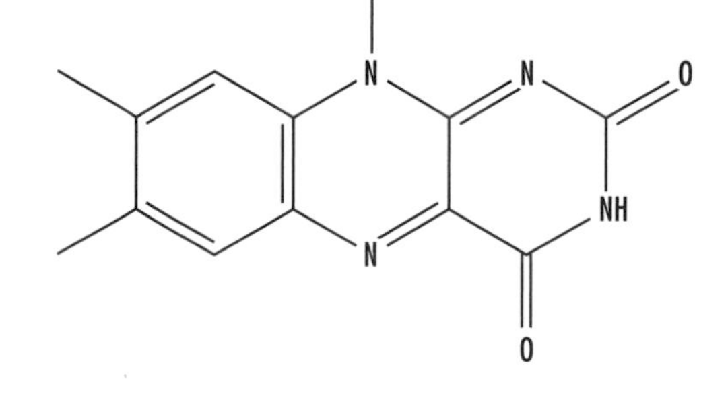

flocculation Coagulation of colloidal particles due to the ions in solution.

flotation A method to separate water-repelling (HYDROPHOBIC) ore particles from water-attracting (HYDROPHILIC) particles.

fluids All substances that flow freely, such as gases and liquids.

fluorescence When colors absorb radiant power at one wavelength and immediately reemit it at another (usually longer) wavelength. The property of giving off light at a particular wavelength (emission wavelength) after it is illuminated by light of a different wavelength (excitation wavelength).

fluorination A process in which a thermoplastic container or closure is exposed to fluorine gas to improve barrier properties and reduce solvent absorption and permeation. Also refers to the introduction of a fluorine atom into a molecule.

flux A substance added to react with a material to lower a melting point. A material used to remove oxides that form on metal surfaces exposed to oxygen in the air. In nuclear terms, the amount of some type of particle (neutrons, alpha radiation, etc.) or energy (photons, heat, etc.) crossing a unit area per unit time. The unit of flux is the number of particles, energy, etc. per square centimeter per second.

fluxional In inorganic chemistry, this term is used to designate positional changes among LIGANDs. A fluxional chemical species undergoes rapid (degenerate) rearrangements, generally detectable by methods that observe the behavior of individual nuclei in a rearranged chemical species.

foam A colloidal suspension of a gas in a liquid.

folate coenzymes A group of heterocyclic compounds that are based on the 4-(2-amino-3,4-dihydro-4-oxopteridin-6-ylmethylamino)benzoic acid (pteroic acid) and conjugated with one or more L-glutamate units. Folate derivatives are important in DNA synthesis and erythrocyte formation. Folate deficiency leads to ANEMIA.

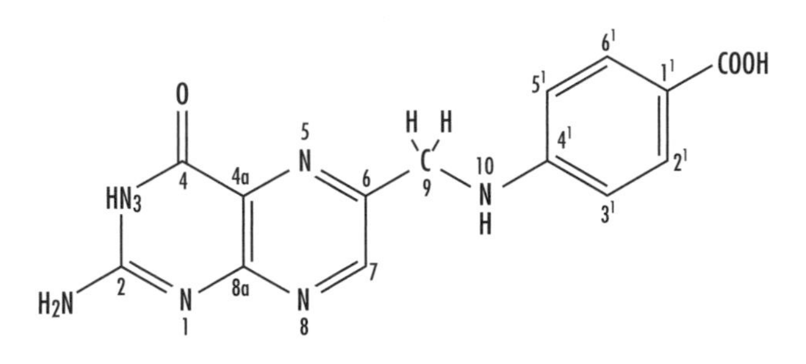

force constant A bond property that determines the steepness of the potential energy parabola confining the two atoms.

force-field calculations *See* MOLECULAR MECHANICS CALCULATION.

formal charge A charge given to an atom in a molecule or polyatomic ion derived from a specific set of rules.

formation constant *See* STABILITY CONSTANT.

formula An exact representation of the structure of a molecule, ion, or compound showing the proportion of atoms that comprise the material, e.g., H_2O.

formula unit The smallest possible integer number of different ions in an ionic compound.

formula weight The sum of the atomic weights of all atoms found in one formula unit of an ionic compound.

fossil fuel Mostly hydrocarbon material that is derived from decay of organic materials under geological conditions of high pressure and temperature (metamorphism), i.e., coal, petroleum, natural gas, peat, and oil shale. Their combustion is considered part of the global warming problem.

fractional distillation Using a fractionating column in DISTILLATION to separate a liquid mixture into component parts that have different boiling points. When a mixture of liquids is boiled, the vapor will be richer in the component with the lower boiling point, but it will still be a mixture. By successive CONDENSATION and revaporization steps (usually in a column called a fractionating column), the vapor becomes increasingly rich in the lower-boiling component and can eventually be collected in pure form.

fractional precipitation Removing ions by precipitation in a solution while leaving other ions that have similar properties in the solution.

fractionation factor, isotopic The ratio $(x_1/x_2)_A/(x_1/x_2)_B$, where x is the abundance, expressed as the atom fraction of the isotope distinguished by the subscript numeral when the two isotopes are equilibrated between two different CHEMICAL SPECIES A and B (or between specific sites A and B in the same or different chemical species). The term is most commonly met in connection with deuterium solvent ISOTOPE EFFECTS, when the fractionation factor φ expresses the ratio

$$\phi = (x_D/x_H)_{solute}/(x_D/x_H)_{solvent}$$

for the exchangeable hydrogen atoms in the chemical species (or sites) concerned. The concept is also applicable to TRANSITION STATES.

fragmentation (1) The heterolytic cleavage of a molecule according to the general reaction

$$a\text{–}b\text{–}c\text{–}d\text{–}X \rightarrow (a\text{–}b)^+ + c{=}d + X^-$$

where a–b is an ELECTROFUGE and X is a NUCLEOFUGE (which may emerge from the reaction in combined form), and the middle group affords the unsaturated fragment c=d. For example,

$$Ph_3C\text{–}CO_2H + H^+ \rightarrow Ph_3C^+ + C{=}O + H_2O$$

(2) The breakdown of a RADICAL into a diamagnetic molecule or ion and a smaller radical, e.g.,

$$(CH_3)_3C\text{–}O\cdot \rightarrow (CH_3)_2C{=}O + H_3C\cdot$$
$$[ArBr]^{\cdot -} \rightarrow Ar\cdot + Br^- \text{ (solution)}$$

(3) The breakdown of a RADICAL ION in a mass spectrometer or in solution, forming an ion of lower molar mass and a radical, e.g.,

$$[(CH_3)_3C\text{–}OH]^{\cdot +} \rightarrow (CH_3)_2C{=}OH^+ + H_3C\cdot \text{ (mass spectrometer)}$$

See also HETEROLYTIC BOND-DISSOCIATION ENERGY.

Frasch process Named for the German-American chemist Herman Frasch (1851–1914), a method to extract elemental sulfur by melting the sulfur with superheated water (at 170°C under high pressure) and forcing it to the surface of the Earth as a slurry.

free energy Energy readily available for producing change in a system.

See also GIBBS FREE ENERGY.

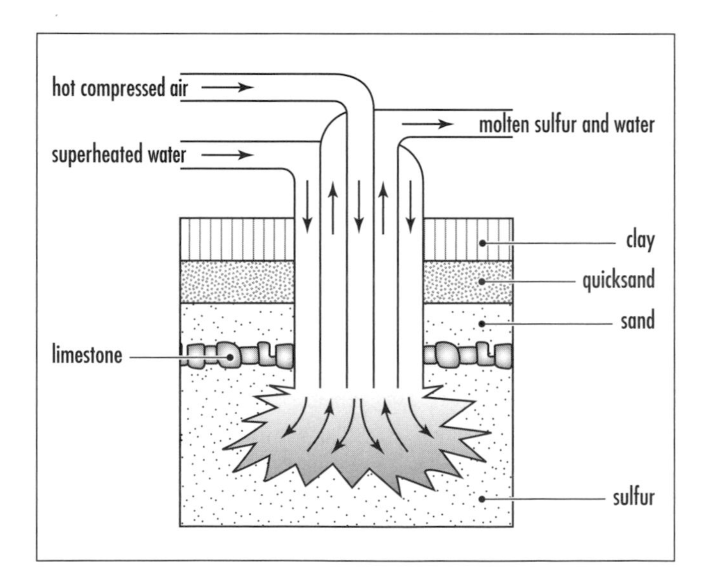

Frasch process

free-energy change Determines the direction of spontaneity of a chemical reaction at constant temperature. A reaction will proceed to decrease the free energy of the system.

free radical A molecule that contains at least one unpaired electron; a highly reactive chemical that usually exists for a short time. Formed in the body during oxidation, a normal by-product of metabolism, they can bind with electrons from other molecules and cause cellular damage by disrupting normal cellular processes. They can be kept in check by antioxidants such as certain enzymes or vitamins (C and E).

freezing-point depression The decrease in the freezing point of a solvent caused by the presence of a solute. The lower the molecular weight, the greater is the ability of a molecule to depress the freezing point for any given concentration by weight.

frequency The rate at which a periodic event occurs. The number of repeating corresponding points on a wave that pass a given observation point per unit time; the rate at which the waves of electromagnetic radiation pass a point.

frontier orbitals The highest-energy occupied molecular orbital (HOMO) (filled or partly filled) and lowest-energy unoccupied molecular orbital (LUMO) (completely or partly vacant) of a MOLECULAR ENTITY. Examination of the mixing of frontier molecular orbitals of reacting molecular entities affords an approach to the interpretation of reaction behavior. This constitutes a simplified perturbation MOLECULAR ORBITAL theory of chemical behavior.

See also LOWEST UNOCCUPIED MOLECULAR ORBIT; SOMO; SUBJACENT ORBITAL.

fuel cell A voltaic cell where reactants are continually supplied to convert chemical energy to electrical energy. Typically, the reaction uses H_2 as fuel, reacting electrochemically with oxygen to produce water and electrical energy.

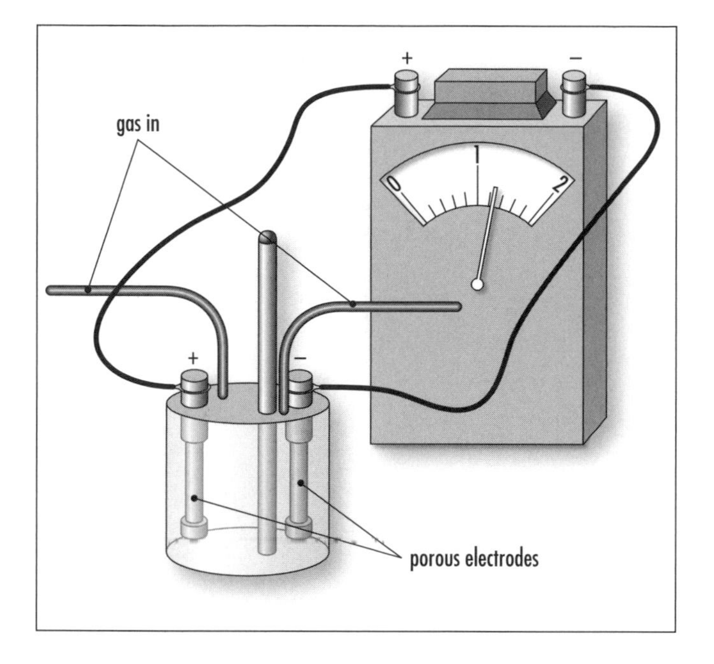

Fuel cell. A voltaic cell where reactants are continually supplied to convert chemical energy to electrical energy

fullerene An ALLOTROPE of carbon containing five- and six-membered rings, whose structure is based on that of C_{60}, BUCKMINSTERFULLERENE.

functional group Organic compounds are thought of as consisting of a relatively unreactive backbone—for example, a chain of sp^3 hybridized carbon atoms—and one or several functional groups. The functional group is an atom (or a group of atoms) that has similar chemical properties whenever it occurs in different compounds. It defines the characteristic physical and chemical properties of families of organic compounds.

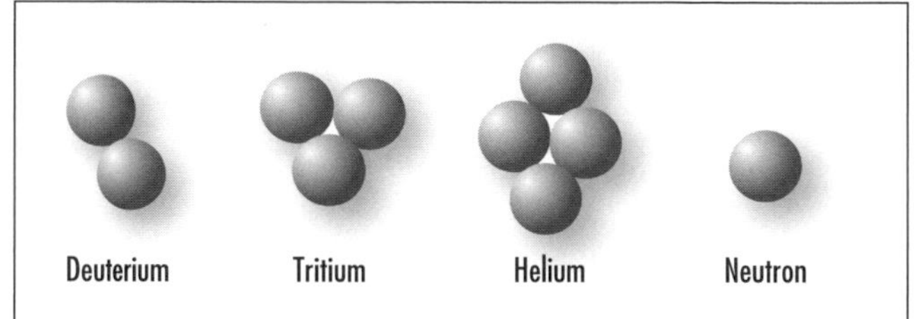

Fusion. A nuclear reaction in which light atomic nuclei combine to form heavier nuclei, typically accompanied by the release of energy

fur The iron-uptake-regulating protein present in PROKARYOTES that binds simultaneously Fe and DNA, thereby preventing the biosynthesis of ENZYMES for the production of SCAVENGER chelates (SIDEROPHORES).

See also CHELATION.

fused-ring compound A fused ring is two rings joined together through one or more atoms. Examples include *trans*-decalin, *cis*-decalin, bicyclo [2.2.2] octane, bicyclo [2.2.1] heptane (norbornane), and adamantane.

fusion A nuclear reaction in which light atomic nuclei combine to form heavier nuclei, typically accompanied by the release of energy. Also used to describe chemical reactions in which rings are formed.

G

galvanizing The process where steel is coated with a layer of zinc, which provides the steel with greater corrosion resistance.

gamma (γ) band *See* SORET BAND.

gamma ray High-energy electromagnetic radiation that is emitted by the nuclei of radioactive atoms.

gangue The term used to describe the nonvaluable or unwanted minerals or rocks associated with a commercial ore.

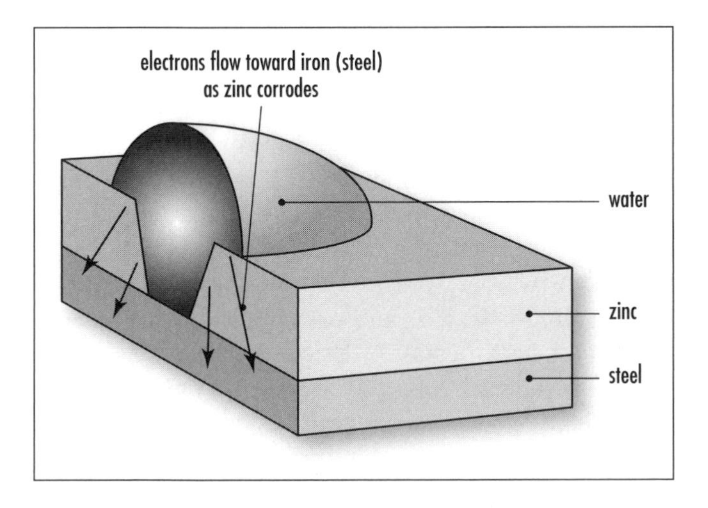

Galvanizing. The process where steel is coated with a layer of zinc, which provides the steel with greater corrosion resistance

gap junction A site between two cells that allows small molecules or ions to cross through and connect between the two cytoplasms; allows electrical potentials between the two cells.

gas Matter that has no definite volume or definite shape and always fills any space given to it.

gas chromatography A type of automated CHROMATOGRAPHY in which the mixture to be analyzed is vaporized and then carried by an inert gas through a special column and further to a detection device.

gas-phase acidity The negative of the Gibbs energy (ΔG_r^o) change for the reaction

$$A\text{–}H \rightarrow A^- + H^+$$

in the gas phase.

See also GIBBS FREE ENERGY.

gas-phase basicity The negative of the Gibbs energy (ΔG_r^o) change associated with the reaction

$$B + H^+ \rightarrow BH^+$$

in the gas phase. Also called absolute or intrinsic basicity.

See also GIBBS FREE ENERGY; PROTON AFFINITY.

gated ion channel A specific ion channel that opens and closes to allow the cell to alter its membrane potential. An ion channel is a membrane protein that forms an aqueous pore so that charged ions can cross through the membrane. There are several types of ion channels. For example, a LIGAND-gated ion channel is where gating is controlled by binding of a chemical signal (the ligand) to a specific binding site on the channel protein. Other ion channels are voltage gated and mechanically gated.

GC-MS Gas chromatograph mass spectroscopy.

See also GAS CHROMATOGRAPHY; MASS SPECTROMETER.

Geiger counter A radiation-detection and -measuring instrument. Consists of a gas-filled tube that discharges electrically when ionizing radiation passes through it. It was named for Hans Geiger and W. Mueller, who invented it in the 1920s.

gel A semisolid or highly viscous colloidal suspension of a solid dispersed in a liquid.

gel electrophoresis The analytical laboratory process to separate molecules according to their size. The sample is put on an end of a slab of polymer gel, a lyophilic colloid that has coagulated to a jelly. An electric field is then applied through the gel, which separates the molecules; small molecules pass easily and move toward the other end faster than corresponding larger ones. Eventually all sizes get sorted, since molecules with similar electric charge and density will migrate together at the same rate. There are several types of gel composition, and various chemicals can be added to help separation.

gem-dimethyl group Two methyl groups of the same carbon atom. A methyl group is the radical –3CH that exists only in organic compounds.

geminate pair Pair of MOLECULAR ENTITIES IN close proximity in solution within a solvent cage and resulting from reaction (e.g., bond scission, electron transfer, group transfer) of a precursor that constitutes a single kinetic entity.

See also ION PAIR; RADICAL PAIR.

geminate recombination The reaction with each other of two transient molecular entities produced from a common precursor in solution. If reaction occurs before any separation by diffusion has occurred, this is termed *primary geminate recombination.* If the mutually reactive entities have been separated, and come together by diffusion, this is termed *secondary geminate recombination.*

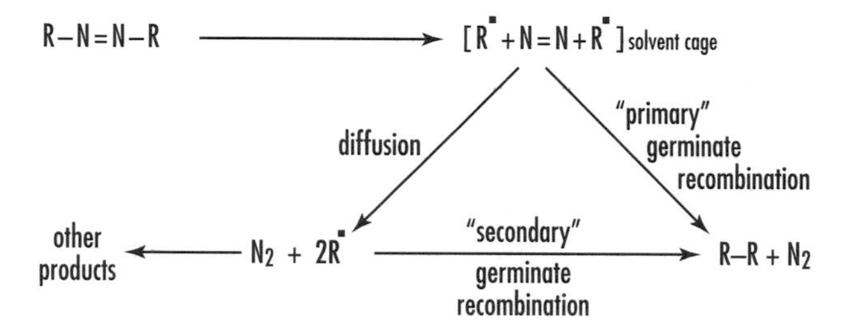

gene Structurally, a basic unit of hereditary material; an ordered SEQUENCE of NUCLEOTIDE bases that encodes one polypeptide chain (via MRNA). The gene includes, however, regions preceding and following the coding region (leader and trailer) as well as (in EUKARYOTEs) intervening sequences (INTRONs) between individual coding segments (EXONs). Functionally, the gene is defined by the *cis–trans* test that determines whether independent MUTATIONs of the same phenotype occur within a single gene or in several genes involved in the same function.

general acid catalysis The CATALYSIS of a chemical reaction by a series of BRONSTED ACIDS (which may include the solvated hydrogen ion) so that the rate of the catalyzed part of the reaction is given by $\Sigma k_{HA}[HA]$ multiplied by some function of SUBSTRATE concentrations. (The acids HA are unchanged by the overall reaction.) General catalysis by acids can be experimentally distinguished from SPECIFIC CATALYSIS by hydrogen cations (HYDRONS) by observation of the RATE OF REACTION as a function of buffer concentration.

See also CATALYSIS; CATALYTIC COEFFICIENT; INTRAMOLECULAR CATALYSIS; PSEUDOCATALYSIS; SPECIFIC CATALYSIS.

general base catalysis The catalysis of a chemical reaction by a series of BRONSTED BASES (which may include the LYATE ION) so that the rate of the catalyzed part of the reaction is given by $\Sigma k_B[B]$ multiplied by some function of substrate concentration.

See also GENERAL ACID CATALYSIS.

genetic code The language of genetics. The instructions in a gene that tell the cell how to make a specific protein. The code defines the series of NUCLEOTIDES in DNA, read as triplets called CODONs, that specifies the sequence of AMINO ACIDs in a protein. The set comprises 64 nucleotide triplets (codons) that specify the 20 amino acids and termination codons (UAA, UAG, UGA).

The code is made up of adenine (A), thymine (T), guanine (G), and cytosine (C), the nucleotide bases of DNA. Each gene's code combines them in various ways to spell out three-letter triplets (codons) that specify which amino acid is needed at each step in making a protein.

genome The complete assemblage of chromosomes and extrachromosomal genes of a cell, organelle, organism, or virus; the complete DNA portion of an organism. The complete set of genes shared by members of any reproductive body such as a population or species.

geological time The span of time that has passed since the creation of the Earth and its components; a scale use to measure geological events millions of years ago. Measured in chronostratic or relative terms, where subdivisions of the Earth's geology are set in an order based on relative age relationships derived from fossil composition and stratigraphic position, or in chronometric or absolute time, where the use of radiometric dating techniques gives numerical ages.

geometrical isomers **(position isomers)** Isomers differing in the way the atoms are oriented in space relative to each other.

geothermal energy Natural heat from within the Earth.

geotropism A plant's response to gravitational effects. A plant's roots grow downward toward the gravitational pull, which is called positive geotropism, while shoots grow upward against gravitational pull (negative geotropism). Also called gravitropism.

g-factor *See* ELECTRON PARAMAGNETIC RESONANCE SPECTROSCOPY.

Gibbs energy diagram A diagram showing the relative standard Gibbs energies of reactants, TRANSITION STATEs, reaction INTERMEDIATEs, and products in the same sequence as they occur in a CHEMICAL REACTION. These points are often connected by a smooth curve (a "Gibbs energy profile," still commonly referred to as a "free energy profile"), but experimental observation can provide information on relative standard Gibbs energies only at the maxima and minima and not at the configurations between them. The abscissa expresses the sequence of reactants, products, reaction intermediates, and transition states and is usually undefined or only vaguely defined by the REACTION COORDINATE (extent of bond breaking or bond making). In some adaptations, however, the abscissas are explicitly defined as BOND ORDERs, Bronsted exponents, etc. Contrary to statements in many textbooks, the highest point on a Gibbs energy diagram does not necessarily correspond to the transition state of the RATE-LIMITING STEP. For example, in a STEPWISE REACTION consisting of two reaction steps

$$1.\ A + B \rightleftarrows C$$
$$2.\ C + D \rightarrow E$$

one of the transition states of the two reaction steps must (in general) have a higher standard Gibbs energy than the other, whatever the concentration of D in the

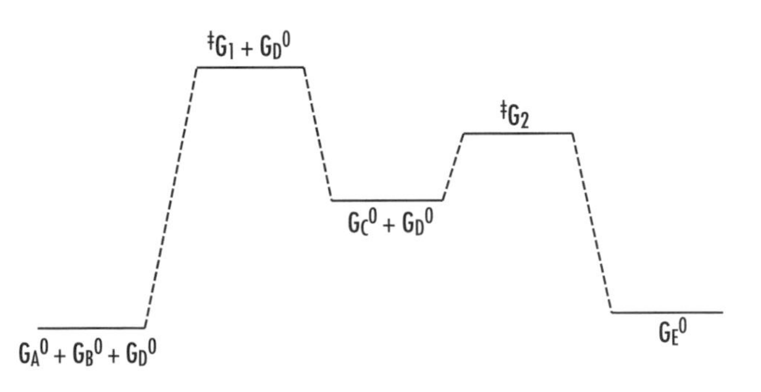

system. However, the value of that concentration will determine which of the reaction steps is rate-limiting. If the particular concentrations of interest, which may vary, are chosen as the standard state, then the rate-limiting step is the one with the highest Gibbs energy.

See also POTENTIAL-ENERGY PROFILE; POTENTIAL-ENERGY (REACTION) SURFACE; REACTION COORDINATE.

Gibbs energy of activation (standard free energy of activation), $\Delta^{‡}G^{o}$ (SI unit: kJ mol^{-1}) The standard Gibbs energy difference between the TRANSITION STATE of a reaction (either an ELEMENTARY REACTION or a STEPWISE REACTION) and the ground state of the reactants. It is calculated from the experimental rate constant k via the conventional form of the absolute rate equation:

$$\Delta^{‡}G = RT\,[\ln(k_B/h) - \ln(k/T)]$$

where k_B is the Boltzmann constant and h the Planck constant ($k_B/h = 2.08358 \times 10^{10}$ K^{-1} s^{-1}). The values of the rate constants, and hence Gibbs energies of activation, depend upon the choice of concentration units (or of the thermodynamic standard state).

See also ENTHALPY OF ACTIVATION; ENTROPY OF ACTIVATION.

Gibbs free energy Energy liberated or absorbed in a reversible process at constant pressure and constant temperature.

Gilbert, Walter (1932–) American *Physicist, biochemist* Walter Gilbert was born on March 21, 1932, in Boston, Massachusetts, to Richard V. Gilbert, an economist at Harvard University, and Emma Cohen, a child psychologist. Educated at home for the first few years, Gilbert moved with the family to Washington, D.C., in 1939, and he was educated there in public schools, later attending the Sidwell Friends high school.

He attended college at Harvard and majored in chemistry and physics, then went to the University of Cambridge for two years, receiving a doctorate degree in physics in 1957 for his work on dispersion relations for elementary particle scattering. He returned to Harvard and, after a postdoctoral year and a year as Julian Schwinger's assistant, became an assistant professor of biophysics in 1964 and professor of biochemistry in 1968. In 1974 he became American Cancer Society professor of molecular biology at Harvard. During the late 1950s and early 1960s, he focused on theoretical physics and worked with graduate students on problems in theory. After a few years, he shifted from the mathematical formulations of theoretical physics to an experimental field.

In the 1970s he developed a widely used technique of using gel electrophoresis to read nucleotide sequences of DNA segments. The same method was developed independently by Frederick Sanger, and they both won the Nobel Prize in chemistry in 1980 "for their contributions concerning the determination of base sequences in nucleic acids."

In 1979 he joined a group of other scientists and businessmen to form Biogen, a commercial genetic-engineering research corporation, but resigned from Biogen in 1984. He went back to Harvard and became a chief proponent of the Human Genome Project, a government-funded effort to compile a complete map of the gene sequences in human DNA, where he continues to work today. He is married to poet Celia Gilbert, and they have two children.

glass A homogeneous material with a random, liquidlike molecular structure.

glass electrode An electrode for measuring pH when it is dipped into an aqueous solution containing H^+ ions.

glass transition temperature The temperature where a polymer changes from hard and brittle to soft and pliable.

glycogen A glucose polymer (also known as animal starch) stored in animal tissue.

glycoprotein Glycoproteins are complexes in which carbohydrates are attached covalently to asparagine (N-glycans) or serine/threonine (O-glycans) residues of peptides. Also known as a conjugated protein. A protein coated with a sugar is termed *glycosylated* and is described or named with the initials "gp" along with its molecular weight, e.g., gp160. Several gps are associated with HIV infection, since they are the outer-coat proteins of HIV: gp41 plays a key role in HIV's infection of CD4+ T cells by facilitating the fusion of the viral and cell membranes. GP120 is one of the proteins that forms the envelope of HIV, and projects from the surface of HIV and binds to the CD4 molecule on helper T cells. GPs are found in mucus and mucins, y-globulins, al-globulins, a2-globulins, and transferrin, an ion-transporting protein. They act as receptors for molecular signals originating outside the cell. Attachment of oligosaccharides to peptides increases solubility, covers the antigenic domains, and protects peptide backbones against proteases.

gold drugs Gold COORDINATION compounds used in the treatment of rheumatoid arthritis, examples being auranofin, (tetraacetylthioglucosato-S)(triethylphosphane)gold(I), and myocrysin, disodium thiomalonatogold(I).

Graham's law The rate of diffusion of a gas is inversely proportional to the square root of its density or molecular weight.

Greek letters used

α: *See* HELIX (for alpha helix); CYTOCHROME
β: *See* BETA SHEET; BETA STRAND; BETA TURN; CYTOCHROME
γ: *See* SORET BAND (for gamma band)
η: *See* HAPTO; ASYMMETRY PARAMETER
κ: *See* DONOR ATOM SYMBOL (for kappa convention)
μ: *See* BRIDGING LIGAND (for mu symbol)

greenhouse effect The warming of an atmosphere by its absorption of infrared radiation while shortwave radiation is allowed to pass through.

Certain gaseous components of the atmosphere, called greenhouse gases, transmit the visible portion of solar radiation but absorb specific spectral bands of thermal radiation emitted by the Earth. The theory is that terrain absorbs radiation, heats up, and emits longer wavelength thermal radiation that is prevented from escaping into space by the blanket of carbon dioxide and other greenhouse gases in the atmosphere. As a result, the climate warms. Because atmospheric and oceanic circulations play a central role in the climate of the Earth, improving our knowledge about their interaction is essential.

See also CARBON DIOXIDE.

Greenhouse effect. The warming of an atmosphere by its absorbing infrared radiation while allowing shortwave radiation to pass through

ground state The state of lowest Gibbs energy of a system.

See also EXCITED STATE; GIBBS ENERGY DIAGRAM.

group A defined linked collection of atoms or a single atom within a MOLECULAR ENTITY. This use of the term in physical, organic, and general chemistry is less restrictive than the definition adopted for the purpose of nomenclature of organic compounds.

Also a vertical column in the periodic table.

See also SUBSTITUENT.

group theory A branch of mathematics concerned with the study of groups.

growth factor A complex family of organic chemicals, especially polypeptides, that bind to cell surface receptors and act to control new cell division, growth, and maintenance by bone marrow.

Synthetic growth factors are being used to stimulate normal white blood cell production following cancer treatments and bone marrow transplants.

Examples of growth factors are: insulin, insulinlike growth factors (IGF), and GF1 and II, which are polypeptides similar to insulin; somatomedin; polypeptides made by the liver and fibroblasts that, when released into the blood (stimulated by somatotropin), help cell division and growth by incorporating sulfates into collagen, RNA, and DNA synthesis; HGH, human growth hormone, also called somatotropin, a protein-like hormone from the pituitary gland that stimulates the liver to produce somatomedins that stimulate growth of bone and muscle; platelet-derived growth factor (PDGF), a glycoprotein that stimulates cell proliferation and chemotaxis in cartilage, bone, and other cell types; fibroblast growth factor, which promotes the proliferation of cells of mesodermal, neuroectodermal, epithelial, or endothelial origin; epidermal growth factor (EGF), important for cell development as it binds to receptors on cell surfaces to create a growth signal; and granulocyte colony-stimulating factor (G-CSF), a growth factor that promotes production of granulocytes, a type of white blood cell.

Grunwald-Winstein equation The LINEAR FREE-ENERGY RELATION

$$\log(k_s/k_0) = mY$$

expressing the dependence of the rate of SOLVOLYSIS of a substrate on ionizing power of the solvent. The rate constant k_0 applies to the reference solvent (ethanol-water, 80:20, v/v) and k_s to the solvent s, both at 25°C. The parameter m is characteristic of the substrate and is assigned the value unity for tert-butyl chloride. The value Y is intended to be a quantitative measure of the IONIZING POWER of the solvent s. The equation was later extended by Winstein, Grunwald, and Jones (1951) to the form

$$\log(k_s/k_0) = mY + lN$$

where N is the NUCLEOPHILICITY of the solvent and l its susceptibility parameter. The equation has also been applied to reactions other than solvolysis.

See also DIMROTH-REICHARDT E_T PARAMETER; POLARITY; Z-VALUE.

guanylate cyclase An ENZYME catalyzing the conversion of guanosine 5′-triphosphate to cyclicguanosine 3′,5′-monophosphate, which is involved in cellular REGULATION processes. One member of this class is a HEME-containing enzyme involved in processes regulated by nitrogen monoxide.

guest An organic or inorganic ion or molecule that occupies a cavity, cleft, or pocket within the molecular structure of A HOST MOLECULAR ENTITY and forms a COMPLEX with it or that is trapped in a cavity within the crystal structure of a host.

See also CROWN; CRYPTAND; INCLUSION COMPOUND.

gypsum A soft, transparent, mineral composed of hydrated calcium sulfate. Burnt gypsum commonly used by artists is known as plaster of paris.

H

Haber, Fritz (1868–1934) German *Chemist* Fritz Haber was born on December 9, 1868, in Breslau, Germany, to Siegfried Haber, a merchant and a member of one of the oldest families of the town. He went to school at the St. Elizabeth classical school at Breslau and conducted many chemistry experiments while still a child.

Haber studied chemistry at the University of Heidelberg from 1886 until 1891, followed by stints at the University of Berlin and at the Technical School at Charlottenberg. He finally decided to devote himself to chemistry in 1894 by accepting an assistantship at Karlsruhe, where he remained until 1911.

In 1896 Haber qualified as a Privatdozent (lecturer) with a thesis on experimental studies of the decomposition and combustion of hydrocarbons. In 1906 he became professor of physical chemistry and electrochemistry and director of an institute established at Karlsruhe to study these subjects.

In 1911 he became director of the Institute for Physical and Electrochemistry at Berlin-Dahlem, and he stayed there until 1933, when Nazi race laws forced most of his staff to resign. Haber also resigned in protest. He had a brief stint at Cambridge, England, then moved to Switzerland.

In 1898 Haber published a textbook on electrochemistry and promoted his research to relate chemical research to industrial processes, showing his results on electrolytic oxidation and reduction. That same year he explained the reduction of nitrobenzene in stages at the cathode, which became the model for other similar reduction processes. He continued electrochemical research for the next decade, working on the electrolysis of solid salts (1904), the establishment of the quinone–hydroquinone equilibrium at the cathode, and inventing the glass electrode that led him to make the first experimental investigations of the potential differences that occur between solid electrolytes and their aqueous solutions.

Haber's work on the fixation of nitrogen from air ("for the synthesis of ammonia from its elements") earned him the Nobel Prize in chemistry for 1918.

Haber died on January 29, 1934, at Basle. The Institute for Physical and Electrochemistry at Berlin-Dahlem was renamed the Fritz Haber Institute after his death.

Haber process An industrial process for the catalyzed production of ammonia from N_2 and H_2 at high temperature and pressure.

Haber-Weiss reaction The Haber-Weiss cycle consists of the following two reactions:

$$H_2O_2 + OH^{\cdot} \rightarrow H_2O + O_2^- + H^+$$
$$H_2O_2 + O_2^- \rightarrow O_2 + OH^- + OH^{\cdot}$$

The second reaction achieved notoriety as a possible source of hydroxyl radicals. However, it has a negligible rate constant. It is believed that iron(III) complexes can catalyze this reaction: first Fe(III) is reduced by superoxide, followed by oxidation by dihydrogenperoxide.

See also FENTON REACTION.

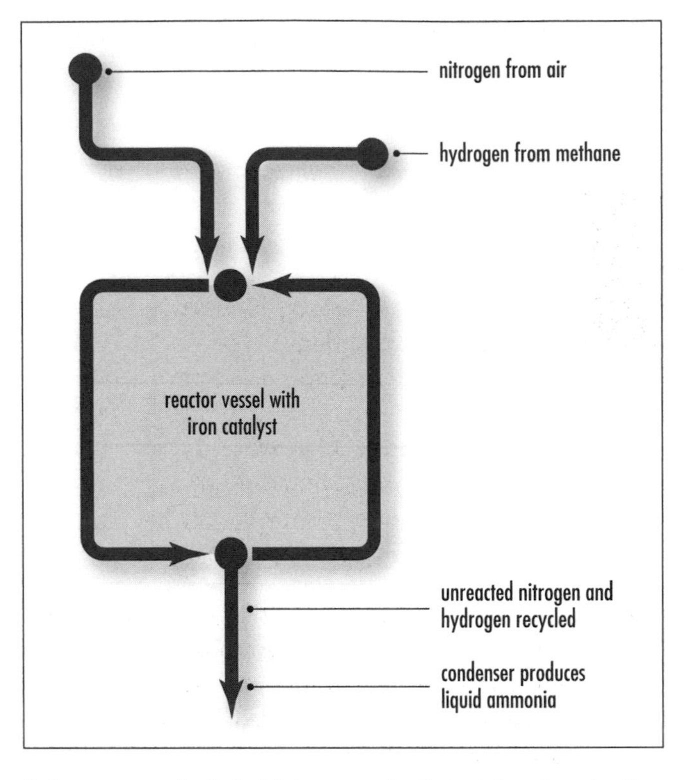

Haber process. An industrial process for the catalyzed production of ammonia from N_2 and H_2 at high temperature and pressure

Hahn, Otto (1879–1968) German *Chemist* Otto Hahn was born on March, 8, 1879, in Frankfurt-on-Main and attended and graduated from the city's secondary high school. In 1897 he began studying chemistry at Marburg and Munich, receiving a doctorate in 1901 at Marburg for a thesis on organic chemistry.

He became assistant in the Chemical Institute at Marburg for two years, and in 1904 moved on to University College, London, working under Sir William Ramsay. Here he discovered a new radioactive substance, radiothorium, while working on the preparation of pure radium salts.

From 1905 to 1906, he worked at the Physical Institute of McGill University, Montreal (Canada), and discovered radioactinium and conducted investigations with Rutherford on alpha rays of radiothorium and radioactinium.

Hahn moved to Berlin, to the Chemical Institute of the University, as a university lecturer in the spring of 1907 and discovered mesothorium. At the end of 1907, Hahn began a 30-year collaboration with Dr. Lise Meitner, who came to Berlin from Vienna. They worked on investigations on beta rays, discovered protactinium, and Hahn discovered the fission of uranium and thorium. In 1944 he was awarded the Nobel Prize in chemistry "for his discovery of the fission of heavy nuclei."

He died on July 28, 1968, in Gottingen, West Germany, after a fall.

half-cell Compartment or location (single electrode) in an electrolytic cell or voltaic cell in which the oxidation or reduction half-reaction occurs. It is oxidation at the anode, reduction at the cathode. A half-cell reaction refers to the chemical equation that describes only the oxidation or reduction part of a redox reaction.

half-life For a given reaction, the half-life ($t_{1/2}$) of a reactant is the time required for its concentration to reach a value that is the arithmetic mean of its initial and final (equilibrium) value. For a reactant that is entirely consumed, it is the time taken for the reactant concentration to fall to one-half of its initial value. For a first-order reaction, the half-life of the reactant may be called the half-life of the reaction. In nuclear chemistry, (radioactive) half-life is defined, for a simple

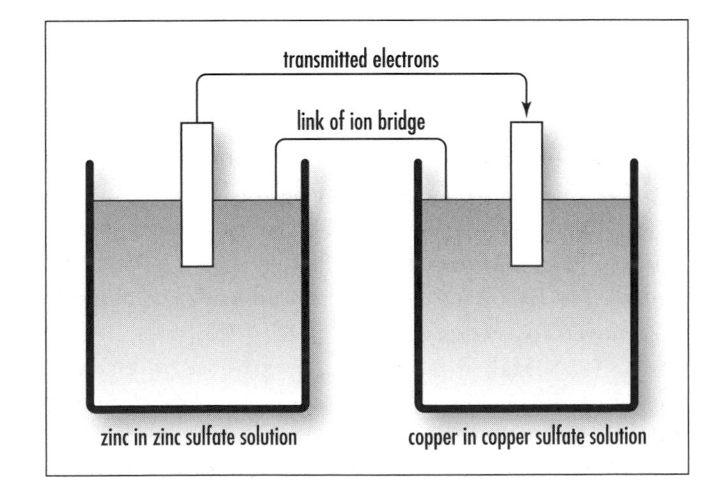

Half-cell. Compartment or location (single electrode) in an electrolytic cell or voltaic cell in which the oxidation or reduction half-reaction occurs

radioactive decay process, as the time required for the activity to decrease to half its value by that process.

See also BIOLOGICAL HALF-LIFE; GEOLOGICAL TIME.

half-reaction Refers to one-half state of the redox reaction, either the oxidation part (loss of electrons) or the reduction part (uptake of electrons).

Hall process The process invented in 1886 by Charles Martin Hall wherein he produced globules of aluminum metal by the electrolysis of aluminum oxide dissolved in a cryolite-aluminum fluoride mixture. Today, alumina is dissolved in an electrolytic bath of molten cryolite (sodium aluminum fluoride) within a large carbon- or graphite-lined steel container (a "pot"). An electric current is passed through the electrolyte at low voltage but very high current, typically 150,000 amperes, and the electric current flows between a carbon anode (positive), made of petroleum coke and pitch, and a cathode (negative), formed by the thick carbon or graphite lining of the pot. Molten aluminum is deposited at the bottom of the pot and is siphoned off.

halocarbon One of the various compounds of carbon combined with any of the halogens; e.g., fluorocarbon has had some of the hydrogen replaced by fluorine.

halochromism Halochromism refers to the color change that occurs upon addition of an ACID (or BASE or a salt) to a solution of a compound. A chemical reaction (e.g., ion formation) transforms a colorless compound into a colored one.

halogens Group VII elements are five nonmetallic elements: fluorine, chlorine, bromine, iodine, and astatine. Halogen means "salt-former," and their compounds are called salts. (They are labeled Group 17 in the newer labeling convention of the PERIODIC TABLE.)

haloperoxidase A PEROXIDASE that catalyzes the oxidative transformation of halides to XO^- (X being Cl, Br, or I) or to organic halogen compounds. Most are HEME proteins, but some bromoperoxidases from algae are vanadium-containing ENZYMEs.

Hammett acidity function *See* ACIDITY FUNCTION.

Hammett equation (Hammett relation) The equation in the form

$$\lg(k/k_o) = \rho\sigma$$

or

$$\lg(K/K_o) = \rho\sigma$$

applied to the influence of meta- or parasubstituents X on the reactivity of the functional group Y in the benzene derivative m- or p-XC_6H_4Y. K and k are the equilibrium and rate constant, respectively, for the given reaction of m- or p-XC_6H_4Y; k_o or K_o refers to the reaction of C_6H_5Y, i.e., X = H; σ is the substituent constant characteristic of m- or p-X; ρ is the reaction constant characteristic of the given reaction of Y. The equation is often encountered in a form with lg k_o or lg K_o written as a separate term on the right-hand side, e.g.,

$$\lg k = \rho\sigma + \lg k_o$$

or

$$\lg K = \rho\sigma + \lg K_o$$

It then signifies the intercept corresponding to X = H in a regression of lg k or lg K on σ.

See also RHO (ρ) VALUE; SIGMA (σ) CONSTANT; TAFT EQUATION; YUKAWA-TSUNO EQUATION.

Hammond principle (Hammond postulate) The hypothesis that, when a TRANSITION STATE leading to an unstable REACTION intermediate (or product) has nearly the same energy as that INTERMEDIATE, the two are interconverted with only a small reorganization of molecular structure. Essentially the same idea is sometimes referred to as "Leffler's assumption," namely, that the transition state bears the greater resemblance to the less stable species (reactant or reaction intermediate/product). Many textbooks and physical organic chemists, however, express the idea in Leffler's form but attribute it to Hammond.

As a corollary, it follows that a factor stabilizing a reaction intermediate will also stabilize the transition state leading to that intermediate.

The acronym *Bemahapothle* (*Be*ll, *Ma*rcus, *Ha*mmond, *Po*lanyi, *Th*ornton, *Le*ffler) is sometimes used in recognition of the principal contributors toward expansion of the original idea of the Hammond postulate.

Hansch analysis The investigation of the quantitative relationship between the biological activity of a series of compounds and their physicochemical substituent or global parameters representing hydrophobic, electronic, steric, and other effects using multiple-regression correlation methodology.

Hansch constant A measure of the capability of a solute for HYDROPHOBIC (LIPOPHILIC) interaction based on the partition coefficient P for distribution of the solute between octan-1-ol and water. The most general way of applying P in CORRELATION ANALYSIS, QSAR, etc., is as log P, but the behavior of substituted benzene derivatives may be quantified by a substituent constant scale, π, which is defined in a way analogous to the Hammett σ scale. There are various π scales, depending on the substrate series used as reference.

hapten A molecule (usually a small organic molecule) that can be bound to an ANTIGENIC determinant/epitope. Usually they are too small to give a response of their own. They become antigenic if they are coupled to a suitable macromolecule, such as a protein.

hapto The hapto symbol, η (Greek eta) with numerical superscript, provides a topological description for the bonding of hydrocarbons or other π-electron systems to metals, by indicating the connectivity between the LIGAND and the CENTRAL ATOM. The symbol is prefixed to the ligand name, or to that portion of the ligand name most appropriate. The right superscript numerical index indicates the number of COORDINATING atoms in the ligand that bind to the metal. Examples:

$[PtCl_2(C_2H_4)(NH_3)]$ amminedichloro(η^2-ethene)platinum
$[Fe(\eta^5\text{-}C_5H_5)_2]$ bis(η^5-cyclopentadienyl)iron (ferrocene)

hard acid A LEWIS ACID with an acceptor center of low polarizability. It preferentially associates with HARD BASEs rather than with soft bases, in a qualitative sense (sometimes called the HSAB [hard and soft acid and base] rule). Conversely, a soft acid possesses an acceptor center of high polarizability and exhibits the reverse preference for a partner for COORDINATION.

hard base A LEWIS BASE with a donor center of low polarizability; the converse applies to soft bases.

See also HARD ACID.

hard drug A nonmetabolizable compound, characterized either by high lipid solubility and accumulation in adipose tissues and organelles, or by high water solubility. In the lay press, the term *hard drug* refers to a powerful DRUG of abuse such as cocaine or heroin.

hard water A property of water that causes the formation of an insoluble residue when used with soap or scales in vessels in which water has been allowed to evaporate. Properties are due primarily to the presence of ions of calcium, magnesium, and iron.

HCP (hexagonal close packing) A type of crystal lattice structure found in zinc, titanium, and cobalt, for example.

heat Kinetic energy in the process of being transferred from one object to another due to a temperature difference. It moves in one of three ways: radiation, conduction, or convection.

heat capacity The amount of heat required to change a unit mass (or unit quantity, such as a mole) of a substance by one degree in temperature.

heat capacity of activation, $\Delta^{\ddagger}C_p^{o}$ (SI unit: J mol^{-1} K^{-1}) A quantity related to the temperature coefficient of $\Delta^{\ddagger}H$ (ENTHALPY OF ACTIVATION) and $\Delta^{\ddagger}S$ (ENTROPY OF ACTIVATION) according to the equations:

$$\Delta^{\ddagger}Cp = (\partial\Delta^{\ddagger} H/\partial T)_p = T(\partial\Delta^{\ddagger}S/\partial T)_p$$

If the rate constant is expressible in the form ln k = a/T + b + c ln T + dT, then

$$\Delta^{\ddagger}C_p = (c-1) R + 2dRT$$

heat of atomization The amount of energy required to dissociate 1 mole of a given substance into atoms.

heat of condensation Heat released when vapors change state to liquid. It is the amount of heat that must be removed from 1 gram of a vapor (specific heat of condensation) at its condensation point to condense the vapor with no temperature change. The molar heat of condensation refers to the quantity of heat released by a mole of vapor.

heat of crystallization The amount of heat that must be removed when 1 mole of a given substance crystallizes from a saturated solution of the same substance.

heat of fusion The amount of heat required to melt 1 gram of solid at its melting point with no change in temperature.

heat of reaction The amount of heat absorbed from the complete chemical reaction of molar amounts of the reactants.

heat of solution The amount of energy required or released when 1 mole of solute dissolves in a solvent. The value is positive if heat is absorbed (endothermic) and negative if heat is released (exothermic).

heat of sublimation The amount of energy required to convert 1 mole of a substance from the solid to the gas state (sublimation) without the appearance of the liquid state.

heat of vaporization The heat energy required to convert 1 gram of liquid to vapor without a change in temperature of the substance being vaporized.

heat-shock proteins (HSPs) A family of closely related proteins, widely distributed in virtually all organisms from plants, animals, microorganisms, and humans. Even though they are found in widely different sources, they show structural similarity. HSP expression increases in response to physiological stresses such as rise in temperature, pH changes, and oxygen deprivation. Many of these stresses can disrupt the three-dimensional structure, or folding, of a cell's proteins, and HSPs bind to those damaged proteins, helping them refold back into their proper shapes. They also help newly synthesized polypeptides fold and prevent premature interactions with other proteins. HSPs, also called chaperones, aid in the transport of proteins throughout the cell's various compartments and aid in the destruction of peptides specific to tumors or pathogens.

heavy water Water in which hydrogen atoms are replaced by deuterium, a heavy isotope of hydrogen.

Heisenberg uncertainty principle It states that it is not possible to determine accurately both the momentum and position of an electron simultaneously.

helium An inert gas. An ELEMENT with atomic number 2. Helium is produced in stars and is the second most abundant element in the universe. Its atom contains two protons, two neutrons, and two electrons, although there is another stable but very rare isotope with only 1 neutron.

helix A particular rigid left- or right-handed arrangement of a polymeric chain, characterized by the number of strands, the number (n) of units per turn, and its pitch (p), which is the distance the helix rises along its axis per full turn. Examples of single-stranded helices are the protein helices: α-helix: n = 3.6, p = 540 pm; ^{3}10-helix: n = 3.0, p = 600 pm; π-helix: n = 4.4, p = 520 pm.

See also DOUBLE HELIX.

heme A near-planar COORDINATION complex obtained from iron and the dianionic form of PORPHYRIN.

Derivatives are known with substitutes at various positions on the ring named a, b, c, d, etc. Heme b, derived from PROTOPORPHYRIN IX, is the most frequently occurring heme.

hemerythrin A dioxygen-carrying protein from marine invertebrates, containing an oxo-bridged DINUCLEAR iron center.

hemileptic *See* HOMOLEPTIC.

hemochromatosis A genetic condition of massive iron overload leading to cirrhosis or other tissue damage, attributable to iron.

hemocyanin A dioxygen-carrying protein (from invertebrates, e.g., arthropods and mollusks), containing dinuclear type 3 copper sites.

See also NUCLEARITY; TYPE 1, 2, 3 COPPER.

hemoglobin A dioxygen-carrying HEME protein of red blood cells, generally consisting of two alpha and two beta SUBUNITs, each containing one molecule of PROTOPORPHYRIN IX.

Henderson-Hasselbach equation An equation of the form

$$pH = pK_a - lg([HA]/[A^-])$$

for calculation of the pH of solutions where the ratio $[HA]/[A^-]$ is known.

Henry's law At equilibrium, the concentration of a gas dissolved in any solvent is proportional to its partial pressure.

herbicide A chemical pesticide that controls or destroys unwanted plants, weeds, or grasses.

hertz A unit for expressing frequency (f). One hertz equals one cycle per second.

Herzberg, Gerhard (1904–1999) German *Chemist* Gerhard Herzberg was born in Hamburg, Germany, on December 25, 1904. He received early schooling in Hamburg and then studied physics at the Darmstadt Institute of Technology, where he obtained a doctoral degree in 1928. From 1928 to 1930 he conducted postdoctorate work at the University of Göttingen and the University of Bristol. He was married in 1929 to Luise Herzberg (neé Oettinger) (widowed in 1971) and had two children. In 1930 he was appointed Privatdozent (lecturer) and senior assistant in the physics department of the Darmstadt Institute of Technology.

After leaving Germany in 1935, he became a guest lecturer at the University of Saskatchewan (Saskatoon, Canada) with financial support by the Carnegie Foundation, and he later became research professor of physics, staying there until 1945.

From 1945 to 1948 he served as a professor of spectroscopy at the Yerkes Observatory of the University of Chicago, but he returned to Canada in 1948 as principal research officer and then as director of the division of physics at the National Research Council. In 1955 he became director of the division of pure physics, a position he held until 1969, when he was appointed as distinguished research scientist in the recombined division of physics.

Herzberg contributed to the field of atomic and molecular spectroscopy, where he and his colleagues determined the structures of a large number of diatomic and polyatomic molecules, the structures of free radicals, and the identification of certain molecules in planetary atmospheres, in comets, and in interstellar space. In 1971 he was awarded the Nobel Prize in chemistry "for his contributions to the knowledge of electronic structure and geometry of molecules, particularly free radicals."

He served in many organizations and received several awards. He was vice president of the International Union of Pure and Applied Physics (1957–63), president of the Canadian Association of Physicists (1956–57), and president of the Royal Society of Canada (1966–67). He was elected a fellow of the Royal Society of Canada (1939) and Royal Society of London (1951). He was Bakerian Lecturer of the Royal Society of London (1960) and recipient of their Royal Medal in 1971. He was George Fischer Baker Non-Resident Lecturer in Chemistry at Cornell University (1968) and Faraday Medallist and Lecturer of the Chemical Society of London (1970).

He died on March 3, 1999, at age 94, at his home in Ottawa, Canada, and left a second wife, Monika, and son and daughter.

Hess's law of heat summation If a reaction goes through two or more steps, the enthalpy of reaction is the sum of the enthalpies of all the steps; the enthalpy change is the same regardless of the number of steps.

heterobimetallic complex A metal complex having two different metal atoms.

heteroconjugation (1) Association between a base and the conjugate acid of a different base through a HYDROGEN BOND ($B' \cdots HB^+$ or $A'H \cdots A^-$). The term has its origin in the CONJUGATE ACID-BASE PAIR and is in no way related to conjugation of ORBITALS. Heteroassociation is a more appropriate term.

(2) Some authors refer to CONJUGATED SYSTEMS containing a heteroatom, e.g., pyridine, as "heteroconjugated systems." This usage is discouraged, since it inappropriately suggests an analogy to HOMOCONJUGATION (2) and conflicts with the currently accepted definition of that term.

heterocyclic amine An amine in which nitrogen is part of a ring.

heterocyclic compound A cyclic organic compound containing one or more atoms other than carbon in its ring.

heterogeneous catalyst A catalyst that exists in a different phase (solid, liquid, or gas) from the reactants; a contact catalyst that furnishes a surface at which a reaction can occur.

heterogeneous equilibrium Equilibria involving species in more than one phase.

heterogeneous mixture A mixture that does not have uniform composition and properties throughout, where parts differ in composition or state of matter.

heteroleptic Transition-metal or main group compounds having more than one type of LIGAND.

See also HOMOLEPTIC.

heterolysis (heterolytic reaction) The cleavage of a COVALENT BOND so that both bonding electrons remain with one of the two fragments between which the bond is broken, e.g., $A{-}B \rightarrow A^+ + B^-$. Heterolytic bond fission is a feature of many BIMOLECULAR REACTIONs in solution (e.g., ELECTROPHILIC substitution and NUCLEOPHILIC substitution).

See also HETEROLYTIC BOND-DISSOCIATION ENERGY; HOMOLYSIS.

heterolytic bond-dissociation energy The energy required to break a given BOND of some specific compound by HETEROLYSIS. For the DISSOCIATION of a neutral molecule AB in the gas phase into A^+ and B^-, the heterolytic BOND-DISSOCIATION ENERGY $D(A^+B^-)$ is the sum of the bond-dissociation energy, D(A–B), and the adiabatic ionization energy of the radical $A^\cdot$ minus the electron affinity of the radical $B^\cdot$.

heteronuclear Consisting of different elements.

heteroreceptor A RECEPTOR regulating the synthesis or the release of mediators other than its own ligand.

See also AUTORECEPTOR.

heterotrophic organisms Organisms that are not able to synthesize cell components from carbon dioxide as a sole carbon source. Heterotrophic organisms use preformed oxidizable organic SUBSTRATEs, such as glucose, as carbon and energy sources, while energy is gained through chemical processes (chemoheterotrophy) or through light sources (photoheterotrophy).

heterovalent hyperconjugation *See* HYPERCONJUGATION.

Heyrovsky, Jaroslav (1890–1967) Czechoslovakian *Electrochemist* Jaroslav Heyrovsky was born in Prague on December 20, 1890, the son of Leopold Heyrovsky, professor of Roman law at the Czech University of Prague, and his wife Clara (née Hanl). He went to secondary school until 1909, when he began studying at the Czech University, Prague, in the fields of chemistry, physics, and mathematics. He continued his studies at University College, London, from 1910 to 1914, receiving a B.Sc. in 1913. He received a Ph.D. degree in Prague in 1918, and a D.Sc. in London in 1921 and started his university career as assistant to Professor B. Brauner in the Institute of Analytical Chemistry of the Charles University, Prague. He later was promoted to associate professor in 1922, and four years later became the first professor of physical chemistry at this university.

In 1922 he invented the polarographic method in electrochemistry (polarography is based on electrolysis using a dropping mercury electrode), and he continued development of this new branch of electrochemistry for decades. In 1926 he married Marie Koranová and had two children. In 1950 he was appointed director of the newly established Polarographic Institute (incorporated into the Czechoslovak Academy of Sciences in 1952). In 1959 he was awarded the Nobel Prize in chemistry "for his discovery and development of the polarographic methods of analysis."

He was a member of most of the leading scientific organizations and had honorary doctorates from many international universities. He died on March 27, 1967, in Prague, Czechoslovakia.

hidden return *See* ION-PAIR RETURN.

high-density polyethylene (HDPE) Polyethylene is a plastic polymer composed of carbon and hydrogen atoms with the formula $-(CH_2-CH_2)n-$ made by polymerizing ethylene. In the high-density form, the polymer chains are unbranched (linear), producing a material of improved strength.

highest occupied molecular orbital (HOMO) The highest-energy molecular orbital of an atom or molecule containing an electron. Most likely the first orbital, from which an atom will lose an electron.

high-performance liquid chromatography (HPLC) An analytical technique for the separation and determination of solutes in any sample (such as biological, pharmaceutical, environmental, etc.). During the process, a liquid (the eluant) is pumped (usually at high pressure) through a porous, solid, stationary phase, which separates the solute species, and then into a flow-through detector.

high spin *See* LOW SPIN.

high-temperature superconductor Four classes of superconductors that have been discovered since 1986 have much higher transition temperatures than previously known superconductors. These resistance-free conductors are made of ceramic materials that exhibit superconducting properties at temperatures from 20 to 130 K (–423 to –225°F) and require less-expensive cooling systems than those needed for low-temperature superconductors (<10 K, –441°F).

Hildebrand parameter A parameter measuring the cohesion of a solvent (energy required to create a cavity in the solvent).

Hill, Archibald Vivian (1886–1977) British *Physiologist* Archibald Vivian Hill was born in Bristol on September 26, 1886. After an early education at Blundell's School, Tiverton, he entered Trinity College, Cambridge, with scholarships. He studied mathematics but was urged to go into physiology by one of his teachers, Walter Morley Fletcher.

In 1909 he began to study the nature of muscular contraction and the dependence of heat production on the length of muscle fiber. During the years 1911–14 until the start of World War I, he continued his work on the physiology of muscular contraction at Cambridge as well as on other studies of nerve impulse, hemoglobin, and calorimetry.

In 1926 he was appointed the Royal Society's Foulerton Research Professor and was in charge of the Biophysics Laboratory at University College until 1952.

His work on muscle function, especially the observation and measurement of thermal changes associated with muscle function, was later extended to similar studies on the mechanism of the passage of nerve impulses. He coined the term *oxygen debt* after his own interests in recovery after exercise.

He discovered and measured heat production associated with nerve impulses and analyzed physical and chemical changes associated with nerve excitation, among other studies. In 1922 he won the Nobel Prize in physiology or medicine (with OTTO MEYERHOF) for work on chemical and mechanical events in muscle contraction, such as the production of heat in muscles. This research helped establish the origin of muscular force in the breakdown of carbohydrates while forming lactic acid in the muscle.

His important works include *Muscular Activity* (1926) and *Muscular Movement in Man* (1927), as well as *Living Machinery* (1927), *The Ethical Dilemma of Science and Other Writings* (1960), and *Traits and Trials in Physiology* (1965).

He was a member of several scientific societies and was elected a fellow of the Royal Society in 1918, serving as secretary for the period 1935–45 and as foreign secretary in 1946. Hill died on June 3, 1977.

HiPIP Formerly used abbreviation for high-potential IRON-SULFUR PROTEIN, now classed as a FERREDOXIN. An ELECTRON-TRANSFER PROTEIN from photosynthetic and other bacteria, containing a [4FE-4S] CLUSTER which undergoes oxidation-reduction between the $[4Fe\text{-}4S]^{2+}$ and $[4Fe\text{-}4S]^{3+}$ states.

See also PHOTOSYNTHESIS.

hirudin A nonenzymatic chemical secreted from the leech that prevents blood clotting. Now genetically engineered as lepirudin, desirudin, and a synthetic bivalirudin and used as anticoagulants.

histamine A hormone and chemical transmitter found in plant and animal tissues. In humans it is involved in local immune response that causes blood vessels to dilate during an inflammatory response; also regulates stomach acid production, dilates capillaries, and decreases blood pressure. It increases permeability of the walls of blood vessels by vasodilation when released from mast cells and causes the common symptoms of allergies such as running nose and watering eyes. It will also shut the airways in order to prevent allergens from entering, making it difficult to breathe. Antihistamines are used to counteract this reaction.

histone A basic unit of CHROMATIN structure. Several types of protein are characteristically associated with the DNA in chromosomes in the cell nucleus of EUKARYOTES. They function to coil DNA into nucleosomes that are a combination of eight histones (a pair each of H2A, H2B, H3, and H4) wrapped by two turns of a DNA molecule. A high number of positively charged amino acids bind to the negatively charged DNA.

Hodgkin, Dorothy Crowfoot (1910–1994) British *Crystallographer* Dorothy Crowfoot was born in Cairo on May 12, 1910, the daughter of John Winter Crowfoot from the Egyptian Education Service, and Grace Mary Crowfoot (née Hood), an archaeologist and botanist. Dorothy became interested in chemistry at age 10 but almost gave it up for a career in archaeology.

She attended Oxford and Somerville College from 1928 to 1932, combining archaeology and chemistry, but after attending a special course in crystallography, she turned her interests to X-ray crystallography.

In 1933 after a brief stint at Cambridge and Oxford, she returned to Somerville and Oxford in 1934 and remained there for most of her life teaching chemistry. In 1934 she crystallized and X-ray photographed insulin, only the second protein to be studied. She went on to map the molecular structure of penicillin (1947) and vitamin B12 (1956). In the late 1960s, she created a three-dimensional map of insulin.

In 1937, the same year she received a doctorate from Cambridge, she married Thomas Hodgkin, with whom she had three children.

In 1964 she was awarded the Nobel Prize in chemistry "for her determinations by X-ray techniques of the structures of important biochemical substances."

She was a founding member of Pugwash in 1957, the international organization of scientists who, during the Cold War, tried to further communication between scientists on both sides of the Iron Curtain.

Hodgkin was elected a fellow of the Royal Society in 1947 (in 1956 she received the Royal Medal) and a foreign member of the Royal Netherlands Academy of Sciences in 1956 and of the American Academy of Arts and Sciences (Boston) in 1958. She was the first woman since Florence Nightingale to become a member of the Order of Merit, the most prestigious of Britain's royal orders. Between 1977 and 1978 she was president of the British Association for the Advancement of Science (awarded the Longstaff Medal in 1978). In 1982 she received the Lomonosov Gold Medal because of her respect within the Soviet scientific community, and in 1987 she received the Lenin Peace Prize for her commitment to the Soviet cause and her efforts toward easing tensions between the East and the West while president of the Pugwash. She died at her home from a stroke on July 28, 1994, at age 84.

Hofmann rule "The principal alkene formed in the decomposition of quaternary ammonium hydroxides that contain different primary alkyl groups is always ethylene, if an ethyl group is present." Originally given in this limited form by A.W. Hofmann, the rule has since been extended and modified as follows: "When two or more alkenes can be produced in a β-elimination reaction, the alkene having the smallest number of alkyl groups attached to the double bond carbon atoms will be the predominant product." This orientation described by the Hofmann rule is observed in elimination reactions of quaternary ammonium salts and tertiary sulfonium salts, and in certain other cases.

holoenzyme An ENZYME containing its characteristic PROSTHETIC GROUP(s) or metal(s).

HOMO (1) An acronym for HIGHEST OCCUPIED MOLECULAR ORBITAL. (*See* FRONTIER ORBITALS.) (2) A prefix (consisting of lower case letters, "homo") used to indicate a higher homologue of a compound.

homoaromatic Whereas in an AROMATIC molecule there is continuous overlap of *P* ORBITALS over a cyclic array of atoms, in a homoaromatic molecule there is a formal discontinuity in this overlap resulting from the presence of a single sp^3 hybridized atom at one or several positions within the ring; *p*-orbital overlap apparently bridges these sp^3 centers, and features associated with aromaticity are manifest in the properties of the compound. Pronounced homoaromaticity is not normally associated with neutral molecules, but mainly with species bearing an electrical charge, e.g., the homotropylium cation, $C_8H_9^+$:

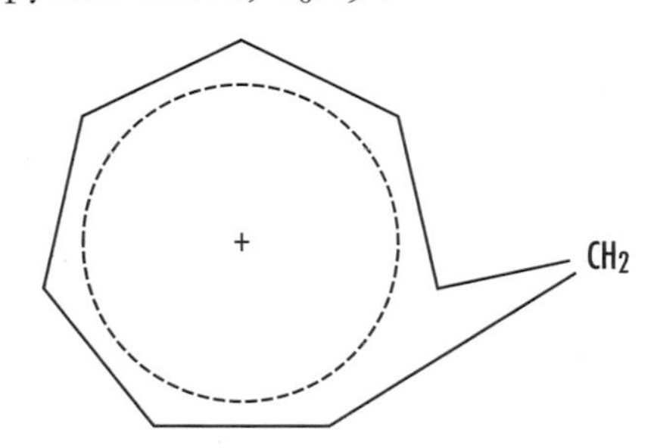

In bis, tris, etc. homoaromatic species, two, three, etc. single sp^3 centers separately interrupt the pi-electron system.

See also HOMOCONJUGATION (2).

homoconjugation (1) Association between a base and its CONJUGATE ACID through a HYDROGEN BOND ($B\cdots HB^+$ or $AH\cdots A^-$). *Homoassociation* is a more appropriate term for this phenomenon.

(2) The orbital overlap of two pi systems separated by a nonconjugating group, such as CH_2.

See also CONJUGATE ACID-BASE PAIR; CONJUGATED SYSTEM; HOMOAROMATIC.

homogeneous catalyst A catalyst existing in the same phase (solid, liquid, or gas) as the reactants.

homogeneous equilibrium Equilibria involving one species in a single phase, e.g., all gases, all liquids, or all solids; an equilibrium within a single phase.

See also HETEROGENEOUS EQUILIBRIUM.

homogeneous mixture Has uniform chemical composition, appearance, and properties throughout. Air is an example.

See also HETEROGENEOUS MIXTURE.

homoleptic Transition-metal or main group compounds having only one type of LIGAND are said to be homoleptic, e.g., $TaMe_5$.

See also HETEROLEPTIC.

homologous series A series of organic compounds in which each member differs from the next by a specific number and kind of atoms (structures differ by some regular increment) and shares the same basic formula, for example, the alkane series: methane (CH_4), ethane (C_2H_6), propane (C_3H_8), butane (C_4H_{10}), etc.

homologue Used to describe a compound belonging to a series of compounds differing from each other by a repeating unit, such as a methylene group, a peptide residue, etc.

homolysis The cleavage of a bond (homolytic cleavage or homolytic fission) so that each of the molecular fragments between which the bond is broken retains one of the bonding electrons.

homonuclear Consisting of only one element.

Hopkins, Frederick Gowland (1861–1947) British *Biochemist* Frederick Gowland Hopkins was born on June 20, 1861, at Eastbourne, England, to a bookseller in Bishopsgate Street, London, who died when Frederick was an infant.

In 1871 he attended the City of London School, and at the early age of 17, he published a paper in *The Entomologist* on the bombardier beetle. He went to University College, London, where he became the assistant to Sir Thomas Stevenson, an expert on poisoning. In 1888 he became a medical student at Guy's Hospital, London.

In 1894 he graduated with a degree in medicine and taught physiology and toxicology at Guy's Hospital for four years, and in 1898 he moved to Cambridge. He was appointed fellow and tutor at Emmanuel College, Cambridge.

Hopkins established biochemistry as a field in Great Britain. He discovered how to isolate the amino acid tryptophan and identified its structure, discovered enzymes, and isolated glutathione. For his research on discovering growth-stimulating vitamins, which he called "accessory substances," he was awarded the Nobel Prize in 1929 in medicine or physiology. He actually isolated vitamins C, A, and D.

Hopkins was knighted in 1925 and received the Order of Merit in 1935. Hopkins died in 1947, at the age of 86. The Sir Frederick Gowland Hopkins Memorial Lecture of the Biochemical Society, named in his honor, is presented by a lecturer to assess the impact of recent advances in his or her particular field on developments in biochemistry. The award is made every two or three years, and the lecturer is presented with a medal and £1,000.

hormone A substance produced by endocrine glands, released in very low concentration into the bloodstream, and which exerts regulatory effects on specific organs or tissues distant from the site of secretion.

host A MOLECULAR ENTITY that forms COMPLEXes with organic or inorganic guests, or a chemical species that can accommodate guests within cavities of its crystal structure. Examples include CRYPTANDS and CROWNS (where there are ion-dipole attractions between heteroatoms and positive ions), HYDROGEN-BONDED molecules that form clathrates (e.g., hydroquinone and water), and host molecules of INCLUSION COMPOUNDS (e.g., urea or thiourea). VAN DER WAALS FORCES and HYDROPHOBIC INTERACTIONS bind the guest to the host molecule in clathrates and inclusion compounds.

Hückel (4n + 2) rule Monocyclic planar (or almost planar) systems of trigonally (or sometimes digonally) hybridized atoms that contain $(4n + 2)$ π-electrons (where n is a nonnegative integer) will exhibit aromatic character. The rule is generally limited to $n = 0–5$.

This rule is derived from the Hückel MO (molecular orbital) calculation on planar monocyclic conjugated hydrocarbons $(CH)_m$ where m is an integer equal to or greater than 3, according to which $(4n + 2)$ π-electrons are contained in a closed-shell system. Examples of systems that obey the Hückel rule include:

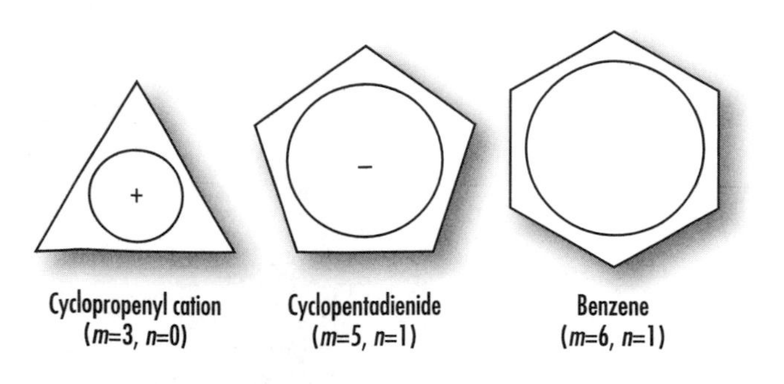

Systems containing $4n$ π-electrons (such as cyclobutadiene and the cyclopentadienyl cation) are antiaromatic.

See also MÖBIUS AROMATICITY.

humic acid An acid substance of indeterminate composition found in the organic materials in soils (humic matter) and which is insoluble in acid, methyl ethyl ketone, and methyl alcohol, but is soluble in alkali.

Hund's rule All orbitals of a given sublevel must be occupied by single electrons before doubling up begins.

hybridization Linear combination of ATOMIC ORBITALS on an atom. Hybrid orbitals are often used in organic chemistry to describe the bonding molecules containing tetrahedral (sp^3), trigonal (sp^2), and digonal (sp) atoms.

hydrate A solid substance that contains a specific percentage of bound water in it.

hydration Addition of water or the elements of water (i.e., H and OH) to a molecular entity. The term is also used in a more restricted sense for the process: A (gas) → A (aqueous solution). The term is also used in inorganic and physical chemistry to describe the state of ions of an aqueous electrolyte solution.

See also AQUATION; SOLVATION.

hydration energy Energy change that accompanies the hydration of a mole of gases or ions.

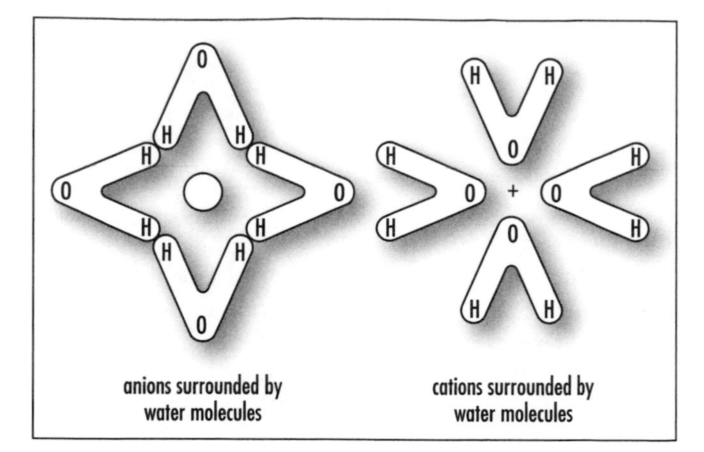

Hydration. Addition of water or the elements of water (i.e., H and OH) to a molecular entity

hydride A binary compound of hydrogen.

hydrocarbon Compound containing only carbon and hydrogen.

hydrogenase An ENZYME, dihydrogen:acceptor OXIDOREDUCTASE, that catalyzes the formation or oxidation of H_2. Hydrogenases are of various types. One class ([Fe]-hydrogenases) contains only IRON-SULFUR CLUSTERS. The other major class ([NiFe]-hydrogenases) has a nickel-containing center and iron-sulfur clusters; a variation of the latter type ([NiFeSe]-hydrogenases) contains selenocysteine.

hydrogenation A reaction where hydrogen is added across a double or triple bond, usually with the assistance of a catalyst.

hydrogen bond A form of ASSOCIATION between an electronegative atom and a hydrogen atom attached to a second, relatively electronegative atom. It is best considered to be an electrostatic interaction, heightened by the small size of hydrogen, which permits the proximity of the interacting dipoles or charges. Both electronegative atoms are usually (but not necessarily) from the first row of the periodic table (i.e., N, O, or F). Hydrogen bonds may be INTERMOLECULAR or INTRAMOLECU-

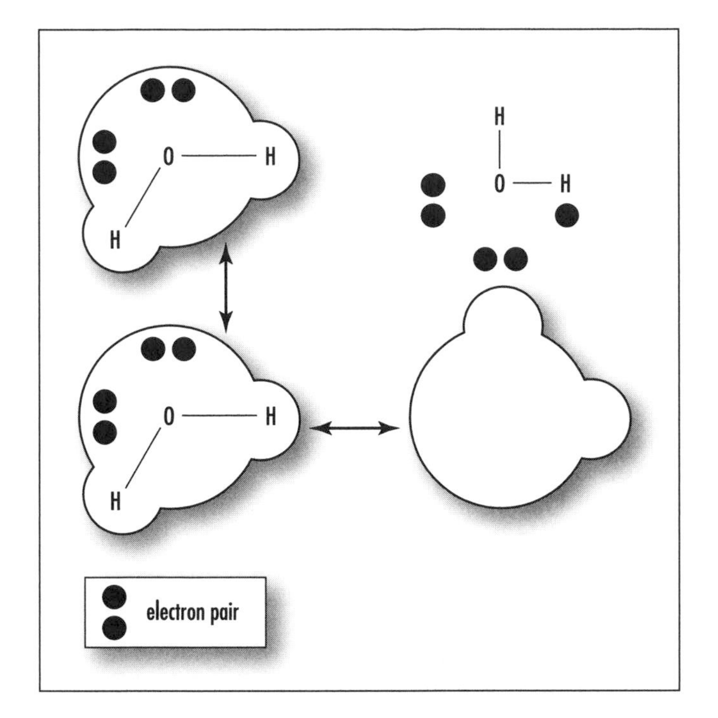

The hydrogen bond is a form of association between an electronegative atom and a hydrogen atom attached to a second, relatively electronegative atom.

LAR. With a few exceptions, usually involving fluorine, the associated energies are less than 20–25 kJ mol^{-1} (5–6 kcal mol^{-1}).

A type of bond formed when the partially positive hydrogen atom of a polar covalent bond in one molecule is attracted to the partially negative atom of a polar covalent bond in another.

hydrogen ion A single proton with a charge of +1. Also called a HYDRON.

hydrogen-oxygen fuel cell A cell in which hydrogen is the fuel (reducing agent) and oxygen is the oxidizing agent. Francis Thomas Bacon (1904–92), a British engineer, developed the first practical hydrogen-oxygen fuel cell.

hydrolase An ENZYME of EC Class 3, also known as a hydro-LYASE, that catalyzes the HYDROLYSIS of a SUBSTRATE.

hydrolysis SOLVOLYSIS by water.

hydrolysis constant An equilibrium constant for a hydrolysis reaction (reaction of a substance with water or its ions).

hydrometallurgy The extraction of metals and metallic compounds such as gold, alumina, nickel, copper, and zinc from minerals by solutions.

hydrometer A device used to measure the densities of liquids and solutions; determines specific gravity of the solution. Invented by the French scientist Antoine Baumé, although Leonardo da Vinci is also given credit.

hydron General name for the ion H^+ either in natural abundance, or where it is not desired to distinguish between the isotopes, as opposed to proton for $^1H^+$, deuteron for $^2H^+$, and triton for $^3H^+$. Not in general use.

hydronium ion A water molecule (H_2O) with an added hydrogen ion, making it H_3O^+.

hydrophilic Water loving. The capacity of a MOLECULAR ENTITY or of a SUBSTITUENT to interact with polar solvents, in particular with water, or with other polar groups. Hydrophilic molecules dissolve easily in water, but not in fats or oils.

hydrophilic colloids Macromolecules that interact strongly with water and form colloids.

hydrophilicity The tendency of a molecule to be solvated by water.

See also SOLVATION.

hydrophobic Fear of water. The tendency to repel water.

See also HYDROPHILIC.

hydrophobic interaction The tendency of hydrocarbons (or of lipophilic hydrocarbonlike groups in solutes) to form intermolecular aggregates in an aqueous medium, and analogous intramolecular interactions. The name arises from the attribution of the phenomenon to the apparent repulsion between water and hydrocarbons. Use of the misleading alternative term *hydrophobic bond* is discouraged.

hydrophobicity The association of nonpolar groups or molecules in an aqueous environment that arises from the tendency of water to exclude nonpolar molecules.

See also LIPOPHILICITY.

hydroxyl group A functional group that has a hydrogen atom joined to an oxygen atom by a polar covalent bond (–OH).

hydroxyl ion One atom each of oxygen and hydrogen bonded into an ion (OH^-) that carries a negative charge.

See also ION.

hydroxyl radical A radical consisting of one hydrogen atom and one oxygen atom. It normally does not exist in a stable form.

hyperconjugation In the formalism that separates bonds into σ- and π-types, hyperconjugation is the interaction of σ-bonds (e.g. C–H, C–C, etc.) with a π-network. This interaction is customarily illustrated by CONTRIBUTING STRUCTURES, e.g., for toluene (below), sometimes said to be an example of heterovalent or sacrificial hyperconjugation, so named because the contributing structure contains one two-electron bond less than the normal LEWIS FORMULA for toluene:

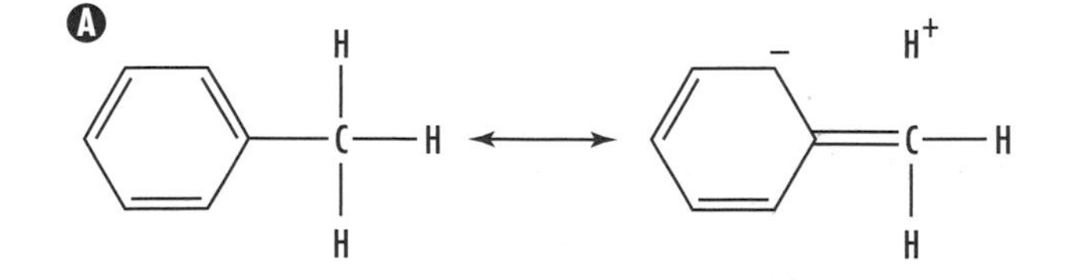

At present, there is no evidence for sacrificial hyperconjugation in neutral hydrocarbons. The concept of hyperconjugation is also applied to CARBENIUM IONS and RADICALS, where the interaction is now between σ-bonds and an unfilled or partially filled π or *p*-ORBITAL. A contributing structure illustrating this for the tert-butyl cation is:

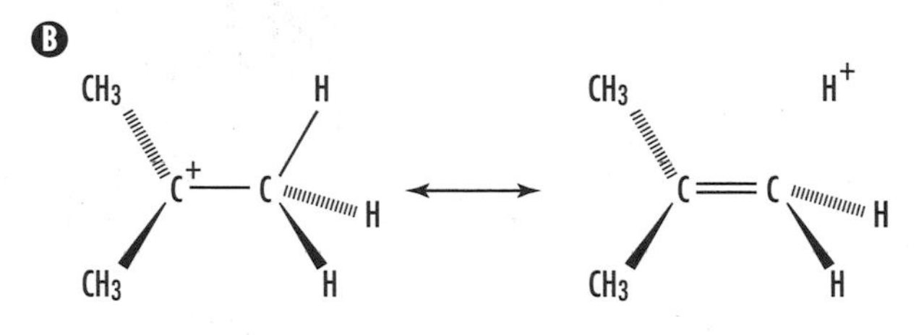

This latter example is sometimes called an example of isovalent hyperconjugation (the contributing structure containing the same number of two-electron bonds as the normal Lewis formula).

Both structures on the right-hand side are also examples of double-bond no-bond resonance.

The interaction between filled π or *p* orbitals and adjacent antibonding σ* orbitals is referred to as negative hyperconjugation, as for example in the fluoroethyl anion:

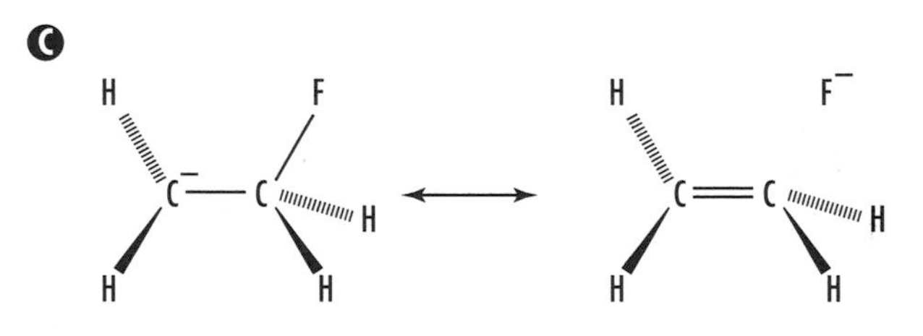

See also PI (π) BOND; SIGMA (σ) BOND; SIGMA PI; N-σ DELOCALIZATION.

hyperfine *See* ELECTRON PARAMAGNETIC RESONANCE SPECTROSCOPY.

hyperpolarization An electrical state where the inside of a cell is made more negative relative to the outside than was the case at resting potential of about –70mV.

hypertonic solution A solution whose solute concentration is high enough to cause water to move out of cells via osmosis.

hypoosmotic solution A solution whose osmotic pressure is less than that of another solution.

hypothesis The formal declaration of the possible explanation of a set of observations that needs to be tested and proved.

hypotonic solution A solution where solute concentration is low enough to cause water to move into cells via osmosis.

hypsochromic shift Shift of a spectral band to higher frequency or shorter wavelength upon substitution or change in medium. It is informally referred to as "blue shift."

See also BATHOCHROMIC SHIFT.

I

ideal gas A hypothetical gas that appears to obey perfectly the ideal gas law as the pressure nears zero; a gas characterized by a complete absence of cohesive forces between molecules; no gas has met this requirement.

ideal gas law A gas that conforms in its physical behavior to an idealized relation between pressure, volume, and temperature.

ideal solution A solution where each component behaves independently of other components; solvent-solvent and solvent-solute interactions are the same.

identity reaction A CHEMICAL REACTION whose products are chemically identical with the reactants, e.g., the bimolecular self exchange reaction of CH_3I with I^-.

See also DEGENERATE REARRANGEMENT.

igneous rock Formed by the cooling and crystallization of molten rock. The term is derived from *ignius,* the Latin word for fire.

imaging A medical diagnostic technique by which useful organ images are obtained from the radiation emitted by RADIONUCLIDEs that are introduced into organs, or from radiation absorbed by atomic nuclei within the organs. Typical examples are imaging obtained by recording the radiation emitted by a radionuclide such as ^{99m}Tc and the ^{1}H-NMR imaging obtained by whole-body NUCLEAR MAGNETIC RESONANCE measurements.

See also BONE IMAGING; BRAIN IMAGING; MAGNETIC RESONANCE IMAGING.

imbalance The situation in which REACTION parameters that characterize different bond-forming or bond-breaking processes in the same reaction have developed to different extents as the TRANSITION STATE is approached along some arbitrarily defined reaction coordinate. For example, in the nitroalkane anomaly, the Bronsted$^{\beta}$ exponent for proton removal is smaller than the Bronsted$^{\alpha}$ for the nitroalkane because of imbalance between the amount of bond breaking and resonance delocalization in the transition state. Imbalance is common in elimination, addition, and other complex reactions that involve proton (hydron) transfer.

See also SYNCHRONIZATION; SYNCHRONOUS.

imene *See* NITRENE.

imidogen *See* NITRENE.

imidonium ion *See* NITRENIUM ION.

imin *See* NITRENE.

imine radical *See* NITRENE.

immiscibility When two or more substances or liquids do not dissolve into one another, e.g., oil and water.

immune response The process by the body of an organism to recognize and fight invasion of microorganisms, viruses, and other substances (antigens) that may be harmful to the body; the total time from recognition of the intrusion to attack or tolerance of the antigen.

immunoglobulin (**Ig**) Also known as antibodies, these are proteins created by plasma cells and B cells that are designed to control the body's immune response by binding to antigens. There are more than 1,000 possible ANTIBODY variations and five major types, and each is specific to a particular antigen. Of the five main types, IgA, IgD, IgE, IgG, and IgM, the most common are IgA, IgG, and IgM.

immunogold A method for visualizing proteins in electron microscopy within a cell using gold particles attached to an ANTIBODY that binds specifically to that protein.

inclusion compound (**inclusion complex**) A COMPLEX in which one component (the HOST) forms a cavity or, in the case of a crystal, a crystal lattice containing spaces in the shape of long tunnels or channels in which molecular entities of a second CHEMICAL SPECIES (the GUEST) are located. There is no covalent bonding between guest and host, the attraction being generally due to VAN DER WAALS FORCES. If the spaces in the host lattice are enclosed on all sides so that the guest species is "trapped" as in a CAGE, such compounds are known as clathrates or cage compounds.

IND (**investigational new drug**) Regulations define IND as a new drug, antibiotic drug, or biological drug that is used in clinical investigation. The term also includes biological products used in vitro for diagnostic purposes. Required when a new indication or marketing/labeling change of an approved drug is being sought.

indicators Compounds that exhibit different colors in solutions of different acidities or bases. Compounds whose color depends upon the pH; typically, they change color over a narrow pH range and are useful in

Demonstration of an acid-base titration that uses the indicator phenolphthalein. An alkaline solution is in the flask, with a few drops of the indicator added to give a purple color. Acid is slowly added from the burette above. When the acid has completely neutralized the alkali in the solution, the indicator turns colorless. This technique is used to gather data for many investigations, such as determining the concentration of the acid or alkali. *(Courtesy of Jerry Mason/Science Photo Library)*

determining the pH of a solution (indicator paper, mixed indicators) and in analytical chemistry in finding the end point of an acid-base titration. Redox indicators similarly have colors determined by the potential of a solution. Complexometric indicators are used to find the end points in titrations of metal ions by complexing agents.

induced dipole moment Polar molecules have bonds that have positive and negative ends, or poles, and the molecules are often said to have a DIPOLE MOMENT, a force of attraction between oppositely charged particles. When an atom comes close to a polar molecule, the electrons can shift to one side of the nucleus to produce a very small dipole moment that lasts for only an instant.

induction period The initial slow phase of a CHEMICAL REACTION that later accelerates. Induction periods are often observed with radical reactions, but they may also occur in other systems (for example, before steady-state concentration of the reactants is reached).

inductive effect In strict definition, an experimentally observable effect (on rates of reaction, etc.) of the transmission of charge through a chain of atoms by electrostatic induction. A theoretical distinction may be made between the FIELD EFFECT and the inductive effect as models for the Coulomb interaction between a given site within a MOLECULAR ENTITY and a remote unipole or dipole within the same entity. The experimental distinction between the two effects has proved difficult, except for molecules of peculiar geometry, which may exhibit "reversed field effects." Ordinarily, the inductive effect and the field effect are influenced in the same direction by structural changes in the molecule, and the distinction between them is not clear. This situation has led many authors to include the field effect in the term *inductive effect.* Thus the separation of σ-values into inductive and RESONANCE components does not imply the exclusive operation of a through-bonds route for the transmission of the nonconjugative part of the substituent effect. To indicate the all-inclusive use of the term *inductive,* the phrase "so-called inductive effect" is sometimes used. Certain modern theoretical approaches suggest that the so-called inductive effect reflects a field effect rather than through-bonds transmission.

See also FIELD EFFECT; MESOMERIC EFFECT; POLAR EFFECT.

inductomeric effect A molecular polarizability effect occurring by the inductive mechanism of electron displacement. The consideration of such an effect and the descriptive term have been regarded as obsolescent or even obsolete, but in recent years theoretical approaches have reintroduced substituent polarizability as a factor governing reactivity, etc., and its parameterization has been proposed.

inert STABLE and unreactive under specified conditions.

See also LABILE.

inert *s*-pair effect A tendency of the outermost *s* electrons to remain nonionized or unshared in compounds, characteristic of the post-transition minerals.

infrared radiation Long-wavelength radiation that can be felt as heat but not seen.

infrared spectroscopy Infrared (IR) spectroscopy is a type of vibrational spectroscopy. It is the study of the interaction of substances with infrared electromagnetic radiation. IR spectroscopy can be used to determine the concentration of the sample under study or to study the spectral characteristics of the sample. The frequencies of the vibrations excited by the infrared radiation are characteristic of the functional groups in a molecule and so provide information on molecular structure and identification.

inhibition The decrease in RATE OF REACTION brought about by the addition of a substance (inhibitor) by virtue of its effect on the concentration of a reactant, CATALYST, or reaction INTERMEDIATE. For example, molecular oxygen and p-benzoquinone can react as inhibitors in many reactions involving

RADICALs as intermediates by virtue of their ability to act as SCAVENGERs toward these radicals.

If the rate of a reaction in the absence of inhibitor is v_o and that in the presence of a certain amount of inhibitor is v, the degree of inhibition *(i)* is given by

$$i = (v_o - v)/v_o$$

See also MECHANISM-BASED INHIBITION.

inhibitor A substance that decreases the rate of catalysis by an ENZYME or of some other chemical reaction.

inhibitory catalyst One that decreases the rate of reaction.

inhibitory postsynaptic potential (IPSP) A small electrical charge of a few millivolts creating a local hyperpolarization (increase in membrane potential on the negativity of the inside of the neuron) in the membrane of a postsynaptic neuron. Occurs when an inhibitory neurotransmitter from a presynaptic cell binds to a postsynaptic receptor; makes it difficult for a postsynaptic neuron to generate an action potential.

initiation A reaction or process generating free RADICALs (or some other REACTIVE reaction intermediates), which then induce a CHAIN REACTION. For example, in the chlorination of alkanes by a radical mechanism, the initiation step is the DISSOCIATION of molecular chlorine.

inner-sphere electron transfer Historically, an electron transfer between two metal centers sharing a ligand or atom in their respective coordination shells. The definition has more recently been extended to any situation in which the interaction between the donor and acceptor centers in the TRANSITION STATE is significant (>20 kJ mol^{-1}).

See also OUTER-SPHERE ELECTRON TRANSFER.

inorganic chemistry The study of inorganic compounds and their structure, reactions, catalysis, and mechanism of action.

insertion A CHEMICAL REACTION or TRANSFORMATION of the general type

$$X\text{–}Z + Y \rightarrow X\text{–}Y\text{–}Z$$

in which the connecting atom or GROUP Y replaces the bond joining the parts X and Z of the reactant XZ. An example is the CARBENE insertion reaction

$$R_3C\text{–}H + H_2C: \rightarrow R_3C\text{–}CH_3$$

The reverse of an insertion is called an EXTRUSION.

See also ALPHA (α) ADDITION.

insertion reaction A chemical reaction or transformation of the general type X–Z + Y → X–Y–Z in which the connecting atom or group Y replaces the bond joining the parts X and Z of the reactant XZ.

insoluble compound One that does not dissolve to a significant extent in the specified solvent.

insulator A material that poorly conducts heat or electricity.

insulin A protein hormone produced in the pancreas by beta cells, located in the islets of Langerhans, that stimulates cellular utilization of glucose by body cells by converting glucose and other carbohydrates to energy, and that helps control blood sugar levels by acting antagonistically with glucagons, the chief source of stored fuel in the liver. It is released by various signals that are sensitive to the intake and digestion of food. It also acts as an important regulator of protein and lipid metabolism. Insulin is used as a drug to control insulin-dependent diabetes mellitus, a disorder that is caused by the insufficient production of insulin. Without insulin, cells do not absorb glucose. Diabetic individuals may have type I diabetes (juvenile), comprising about 10 percent of the population, or type II diabetes (adult). Presently some 16 million Americans have diabetes, with 1,700 new cases being diagnosed daily. Diabetes has been linked to the development of a variety of diseases, including heart disease, stroke, peripheral vascular disease, and neurological disorders.

integrated rate laws Integrated rate laws are used to relate concentrations and time. The rate law provides an equation that relates the reaction rate to the concentration of a species in the reaction. The integrated rate law has a very different form for first-, second-, and zero-order reactions.

See also RATE OF REACTION.

inter- A prefix meaning between or among.

intercalation compounds Compounds resulting from inclusion, usually without covalent bonding, of one kind of molecule (the GUEST molecule) in a matrix of another compound (the HOST compound), which has a layered structure. The host compound, with a rather rigid structure, may be macromolecular, crystalline, or amorphous.

interferon A chemical messenger of the immune system, composed of a group of cytokine proteins that have antiviral characteristics that are capable of helping the immune response. Three main types of interferon—alpha, beta, and gamma—are produced by virus-infected cells and are released to coat uninfected cells, thus preventing them from becoming infected. Alpha interferon is produced by virus-infected monocytes and lymphocytes, while beta is produced by virus-infected fibroblasts. Gamma is produced by stimulated T and NK (natural killer) cells.

intermediate A MOLECULAR ENTITY with a LIFETIME appreciably longer than a molecular vibration (corresponding to a local potential energy minimum of depth greater than reagent type) that is formed (directly or indirectly) from the reactants and reacts further to give (either directly or indirectly) the products of a chemical reaction as well as the corresponding CHEMICAL SPECIES.

See also ELEMENTARY REACTION; REACTION STEP; STEPWISE REACTION.

intermolecular (1) Descriptive of any process that involves a transfer (of atoms, GROUPs, electrons, etc.) or interactions between two or more MOLECULAR ENTITIES.

(2) Relating to a comparison between different molecular entities.

See also INTRAMOLECULAR.

intermolecular forces Forces of attraction that exist between particles (atoms, molecules, ions) in a compound.

internal conversion Described as an alternative process to gamma-ray emission that frequently occurs in metastable NUCLIDEs. An excited nucleus transfers its excitation energy to an orbital electron (called the conversion electron), which is then ejected from the atom. The excess energy of the electron-binding energy travels with the electron as kinetic energy. The orbital vacancy is filled by another shell electron, giving rise to the emission of characteristic X-rays or Auger electrons.

internal energy Energy that is associated with the random, disordered motion of molecules. The internal energy of a system is defined as the sum of the energies of all the constituent particles.

internal return *See* ION-PAIR RETURN.

interstitial One of the three classes of HYDRIDEs: covalent, interstitial, and ionic. A hydride is a binary compound containing hydrogen. The hydride ion, H-, occurs in ionic hydrides. More generally, applied to solid solutions in which the (small) solute atoms occupy interstitial positions between the (larger) atoms of the solvent. Interstitial positions are the spaces that are empty when spherical particles such as atoms are packed together. Octahedral and tetrahedral holes are examples on interstitial sites in close packed lattices.

intimate ion pair *See* ION PAIR.

intra- A prefix meaning within or inside.

intramolecular (1) Descriptive of any process that involves a transfer (of atoms, GROUPS, electrons, etc.) or interactions between different parts of the same MOLECULAR ENTITY.

(2) Relating to a comparison between atoms or groups within the same molecular entity.

See also INTERMOLECULAR.

intramolecular catalysis The acceleration of a chemical transformation at one site of a MOLECULAR ENTITY through the involvement of another FUNCTIONAL ("catalytic") GROUP in the same molecular entity, without that group appearing to have undergone change in the reaction product. The use of the term should be restricted to cases for which analogous INTERMOLECULAR CATALYSIS by CHEMICAL SPECIES bearing that catalytic group is observable. Intramolecular catalysis can be detected and expressed in quantitative form by a comparison of the reaction rate with that of a comparable model compound in which the catalytic group is absent, or by measurement of the EFFECTIVE MOLARITY of the catalytic group.

See also CATALYSIS; NEIGHBORING-GROUP PARTICIPATION.

intrinsic activity The maximal stimulatory response induced by a compound in relation to that of a given reference compound.

This term has evolved with common usage. It was introduced by Ariëns as a proportionality factor between tissue response and RECEPTOR occupancy. The numerical value of intrinsic activity (alpha) could range from unity (for full AGONISTs, i.e., agonists inducing the tissue maximal response) to zero (for ANTAGONISTs), the fractional values within this range denoting PARTIAL AGONISTs. Ariëns's original definition equates the molecular nature of alpha to maximal response only when response is a linear function of receptor occupancy. This function has been verified. Thus, intrinsic activity, which is a DRUG and tissue parameter, cannot be used as a characteristic drug parameter for classification of drugs or drug receptors. For this purpose, a proportionality factor derived by null methods, namely, relative EFFICACY, should be used.

Finally, the term *intrinsic activity* should not be used instead of *intrinsic efficacy.* A "partial agonist" should be described as "agonist with intermediate intrinsic efficacy" in a given tissue.

See also PARTIAL AGONIST.

intrinsic barrier The GIBBS ENERGY OF ACTIVATION ($\Delta^{\ddagger}G$) in the limiting case where $\Delta G^{o} = 0$, i.e., when the effect of thermodynamic driving force is eliminated. According to the MARCUS EQUATION, the intrinsic barrier is related to the REORGANIZATION ENERGY, λ, of the reaction by the equation

$$\Delta^{\ddagger}G = \lambda/4$$

intron An intervening section of DNA that occurs almost exclusively within a eukaryotic GENE, but which is not translated to amino-acid SEQUENCEs in the gene product. The introns are removed from the pre-mature mRNA through a process called splicing, which leaves the EXONs untouched, to form an active mRNA.

See also EUKARYOTE.

inverse agonist A DRUG that acts at the same RECEPTOR as that of an AGONIST, yet produces an opposite effect. Also called negative ANTAGONIST.

inverse kinetic isotope effect *See* ISOTOPE EFFECT.

inverted micelle The reversible formation of association colloids from surfactants in nonpolar solvents leads to aggregates termed inverted (or inverse, reverse, or reversed) MICELLEs. Such association is often of the type

$$\text{Monomer} \rightleftharpoons \text{Dimer} \rightleftharpoons \text{Trimer} \rightleftharpoons \dots n\text{-mer}$$

and the phenomenon of CRITICAL MICELLE CONCENTRATION (or an analogous effect) is consequently not observed.

In an inverted micelle, the polar GROUPs of the surfactants are concentrated in the interior, and the LIPOPHILIC groups extend toward and into the nonpolar solvent.

ion An atom or group of atoms that acquires a charge by either gaining or losing one or more electrons.

ion channel Enables ions to flow rapidly through membranes in a thermodynamically downhill direction after an electrical or chemical impulse. Their structures usually consist of four to six membrane-spanning DOMAINS. This number determines the size of the pore and thus the size of the ion to be transported.

See also ION PUMP.

ion exchange A reversible process where ions are released from an insoluble permanent material in exchange for other ions in a surrounding solution. The direction of the exchange depends on (a) the affinities of the ion exchanger for the ions present and (b) the concentrations of the ions in the solution. In water treatment, this process adds and removes ions from water. In nuclear chemistry, it is a common method for concentrating uranium from a solution. The uranium solution is passed through a resin bed, where the uranium-carbonate complex ions are transferred to the resin by exchange with a negative ion like chloride. After buildup of the uranium complex on the resin, the uranium is eluted with a salt solution, and the uranium is precipitated in another process.

ionic bond A chemical bond or link between two atoms due to an attraction between oppositely charged (positive-negative) ions.

ionic bonding Chemical bonding that results when one or more electrons from one atom or a group of atoms is transferred to another. Ionic bonding occurs between charged particles.

ionic compounds Compounds where two or more ions are held next to each other by electrical attraction.

ionic radius The radius of an ion.

ionic strength I (SI unit: mol dm^{-3}) In a solution of fully dissociated electrolytes, the ionic strength is defined as $I = 0.5\ \Sigma_i c_i Z_i^2$, in which c_i is the concentration and Z_i the charge number of ionic species i. μ is also defined as $I_m = 0.5\ \Sigma_i m_i Z_i^2$, where m_i is the molality.

ionization The generation of one or more ions. It can occur by loss of an electron from a neutral MOLECULAR ENTITY, by the UNIMOLECULAR heterolysis of such an entity into two or more ions, or by a heterolytic SUBSTITUTION REACTION involving neutral molecules, such as

$$CH_3CO_2H + H_2O \rightarrow H_3O^+ + CH_3CO_2^-$$

$$Ph_3CCl + AlCl_3 \rightarrow Ph_3C^+ + AlCl_4^-\ \text{(electrophile assisted)}$$

$$Ph_3CCl \rightarrow Ph_3C^+\ Cl^-\ \text{(ion pair, in benzene)}$$

The loss of an electron from a singly, doubly, etc., charged cation is called second, third, etc., ionization. This terminology is used especially in mass spectroscopy.

See also DISSOCIATION; HETEROLYSIS; IONIZATION ENERGY.

ionization constant The equilibrium constant for the ionization of a weak electrolyte. An example is 1.75×10^{-5} for acetic acid (CH_3COOH).

ionization energy E_i (SI unit: kJ mol^{-1} or J per molecule) The minimum energy required to remove an electron from an isolated MOLECULAR ENTITY (in its vibrational GROUND STATE) in the gaseous phase. If the resulting molecular entity is considered to be in its vibrational ground state, one refers to the energy as the "adiabatic ionization energy." If the molecular entity produced possesses the vibrational energy determined by the Franck-Condon principle (according to which the electron ejection takes place without an accompanying change in molecular geometry), the energy is called the "vertical ionization energy." The name "ionization energy" is preferred to the somewhat misleading earlier name "ionization potential."

See also IONIZATION.

ionizing power A term to denote the tendency of a particular solvent to promote IONIZATION of an uncharged or, less often, charged solute. The term has been used both in a kinetic and in a thermodynamic context.

See also DIMROTH-REICHARDT E_T PARAMETER; GRUNWALD-WINSTEIN EQUATION; Z-VALUE.

ionizing radiation Radiation capable of producing ions or charged particles. Ionizing radiation includes alpha, beta, gamma, and X rays. Nonionizing radiation is radiation without enough energy to remove tightly bound electrons from their orbits around atoms (e.g., microwaves and visible light).

ionophore A compound that can carry specific ions through membranes of cells or organelles.

ion pair (ionic bond) A pair of oppositely charged ions held together by Coulomb attraction without formation of a COVALENT BOND. Experimentally, an ion pair behaves as one unit in determining conductivity, kinetic behavior, osmotic properties, etc.

Following Bjerrum, oppositely charged ions with their centers closer together than a distance

$$q = 8.36 \times 10^6 Z^+Z^-/(\varepsilon_r T) \text{ pm}$$

are considered to constitute an ion pair (Bjerrum ion pair). [Z^+ and Z^- are the charge numbers of the ions, and ε_r is the relative permittivity (or dielectric constant) of the medium.]

An ion pair, the constituent ions of which are in direct contact (and not separated by an intervening solvent or other neutral molecule), is designated as a "tight ion pair" (or "intimate" or "contact ion pair"). A tight ion pair of X^+ and Y^- is symbolically represented as X^+Y^-.

In contrast, an ion pair whose constituent ions are separated by one or several solvent or other neutral molecules is described as a "loose ion pair," symbolically represented as $X^+||Y^-$. The members of a loose ion pair can readily interchange with other free or loosely paired ions in the solution. This interchange may be detectable (e.g., by isotopic labeling), thus affording an experimental distinction between tight and loose ion pairs.

A further conceptual distinction has sometimes been made between two types of loose ion pairs. In "solvent-shared ion pairs," the ionic constituents of the pair are separated by only a single solvent molecule, whereas in "solvent-separated ion pairs," more than one solvent molecule intervenes. However, the term *solvent-separated ion pair* must be used and interpreted with care, since it has also been widely used as a less-specific term for loose ion pair.

See also COMMON-ION EFFECT; DISSOCIATION; ION-PAIR RETURN; SPECIAL SALT EFFECT.

ion-pair return The recombination of a pair of ions, R^+ and Z^-, formed from ionization of RZ. If the ions are paired as a tight ION PAIR and recombine without prior separation into a loose ion pair, this is called "internal ion-pair return":

$$R^+Z^- \text{ (tight ion pair)} \rightleftharpoons RZ \text{ (covalent molecule)}$$

It is a special case of primary GEMINATE RECOMBINATION.

If the ions are paired as a loose ion pair and form the covalent chemical species via a tight ion pair, this is called external ion-pair return:

$$R^+||Z^- \text{ (loose ion pair)} \rightleftharpoons R^+Z^- \text{ (tight ion pair)} \rightleftharpoons RZ \text{ (covalent molecule)}$$

It is a special case of secondary geminate recombination.

When the covalent molecule RZ is re-formed without direct evidence of prior partial racemization or without other direct evidence of prior formation of a tight ion pair (e.g., without partial racemization if the group R is suitably chiral), the internal ion-pair return is sometimes called a hidden return.

External (unimolecular) ion-pair return should be distinguished from external (bimolecular) ion return, the (reversible) process whereby dissociated ions are converted into loose ion pairs:

$$R^+ + Z^- \rightleftharpoons R^+||Z^-$$

ion pump Enables ions to flow through membranes in a thermodynamically uphill direction by the use of an energy source such as ATP or light. An ion pump consists of sugar-containing heteropeptide assemblies that open and close upon the binding and subsequent HYDROLYSIS of ATP, usually transporting more than one ion toward the outside or the inside of the membrane.

See also ION CHANNEL.

ion-selective electrode An electrode, the potential of which depends on the concentration of a specific ion with which it is in contact. An example is the glass electrode that is used to measure hydrogen ion concentrations (pH). An ion-selective electrode is used in a wide variety of applications for determining the concentrations of various ions in aqueous solutions.

ipso attack The attachment of an entering group to a position in an aromatic compound already carrying a SUBSTITUENT group (other than hydrogen). The entering group may displace that substituent group but may also itself be expelled or migrate to a different position in a subsequent step. Ipso is a position on a phenyl ring. The term *ipso substitution* is not used, since it is synonymous with substitution.

For example:

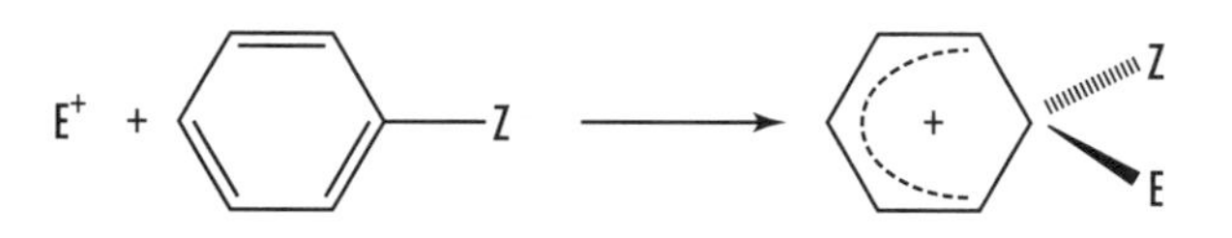

where E^+ is an ELECTROPHILE and Z is a substituent (other than hydrogen).

See also CINE-SUBSTITUTION; TELE-SUBSTITUTION.

iron-responsive element A specific base SEQUENCE in certain MESSENGER RNAs that code for various proteins of iron METABOLISM, which allows REGULATION at translational level by the IRON-RESPONSIVE PROTEIN.

iron-responsive protein (IRP) A protein that responds to the level of iron in the cell and regulates the biosynthesis of proteins of iron METABOLISM by binding to the IRON-RESPONSIVE ELEMENT on MESSENGER RNA.

See also REGULATION.

iron-sulfur cluster A unit comprising two or more iron atoms and BRIDGING sulfide LIGANDs in an IRON-SULFUR PROTEIN. The recommended designation of a CLUSTER consists of the iron and sulfide content in square brackets, for example [2Fe-2S], [3Fe-4S]. The possible oxidation levels are indicated by the net charge excluding the ligands, for example a $[4Fe\text{-}4S]^{2+}$; $[4Fe\text{-}4S]^{1+}$ (or $[4Fe\text{-}4S]^{2+;1+}$) cluster.

iron-sulfur protein Proteins in which non-HEME iron is coordinated with cysteine sulfur and, usually, with inorganic sulfur. Divided into three major categories: RUBREDOXINs; simple iron-sulfur proteins, containing only IRON-SULFUR CLUSTERs; and complex iron-sulfur proteins, containing additional active redox centers such as FLAVIN, molybdenum, or heme. In most iron-sulfur proteins, the clusters function as ELECTRON TRANSFER groups, but in others they have other functions such as catalysis of hydratase/dehydratase reactions, maintenance of protein structure, or REGULATION of activity.

See also COORDINATION.

IRP *See* IRON-RESPONSIVE PROTEIN.

ischemia Local deficiency of blood supply and dioxygen to an organ or tissue owing to constriction of the blood vessels or to an obstruction.

isobacteriochlorin (2,3,7,8-Tetrahydroporphyrin) A reduced PORPHYRIN with two pairs of confused saturated carbon atoms (C-2, C-3 and C-7, C-8) in two of the pyrrole rings.

See also CHLOROPHYLL.

isochore A line or surface of constant volume on a graphical representation of a physical system. A contour line that corresponds to values measured at identical volumes.

isodesmic reaction A reaction (actual or hypothetical) in which the types of bonds that are made in forming the products are the same as those that are broken in the reactants, e.g.,

$$PhCOOH + p\text{-}ClC_6H_4COO^- \rightarrow PhCOO^- + p\text{-}ClC_6H_4COOH$$

$$ClCH{=}CH_2 + ClCH_2CH_2Cl \rightarrow CH_2{=}CH_2 + Cl_2CHCH_2Cl$$

Such processes have advantages for theoretical treatment. The HAMMETT EQUATION as applied to equilibria essentially deals with isodesmic processes.

isoelectronic Two or more MOLECULAR ENTITIES are described as isoelectronic if they have the same number of valence electrons and the same structure, i.e., number and CONNECTIVITY of atoms, but differ in some of the elements involved. Thus

CO, N_2, and NO^+ are isoelectronic
$CH_2{=}C{=}O$ and $CH_2{=}N{=}N$ are isoelectronic

CH_3COCH_3 and $CH_3N{=}NCH_3$ have the same number of electrons but have different structures; hence they are not described as isoelectronic.

isoentropic A reaction series is said to be isoentropic if the individual reactions of the series have the same standard ENTROPY OF ACTIVATION.

isoenzymes Multiple forms of ENZYMES arising from genetically determined differences in PRIMARY STRUCTURE. The term does not apply to isoenzymes derived by modification of the same primary SEQUENCE.

isoequilibrium relationship A relationship analogous to the ISOKINETIC RELATIONSHIP but applied to equilibrium data. The equation defining the isoequilibrium temperature β is

$$\Delta_r H - \beta \Delta_r S = \text{constant}$$

where ΔH and ΔS are enthalpy and entropy of reaction, respectively.

See also ISOKINETIC RELATIONSHIP.

isokinetic relationship When a series of structurally related substrates undergo the same general reaction, or when the reaction conditions for a single substrate are changed in a systematic way, the ENTHALPIES and ENTROPIES OF ACTIVATION sometimes satisfy the relation

$$\Delta^{\ddagger}H - \beta \Delta^{\ddagger}S = \text{constant}$$

where the parameter β is independent of temperature. This equation (or some equivalent form) is said to represent an isokinetic relationship. The temperature T = β (at which all members of a series obeying the isokinetic relationship react at the same rate) is termed the isokinetic temperature.

Supposed isokinetic relationships as established by direct correlation of $\Delta^{\ddagger}H$ with $\Delta^{\ddagger}S$ are often spurious, and the calculated value of β is meaningless, because errors in $\Delta^{\ddagger}H$ lead to compensating errors in $\Delta^{\ddagger}S$. Satisfactory methods of establishing such relationships have been devised.

See also COMPENSATION EFFECT; ISOEQUILIBRIUM RELATIONSHIP; ISOSELECTIVE RELATIONSHIP.

isolobal The term is used to compare molecular fragments with each other and with familiar species from organic chemistry. Two fragments are isolobal if the number, symmetry properties, approximate energy, and shape of the FRONTIER ORBITALS and the number of electrons in them are similar.

See also ISOELECTRONIC.

isomerase An ENZYME of EC class 5 that catalyzes the isomerization of a SUBSTRATE.

See also EC NOMENCLATURE FOR ENZYMES.

isomerization A CHEMICAL REACTION, the principal product of which is isomeric with the principal reactant. An INTRAMOLECULAR isomerization that involves the breaking or making of bonds is a special case of a MOLECULAR REARRANGEMENT.

Isomerization does not necessarily imply molecular rearrangement (e.g., in the case of the interconversion of conformational isomers).

isomers Compounds that have the same number and types of atoms—same molecular formula—but differ in the way they are combined with each other. They can differ by the bonding sequence—called structural or constitutional isomerism—or the way their atoms are arranged spatially, called stereoisomerism. Other types include conformational and configurational isomers,

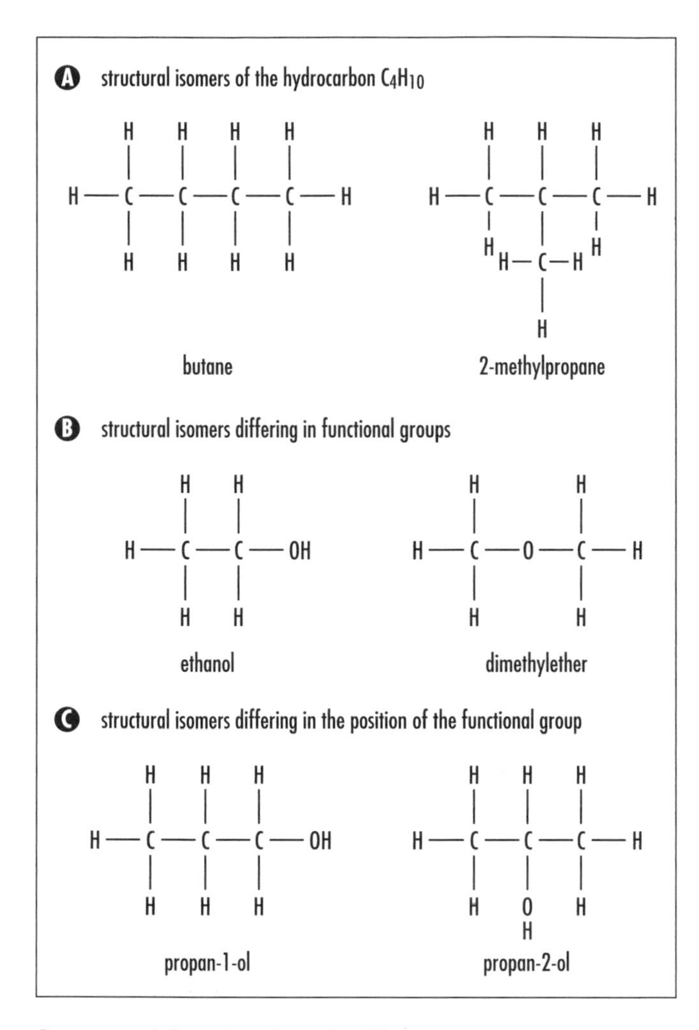

A compound that arises from the different ways atoms are grouped in a given molecular formula

GEOMETRICAL ISOMERS, OPTICAL ISOMERS, ENANTIOMERS, and DIASTEREOMERS.

isomorphous Refers to crystals having the same atomic arrangement and the same crystal form, except for the presence of different elements. An example is calcite ($CaCO_3$) and siderite ($FeCO_3$).

isosbestic point This term is usually employed with reference to a set of absorption spectra that are plotted on the same chart for a set of solutions in which the sum of the concentrations of two principal absorbing components, A and B, are constant. The curves of absorbance against wavelength (or frequency) for such a set of mixtures often all intersect at one or more points, called isosbestic points.

Isosbestic points are commonly met when electronic spectra are taken (a) on a solution in which a CHEMICAL REACTION is in progress (in which case the two absorbing components concerned are a reactant and a product, A + B) or (b) on a solution in which the two absorbing components are in equilibrium and their relative proportions are controlled by the concentration of some other component, typically the concentration of hydrogen ions, e.g., an acid-base indicator equilibrium.

$$A \rightleftharpoons B + H^+_{aq}$$

The effect may also appear (c) in the spectra of a set of solutions of two unrelated noninteracting components having the same total concentration. In all these examples, A (or B) may be either a single CHEMICAL SPECIES or a mixture of chemical species present in invariant proportion.

If A and B are single chemical species, isosbestic points will appear at all wavelengths at which their molar absorption coefficients (formerly called extinction coefficients) are the same. (A more involved identity applies when A and B are mixtures of constant proportion.)

If absorption spectra of the types considered above intersect not at one or more isosbestic points, but over progressively changing wavelengths, this is prima facie evidence in case (a) for the formation of a REACTION INTERMEDIATE in substantial concentration ($A \rightarrow C \rightarrow B$); in case (b) for the involvement of a third absorbing species in the equilibrium, e.g.,

$$A \rightleftharpoons B + H^+_{aq} \rightleftharpoons C + 2\,H^+_{aq}$$

or in case (c) for some interaction of A and B, e.g.,

$$A + B \rightleftharpoons C$$

isoselective relationship A relationship analogous to the ISOKINETIC RELATIONSHIP, but applied to SELECTIVITY data of reactions. At the isoselective temperature, the selectivities of the series of reactions following the relationship are identical.

See also ISOEQUILIBRIUM RELATIONSHIP.

isosteres Molecules or ions of similar size containing the same number of atoms and valence electrons, e.g., O^{2-}, F^-, Ne.

isotonic solutions Solutions having identical osmotic pressures, i.e., solutions where cells do not swell or shrink.

isotope Different forms of a single element that have the same number of protons but have different numbers of neutrons in their nuclei. Radioactive isotopes are unstable and break down until they become stable. Carbon-14 is a radioactive isotope of carbon that is used to date fossilized organic matter.

isotope effect The effect on the rate or equilibrium constant of two reactions that differ only in the isotopic composition of one or more of their otherwise chemically identical components is referred to as a kinetic isotope effect (*see* ISOTOPE EFFECT, KINETIC) or a thermodynamic (or equilibrium) isotope effect (*see* ISOTOPE EFFECT, THERMODYNAMIC), respectively.

isotope effect, equilibrium *See* ISOTOPE EFFECT, THERMODYNAMIC.

isotope effect, heavy atom An ISOTOPE EFFECT due to isotopes other than those of hydrogen.

isotope effect, intramolecular A kinetic ISOTOPE EFFECT observed when a single substrate, in which the isotopic atoms occupy equivalent reactive positions, reacts to produce a nonstatistical distribution of ISOTOPOLOGUE products. In such a case, the isotope effect will favor the pathway with lower force constants for displacement of the isotopic nuclei in the TRANSITION STATE.

isotope effect, inverse A kinetic ISOTOPE EFFECT wherein $k^l/k^h < 1$, i.e., the heavier substrate, reacts more rapidly than the lighter one, as opposed to the usual normal isotope effect, in which $k^l/k^h > 1$. The isotope effect is typically "normal" when the frequency differences between the isotopic TRANSITION STATEs are smaller than in the reactants. Conversely, an inverse isotope effect can be taken as evidence of an increase in the corresponding force constants upon passing from the reactant to the transition state.

isotope effect, kinetic The effect of isotopic substitution on a rate constant is referred to as a kinetic isotope effect.

For example in the reaction

$$A + B \rightarrow C$$

the effect of isotopic substitution in reactant A is expressed as the ratio of rate constants k^l/k^h, where the superscripts l and h represent reactions in which the molecules A contain the light and heavy isotopes, respectively.

Within the framework of TRANSITION STATE theory, in which the reaction is rewritten as

$$A + B \rightleftharpoons [TS]^{\ddagger} \rightarrow C$$

and with neglect of isotopic mass on TUNNELING and the TRANSMISSION COEFFICIENT, k^l/k^h can be regarded as if it were the equilibrium constant for an isotope exchange reaction between the transition state $[TS^{\ddagger}]$ and the isotopically substituted reactant A, and can be calculated from their vibrational frequencies, as in the case of a thermodynamic isotope effect.

Isotope effects like the above, involving a direct or indirect comparison of the rates of reaction of ISOTOPOLOGUEs, are called intermolecular, in contrast to intramolecular isotope effects, in which a single substrate reacts to produce a nonstatistical distribution of isotopologue product molecules.

See also ISOTOPE EFFECT, INTRAMOLECULAR.

isotope effect, primary A kinetic isotope effect attributable to isotopic substitution of an atom to which a bond is made or broken in the RATE-CONTROLLING STEP or in a PRE-EQUILIBRIUM step of a specified reaction. The corresponding isotope effect on the equilibrium constant of a reaction in which one or more bonds to isotopic atoms are broken is called a primary equilibrium isotope effect.

See also ISOTOPE EFFECT, SECONDARY.

isotope effect, secondary A kinetic isotope effect that is attributable to isotopic substitution of an atom to which bonds are neither made nor broken in the RATE-CONTROLLING STEP or in a PRE-EQUILIBRIUM step of a specified reaction, and is therefore not a primary isotope effect, is termed a secondary isotope effect. One speaks of α, β, etc. secondary isotope effects, where α, β, etc. denote the position of isotopic substitution relative to the reaction center. The corresponding isotope effect on the equilibrium constant of such a reaction is called a "secondary equilibrium isotope effect."

Secondary isotope effects have been discussed in terms of the conventional electronic effects of physical organic chemistry, e.g., induction, HYPERCONJUGATION, HYBRIDIZATION, etc., since these properties are determined by the electron distribution that depends on vibrationally averaged bond lengths and angles that vary slightly with isotopic substitution. While this usage is legitimate, the term *electronic isotope effect* should be avoided because of the misleading implication that such an effect is electronic rather than vibrational in origin.

See also ISOTOPE EFFECT, PRIMARY; ISOTOPE EFFECT, STERIC.

isotope effect, solvent A kinetic or equilibrium isotope effect resulting from a change in the isotopic composition of the solvent.

isotope effect, steric A secondary isotope effect attributed to the different vibrational amplitudes of ISOTOPOLOGUES. For example, both the mean and mean-square amplitudes of vibrations associated with C–H bonds are greater than those of C–D bonds. The greater effective bulk of molecules containing the former may be manifested by a STERIC EFFECT on a RATE CONSTANT or an EQUILIBRIUM CONSTANT.

See also ISOTOPE EFFECT, SECONDARY.

isotope effect, thermodynamic The effect of isotopic substitution on an EQUILIBRIUM CONSTANT is referred to as a thermodynamic (or equilibrium) isotope effect.

For example, the effect of isotopic substitution in reactant A that participates in the equilibrium:

$$A + B \rightleftharpoons C$$

is the ratio K^l/K^h of the equilibrium constant for the reaction in which A contains the light isotope to that in which it contains the heavy isotope. The ratio can be expressed as the equilibrium constant for the isotopic exchange reaction:

$$A^l + C^h \rightleftharpoons A^h + C^l$$

in which reactants such as B that are not isotopically substituted do not appear.

The potential energy surfaces of isotopic molecules are identical to a high degree of approximation, so thermodynamic isotope effects can only arise from the effect of isotopic mass on the nuclear motions of the reactants and products, and this can be expressed quantitatively in terms of partition function ratios for nuclear motion:

$$\frac{K^l}{K^h} = \frac{(Q^l_{nuc} / Q^h_{nuc})_C}{(Q^l_{nuc} / Q^h_{nuc})_A}$$

Although the nuclear partition function is a product of the translational, rotational, and vibrational partition functions, the isotope effect is determined almost entirely by the latter, specifically by vibrational modes involving motion of isotopically different atoms. In the case of light atoms (i.e., protium vs. deuterium or tritium) at moderate temperatures, the isotope effect is dominated by ZERO-POINT ENERGY differences.

See also FRACTIONATION FACTOR, ISOTOPIC.

isotope exchange A CHEMICAL REACTION in which the reactant and product CHEMICAL SPECIES are chemically identical but have different isotopic composition. In such a reaction, the isotope distribution tends toward equilibrium (as expressed by fractionation factors) as a result of transfers of isotopically different atoms or groups. For example:

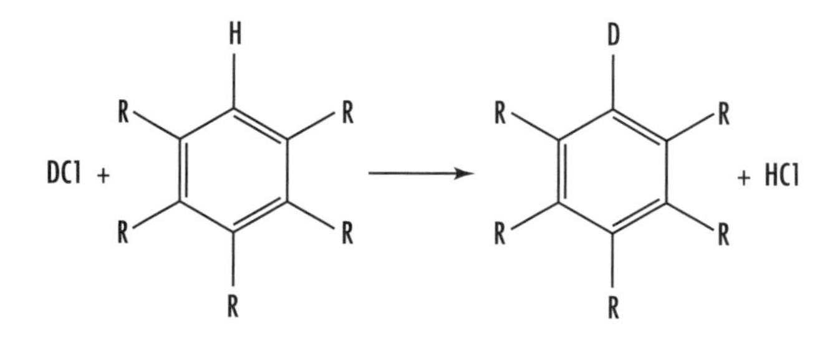

isotopic perturbation, method of NMR shift-difference measurement of the isotope effect on a fast (degenerate) equilibrium between two (except for isotopic substitution) species that are equivalent. This can be used to distinguish a rapidly equilibrating mixture with time-averaged symmetry from a single structure with higher symmetry.

See also NUCLEAR MAGNETIC RESONANCE SPECTROSCOPY.

isotopic scrambling The achievement, or the process of achieving, an equilibrium distribution of isotopes within a specified set of atoms in a CHEMICAL SPECIES or group of chemical species. For example:

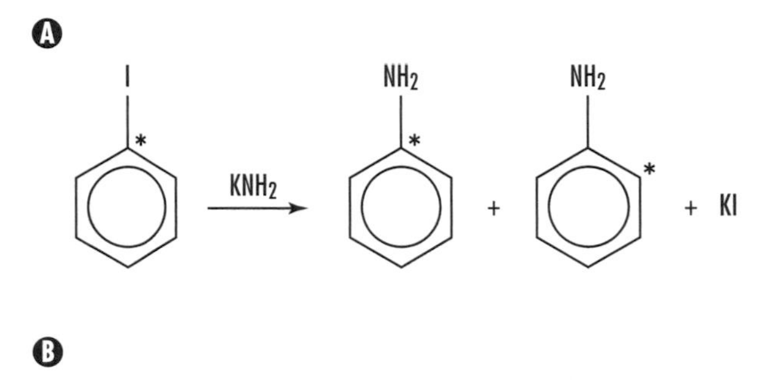

(* denotes position of an isotopically different atom.)

See also FRACTIONATION FACTOR, ISOTOPIC.

isotopologue A MOLECULAR ENTITY that differs only in isotopic composition (number of isotopic substitutions), e.g., CH_4, CH_3D, CH_2D_2....

isotopomer Isomers having the same number of each isotopic atom but differing in their positions. The term is a contraction of "isotopic isomer."

Isotopomers can be either constitutional isomers (e.g., $CH_2DCH{=}O$ and $CH_3CD{=}O$) or isotopic stereoisomers (e.g., (R)– and (S)–CH_3CHDOH or (Z)– and (E)–$CH_3CH{=}CHD$).

isotropy Lack of ANISOTROPY; the property of molecules and materials of having identical physical properties in all directions.

isovalent hyperconjugation *See* HYPERCONJUGATION.

J

Joliot-Curie, Irène (1897–1956) French *Nuclear physicist* Irène Curie was born in Paris on September 12, 1897, the daughter of the Nobel laureate physicists Pierre and Marie Curie. Growing up in the Curie family, Irène had no doubt that she would follow in her famous parents' footsteps. First home-schooled, she finished high school at Collège Sévigné, an independent school in the center of Paris, and then received a *baccalauréat* from the same academy in 1914 and a doctorate of science from the Sorbonne in 1925 for her thesis on the alpha rays of polonium. She served in World War I as a nurse radiographer. After the war, she joined her mother as an assistant at the Institute of Radium.

In 1926 she met and married Frédéric Joliot, also an assistant at the Institute of Radium, and had a daughter, Hélène, and a son, Pierre. The couple both (and singularly) worked on natural and artificial radioactivity and nuclear physics, their joint papers published during the years 1932–34. They confirmed the discovery of the positron in 1932. In 1933 they bombarded alpha particles at the stable element boron that created a radioactive compound of nitrogen. They published the results in the 1934 paper, "*Production artificielle d'éléments radioactifs. Preuve chimique de la transmutation des éléments*" (Artificial production of radioactive elements. Chemical proof of the transmutation of the elements.). The Nobel Committee recognized the profound implications of their discovery by awarding both Joliot-Curies the Nobel Prize in chemistry for 1935 "in recognition of their synthesis of new radioactive elements."

The Faculty of Science in Paris had appointed her as a lecturer in 1932, and in 1937 it conferred on her the title of professor. The year before, the French government had named her undersecretary of state for scientific research. In 1938 her research on the action of neutrons on the heavy elements was important in the discovery of uranium fission. The next year, the Legion of Honour inducted her as an officer. During World War II, Frédéric led the underground resistance movement as the president of the Front National.

After the war, Irène succeeded her mother as the director of the Institute of Radium. Also in 1946, she was named a commissioner for atomic energy, a position she retained for six years, during which France amassed its first atomic stockpile. After the war, she worked to promote peace as a member of the World Peace Council. She contributed her energy to women's rights as a member of the Comité National de l'Union des Femmes Françaises.

In 1948 while on a planned fund-raising tour in the United States to raise money for Spanish refugees, it was the time of McCarthyism, and she was refused entry (she was a socialist) and kept in a detention center on Ellis Island until the French embassy in Washington could get her out.

As with her mother, Marie Curie, who died from sustained overexposure to radiation, Irène Joliot-Curie also died from overexposure to radiation throughout her lifetime. She contracted leukemia and died in Paris on March 17, 1956. Her husband succeeded her as director of the Institute of Radium. He also continued to oversee the construction of the 160-MeV synchrocyclotron at the new center for nuclear physics in Orsay, a project she commenced before dying.

Jones reductor A column packed with amalgamated zinc and used for prereduction of an analyte.

Josephson junction A type of electronic circuit capable of switching at very high speeds when operated at temperatures approaching absolute zero. Composed of two superconductors that are separated by a thin nonsuperconducting layer in which electrons can cross through the insulating barrier, it was discovered by Brian David Josephson, who won the Nobel Prize in physics in 1973 with Leo Esaki and Ivar Giaever.

joule The international unit of energy or work. The amount of work done when a force of 1 Newton is applied through a distance of 1 meter. One joule is 1 $kg{\cdot}m^2/s^2$ or 0.2390 calorie.

Joule-Thompson effect The physical process in which the temperature of a gas is changed by allowing the gas to expand. Depending on the gas, the pressure, and the temperature, the change can be positive or negative.

Kamlet-Taft solvent parameters Parameters of the Kamlet-Taft solvatochromic relationship that measure separately the hydrogen bond donor (α), hydrogen bond acceptor (β), and dipolarity/POLARIZABILITY (π*) properties of solvents as contributing to overall solvent POLARITY.

kappa convention *See* DONOR ATOM SYMBOL.

K capture Absorption of a K shell (n = 1) electron by an atomic nucleus, where it combines with a proton as it is converted to a neutron.

Kekulé structure (for aromatic compounds) A representation of an AROMATIC MOLECULAR ENTITY (such as benzene), with fixed alternating single and double bonds, in which interactions between multiple bonds are assumed to be absent.

For benzene:

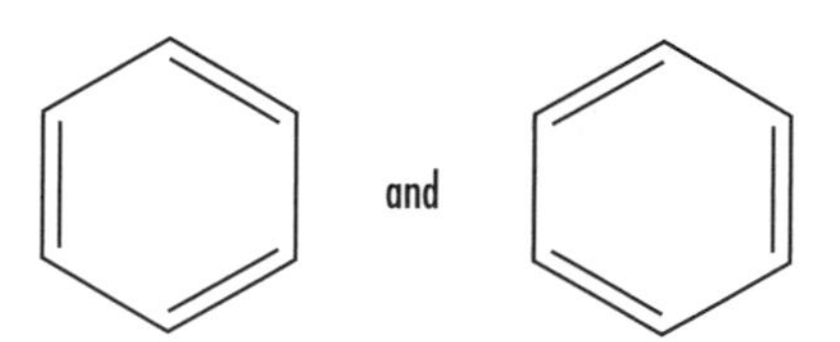

are the Kekulé structures.

Kelvin The standard unit of thermodynamic temperature. It is defined as 1/273.16 of the temperature of the triple point of water above absolute zero. The symbol for this is K. Kelvin is measured by the same temperature steps as Celsius but is shifted downward so that 0 K is absolute zero; water freezes at 273 K and boils at 373 K.

Kendall, Edward Calvin (1886–1972) American *Biochemist* Edward Calvin Kendall was born on March 8, 1886, in South Norwalk, Connecticut. He was educated at Columbia University and obtained a bachelor of science degree in 1908, a master's degree in chemistry in 1909, and a Ph.D. in chemistry in 1910.

From 1910 until 1911 he was a research chemist for Parke, Davis and Co., in Detroit, Michigan, and conducted research on the thyroid gland, continuing the work from 1911 until 1914 at St. Luke's Hospital, New York City.

In 1914 he was appointed head of the biochemistry section in the graduate school of the Mayo Foundation, Rochester, part of the University of Minnesota. In 1915 he was appointed director of the division of biochemistry and subsequently professor of physiological chemistry. In 1951 he retired from the Mayo Foundation and accepted the position of visiting professor in the department of biochemistry at Princeton University.

In 1914 he isolated thyroxine, the active principle of the thyroid gland. Kendall isolated and identified a series of compounds from the adrenal gland cortex and, with Merck & Co., Inc., prepared cortisone by partial synthesis. He also investigated the effects of cortisone and of adrenocorticotropic hormone (ACTH) on rheumatoid arthritis with Philip S. Hench, H. F. Polley, and C. H. Slocumb. Kendall and Hench, along with Tadeus Reichstein, shared the Nobel Prize in physiology or medicine in 1950 for this work. Kendall received many awards and honors. He died on May 4, 1972.

keratin A tough, insoluble, fibrous protein with high sulfur content that forms the main structure and protective barrier or cytoskeleton of epidermal cells and is the chief constituent of skin, hair and nails, and enamel of the teeth. It is produced by keratinocytes, the most abundant cells in the epidermis (95 percent). Keratin makes up 30 percent of the cellular protein of all living epidermal cells. The high amount of sulfur content is due to the presence of the amino acid cystine.

ketone Organic compound in which a carbonyl group is bound to two carbon atoms.

kilocalorie (kcal) One kilocalorie is equal to 1,000 calories, which is based on the amount of heat energy required to raise the temperature of 1 kilogram of water by 1°C; used to measure the energy value in food and labor. Usually just called calorie; 1 kilocalorie (kcal) = 1 Calorie (Cal) = 1,000 calories (cal). However, in the International System of Units (ISU), the universal unit of energy is the joule (J). One kilocalorie = 4.184 kilojoules (kJ).

kilogram The basic unit of mass (not of weight or of force). A kilogram is equal to the mass of 1.000028 cubic decimeters of water at the temperature of its maximum density.

kinetic ambiguity *See* KINETIC EQUIVALENCE.

kinetic control (of product composition) The term characterizes conditions (including reaction times) that lead to reaction products in a proportion governed by the relative rates of the parallel (forward) reactions in which the products are formed, rather than by the respective overall equilibrium constants.

See also THERMODYNAMIC CONTROL.

kinetic electrolyte effect (kinetic ionic-strength effect) The general effect of an added electrolyte (i.e., an effect other than, or in addition to, that due to its possible involvement as a reactant or catalyst) on the observed RATE CONSTANT of a reaction in solution. At low concentrations (when only long-range coulombic forces need to be considered), the effect on a given reaction is determined only by the IONIC STRENGTH of the solution and not by the chemical identity of the ions. For practical purposes, this concentration range is roughly the same as the region of validity of the Debye-Hückel limiting law for activity coefficients. At higher concentrations, the effect of an added electrolyte depends also on the chemical identity of the ions. Such specific action can usually be interpreted as the incursion of a REACTION PATH involving an ion of the electrolyte as reactant or catalyst, in which case the action should not be regarded as simply a kinetic electrolyte effect.

Kinetic electrolyte effects are usually (too restrictively and therefore incorrectly) referred to as "kinetic salt effects."

A kinetic electrolyte effect ascribable solely to the influence of the ionic strength on activity coefficients of ionic reactants and transition states is called a "primary kinetic electrolyte effect." A kinetic electrolyte effect arising from the influence of the ionic strength of the solution upon the PRE-EQUILIBRIUM concentration of an ionic species that is involved in a subsequent RATE-LIMITING STEP of a reaction is called a "secondary kinetic electrolyte effect." A common case encountered in practice is the effect on the concentration of a hydrogen ion (acting as catalyst) produced from the IONIZATION of a weak acid in a buffer solution.

See also COMMON-ION EFFECT; ORDER OF REACTION.

kinetic energy Energy of motion; kinetic energy depends on the object's mass and velocity and can be described mathematically as K.E. = $^{1}/_{2}$ mv^2. Moving matter, be it a rolling rock, flowing water, or falling ball, transfers a portion of its kinetic energy to other matter. For example, an inelastic collision is one in which at least a portion of the kinetic energy of the colliding particles is lost through conversion to some other form of energy. Potential energy, energy stored in a body, can be converted to kinetic energy.

kinetic equivalence Two reaction schemes are kinetically equivalent if they imply the same RATE LAW. For example, consider the two schemes (i) and (ii) for the formation of *C* from *A*:

$$\text{(i) A} \underset{k_{-1}\cdot \text{OH}^-}{\overset{k_1\cdot \text{OH}^-}{\rightleftharpoons}} \text{B} \xrightarrow{k_2} \text{C}$$

Providing that *B* does not accumulate as a reaction intermediate

$$\frac{d[C]}{dt} = \frac{k_1 k_2 [A][OH^-]}{k_2 + k_{-1}[OH^-]} \quad (1)$$

$$\text{(ii) A} \underset{k_{-1}\cdot \text{OH}^-}{\overset{k_1\cdot \text{OH}^-}{\rightleftharpoons}} \text{B} \xrightarrow[\text{OH}^-]{k_2} \text{C}$$

Providing that *B* does not accumulate as a reaction intermediate

$$\frac{d[C]}{dt} = \frac{k_1 k_2 [A][OH^-]}{k_1 + k_2[OH^-]} \quad (2)$$

Both equations for d[C]/dt are of the form

$$\frac{d[C]}{dt} = \frac{r[A][OH^-]}{1 + s[OH^-]} \quad (3)$$

where *r* and *s* are constants (sometimes called "coefficients in the rate equation"). The equations are identical in their dependence on concentrations and do not distinguish whether OH^- catalyzes the formation of *B*, and necessarily also its reversion to *A*, or is involved in its further transformation to *C*. The two schemes are therefore kinetically equivalent under conditions to which the stated provisos apply.

kinetic isotope effect *See* ISOTOPE EFFECT, KINETIC.

kinetic-molecular theory An ideal gas is composed of tiny particles (molecules) in constant motion.

kingdom Taxonomic name used to organize, classify, and identify plants and animals. There are five taxonomic kingdoms: Monera, Protista, Plantae, Fungi, and Animalia. Only the domain is higher in ranking. This system of ranking, called the Linnaean system, was developed by the Swedish scientist Carolus Linnaeus (1707–78), who developed a two-name system, binomial nomenclature (genus and species), for identifying and classifying all living things. The system is based on a hierarchical structure in which organisms are sorted by kingdom, phylum, class, order, family, genus, and species. Organisms belonging to the same kingdom do not have to be very similar, but organisms belonging to the same species are and can reproduce and create offspring.

klinotaxis A movement in a specific direction relative to a given stimulus, either directly toward or away from the source.

Koppel-Palm solvent parameters Parameters to measure separately the ability of a solvent to enter into nonspecific solvent-solute interactions (permittivity, ε, and refractive index, n_D) and specific solvent-solute interaction (solvent basicity or NUCLEOPHILICITY B and solvent acidity or ELECTROPHILICITY E) as contributing to overall solvent POLARITY.

Kosower Z-value *See* Z-VALUE.

Kossel, Ludwig Karl Martin Leonhard Albrecht (1853–1927) German *Chemist* Ludwig Karl Martin Leonhard Albrecht Kossel was born in Rostock on September 16, 1853, the eldest son of Albrecht Kossel, the merchant and Prussian consul, and his wife Clara Jeppe. He attended the secondary school in Rostock and went to the newly founded University of Strassburg in 1872 to study medicine. He received his doctorate of medicine in 1878.

Kossel specialized in chemistry of tissues and cells (physiological chemistry), and by the 1870s, he had begun his investigations into the constitution of the cell nucleus. He isolated nucleoproteins from the heads of fish sperm cells in 1879. By the 1890s he had focused his study on proteins. In 1910 he received the Nobel Prize in physiology or medicine for his contributions in cell chemistry and work on proteins.

Among his important publications are *Untersuchungen über die Nukleine und ihre Spaltungsprodubte* (Investigations into the nucleins and their cleavage products), 1881; *Die Gewebe des menschlichen Körpers und ihre mikroskopische Untersuchung* (The tissues in the human body and their microscopic investigation), 1889–91, in two volumes, with Behrens and Schieerdecker; and the *Leitfaden für medizinisch-chemische Kurse* (Textbook for medical-chemical courses), 1888. He was also the author of *Die Probleme der Biochemie* (The problems of biochemistry), 1908; and *Die Beziehungen der Chemie zur Physiologie* (The relationships between chemistry and physiology), 1913.

Kossel had honorary doctorates from the Universities of Cambridge, Dublin, Ghent, Greifswald, St. Andrews, and Edinburgh, and he was a member of various scientific societies, including the Royal Swedish Academy of Sciences and the Royal Society of Sciences of Uppsala. Albrecht Kossel died on July 5, 1927.

Portrait of Sir Hans Krebs (1900–81), German-British biochemist and Nobel laureate. Krebs trained, like his father, in medicine. During 1932 he discovered the ornithine cycle, in which the liver converts amino acids to nitrogen and urea. The following year he fled from Nazism to Britain. He eventually settled in Sheffield, and it was there that he discovered the Krebs cycle. This describes how the body breaks down glucose into carbon dioxide, water, and energy. This is central to energy production in the mitochondria of most cells and generates energy for whole organisms. For this work he shared the 1953 Nobel Prize with F. Lipmann. ***(Courtesy of Science Photo Library)***

Krebs, Sir Hans Adolf (1900–1981) German/British *Biochemist* Sir Hans Adolf Krebs was born at Hildesheim, Germany, on August 25, 1900, to Georg Krebs, M.D., an ear, nose, and throat surgeon of that city, and his wife Alma (née Davidson).

Krebs was educated at the Gymnasium Andreanum at Hildesheim. Between 1918 and 1923 he studied medicine at the Universities of Göttingen, Freiburg-im-Breisgau, and Berlin. He received a M.D. degree at the University of Hamburg in 1925. In 1926 he was appointed assistant to professor Otto Warburg at the Kaiser Wilhelm Institute for Biology at Berlin-Dahlem, where he remained until 1930. He was forced to leave Germany in 1933 because of his Jewish background.

In 1934 he was appointed demonstrator of biochemistry in the University of Cambridge, and the following year he was appointed lecturer in pharmacology at the University of Sheffield. In 1938 he became the newly founded lecturer-in-charge of the department of biochemistry. In 1939 he became an English citizen. By 1945 he was professor and director of the Medical Research Council's research unit established in the department. In 1954 he was appointed Whitley Professor of Biochemistry in the University of Oxford, and the Medical Research Council's Unit for Research in Cell Metabolism was transferred to Oxford.

At the University of Freiburg (1932), he discovered a series of chemical reactions (now known as the urea cycle) by which ammonia is converted to urea in mammalian tissue. For his discoveries of chemical reactions in living organisms now known as the citric acid or KREBS CYCLE, he was awarded the 1953 Nobel Prize for physiology or medicine. These reactions involve the conversion—in the presence of oxygen—of substances that are formed by the breakdown of sugars, fats, and protein components to carbon dioxide, water, and energy-rich compounds.

He was a member of many scientific societies, winning many awards and citations for his work and published works, including *Energy Transformations in Living Matter* (1957) with British biochemist Hans Kornberg. He was knighted in 1958. He died on November 22, 1981, in Oxford, England.

Krebs cycle A biochemical cycle in the second stage of cellular respiration involving eight steps that complete the metabolic breakdown of glucose molecules to carbon dioxide. Acetyl CoA is combined with oxaloacetate to form citric acid. Citric acid is then converted into a number of other chemicals, and carbon dioxide is released. The process takes place within the mitochondrion. Also called citric acid cycle or tricarboxylic acid (TCA) cycle. Conceived and published by British scientist SIR HANS ADOLF KREBS in 1957.

L

labeled compound A compound consisting of radioactively labeled molecules that can be observed as it passes through physical, chemical, or biological processes.

labile The term has loosely been used to describe either a relatively unstable and transient chemical species or a relatively STABLE but reactive species.

See also INERT.

laccase A copper-containing ENZYME, 1,4-benzenediol oxidase, found in higher plants and microorganisms. Laccases are MULTICOPPER OXIDASEs of wide specificity that carry out one-electron oxidation of phenolic and related compounds and reduce O_2 to water. The enzymes are polymeric and generally contain one each of TYPE 1, TYPE 2, TYPE 3 COPPER centers per SUBUNIT, where the type 2 and type 3 are close together, forming a trinuclear copper CLUSTER.

See also NUCLEARITY.

lactate Alternate name for lactic acid, a chemical created from sugars when broken down for energy in the absence of oxygen. Strictly, it refers to the deprotonated form of lactic acid as it exists as the anion or in salts and esters, for example.

lactoferrin An iron-binding protein from milk, structurally similar to the TRANSFERRINs.

Landsteiner, Karl (1868–1943) Austrian *Biochemist* Karl Landsteiner was born in Vienna on June 14, 1868, to Leopold Landsteiner, a journalist and newspaper publisher, and Fanny Hess. Landsteiner studied medicine at the University of Vienna, graduating in 1891.

From 1898 until 1908 he held the post of assistant in the university department of pathological anatomy in Vienna. In 1908 he received the appointment as prosector in the Wilhelminaspital in Vienna, and in 1911 he became professor of pathological anatomy at the University of Vienna.

With a number of collaborators, he published many papers on his findings in anatomy and immunology, such as the immunology of syphilis and the Wassermann reaction, and he discovered the immunological factors, which he named HAPTENs. He also laid the foundations of the cause and immunology of poliomyelitis.

His discovery of the major blood groups and development of the ABO system of blood typing in 1901 and, in 1909, the classification of the bloods of human beings into the now well-known A, B, AB, and O groups (as well as the M and N groups), which made blood transfusion a routine medical practice, resulted in his receiving the Nobel Prize for physiology or medicine in 1930. In 1936 he wrote *The Specificity of Serological Reactions,* a classic text that helped to establish the science of immunochemistry. In 1940 he discovered the Rh factor, the protein on the surface of red blood cells that determines if the blood type is positive (Rh-positive) or negative (Rh-negative). If the mother has a negative Rh factor (Rh-negative) and the father and fetus are Rh-positive, the mother can become Rh-sensitized and produce antibodies to combat fetal blood cells that cross the placenta into her bloodstream. These antibodies can destroy the fetus's Rh-positive blood cells, putting it at serious risk of anemia.

In 1939 he became professor emeritus at the Rockefeller Institute. On June 24, 1943, he had a heart attack in his laboratory and died two days later.

Langmuir, Irving (1881–1957) American *Chemist* Irving Langmuir was born in Brooklyn, New York, on January 31, 1881, the third of four sons to Charles Langmuir and Sadie Comings Langmuir. Langmuir's early education was scattered among various schools in the United States and Paris, and he finally graduated from the Pratt Institute's Manual Training High School in Brooklyn. He attended Columbia University in New York City, where he received a bachelor's degree in metallurgical engineering from the university's school of mines in 1903. He attended graduate school in Göttingen University in Germany, working with Walther Nernst, a theoretician, inventor, and Nobel laureate, and he received his master's degree and Ph.D. in physical chemistry under Nernst in 1906.

Langmuir returned to America and became an instructor in chemistry at the Stevens Institute of Technology in Hoboken, New Jersey, where he taught until July 1909. He next took a job at the General Electric Company Research Laboratory in Schenectady, New York, where he eventually became associate director of research and development. In 1912 he married Marion Mersereau, and they had two children.

While his studies included chemistry, physics, and engineering, he also became interested in cloud physics. He investigated the properties of adsorbed films and the nature of electric discharges in high vacuum and in certain gases at low pressures, and his research on filaments in gases led directly to the invention of the gas-filled incandescent lamp and the discovery of atomic hydrogen. He used his discovery of hydrogen in the development of the atomic hydrogen welding process. He formulated a general theory of adsorbed films after observing the very stable adsorbed monatomic films on tungsten and platinum filaments, and after experiments with oil films on water. He also studied the catalytic properties of such films.

In chemistry, his interest in reaction mechanisms led him to study structure and valence, and he contributed to the development of the Lewis theory of shared electrons. In 1927 he invented the term *plasma* for an ionized gas. In 1932 he won the Nobel Prize for his studies on surface chemistry. While at GE, he invented the mercury-condensation vacuum pump, the nitrogen-argon-filled incandescent lamp, and an entire family of high-vacuum radio tubes. He had a total of 63 patents at General Electric. He also worked with Vincent Schaefer, Bernard Vonnegut, and Duncan Blanchard on a number of experiments, including the first successful cloud-seeding project (making rain) and the development of smoke generators for the Second World War effort. Their smoke generator was 400 times more efficient than anything the military had and filled the entire Schoharie Valley within one hour during a demonstration.

He was given many awards and honors including: Nichols Medal (1915 and 1920); Hughes Medal (1918); Rumford Medal (1921); Cannizzaro Prize (1925); Perkin Medal (1928); School of Mines Medal (Columbia University, 1929); Chardler Medal (1929); Willard Gibbs Medal (1930); Popular Science Monthly Award (1932); Franklin Medal and Holly Medal (1934); John Scott Award (1937); "Modern Pioneer of Industry" (1940); Faraday Medal (1944); and Mascart Medal (1950). He was a foreign member of the Royal Society of London, a fellow of the American Physical Society, and an honorary member of the British Institute of Metals and the Chemical Society (London). He served as president of the American Chemical Society and as president of the American Association for the Advancement of Science. He received over a dozen honorary degrees.

Langmuir was an avid outdoorsman and skier, and in his early years he was associated with the Boy Scout movement, where he organized and served as scoutmaster of one of the first troops in Schenectady, New York. After a heart attack, he died on August 16, 1957, in Falmouth, Massachusetts. In 1975 his son Kenneth Langmuir bequeathed the residue of his estate to the Irving Langmuir Laboratory for Atmospheric Research, where a great deal of lightning research takes place. The bequest supports the laboratory, Langmuir Fellowships at New Mexico Institute of Mining and Technology, and an annual research award.

Langmuir's discoveries helped shape the establishment of modern radio and television broadcasting, safeguarded the lives of soldiers in war, and provided the framework that allowed his research team to develop a key to possibly control the weather. Bernard Vonnegut's brother, the writer Kurt Vonnegut, made Langmuir a character, Dr. Felix Hoenikker, in his novel *Cat's Cradle*. Vonnegut claims that the absentminded scientist really did leave a tip for his wife after breakfast one time and abandoned his car in the middle of a traffic jam.

lanthanide contraction The ionic radii decrease smoothly across the lanthanide series in the PERIODIC TABLE. Caused by the increase in effective nuclear charge across the series due to the poor shielding ability of 4*f* electrons.

lanthanides The elements 58 to 71 (after lanthanum) that are very reactive and electropositive. Sometimes includes lanthanum (La). They are characterized by having a partially filled set of *f* orbitals in the $n = 4$ shell.

laser Acronym for *l*ight *a*mplification by *s*timulated *e*mission of *r*adiation. A device that has mirrors at the ends and is filled with material such as crystal, glass, liquid, gas, or dye that has atoms, ions, or molecules capable of being excited to a metastable state by light, electric discharge, or other stimulus. The transition from the metastable state back to the normal ground state is accompanied by the emission of photons that form a straight, coherent beam. Laser light is directional and covers a narrow range of wavelengths. It is more coherent than ordinary light.

latentiated drug *See* DRUG LATENTIATION.

latex A polymer of *cis*-1-4 isoprene; milky sap from the rubber tree *Hevea brasiliensis*.

lattice The positioning of atoms in crystalline solids.

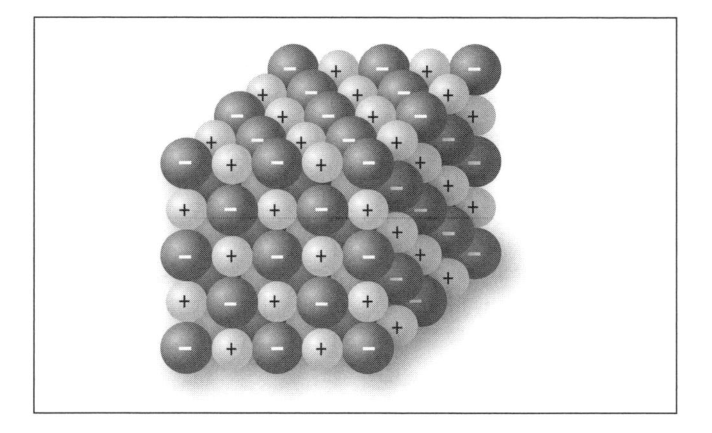

Lattice. The positioning of atoms in crystalline solids

lattice energy The amount of energy required to separate ions or molecules in a mole of a crystalline compound.

law of combining volumes **(Gay-Lussac's law)** At constant temperature and pressure, the volumes of reacting gases are expressed as ratios of small whole numbers.

law of conservation of energy Energy cannot be created or destroyed, just transformed.

law of conservation of matter Matter cannot be created or destroyed in an isolated system.

law of conservation of matter and energy The total amount of matter and energy available in the universe is fixed.

law of definite proportions **(law of constant composition)** Different samples of a pure compound will always contain the same elements in the same proportions by mass.

law of partial pressures **(Dalton's law)** Total pressure exerted by a mixture of gases is the sum of the partial pressures of the individual gases.

LCAO (linear combination of atomic orbitals) A form of expression in showing molecular orbitals. Written as linear combinations of atomic orbital wave functions.

lead discovery The process of identifying active new chemical entities that, by subsequent modification, may be transformed into a clinically useful DRUG.

lead generation The term applied to strategies developed to identify compounds possessing a desired but nonoptimized biological activity.

lead optimization The synthetic modification of a biologically active compound to fulfill all stereoelectronic, physicochemical, pharmacokinetic, and toxicologic requirements for clinical usefulness.

lead storage battery Secondary voltaic cell used in most automobiles, where the anode is lead and the cathode is lead coated with lead dioxide.The electrolyte is sulfuric acid.

least nuclear motion, principle of The hypothesis that, for given reactants, the reactions involving the smallest change in nuclear positions will have the lowest ENERGY OF ACTIVATION. (It is also often simply referred to as the principle of least motion.)

leaving group An atom or GROUP (charged or uncharged) that becomes detached from an atom in what is considered to be the residual or main part of the SUBSTRATE in a specified reaction. For example, in the heterolytic SOLVOLYSIS of benzyl bromide in acetic acid

$$PhCH_2Br + AcOH \rightarrow PhCH_2OAc + HBr$$

the leaving group is Br^-; in the reaction

$$MeS^- + PhCH_2N^+Me_3 \rightarrow MeSCH_2Ph + NMe_3$$

the leaving group is NMe_3; in the ELECTROPHILIC nitration of benzene, it is H^+. The term has meaning only in relation to a specified reaction. The leaving group is not, in general, the same as the SUBSTITUENT group present in the substrate (e.g., bromo and trimethylammonio in the substrates of the first two examples above).

A slightly different usage of the term prevails in the (nonmechanistic) naming of TRANSFORMATIONs, where the actual substituent group present in the substrate (and also in the product) is referred to as the leaving group.

See also ELECTROFUGE; ENTERING GROUP; NUCLEOFUGE.

Le Châtelier's principle In 1888 French chemist Henri-Louis Le Châtelier stated that a system at equilibrium, or striving to attain equilibrium, responds in such a way as to counteract any stress placed upon it, and a new equilibrium position will be reached.

Leclanché cell A common name for a type of dry cell and one of the earliest practical nonrechargeable batteries created by French scientist Georges-Lionel Leclanché (1839–82). It consists of a zinc ANODE (negative electrode) and a manganese dioxide CATHODE (positive electrode) with ammonium chloride solution as electrolyte. Became the first DRY CELL.

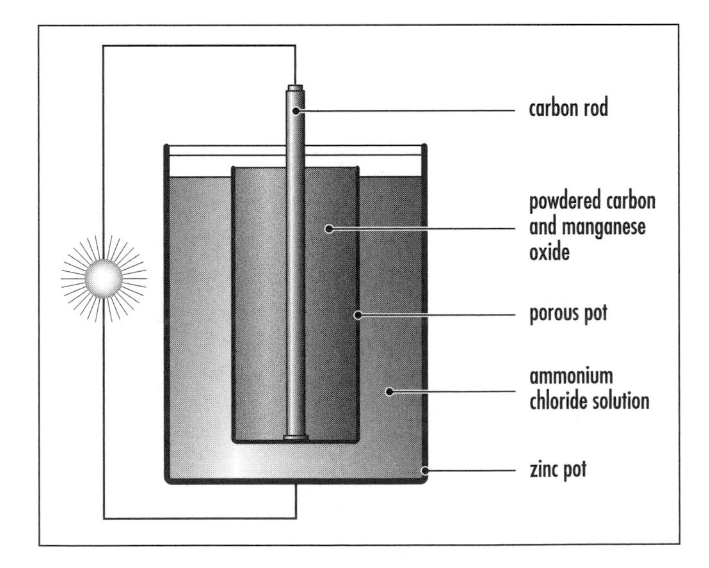

Leclanché cell. A common name for a type of dry cell and one of the earliest practical nonrechargeable batteries created by French scientist Georges-Lionel Leclanché (1839–82)

Leffler's assumption *See* HAMMOND PRINCIPLE.

left-to-right convention Arrangement of the structural formulae of the reactants so that the bonds to be made or broken form a linear array in which the electrons move from left to right.

See also LINE FORMULA.

leghemoglobin A monomeric HEMOGLOBIN synthesized in the root nodules of leguminous plants that are host to nitrogen-fixing bacteria. Has a high affinity for dioxygen and serves as an oxygen supply for the bacteria.

See also NITROGEN FIXATION.

Leloir, Louis F. (1906–1987) Argentinian/French *Chemist* Luis F. Leloir was born in Paris on September 6, 1906, to Argentine parents but lived in Buenos Aires from the age of two.

He graduated with an M.D. from the University of Buenos Aires in 1932 and started his scientific career at the Institute of Physiology researching adrenalin carbohydrate metabolism.

In 1936 he worked at the biochemical laboratory of Cambridge, England, but returned to Buenos Aires to conduct research on the oxidation of fatty acids in the liver.

He was married in 1943 to Amelia Zuberbuhler and had a daughter.

The following year he moved to the United States and worked as a research assistant in the laboratory of Dr. Carl F. Cori in St. Louis, and then with D.E. Green in the College of Physicians and Surgeons at Columbia University in New York City. He was director of the Instituto de Investigaciones Bioquímicas, Fundación Campomar in Argentina and isolated glucose 1,6-diphosphate and uridine diphosphate glucose, uridine diphosphate acetylglucosamine, and guanosine diphosphate mannose.

He received the Nobel Prize in chemistry in 1970 "for his discovery of sugar nucleotides and their role in the biosynthesis of carbohydrates." Leloir had elucidated the biosynthesis of glycogen, the chief sugar reserve in animals and many microorganisms.

He received the following awards: Argentine Scientific Society, Helen Hay Whitney Foundation (United States), Severo Vaccaro Foundation (Argentina), Bunge and Born Foundation (Argentina), Gairdner Foundation (Canada), Louisa Gross Horowitz (United States), and Benito Juarez (Mexico). Leloir was also president of the Pan-American Association of Biochemical Societies. He died in 1987.

leveling effect The tendency of a solvent to make all BRONSTED ACIDS whose ACIDITY exceeds a certain value appear equally acidic. It is due to the complete transfer to a PROTOPHILIC SOLVENT of a HYDRON from a dissolved acid stronger than the CONJUGATE ACID of the solvent. The only acid present to any significant extent in all such solutions is the LYONIUM ION. For example, the solvent water has a leveling effect on the acidities of $HClO_4$, HCl, and HI: aqueous solutions of these acids at the same (moderately low) concentrations have the same acidities. A corresponding leveling effect applies to strong bases in PROTOGENIC SOLVENTS.

levorotatory Refers to an optically active compound that rotates plane polarized light counterclockwise (to the left) when viewed in the direction of the light source. Opposite of DEXTROROTATORY, which rotates it to the right (clockwise).

Lewis acid A molecular entity that is an ELECTRON-PAIR ACCEPTOR and therefore is able to react with a LEWIS BASE to form a LEWIS ADDUCT by sharing the electron pair furnished by the Lewis base.

Lewis acidity The thermodynamic tendency of a substrate to act as a LEWIS ACID. Comparative measures of this property are provided by the equilibrium constants for LEWIS ADDUCT formation of a series of Lewis acids with a common reference LEWIS BASE.

See also ACCEPTOR NUMBER; ELECTROPHILICITY; STABILITY CONSTANT.

Lewis adduct The adduct formed between a LEWIS ACID and a LEWIS BASE. An adduct is formed by the union of two molecules held together by a coordinate COVALENT BOND.

See also COORDINATION.

Lewis base A molecular entity able to provide a pair of electrons and thus capable of COORDINATION to a LEWIS ACID, thereby producing a LEWIS ADDUCT.

Lewis formula (electron dot or Lewis structure) Molecular structure in which the valency electrons are shown as dots placed between the bonded atoms, with one pair of dots representing two electrons or one (single) COVALENT BOND, for example

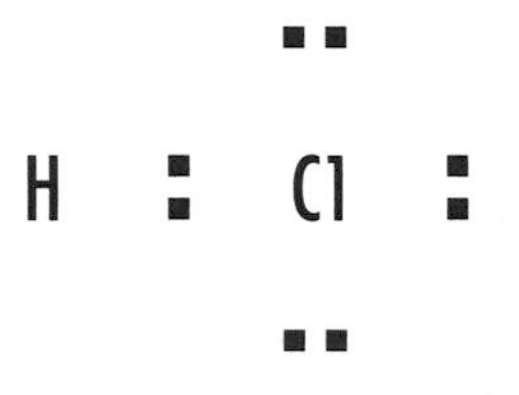

A double bond is represented by two pairs of dots, etc. Dots representing nonbonded outer-shell electrons are placed adjacent to the atoms with which they are associated, but not between the atoms. Formal charges (e.g. +, –, 2+, etc.) are attached to atoms to indicate the difference between the positive nuclear charge (atomic number) and the total number of electrons (including those in the inner shells) on the formal basis that bonding electrons are shared equally between atoms they join. (Bonding pairs of electrons are usually denoted by lines, representing covalent bonds, as in LINE FORMULAe.)

Libby, Willard Frank (1908–1980) American *Physical chemist* Willard Frank Libby was born in Grand Valley, Colorado, on December 17, 1908, to Ora Edward Libby and Eva May (née Rivers). After attending grammar and high schools near Sebastopol, California (1913–26), he attended the University of California at Berkeley from 1927 to 1933, taking his bachelor's and Ph.D. degrees in 1931 and 1933, respectively.

In 1933 he became an instructor in the department of chemistry at California University (Berkeley), eventually becoming associate professor of chemistry. He was awarded a Guggenheim Memorial Foundation Fellowship in 1941, but he went to Columbia University to work on the Manhattan Project, on leave from the Department of Chemistry, California University, until 1945.

After the war Libby became professor of chemistry in the department of chemistry and Institute for Nuclear Studies (Enrico Fermi Institute for Nuclear Studies) of Chicago University. In 1954 he was appointed by President Eisenhower to the U.S. Atomic Energy Commission, staying until 1959, when he became professor of chemistry in the University of California at Los Angeles. He became founding director of the UCLA Institute of Geophysics and Planetary Physics in 1962.

Libby developed carbon dating, a method of using carbon-14 for age determination in archaeology, geology, geophysics, and other branches of science. He was awarded the Nobel Prize in chemistry in 1960 "for his method to use carbon-14 for age determination in archaeology, geology, geophysics, and other branches of science."

He also received the Research Corporation Award for 1951 for the radiocarbon dating technique; the Chandler Medal of Columbia University for outstanding achievement in the field of chemistry (1954); the American Chemical Society Award for Nuclear Applications in Chemistry (1956); the Elliott Cresson Medal of the Franklin Institute (1957); the American Chemical Society's Willard Gibbs Medal Award (1958); the Albert Einstein Medal Award (1959); and the Day Medal of the Geological Society of America (1961).

Libby published *Radiocarbon Dating* (1952), authored numerous scientific articles, and was a member of the editorial board of the *Proceedings of the National Academy of Sciences* and of *Science*.

He died on September 8, 1980, in Los Angeles, California.

lifetime (mean lifetime τ) The lifetime of a CHEMICAL SPECIES that decays in a first-order process is the time needed for a concentration of this species to decrease to 1/e of its original value. Statistically, it represents the mean life expectancy of an excited species. In a reacting system in which the decrease in concentration of a particular chemical species is governed by a first-order RATE LAW, it is equal to the reciprocal of the sum of the (pseudo)unimolecular rate constants of all processes that cause the decay. When the term is used for processes that are not first order, the lifetime depends on the initial concentration of the species, or of a quencher, and should be called "apparent lifetime" instead.

See also CHEMICAL RELAXATION; HALF-LIFE; RATE OF REACTION.

ligand A molecule, ion, hormone or compound that binds to a specific receptor site that binds to another molecule.

The atoms or groups of atoms bound to the CENTRAL ATOM (*see* COORDINATION). The root of the word is sometimes converted into the verb *to ligate,* meaning to coordinate as a ligand, and the derived participles, *ligating* and *ligated.* This use should not be confused with its use to describe the action of LIGASEs (a class of ENZYMEs). The names for anionic ligands, whether inorganic or organic, end in -o. In general, if the anion name ends in -ide, or -ate, the final -e is replaced by -o, giving -ido and -ato, respectively. Neutral and cationic ligand names are used without modification. Ligands bonded by a single carbon atom to metals are regarded as radical substituents, their names being derived from the parent hydrocarbon, from which one hydrogen atom has been removed. In general, the final letter -e of the name is replaced by -yl.

In biochemistry the term *ligand* has been used more widely: if it is possible or convenient to regard part of a polyatomic molecular entity as central, then the atoms or groups or molecules bound to that part may be called ligands.

ligand field Ligand field theory is a modified CRYSTAL FIELD theory that assigns certain parameters as variables rather than taking them as equal to the values found for free ions, thereby taking into account the potential covalent character of the metal-LIGAND bond.

ligand field splitting Removal of a degeneracy of atomic or molecular levels in a molecule or ion with a given symmetry induced by the attachment or removal of ligands to produce reduced symmetries.

ligand-gated ion-channel receptor Ion channels are specialized pores in the cell membrane that help control and transfer electrical impulses (action potentials) in the cell. They regulate the flow of sodium, potassium, and calcium ions into and out of the cell. The ligand-gated ion-channel receptor is a signal receptor protein in a cell membrane that may act as a channel for the passage of a specific ion across the membrane. When activated by a signal molecule, it will allow or block the passage of the ion. This results in a change in ion concentration that often affects cell functioning.

ligase An ENZYME of EC class 6, also known as a synthetase, that catalyzes the formation of a bond between two SUBSTRATE molecules coupled with the HYDROLYSIS of a diphosphate bond of a NUCLEOSIDE triphosphate or similar cosubstrate.

See also EC NOMENCLATURE FOR ENZYMES.

ligating *See* LIGAND.

light microscope A common laboratory instrument that uses optics to bend visible light to magnify images of specimens placed on an attached platform or other viewing area.

light reactions A major component of photosynthesis in which a group of chemical reactions occurs in the thylakoid membranes of chloroplasts that harvest energy from the sun to produce energy-packed chemical bonds of ATP and NADPH and give off oxygen as a by-product.

light water reactor A term used to describe reactors using ordinary water as coolant, including boiling water reactors (BWRs) and pressurized water reactors (PWRs), the most common types used in the United States.

lignin A complex amorphous polymer in the secondary cell wall (middle lamella) of woody plant cells that cements or naturally binds cell walls to help make them rigid. Highly resistant to decomposition by chemical or enzymatic action. It is the major source material for coal. It also acts as support for cellulose fibers.

Cells that contain lignin are fibers, sclerids, vessels, and tracheids.

limestone A sedimentary rock composed of more than 50 percent calcium carbonate ($CaCO_3$).

limewater A colorless, somewhat milky, and strongly alkaline solution of calcium hydroxide produced by slaking lime.

limiting law An equation that is valid only at very low concentrations but reflects the correct dependence upon concentration and charge.

limiting reactant (limiting reagent) In the presence of two or more reactants in such an amount that a reaction proceeded to completion, the limiting reactant would be completely consumed.

linear accelerator A device used for accelerating charged particles along a straight-line path and often used in cancer treatment as well as for the study of nuclear reactions.

linear free-energy relation A linear correlation between the logarithm of a RATE CONSTANT or equilibrium constant for one series of reactions and the logarithm of the rate constant or equilibrium constant for a related series of reactions. Typical examples of such relations (also known as linear Gibbs energy relations) are the BRONSTED RELATION, and the HAMMETT EQUATION.

The name arises because the logarithm of an equilibrium constant (at constant temperature and pressure) is proportional to a standard free energy (Gibbs energy) change, and the logarithm of a rate constant is a linear function of the free energy (Gibbs energy) of activation.

The area of physical organic chemistry that deals with such relations is commonly referred to as linear free-energy relationships.

See also GIBBS ENERGY DIAGRAM; GIBBS ENERGY OF ACTIVATION; GIBBS FREE ENERGY; STABILITY CONSTANT.

linear Gibbs energy relation *See* LINEAR FREE-ENERGY RELATION.

linear solvation energy relationships Equations involving the application of SOLVENT PARAMETERS in linear or multiple (linear) regression expressing the solvent effect on the rate or equilibrium constant of a reaction.

See also DIMROTH-REICHARDT E_T PARAMETER; KAMLET-TAFT SOLVENT PARAMETER; KOPPEL-PALM SOLVENT PARAMETER; STABILITY CONSTANT; Z-VALUE.

line formula A two-dimensional representation of MOLECULAR ENTITIES in which atoms are shown joined by lines representing single or multiple bonds, without any indication or implication concerning the spatial direction of bonds. For example, methanol is represented as

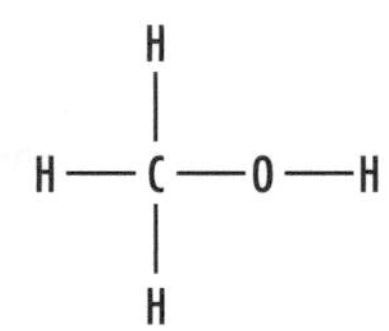

(The term should not be confused with the representation of chemical formulae by the Wiswesser line notation, a method of string notation. Formulae in this notation are also known as "Wiswesser line formulae.")

line-shape analysis Determination of RATE CONSTANTs for a chemical exchange from the shapes of spectroscopic lines of dynamic processes. The method is most often used in NUCLEAR MAGNETIC RESONANCE SPECTROSCOPY.

line spectrum An atomic emission or ABSORPTION SPECTRUM.

Lineweaver-Burk plot *See* MICHAELIS-MENTEN KINETICS.

linkage isomers An isomer where a specific ligand bonds to a metal ion through different donor atoms.

Two or more coordination compounds in which the donor atom of at least one of the ligands is different. Linkage isomers have different physical and chemical properties. Ligands that can form these isomers are CN–, SCN–, and NO_2–.

lipid A large group of HYDROPHOBIC (water insoluble) molecules that are the building blocks of cell membranes and liposomes (lipid vesicles) and contain fatty acids; the principal components of fats, oils, waxes, triglycerides, and cholesterol. They are insoluble in water but soluble in solvents such as alcohol and ether. The phospholipid bilayer of the plasma membrane is a double layer of phospholipid molecules arranged so that the hydrophobic "tails" lie between the HYDROPHILIC "heads." Also known as fat, they easily store in the body and are an important source of fuel for the body.

Lipmann, Fritz Albert (1899–1986) German/American *Biochemist* Fritz Albert Lipmann was born on June 12, 1899, at Koenigsberg, Germany, to Leopold Lipmann and his wife Gertrud Lachmanski.

From 1917 to 1922 he was educated at the Universities of Koenigsberg, Berlin, and Munich, where he studied medicine and received a M.D. degree in 1924 at Berlin. In 1926 he was an assistant in Otto Meyerhof's laboratory at the Kaiser Wilhelm Institute, Berlin, and received a Ph.D. in 1927. He then went with Meyerhof to Heidelberg to conduct research on the biochemical reactions occurring in muscle.

In 1930 Lipmann went back to the Kaiser Wilhelm Institute in Berlin, then to a new institute in Copenhagen in 1932. Between 1931 and 1932, he served as a Rockefeller fellow at the Rockefeller Institute in New York and identified serine phosphate as the constituent of phosphoproteins that contains the phosphate.

He went to Copenhagen in 1932 as research associate in the Biological Institute of the Carlsberg Foundation. In 1939 he came to America and became a research associate in the department of biochemistry at Cornell Medical School, in Ithaca, New York, and in 1941 joined the research staff of the Massachusetts General Hospital in Boston, first as a research associate in the department of surgery, then heading his own group in the biochemical research laboratory of the hospital. In 1944 he became an American citizen. In 1949 he became professor of biological chemistry at Harvard Medical School, Boston. In 1957 he was appointed a member and professor of the Rockefeller Institute, New York City.

In 1947 he isolated and named coenzyme A (or CoA) as well as determining the molecular structure (1953) of this factor that is now known to be bound to acetic acid as the end product of sugar and fat breakdown in the absence of oxygen. It is one of the most important substances involved in cellular metabolism, since it helps convert amino acids, steroids, fatty acids, and hemoglobins into energy. For his discovery of this coenzyme, he was awarded the 1953 Nobel Prize for physiology or medicine. He died on July 24, 1986, in Poughkeepsie, New York.

lipophilic Literally "fat-loving." Applied to MOLECULAR ENTITIES (or parts of molecular entities) having a tendency to dissolve in fatlike (e.g., hydrocarbon) solvents.

See also HYDROPHILIC; HYDROPHOBIC INTERACTION.

lipophilicity Represents the AFFINITY of a molecule or a moiety (portion of a molecular structure) for a LIPOPHILIC (fat soluble) environment. It is commonly measured by its distribution behavior in a biphasic system, either liquid-liquid (e.g., partition coefficient in octan-1-ol/water) or solid-liquid (retention on reversed-phase HIGH-PERFORMANCE LIQUID CHROMATOGRAPHY [RP-HPLC] or thin-layer chromatography [TLC] system).

See also HYDROPHOBICITY.

lipoprotein Since lipids are hydrophobic, or water insoluble, certain lipids like cholesterol and triglycerides are coated or bonded with a protein so they can be carried in the blood. Since it is not possible to determine the exact lipoprotein content in blood due to the variety of lipoproteins, the medical profession talks about low-density lipoproteins (LDLs) and high-density lipoproteins (HDLs), which transport fats and cholesterol through the blood.

lipoxygenase A nonHEME iron ENZYME that catalyzes the INSERTION of O_2 into polyunsaturated fatty acids to form hydroperoxy DERIVATIVEs.

liquefaction When a solid begins to act as a fluid.

liquid aerosol A colloidal suspension of liquid in gas.

liquid crystal An organic molecule that has crystal-like properties but is liquid at normal temperatures.

liquid drop model (of the nucleus) States that NUCLEONs interact strongly with each other, just like molecules in a drop of liquid, which allows scientists to correlate many facts about nuclear masses and energies, as well as providing a useful model for understanding a large class of nuclear reactions, including fission.

littoral zone The shallow shoreward region of a freshwater body, just beyond the breaker zone, where light penetrates to the bottom sediments, giving rise to a zone that is colonized by rooted plants called helophytes; a region of a lake or pond where the water is less than 6 meters deep; in oceanography, the line extending from the high-water line to about 200 meters; also called the intertidal zone where submersion of tides is a normal event. The near-surface open water surrounded by the littoral zone is the limnetic zone, which gets ample light and is dominated by plankton. The littoral system is divided into a eulittoral (lower, middle, and upper) zone and a sublittoral (subtidal or supratidal) zone, the zone exposed to air only at its upper limit by the lowest spring tides. They are separated at a depth of about 50 meters. The term is also frequently used interchangeably with intertidal zone.

Loewi, Otto (1873–1961) Austrian/American *Physician, pharmacologist* Otto Loewi was born on June 3, 1873, in Frankfurt-am-Main, Germany, to Jacob Loewi, a merchant, and Anna Willstätter. He attended the humanistic gymnasium (grammar school) locally in 1881–90 and entered the Universities of Munich and Strassburg as a medical student in 1891. In 1896 he received a doctor's degree at Strassburg University.

After spending a few months working in the biochemical institute of Franz Hofmeister in Strassburg, he became an assistant to Carl von Noorden, clinician at the city hospital in Frankfurt during 1897–98. In 1898 he became an assistant of Professor Hans Horst Meyer, a renowned pharmacologist at the University of Marburg-an-der-Lahn, and professor of pharmacology in Vienna. In 1905 Loewi became associate professor at Meyer's laboratory, and in 1909 he was appointed to the chair of pharmacology in Graz. In 1940 he moved to the United States and became research professor at the School of Medicine of New York University, New York City, where he remained until his death.

His neurological researches during the period 1921–26 provided the first proof that chemicals were involved in the transmission of impulses from one nerve cell to another and from neuron to the responsive organ. It was for his discovery of the chemical transmission of nerve impulses that he received the Nobel Prize in physiology or medicine in 1936, jointly with Sir Henry Dale. Loewi spent his years investigating the physiology and pharmacology of metabolism, the kidneys, the heart, and the nervous system. He became an American citizen in 1946 and died on December 25, 1961.

London forces Attractive forces between apolar molecules due to their mutual polarizability. They are also components of the forces between POLAR MOLECULEs. Also called dispersion forces.

See also VAN DER WAALS FORCES.

lone (electron) pair Two paired electrons localized in the valence shell on a single atom. Lone pairs should be designated with two dots. The term *nonbonding electron pair* is more appropriate and is found in many modern textbooks.

loose ion pair *See* ION PAIR.

lowest unoccupied molecular orbital (LUMO) The orbital that can act as the electron acceptor, since it is the innermost (lowest energy) orbital that has room to accept an electron.

low spin In any COORDINATION entity with a particular d^n $(1 < n < 9)$ configuration and a particular geometry, if the *n* electrons are distributed so that they occupy the lowest possible energy levels, the entity is a low-spin complex. If some of the higher-energy *d* orbitals are occupied before all the lower-energy ones are completely filled, then the entity is a high-spin complex.

luminescent Refers to the act of emitting light without causing heat (called cold light). Bioluminescence is the act of producing light by biological organisms, e.g., fireflies, luminescent bacteria *(Photobacterium phosphoreum),* and foxfire in the form of Clitocybeilludens *(Omphalotus olerius), Panellus stypticus,* and *Armillaria mellea.* The lanthanides are a special group of elements—cerium (Ce, atomic no. 58) through lutetium (Lu, atomic no. 71)—within the PERIODIC TABLE that have trivalent cations that emit light. When these elements are absorbed into materials, the materials can become luminescent after being excited by an electrical current (electroluminescence) or by absorbing light (photoluminescence). As the ions relax to their GROUND STATE, they release light.

luster The appearance of a substance in reflected light.

lyase An ENZYME of EC class 4 that catalyzes the separation of a bond in a SUBSTRATE molecule.

See also EC NOMENCLATURE FOR ENZYMES.

lyate ion The anion produced by HYDRON removal from a solvent molecule. For example, the hydroxide ion is the lyate ion of water.

Lyman series A series of lines in the hydrogen atom spectrum that corresponds to transitions between the GROUND STATE (principal quantum number $n = 1$) and successive EXCITED STATES.

lyonium ion The cation produced by hydronation (PROTONATION) of a solvent molecule. For example, $CH_3OH_2^+$ is the lyonium ion of methanol.

See also ONIUM ION.

M

macromolecule A large molecule of high molecular mass composed of more than 100 repeated monomers (single chemical units of lower relative mass); a polymer. DNA, proteins, and polysaccharides are examples of macromolecules in living systems; a large complex molecule formed from many simpler molecules.

macrophage A type of blood cell that is able to ingest a wide variety of particulate materials. They are a type of PHAGOCYTE.

macroscopic diffusion control *See* MIXING CONTROL.

Madelung constant A term that accounts for the particular structure of an ionic crystal when the lattice energy is evaluated from the coulombic interactions. The value is different for each crystalline structure.

magic acid *See* SUPERACID.

magnetic circular dichroism (MCD) A measurement of CIRCULAR DICHROISM of a material that is induced by a magnetic field applied parallel to the direction of the measuring light beam. Materials that are achiral still exhibit MCD (the Faraday effect), since the magnetic field leads to the lifting of the degeneracy of electronic orbital and spin states and to the mixing of electronic states. MCD is frequently used in combination with absorption and CD studies to effect electronic assignments. The three contributions to the MCD spectrum are the A-term, due to Zeeman splitting of the ground and/or excited degenerate states; the B-term, due to field-induced mixing of states; and the C-term, due to a change in the population of molecules over the Zeeman sublevels of a PARAMAGNETIC ground state. The C-term is observed only for molecules with ground-state paramagnetism and becomes intense at low temperatures; its variation with field and temperature can be analyzed to provide magnetic parameters of the ground state, such as spin, g-factor, and zero-field splitting. Variable-temperature MCD is particularly effective in identifying and assigning electronic transitions originating from paramagnetic CHROMOPHORES.

magnetic equivalence Nuclei having the same resonance frequency in NUCLEAR MAGNETIC RESONANCE SPECTROSCOPY; also, identical spin-spin interactions with the nuclei of a neighboring group are magnetically equivalent. The spin-spin interaction between magnetically equivalent nuclei does not appear, and thus has no effect on the multiplicity of the respective NMR signals. Magnetically equivalent nuclei are necessarily also chemically equivalent, but the reverse is not necessarily true.

magnetic moment The twisting force exerted on a magnet or dipole when placed in a magnetic field. Magnetic moment and spin are interrelated.

magnetic quantum number (ml) The quantum number that signifies the orientation of an orbital around the nucleus; designates the particular orbital within a given set *(s, p, d, f)* in which an electron resides. Orbitals that differ only in their value of ml have the same energy in the absence of a magnetic field but a different energy in its presence.

magnetic resonance imaging (MRI) The visualization of the distribution of nuclear spins (usually water) in a body by using a magnetic-field gradient (NUCLEAR MAGNETIC RESONANCE SPECTROSCOPY). A similar technique, but less widely used, is to visualize the distribution of PARAMAGNETIC centers (ELECTRON PARAMAGNETIC RESONANCE SPECTROSCOPY).

See also IMAGING.

magnetic susceptibility For PARAMAGNETIC materials, the magnetic susceptibility can be measured experimentally and used to give information on the molecular magnetic DIPOLE MOMENT, and hence on the electronic structure of the molecules in the material. The paramagnetic contribution to the molar magnetic susceptibility of a material, χ, is related to the molecular magnetic dipole moment m by the Curie relation: χ = constant m^2/T.

magnetization transfer NMR method for determining kinetics of chemical exchange by perturbing the magnetization of nuclei in a particular site or sites and following the rate at which magnetic equilibrium is restored. The most common perturbations are saturation and inversion, and the corresponding techniques are often called "saturation transfer" and "selective inversion-recovery."

See also SATURATION TRANSFER.

magnetotactic Ability to orient in a magnetic field.

main group The *s* and *p* block elements (Li, Be, Na, Mg, K, Ca, B, C, N, O, F, Ne, Al, Si, P, S, Cl, Ar, Ga, Ge, As, Se, Br, and Kr). Zinc, cadmium, and mercury are often classified as main group elements. The PERIODIC TABLE is divided into blocks. The *s*-block elements have valence configuration *s*1 or *s*2. The *p*-block elements have valence configuration *s*2*p*1 to *s*2*p*6. The *d*-block and *f*-block elements usually have two electrons in the outermost *s*-orbital but have partially filled *d* or *f* subshells in an inner orbital.

malleability The property of metals that allows them to be beaten into thin sheets or extended or shaped or deformed without fracture; having characteristics that permit plastic deformation in compression without rupture.

manometer A two-armed barometer; reads air pressure and pressure of gases and vapors by balancing the pressure against a column of liquid in a U-tube.

marble A metamorphic rock made of calcium carbonate. Marble forms from limestone by metamorphism.

Marcus equation A general expression that correlates the GIBBS ENERGY OF ACTIVATION ($\Delta^{\ddagger}G$) with the driving force ($\Delta_rG^{o\prime}$) of the reaction:

$$\Delta^{\ddagger}G = (\lambda/4)(1 + \Delta_rG^{o\prime}/\lambda)^2$$

where λ is the reorganization energy and $\Delta_rG^{o\prime}$ is the standard free energy of the reaction corrected for the electrostatic work required to bring the reactants together. The INTRINSIC BARRIER of the reaction is $\lambda/4$. Originally developed for OUTER-SPHERE ELECTRON TRANSFER reactions, the Marcus equation has later been applied also to atom and group transfer reactions.

Markownikoff rule "In the addition of hydrogen halides to unsymmetrically constituted (unsaturated) hydrocarbons, the halogen atom becomes attached to the carbon bearing the lesser number of hydrogen atoms." Originally formulated by Markownikoff (Markovnikov) to generalize the orientation in additions of hydrogen halides to simple alkenes, this rule has been extended to polar addition reactions as follows. "In the HETEROLYTIC addition of a polar molecule to an alkene or alkyne, the more electronegative (nucleophilic) atom (or part) of the polar molecule becomes attached to the carbon atom bearing the smaller number of hydrogen atoms."

This is an indirect statement of the common mechanistic observation that the more electropositive (electrophilic) atom (or part) of the polar molecule becomes attached to the end of the multiple bond that would result in the more stable CARBENIUM ION (whether or not a carbenium ion is actually formed as a reaction INTERMEDIATE in the addition reaction). Addition in the opposite sense is commonly called anti-Markownikoff addition.

mass A measure of the amount of matter in an object, usually measured in grams or kilograms.

mass action law The rate of any given chemical reaction is proportional to the product of the activities or concentrations of the reactants. Also known as the law of mass action.

mass-law effect At equilibrium, the product of the activities (or concentrations) of the reacting species is constant. Thus for the equilibrium

$$\alpha A + \beta B \rightleftharpoons \gamma C + \delta D$$
$$K = [C]^{\gamma} [D]^{\delta}/[A]^{\alpha} [B]^{\beta}$$

See also COMMON-ION EFFECT; EQUILIBRIUM.

mass number The sum of the numbers of protons and neutrons in an atom.

mass spectrometer An instrument in which ions are separated according to the quotient mass/charge and in which the ions are measured electrically.

matrix isolation A term that refers to the isolation of a reactive or unstable species by dilution in an inert matrix (argon, nitrogen, etc.), usually condensed on a window or in an optical cell at low temperature to preserve its structure for identification by spectroscopic or other means.

matter Any substance that has inertia and occupies physical space; can exist as solid, liquid, gas, plasma, foam, or Bose-Einstein condensate.

States of matter. Illustration showing three states of matter for water: solid (ice), liquid (water), and gas (steam). The state of matter (or phase) of a substance depends on the ambient temperature and pressure. At any combination, there is a dynamic equilibrium between two or more phases. Water at a temperature of 0.072°C and an ambient pressure of 612 Pa has a dynamic equilibrium between all three phases. This is known as its TRIPLE POINT. A fourth phase, the plasma, exists at extremely high temperatures and is normally seen only in elements. ***(Courtesy of Mehau Kulyk/Science Photo Library)***

MCD *See* MAGNETIC CIRCULAR DICHROISM.

McMillan, Edwin Mattison (1907–1991) American *Physicist* Edwin Mattison McMillan was born on September 18, 1907, at Redondo Beach, California, the son of Dr. Edwin Harbaugh McMillan, a physician, and Anne Marie McMillan (*née* Mattison). He spent his early years in Pasadena, California, obtaining his education.

McMillan attended the California Institute of Technology, where he received a B.Sc. degree in 1928 and a M.Sc. degree the following year. He went to Princeton University for his Ph.D. in 1932.

He attended the University of California at Berkeley as a national research fellow working in the field of molecular beams, in particular the measurement of the magnetic moment of the proton by a molecular beam method. He became a member of the team at the radiation laboratory under Professor E.O. Lawrence, studying nuclear reactions and their products and helping design and construct cyclotrons.

He was a member of the faculty in the department of physics at Berkeley as an instructor in 1935, an assistant professor in 1936 and 1941, and a professor in 1946. In 1940 the creation of element 93, neptunium (symbol Np), was announced by Edwin M. McMillan and Philip H. Abelson. It was the first element heavier than uranium (known as a transuranium element).

He worked on national defense matters from 1940 to 1945, and during 1945 he helped design the synchrotron and synchrocyclotron. He returned to the University of California Radiation Laboratory from 1954 to 1958.

In 1951 McMillan and Glenn T. Seaborg received the Nobel Prize in chemistry "for their discoveries in the chemistry of the transuranium elements." He also received the 1950 Research Corporation Scientific Award and, in 1963, the Atoms for Peace Award along with Professor V. I. Veksler. He retired in 1973.

He was married to Elsie Walford Blumer, a daughter of Dr. George Blumer, dean emeritus of the Yale Medical School, and they had three children. He died on September 7, 1991, in El Cerrito, California.

mean lifetime *See* LIFETIME.

mechanism A detailed description of the process leading from the reactants to the products of a reaction, including a characterization as complete as possible of the composition, structure, energy, and other properties of REACTION INTERMEDIATES, products, and TRANSITION STATES. An acceptable mechanism of a specified reaction (and there may be a number of such alternative mechanisms not excluded by the evidence) must be consistent with the reaction stoichiometry, the RATE LAW, and with all other available experimental data, such as the stereochemical course of the reaction. Inferences concerning the electronic motions that dynamically interconvert successive species along the REACTION PATH (as represented by curved arrows, for example) are often included in the description of a mechanism.

It should be noted that for many reactions, all this information is not available, and the suggested mechanism is based on incomplete experimental data. It is not appropriate to use the term *mechanism* to describe a statement of the probable sequence in a set of stepwise reactions. That should be referred to as a reaction sequence, and not a mechanism.

See also GIBBS ENERGY DIAGRAM.

mechanism-based inhibition Irreversible INHIBITION of an enzyme due to its catalysis of the reaction of an artificial substrate. Also called "suicide inhibition."

mechanoreceptor A specialized sensory receptor that responds to mechanical stimuli, i.e., tension, pressure, or displacement. Examples include the inner-ear hair cells, carotid sinus receptors, and muscle spindles.

mediator modulator (immune modulator; messenger) An object or substance by which something is mediated, such as:

- A structure of the nervous system that transmits impulses eliciting a specific response
- A chemical substance (transmitter substance) that induces activity in an excitable tissue, such as nerve or muscle (e.g., hormones)
- A substance released from cells as the result of an antigen-antibody interaction or by the action of an antigen with a sensitized lymphocyte (e.g., cytokine)

Concerning mediators of immediate hypersensitivity, the most important include histamine, leukotriene e.g., SRS-A (slow-reacting substance of anaphylaxis, ECF-A (eosinophil chemotactic factor of anaphylaxis), PAF (platelet-activating factor), and serotonin. There are also three classes of lipid mediators that are synthesized by activated mast cells through reactions initiated by the actions of phospholipase A2. These are prostaglandins, leukotrienes, and platelet-activating factors (PAF).

medicinal chemistry A chemistry-based discipline, also involving aspects of biological, medical, and pharmaceutical sciences. It is concerned with the invention, discovery, design, identification, and preparation of biologically active compounds; the study of their METABOLISM; the interpretation of their mode of action at the molecular level; and the construction of STRUCTURE-ACTIVITY RELATIONSHIPS.

medium The phase (and composition of the phase) in which CHEMICAL SPECIES and their reactions are studied in a particular investigation.

megapascal (MPa) A unit of pressure. 1 MPa = 1,000,000 Pa (pascals); 1 megapascal (MPa) = 10 bar; 1 bar is approximately equal to 1 atmosphere of pressure.

Meisenheimer adduct A cyclohexadienyl derivative formed as LEWIS ADDUCT from a NUCLEOPHILE (LEWIS BASE) and an AROMATIC or heteroaromatic compound, also called Jackson-Meisenheimer adduct. In earlier usage the term *Meisenheimer complex* was restricted to the typical Meisenheimer alkoxide ADDUCTs of nitro-substituted aromatic ethers, for example

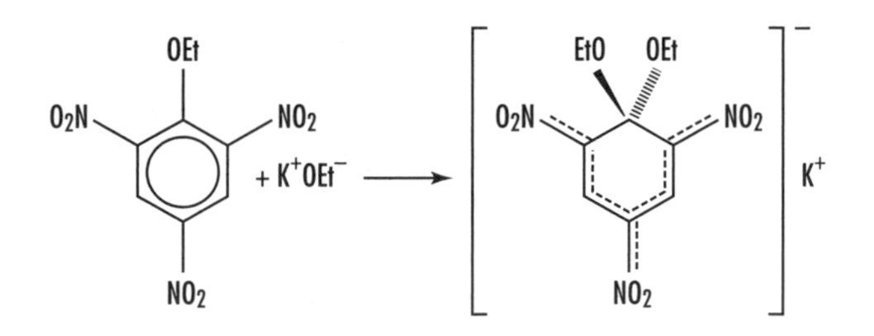

Analogous cationic adducts, such as

are considered to be reaction INTERMEDIATES in ELECTROPHILIC aromatic SUBSTITUTION REACTIONS, are called "Wheland intermediates" and sometimes, inappropriately, σ-complexes.

See also CHEMICAL REACTION; SIGMA (σ) ADDUCT.

melting point The temperature when matter is converted from solid to liquid.

melting point (corrected/uncorrected) The term originally signified that a correction was made (not made) for the emergent stem of the thermometer. In current usage, it often means that the accuracy of the thermometer was (was not) verified. This current usage is inappropriate and should be abandoned.

membrane potential The difference in electrical charge (voltage difference) across the cell membrane due to a slight excess of positive ions on one side and of negative ions on the other; the potential inside a membrane minus the potential outside. A typical membrane potential is –60 mV, where the inside is negative relative to the surrounding fluid, and resting membrane potentials are typically found between –40 and –100 mV.

meniscus The curvature of the surface of a liquid in a vessel at the interface of the liquid with the container wall. If the attractive forces between the molecules of the liquid and the wall are greater than those between the molecules of the liquid itself, the meniscus curves up, and the surface is "wet" by the liquid. The reverse causes the meniscus to curve down (nonwetting).

See also VAN DER WAALS FORCES.

Menkes disease A sex-linked inherited disorder, causing defective gastrointestinal absorption of copper and resulting in copper deficiency early in infancy.

mercury battery No longer used or manufactured in the United States due to pollution potential.

mesolytic cleavage Cleavage of a bond in a RADICAL ION whereby a RADICAL and an ion are formed. The term reflects the mechanistic duality of the process, which can be viewed as homolytic or heterolytic, depending on how the electrons are attributed to the fragments.

See also HETEROLYSIS; HOMOLYSIS.

mesomeric effect The effect (on reaction rates, ionization equilibria, etc.) attributed to a substituent due to overlap of its *p* or pi orbitals with the *p* or pi orbitals of the rest of the MOLECULAR ENTITY. DELOCALIZATION is thereby introduced or extended, and electronic charge may flow to or from the substituent. The effect is symbolized by M.

Strictly understood, the mesomeric effect operates in the ground electronic state of the molecule. When the molecule undergoes electronic excitation or its energy is increased on the way to the TRANSITION STATE of a CHEMICAL REACTION, the mesomeric effect may be enhanced by the ELECTROMERIC EFFECT, but this term is not much used, and the mesomeric and electromeric effects tend to be subsumed in the term RESONANCE EFFECT of a SUBSTITUENT.

See also ELECTRONIC EFFECT; FIELD EFFECT; INDUCTIVE EFFECT.

mesomerism Essentially synonymous with RESONANCE. The term is particularly associated with the picture of pi electrons as less localized in an actual molecule than in a LEWIS FORMULA. The term is intended to imply that the correct representation of a structure is intermediate between two or more Lewis formulae.

See also AROMATIC (2); DELOCALIZATION.

mesophase The phase of a liquid crystalline compound between the crystalline and the isotropic liquid phase.

messenger RNA (mRNA) An RNA molecule that transfers the coding information for protein synthesis from the chromosomes to the ribosomes. Fragments of ribonucleic acid serve as templates for protein synthesis by carrying genetic information from a strand of DNA to ribosomes for translation into a protein. The information from a particular gene or group of genes is transferred from a strand of DNA by constructing a complementary strand of RNA through transcription. Transfer RNA (tRNA), composed of three nucleotide segments attached to specific amino acids, correctly match with a template strand of mRNA, lining up the correct order of amino acids and bonding them, via translation in the ribosome with rRNA (ribosomal RNA), to form a protein.

met- A qualifying prefix indicating the oxidized form of the parent protein, e.g., methemoglobin.

metabolism The entire physical and chemical processes involved in the maintenance and reproduction of life in which nutrients are broken down to generate energy and to give simpler molecules (CATABOLISM) that can be used to form more complex molecules (ANABOLISM).

In the case of HETEROTROPHIC ORGANISMS, the energy evolving from catabolic processes is made available for use by the organism.

In medicinal chemistry the term *metabolism* refers to the BIOTRANSFORMATION of XENOBIOTICS and particularly DRUGS.

metabolite Any intermediate or product resulting from METABOLISM.

metal Metals comprise 80 percent of known elements. Any element below and to the left of the stepwise division (metalloids) in the upper right corner of the PERIODIC TABLE of elements.

metallic bonding The bonding in metallic elements and a few other compounds in which the valence electrons are delocalized over a large number of atoms to produce a large number of molecular orbitals whose energies are close enough together to be considered to make up a continuous band rather than discrete energy

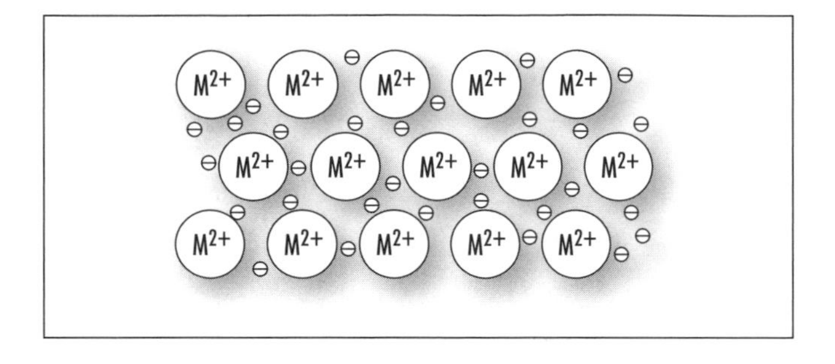

The bonding in metallic elements

levels. The band is not filled, and electrons are free to move in an electric field, giving typical metallic conductivity. Sometimes this is modeled as ions surrounded by a "sea" of electrons.

metallic conduction The conduction of an electrical current through a METAL or along a metallic surface.

metalloenzyme An ENZYME that, in the active state, contains one or more METAL ions that are essential for its biological function.

metalloids Elements with properties intermediate between METALS and nonmetals: boron, silicon, germanium, arsenic, antimony, tellurium, and polonium.

metallo-immunoassay A technique in which ANTIGEN-ANTIBODY recognition is used, with attachment of a METAL ion or metal complex to the antibody. The specific absorption or (radioactive) emission of the metal is then used as a probe for the location of the recognition sites.

See also IMAGING; RADIONUCLIDE.

metallothionein A small, cysteine-rich protein that binds heavy METAL ions such as zinc, cadmium, and copper in the form of CLUSTERS.

metallurgy The science of METALS and their properties at the macroscopic and atomic level; overall processes by which metals are extracted from ores.

metastable *See* STABLE.

metastable (chemical) species *See* TRANSIENT (CHEMICAL) SPECIES.

metathesis A bimolecular process formally involving the exchange of a BOND (or bonds) between similar interacting CHEMICAL SPECIES so that the bonding affiliations in the products are identical (or closely similar) to those in the reactants. For example:

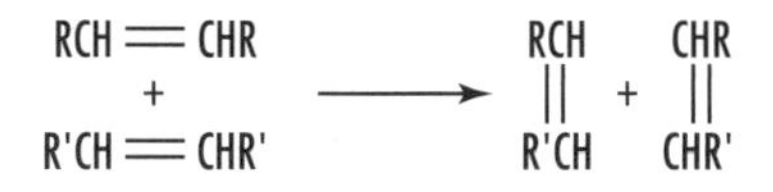

(The term has its origin in inorganic chemistry with a different meaning, but this older usage is not applicable in physical organic chemistry.)

See also BIMOLECULAR REACTION.

meter A unit of metric measure that equals 39.37 in.

methane hydrate A frozen latticelike substance formed when water and methane, CH_4, are combined under low temperatures and high pressures. It is a crystalline combination of a natural gas and water, called a CLATHRATE, and looks like ice but burns like a candle.

methane monooxygenase A METALLOENZYME that converts methane and dioxygen to methanol using NADH as CO-SUBSTRATE. Two types are known, one containing a dinuclear oxo-bridged iron center, the other a copper protein.

See also NUCLEARITY.

methanogen Strictly ANAEROBIC ARCHAEA, able to use a variety of SUBSTRATES (e.g., dihydrogen, formate, methanol, methylamine, carbon monoxide, or acetate) as ELECTRON DONORS for the reduction of carbon dioxide to methane.

methylene *See* CARBENE.

methylidyne *See* CARBYNE.

me-too drug A compound that is structurally very similar to already known DRUGs, with only minor pharmacological differences.

Meyerhof, Otto Fritz (1884–1951) German *Physiologist, chemist* Otto Fritz Meyerhof was born on April 12, 1884, in Hannover to Felix Meyerhof, a merchant, and Bettina May. He went to the Wilhelms Gymnasium (classical secondary school) in Berlin, leaving at age 14 only to have kidney problems two years later that kept him confined for a long period. He eventually studied medicine at Freiburg, Berlin, Strassburg, and Heidelberg and graduated in 1909. From 1912 he worked at the University of Kiel, becoming a professor in 1918.

Meyerhof conducted experiments on the energy changes in cellular respiration. For his discovery of the fixed relationship between the consumption of oxygen and the metabolism of lactic acid in the muscle, he was awarded, together with the English physiologist A.V. Hill, the Nobel Prize for physiology or medicine in 1922. In 1925 Meyerhof successfully extracted the enzymes that convert glycogen to lactic acid from the muscle. He introduced the term *glycolysis* to describe the anaerobic degradation of glycogen to lactic acid, and he showed the cyclic nature of energy transformations in living cells. This metabolic pathway of glycolysis—conversion of glucose to lactic acid—is now known as the Embden-Meyerhof pathway after Meyerhof and Gustav George Embden.

During World War II, he went to the United States and became a research professor of physiological chemistry, a position created for him by the University of Pennsylvania and the Rockefeller Foundation. He died from a heart attack on October 6, 1951.

mica A group of silicate minerals composed of varying amounts of aluminum, potassium, magnesium, iron, and water that forms flat, platelike crystals that cleave into smooth flakes.

micellar catalysis The acceleration of a CHEMICAL REACTION in solution by the addition of a surfactant at a concentration higher than its CRITICAL MICELLE CONCENTRATION so that the reaction can proceed in the environment of surfactant aggregates (MICELLEs). (Rate enhancements may be due, for example, to higher concentration of the reactants in that environment, more favorable orientation and solvation of the species, or enhanced rate constants in the micellar pseudophase of the surfactant aggregate.) Micelle formation can also lead to a decreased reaction rate.

See also CATALYST.

micelle Surfactants in solution are often association COLLOIDs, i.e., they tend to form aggregates of colloidal dimensions that exist in equilibrium with the molecules or ions from which they are formed. Such aggregates are termed micelles.

See also INVERTED MICELLE.

Michaelis-Menten kinetics The dependence of an initial RATE OF REACTION upon the concentration of a SUBSTRATE S that is present in large excess over the concentration of an enzyme or other CATALYST (or reagent) E, with the appearance of saturation behavior following the Michaelis-Menten equation:

$$nu = V[S]/(K_m + [S]),$$

where nu is the observed initial rate, V is its limiting value at substrate saturation (i.e., $[S] >> K_m$), and K_m is the substrate concentration when nu = $V/2$. The definition is experimental, i.e., it applies to any reaction that follows an equation of this general form. The symbols V_{ma} or nu_{ma} are sometimes used for V.

The parameters V and K_m (the Michaelis constant) of the equation can be evaluated from the slope and intercept of a linear plot of nu^{-1} against $[S]^{-1}$ (a LINEWEAVER-BURK PLOT) or from the slope and intercept of a linear plot of nu against h/$[S]$ (Eadie-Hofstee plot).

A Michaelis-Menten equation is also applicable to the condition where E is present in large excess, in which case the concentration [E] appears in the equation instead of [S]. The term has sometimes been used to describe reactions that proceed according to the scheme

$$E + S \underset{k_{-1}}{\overset{k_1}{\rightleftharpoons}} ES \xrightarrow{k_{cat}} \text{Products} + E$$

in which case $K_m = (k_{-1} + k_{cat})/k_1$ (Briggs-Haldane conditions). It has more usually been applied only to the special case in which $k_{-1} >> k_{cat}$ and $K_m = k_{-1}/k_1 = K_s$; in this case, K_m is a true dissociation constant (Michaelis-Menten conditions).

See also RATE-DETERMINING STEP.

micronutrient A compound essential for cellular growth, being present in concentrations less than about 1 m*M* in the growth medium.

microscopic chemical event *See* CHEMICAL REACTION; MOLECULARITY.

microscopic diffusion control (**encounter control**) The observable consequence of the limitation that the rate of a bimolecular CHEMICAL REACTION in a homogeneous medium cannot exceed the rate of encounter of the reacting MOLECULAR ENTITIES.

If (hypothetically) a BIMOLECULAR reaction in a homogeneous medium occurred instantaneously when two reactant molecular entities made an encounter, the RATE OF REACTION would be an ENCOUNTER-CONTROLLED RATE, determined solely by rates of diffusion of reactants. Such a hypothetical fully diffusion-controlled rate is also said to correspond to total microscopic diffusion control and represents the asymptotic limit of the rate of reaction as the RATE CONSTANT for the chemical conversion of the encounter pair into product (or products) becomes large relative to the rate constant for separation (or dissociation) of the encounter pair.

"Partial microscopic diffusion control" is said to operate in a homogeneous reaction when the rates of chemical conversion and of separation are comparable. (The degree of microscopic diffusion control usually cannot be determined with any precision.)

See also MIXING CONTROL.

microscopic reversibility, principle of In a REVERSIBLE REACTION, the mechanism in one direction is exactly the reverse of the mechanism in the other direction. This does not apply to reactions that begin with a photochemical excitation.

See also CHEMICAL REACTION; DETAILED BALANCING.

microstate A microstate describes a specific detailed microscopic configuration of a system. For an atom, it is a specific combination of quantum numbers that the electrons can have in that configuration. For a larger system, it is the state defined by specifying the location and momentum of each molecule and atom in the system.

microwave Any electromagnetic wave having a wavelength from 10 mm to 300 mm (1 GHz to 30 GHz).

microwave spectrum Usually refers to the SHF and EHF frequencies. Super-high frequency (SHF) ranges from 3 to 30 GHz, or free-space wavelengths of 100 to 10 mm. Extremely-high frequency (EHF) ranges from 30 to 300 GHz, or free-space wavelengths of 10 to 1 mm.

migration (1) The (usually INTRAMOLECULAR) transfer of an atom or GROUP during the course of a MOLECULAR REARRANGEMENT.

(2) The movement of a BOND to a new position, within the same MOLECULAR ENTITY, is known as bond migration.

Allylic rearrangements, for example:

$$RCH{=}CHCH_2X \rightarrow RCH(X)CH{=}CH_2$$

exemplify both types of migration.

migratory aptitude The term is applied to characterize the relative tendency of a group to participate in a rearrangement. In nucleophilic rearrangements (MIGRATION to an electron-deficient center), the migratory aptitude of a group is loosely related to its capacity to stabilize a partial positive charge, but exceptions are known, and the position of hydrogen in the series is often unpredictable.

migratory insertion A combination of MIGRATION and INSERTION. The term is mainly used in organometallic chemistry.

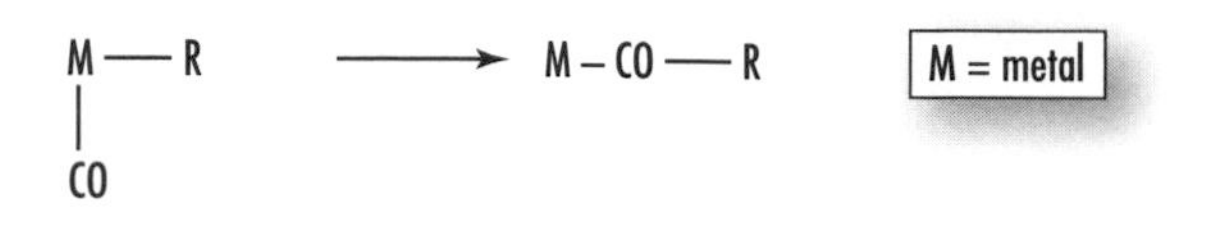

mineral A naturally occurring homogeneous solid, inorganically formed, with a definite chemical composition, usually crystalline in form, and an ordered atomic arrangement, e.g., quartz. Also a naturally occurring inorganic element or compound having an orderly internal structure and characteristic chemical composition, crystal form, and physical properties. The important point is that while a mineral has a characteristic composition, it is not always definite.

minimum structural change, principle of *See* MOLECULAR REARRANGEMENT.

miscibility The ability of one liquid to mix with or dissolve in another liquid to form a uniform blend.

mitochondria CYTOPLASMIC organelles of most eukaryotic cells, they are surrounded by a double membrane and produce ADENOSINE 5′-TRIPHOSPHATE (ATP) as useful energy for the cell by oxidative PHOSPHORYLATION. The proteins for the ATP-generating electron transport of the respiration chain are located in the inner mitochondrial membrane. Mitochondria contain many ENZYMEs of the citric acid cycle and for fatty-acid β-oxidation. They also contain DNA, which encodes some of their proteins, the remainder being encoded by nuclear DNA.

See also EUKARYOTE.

mitosis The cell-division process in eukaryotic cells that replicates chromosomes so that two daughter cells get equally distributed genetic material from a parent cell, making them identical to each other and the parent. It is a five-step process that includes prophase, prometaphase, metaphase, anaphase, and telophase. Interphase is the time in the cell cycle when DNA is replicated in the nucleus.

See also EUKARYOTE.

mixed valency This is one of several names, such as "mixed oxidation state" or "nonintegral oxidation state," used to describe COORDINATION compounds and CLUSTERs, in which a METAL is present in more than one level of OXIDATION. The importance in biology is due to the often-complete DELOCALIZATION of the valence electrons over the cluster, allowing efficient ELECTRON-TRANSFER processes.

See also OXIDATION NUMBER.

mixing control The experimental limitation of the RATE OF REACTION in solution by the rate of mixing of solutions of the two reactants. It can occur even when the reaction rate constant is several powers of 10 less than that for an ENCOUNTER-CONTROLLED rate. Analogous (and even more important) effects of the limitation of reaction rates by the speed of mixing are encountered in heterogeneous (solid-liquid, solid-gas, liquid-gas) systems.

See also MICROSCOPIC DIFFUSION CONTROL; STOPPED FLOW.

mixture Matter composed of two or more substances, each of which retains its identity and properties.

mobile phase Part of an analytical method in GC (GAS CHROMATOGRAPHY) in which a sample is vaporized and injected into a carrier gas (called the mobile phase, usually helium) moving through a column.

Möbius aromaticity A monocyclic array of ORBITALs in which a single out-of-phase overlap (or, more generally, an odd number of out-of-phase overlaps) reveals the opposite pattern of AROMATIC character to Hückel systems; with 4n electrons it is stabilized (aromatic), whereas with 4n + 2 it is destabilized (antiaromatic). In the excited state 4n + 2, Möbius pi-electron systems are stabilized, and 4n systems are destabilized. No examples of GROUND-STATE Möbius pi systems are known, but the concept has been applied to TRANSITION STATEs of PERICYCLIC REACTIONs (*see* AROMATIC [3]).

The name is derived from the topological analogy of such an arrangement of orbitals to a Möbius strip.

See also HÜCKEL (4N + 2) RULE.

Moco *See* MOLYBDENUM COFACTOR.

model A synthetic COORDINATION entity that closely approaches the properties of a METAL ion in a PROTEIN and yields useful information concerning biological structure and function. Given the fact that the term is also loosely used to describe various types of molecular structures (constructed, for example, in the computer), the term BIOMIMETIC is more appropriate.

moderator A substance such as hydrogen, deuterium, oxygen, or paraffin used in a nuclear reactor to slow down the NEUTRON.

moiety In physical organic chemistry, moiety is generally used to signify part of a molecule, e.g., in an ester R^1COOR^2, the alcohol moiety is R^2O. The term should not be used for a small fragment of a molecule.

molality Concentration term expressed as number of moles of solute per kilogram of solvent.

molarity The number of moles of solute dissolved in 1 liter of solution.

molar solubility Number of moles of a solute that dissolve to produce a liter of saturated solution.

mole (mol) An amount of substance that contains as many items such as ions, molecules, etc., as the number of atoms in exactly 12 grams of carbon (C). The number of molecules contained is equal to 6.022×10^{23} (602,200,000,000,000,000,000,000), known as Avogadro's number. Therefore a mole is anything that has Avogadro's number of items in it.

molecular entity Any constitutionally or isotopically distinct atom, MOLECULE, ion, ION PAIR, RADICAL, RADICAL ION, COMPLEX, conformer, etc., identifiable as a separately distinguishable entity.

The term *molecular entity* is used in this glossary as a general term for singular entities, irrespective of their nature, while CHEMICAL SPECIES stands for sets or ensembles of molecular entities. Note that the name of a compound may refer to the respective molecular entity or to the chemical species, e.g., methane may mean a single molecule of CH_4 (molecular entity) or a molar amount—specified or not (chemical species)—participating in a reaction.

The degree of precision necessary to describe a molecular entity depends on the context. For example, "hydrogen molecule" is an adequate definition of a certain molecular entity for some purposes, whereas for others it is necessary to distinguish the electronic state and/or vibrational state and/or nuclear spin, etc., of the hydrogen molecule.

molecular equation Any equation for a chemical reaction where all formulas are written as if all substances exist as molecules.

molecular formula The formula of a compound in which the subscripts give the number of each element in the formula.

molecular geometry The arrangement of atoms around a central atom of a molecule or polyatomic ion; the general shape of a molecule determined by the relative positions of the atomic nuclei.

molecular graphics The visualization and manipulation of three-dimensional representations of molecules on a graphical display device.

molecularity The number of reactant MOLECULAR ENTITIES that are involved in the "microscopic chemical event" constituting an ELEMENTARY REACTION. (For reactions in solution, this number is always taken to exclude molecular entities that form part of the MEDIUM and that are involved solely by virtue of their solvation of solutes.) A reaction with a molecularity of one is called "unimolecular"; one with a molecularity of two is "bimolecular"; and a molecularity of three is "termolecular."

See also CHEMICAL REACTION; ORDER OF REACTION.

molecular mechanics calculation An empirical calculational method intended to give estimates of structures and energies for conformations of molecules. The method is based on the assumption of "natural" bond lengths and angles, deviation from which leads to strain, and the existence of torsional interactions and attractive and/or repulsive VAN DER WAALS and dipolar forces between nonbonded atoms. The method is also called "(empirical) force-field calculations."

molecular metal A nonmetallic material whose properties resemble those of METALS, usually following oxidative doping, e.g., polyacetylene following oxidative doping with iodine.

molecular modeling A technique for the investigation of molecular structures and properties using computational chemistry and graphical visualization techniques in order to provide a plausible three-dimensional representation under a given set of circumstances.

molecular orbital A one-electron wave function describing an electron moving in the effective field provided by the nuclei and all other electrons of a MOLECULAR ENTITY of more than one atom. Such molecular orbitals can be transformed in prescribed ways into component functions to give localized molecular orbitals. Molecular orbitals can also be described, in terms of the number of nuclei (or centers) encompassed, as two-center, multicenter, etc., molecular orbitals, and they are often expressed as a linear combination of ATOMIC ORBITALS.

An ORBITAL is usually depicted by sketching contours on which the wave function has a constant value (contour map) or by indicating schematically the envelope of the region of space in which there is an arbitrarily fixed high (say 96 percent) probability of finding the electron occupying the orbital, giving also the algebraic sign (+ or –) of the wave function in each part of that region.

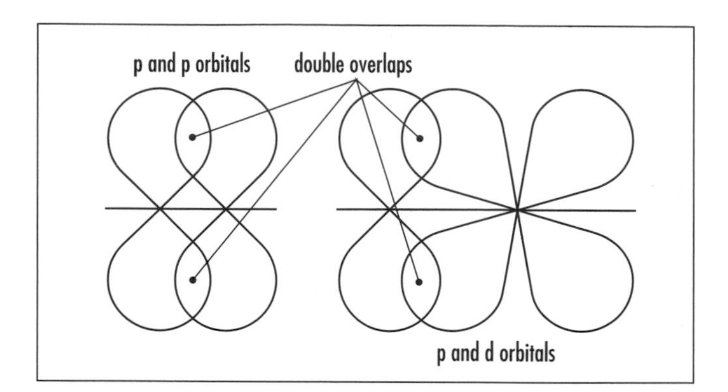

Molecular orbitals. A one-electron wave function describing an electron moving in the effective field provided by the nuclei and all other electrons of a molecular entity of more than one atom.

molecular orbital theory A theory of chemical bonding that describes COVALENT BONDing as ORBITALS that are formed by the combination of atomic orbitals on different atoms.

molecular rearrangement The term is traditionally applied to any reaction that involves a change of connectivity (sometimes including hydrogen) and violates the so-called principle of minimum structural change. According to this oversimplified principle, CHEMICAL SPECIES do not isomerize in the course of a TRANSFORMATION, e.g., SUBSTITUTION, or the change of a functional GROUP of a chemical species into a different functional group is not expected to involve the making or breaking of more than the minimum number of bonds required to effect that transformation. For example, any new substituents are expected to enter the precise positions previously occupied by displaced groups.

The simplest type of rearrangement is an INTRAMOLECULAR reaction in which the product is isomeric with the reactant (one type of intramolecular isomerization). An example is the first step of the Claisen rearrangement:

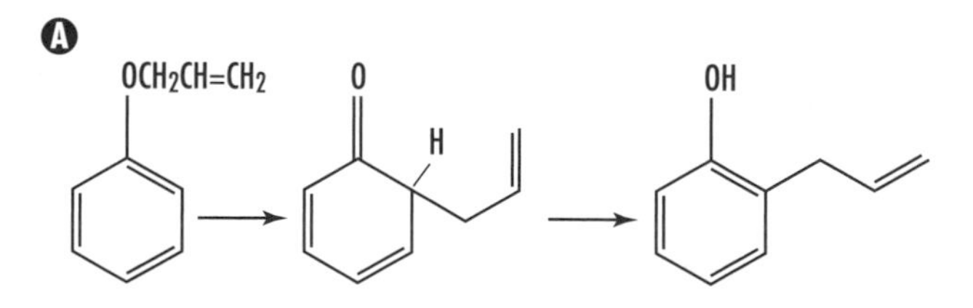

The definition of molecular rearrangement includes changes in which there is a MIGRATION of an atom or

(continued on page 186)

Molecular Modeling

by Karl F. Moschner, Ph.D.

Models, representations of real objects, have long been used to understand, explain, predict, and, ultimately, harness and exploit natural phenomena. They range from simple descriptions or drawings useful for conveying basic concepts to precise mathematical relationships that can be embodied in sophisticated computer programs. Whatever their form, all models are approximations with individual strengths and limitations that must be astutely applied to solve particular problems quickly and properly.

Molecular modeling deals with the representation and prediction of structures, properties, interactions, and reactions of chemical substances. It is intimately linked with experimental investigations of atomic and molecular structure and determinations of physical, chemical, and biological properties; mathematics (including statistics); and computer science and graphics. At its heart is the representation of molecular structure and interactions, especially chemical bonding. Modern molecular modeling has many uses as an effective communication tool, as a means of simulating chemical phenomena that are difficult or impossible to observe experimentally, and, ultimately, as a means of designing new compounds and materials.

Chemists have historically employed various means of representating molecular structure. Two-dimensional drawings of atoms connected by lines are some of the most common molecular representations. Each line represents a chemical bond that, in the simplest case, is a pair of electrons shared between the connected atoms, resulting in a very strong attractive interatomic force. The various interatomic forces define the structure or shape of a molecule, while its chemistry is dependent on the distribution of electrons. A chemical reaction involves a change in the electron distribution, i.e., a change in bonding.

X-ray crystallographic studies demonstrated that bond distances are very uniform and that the three-dimensional arrangements of atoms in a molecule have well-defined geometries. The regularity in molecular structures made it possible to build scale models about 250 million times larger than the molecule. Some of the earliest molecular scale models used standard atom-type wooden balls with holes at appropriate angles that could be connected by ideal bond-length sticks or springs. Such simple models were often a chemist's first opportunity to "see" a molecule, i.e., to develop a concept of its shape or conformation.

Molecular scale models of various types served as important tools for chemists. LINUS PAULING was a proponent for using molecular scale models to better understand the critical influence of three-dimensional structure on molecular properties and reactivities, and models helped Francis Crick and James Watson to elucidate the double helical structure of DNA (WATSON-CRICK MODEL). But they were awkward, fragile, and costly and offered only limited structural information. Indeed, they failed to provide any means of quantitatively comparing conformations of flexible molecules, interactions between molecules, or chemical reactivities. During the second half of the 20th century, chemists sought to address these needs by taking advantage of theoretical advances and emerging computer technology to develop two general approaches to computational molecular modeling based on molecular mechanics (MM) and quantum mechanics (QM).

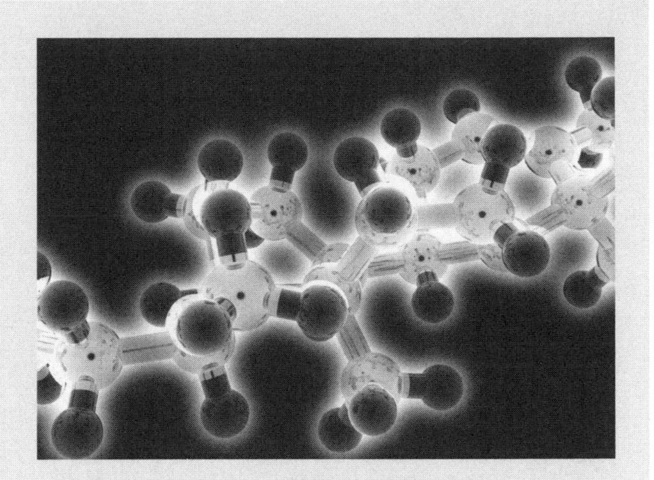

Molecular model of an unidentified chemical. Its atoms (spheres) interact to form chemical bonds (rods) that hold the molecule together. ***(Courtesy of Lawrence Lawry/ Science Photo Library)***

Molecular mechanics computes molecular potential energy using a force field, a series of discrete mathematical functions that reflect measurable intra- and intermolecular forces. In a manner similar to molecular scale models, MM employs "ideal" atom- and bond-types. Distances are based principally on X-ray crystal structures, and forces are derived from vibrational spectra. MM computer programs (e.g., MM2, MM3, SYBYL, CHARMM, and MACROMODEL) are differentiated by the range and specificity of their atom types, the mathematical expressions in their force fields, and their treatment of nonbonding interactions, including electrostatics, hydrogen-bonding, van der Waals forces, and solvation. Some force fields have

(continues)

Molecular Modeling
(continued)

been optimized to better reproduce structures for a specific class of compounds, such as peptides (for proteins and enzymes) or carbohydrates (for sugars, polysaccharides, and cellulose), thereby sacrificing some degree of general utility. Force fields have also been parameterized using QM results, a technique useful to extend MM capabilities when little or no experimental data are available for specific atom or bond types.

MM is inherently limited to studying systems composed of well characterized atom and bond types. It provides molecular geometries in good agreement with experimental values and reliable comparative energies, but it can not model chemical reactions. The biggest advantage of MM is its speed. MM studies can consist of multiple molecules including thousands of atoms. MM force fields can also be used in molecular dynamics and Monte Carlo calculations, which are used to investigate time-dependent phenomena (e.g., protein folding), and in free-energy calculations that are not feasible with QM. Molecular mechanics has been used in a wide range of applications, including simulation of ice crystal growth inhibition by fish antifreeze peptides, comparison of enzyme inhibitors to design improved drugs, investigation of surfactant aggregation in micelles, and studying polymer conformations in solution.

A fundamental postulate of quantum mechanics is that atoms consist of a nucleus surrounded by electrons in discrete atomic orbitals. When atoms bond, their atomic orbitals combine to form molecular orbitals. The redistribution of electrons in the molecular orbitals determines the molecule's physical and chemical properties. QM methods do not employ atom or bond types but derive approximate solutions to the Schrödinger equation to optimize molecular structures and electronic properties. QM calculations demand significantly more computational resources than MM calculations for the same system. In part to address computer-resource constraints, QM calcu-

Molecular model of hydrogen gas. The two white spheres represent individual hydrogen atoms, and the gray bar represents the single bond between them. Two forms of hydrogen exist: orthohydogen (75 percent) and parahydrogen (25 percent). The former's two nuclei spin in parallel; the latter's spin antiparallel. They have slightly different boiling and melting points. Hydrogen is the lightest element and is a widespread constituent of water, minerals, and organic matter. It is produced by electrolysis of water or reactions between acids and metals. Hydrogen is used industrially in hydrogenation of fats and oils and in hydrocarbon synthesis. ***(Courtesy of Adam Hart-Davis/Science Photo Library)***

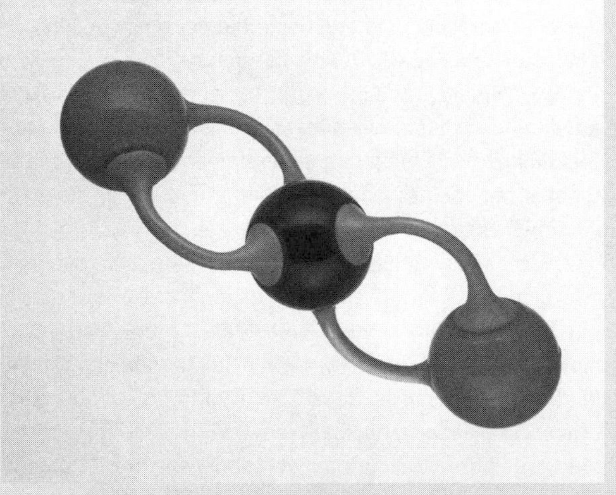

Molecular model of carbon dioxide, CO_2. The black sphere represents an atom of carbon. Gray spheres represent oxygen. The atoms in this linear molecule are held together by two double bonds, each involving a shared pair of electrons. Carbon dioxide is a colorless gas at room temperature. It occurs naturally in the atmosphere and is a waste product of animal and plant respiration. ***(Courtesy of Adam Hart-Davis/Science Photo Library)***

lations may be performed at different levels of approximation that can be divided into two classes: semiempirical and ab initio (from the beginning, i.e., based on first principles). Even so, a geometrical optimization of a molecule composed of 30 atoms that is nearly instantaneous on a personal computer (PC) using MM methods may require several minutes using semiempirical QM methods and an hour or even days using ab initio techniques.

Semiempirical QM methods (e.g., PM3, AM1, and MNDO) employ a variety of simplifications and experimentally derived elemental parameters to speed up calculations versus ab initio methods. All implementations support most of the elements in biologically and commercially important organic compounds. Some programs also support a wide range of transition metals. Semiempirical QM calculations provide very good geometries and associated ground-state properties: atom-centered charges, ionization potential, heats of formation, and some indication of reactivity based on the frontier molecular orbitals (the highest occupied and lowest unoccupied molecular orbitals, HOMO and LUMO). But these methods are generally not suited for studying reaction mechanisms. The limitations of semiempirical QM methods are offset by the ability to conduct QM calculations on systems consisting of hundreds of atoms, including small enzymes.

The ab initio QM methods are based solely on the laws of quantum mechanics and therefore have the broadest applicability. They can be carried out at different levels of approximation in order to balance the required accuracy against the computational demands. The quality of the calculations is principally determined by the selected basis set (functions that describe the atomic orbitals) and the treatment of electron correlation (interaction between electrons). Generally, moderate basis sets are sufficient for accurate ground-state calculations, but large basis sets and proper treatment of electron correlation are required to model excited states, transient species, or chemical reaction mechanisms. Fortunately, modern treatment of electron correlation, based on density functional theory, has made high-quality calculations using a PC feasible for systems containing tens of atoms, sufficient to study enzyme-active sites. Applications of ab initio QM include designing new catalysts, semiconductors, and dyes and studying atmospheric chemistry, such as the impact of greenhouse gases and chlorofluorocarbons (freons) on ozone depletion.

Another area of molecular modeling involves development of quantitative structure-activity or structure-property relationships (QSAR and QSPR). These studies use a range of statistical methods (linear and nonlinear regression, neural nets, clustering, genetic algorithms, etc.) to correlate molecular properties determined experimentally and derived from MM or QM calculations against the known end-use biological activities or physical or chemical properties for a large training set of molecules. Such activity models can then be used to predict the performance of similar molecules, even ones that do not yet exist. A key to success of QSAR studies is that the composition and structure, i.e., chemistry, of the test compound must be represented in the training set, otherwise the predictions can be very misleading. Even with this limitation, it is often possible to generate hundreds or even thousands of ideas that can be rapidly screened for the most promising compounds to advance for laboratory synthesis and testing. Such high-throughput screening (HTS) is rapidly being adopted as standard research practice. In particular, pharmaceutical companies employ ADME (adsorption, digestion, metabolism, and elimination) and TOX (toxicology) models in their screening process. Indeed, regulatory agencies in the United States and European Union also employ QSAR models as part of their review of new materials, and some groups have proposed them as replacements for safety studies involving animals. These same approaches are used to predict protein structure activities (proteomics) and decipher genetic codes (genomics).

The advent of advanced computer graphics workstations during the 1990s dramatically improved the scientific research communities' access to molecular-modeling capabilities. Continued advances are rapidly making computational molecular modeling an integral part of chemistry and its related scientific fields. Chemists, knowledgeable about the available modeling tools, now have the ability to test ideas on their PCs before stepping into the laboratory, thereby maximizing the likelihood of success and eliminating unnecessary work. Chemists once sketched molecules on paper and built molecular-scale models on their desks. Today they assemble them on a three-dimensional computer display, optimize the structure quickly, conduct a conformational search, compute spectral properties, estimate physiochemical properties, and compute and display molecular orbitals or space-filling models with mapped electrostatic charges—all of which can be dynamically rotated, resized, modified, or combined into new models.

—**Karl F. Moschner, Ph.D.**, is an organic chemistry and scientific computing consultant in Troy, New York.

(continued from page 182)

bond (unexpected on the basis of the principle of minimum structural change), as in the reaction

$$CH_3CH_2CH_2Br + AgOAc \rightarrow (CH_3)_2CHOAc + AgBr$$

where the REARRANGEMENT STAGE can formally be represented as the "1,2-shift" of hydride between adjacent carbon atoms in the CARBOCATION

$$CH_3CH_2CH_2^+ \rightarrow (CH_3)_2CH^+$$

Such migrations also occur in radicals, for example:

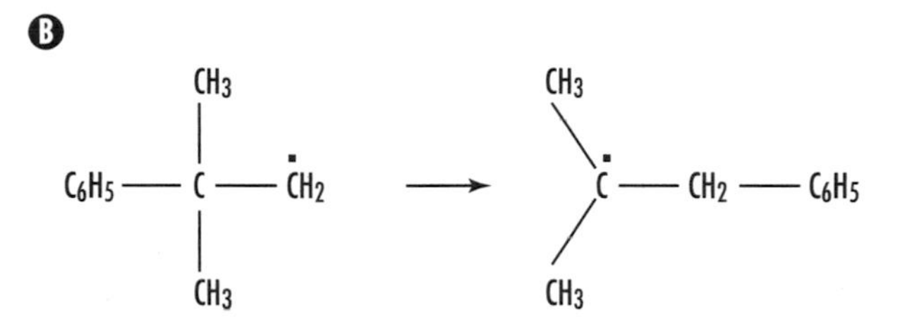

The definition also includes reactions in which an ENTERING GROUP takes up a different position from the LEAVING GROUP, with accompanying bond migration. An example of the latter type is the allylic rearrangement:

$$(CH_3)_2C{=}CHCH_2Br + OH^- \rightarrow$$
$$(CH_3)_2C(OH)CH{=}CH_2 + Br^-$$

A distinction is made between intramolecular rearrangements (or "true" molecular rearrangements) and INTERMOLECULAR rearrangements (or "apparent" rearrangements). In the former case the atoms and groups that are common to a reactant and a product never separate into independent fragments during the rearrangement stage (i.e., the change is intramolecular), whereas in an intermolecular rearrangement, a migrating group is completely free from the parent molecule and is reattached to a different position in a subsequent step, as in the Orton reaction:

$$PhN(Cl)COCH_3 + HCl \rightarrow PhNHCOCH_3 + Cl_2 \rightarrow \text{o-}$$
$$\text{and p-}ClC_6H_4NHCOCH_3 + HCl$$

molecular solid Solids composed of molecules held together by relatively weak INTERMOLECULAR forces; low-melting and tend to dissolve in organic solvents. Sulfur, ice, and sugar (sucrose) are examples.

molecular weight The mass of one mole of molecules of a substance.

molecule The smallest unit in a chemical element or compound that contains the chemical properties of the element or compound. They are made of atoms held together by chemical bonds that form when they share or exchange electrons. They can vary in complexity from a simple sharing or two atoms, such as oxygen, O_2, to a more complex substance such as nitroglycerin, $C_3H_5(NO_3)_3$.

mole fraction Number of moles of a component of a mixture divided by the total number of moles in the mixture.

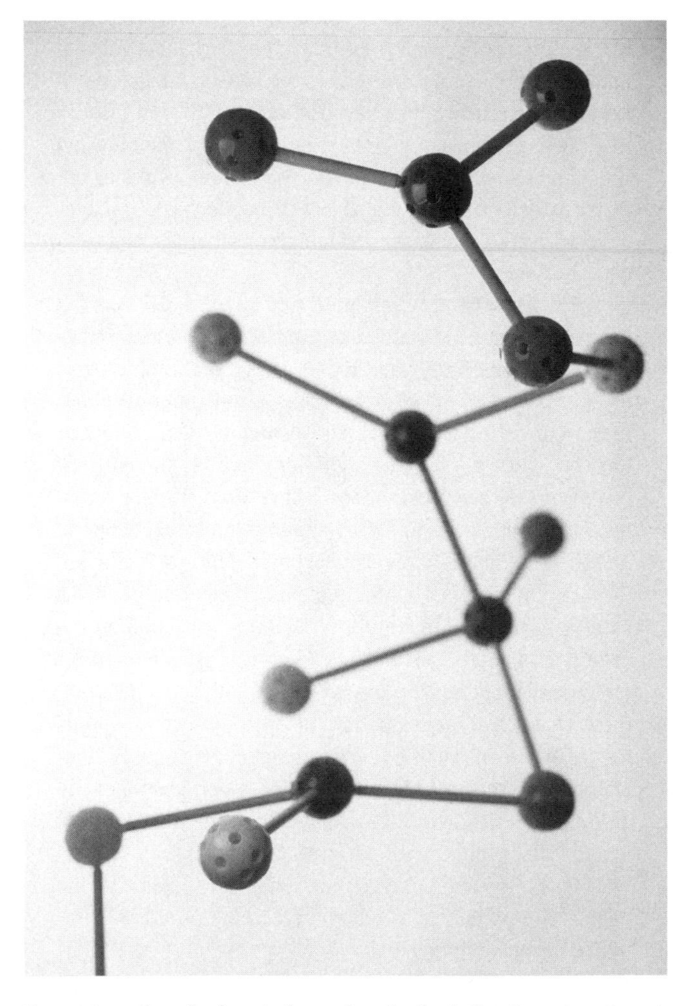

Computer artwork of part of a molecule depicting its arrangement of atoms (balls). The rods holding the balls together represent the chemical bonds between the atoms. ***(Courtesy of Laguna Design/Science Photo Library)***

molybdenum cofactor (Moco) The molybdenum complex of the MOLYBDOPTERIN PROSTHETIC GROUP (LIGAND). In the molybdenum COFACTOR, the minimal COORDINATION of the Mo atom is thought to be provided by the chelating dithiolenato group of the molybdopterin and either two oxo or one oxo and one sulfido ligands.

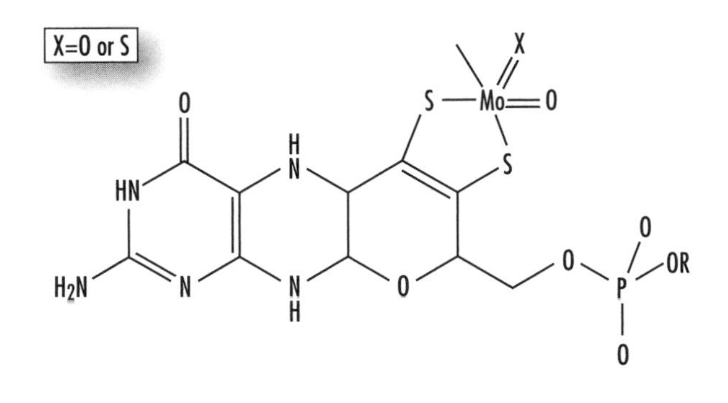

molybdopterin The PROSTHETIC GROUP associated with the Mo atom of the MOLYBDENUM COFACTOR found in all molybdenum-containing ENZYMEs except NITROGENASE. Many of the enzymes catalyze two-electron redox reactions that involve the net exchange of an oxygen atom between SUBSTRATE and water. The molybdopterin prosthetic group contains a pterin ring bound to a dithiolene functional group on the 6-alkyl side chain. In bacterial enzymes a NUCLEOTIDE is attached to the phosphate group.

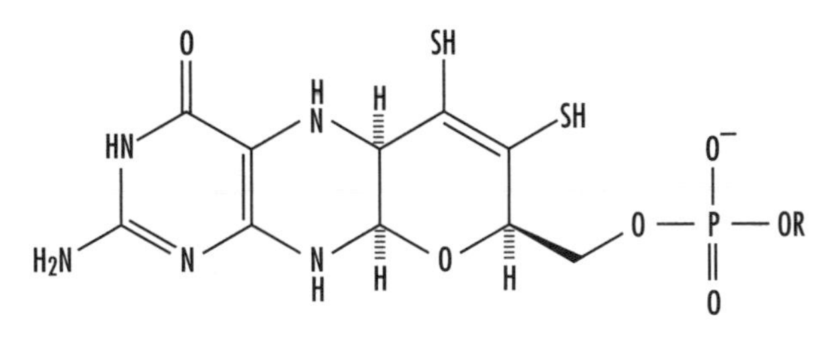

monoamine Small organic molecule containing both a carboxyl group and an amino group bonded to the same carbon atom, e.g., histamine, serotonin, epinephrine, and norepinephrine.

monomer A basic building block or small organic molecule that makes up a polymer when combined with identical or similar monomers through polymerization. Polymers are important substances in organisms, e.g., proteins are polymers.

monooxygenase An ENZYME that catalyzes the INSERTION of one atom of oxygen, derived from O_2, into an aromatic or aliphatic compound. The reaction is coupled to the oxidation of a COSUBSTRATE such as NAD(P)H or 2-oxoglutarate.

monoprotic acid An acid that can donate one H^+. Hydrochloric acid (HCl) is an example.

monosaccharide A simple sugar such as fructose or glucose that cannot be decomposed by hydrolysis; colorless crystalline substances with a sweet taste that have the same general formula, $C_nH_{2n}O_n$. They are classified by size according to the number of carbon atoms in the chain such as dioses, two carbon-ring backbone; trioses, three carbon-ring backbone; heptose, with seven carbon-ring backbone, etc.; further classified as aldoses (when carbonyl group is an aldehyde) or ketoses (contains a carbonyl [keto] group in its straight-chain form).

morphogen A diffusible protein molecule present in embryonic tissues that, through a concentration gradient, can influence the development process of a cell; different morphogen concentrations specify different cell fates.

morphometrics A branch of mathematics that focuses on the study of the metrical and statistical properties of shapes and the changes of geometric objects both organic or inorganic. Biologically relevant when dealing with species that have morphs that appear radically different.

Mössbauer effect Resonance absorption of gamma radiation by specific nuclei arranged in a crystal lattice in such a way that the recoil momentum is shared by many atoms. It is the basis of a form of spectroscopy used for studying coordinated metal ions. The principal application in bioinorganic chemistry is ^{57}Fe. The parameters derived from the Mössbauer spectrum (isomer shift, quadrupole splitting, and the HYPERFINE coupling) provide information about the oxidation, spin, and COORDINATION state of the ion.

mother nuclide The nuclide that undergoes actual nuclear decay.

motif A pattern of amino acids in a protein SEQUENCE that has a specific function, e.g., metal binding.

See also CONSENSUS SEQUENCE.

MRI *See* MAGNETIC RESONANCE IMAGING.

mRNA *See* MESSENGER RNA.

Müller, Paul Hermann (1899–1965) Swiss *Chemist* Paul Hermann Müller was born at Olten, Solothurn, Switzerland, on January 12, 1899. He attended primary school and the Free Evangelical elementary and secondary schools. He began working in 1916 as a laboratory assistant at Dreyfus and Company, followed by a position as an assistant chemist in the Scientific-Industrial Laboratory of their electrical plant. He attended Basel University and received a Ph.D. in 1925. He became deputy director of scientific research on substances for plant protection in 1946.

Müller began his career with investigations of dyes and tanning agents with the J.R. Geigy Company, Basel (1925–65), and he concentrated his research beginning in 1935 to find an "ideal" insecticide, one that had rapid, potent toxicity for the greatest number of insect species but would cause little or no damage to plants and warm-blooded animals. He tested and concluded that dichlorodiphenyltrichloroethane (DDT) was the ideal insecticide.

In 1939 DDT was successfully tested against the Colorado potato beetle by the Swiss government and by the U.S. Department of Agriculture in 1943.

For this discovery of DDT's potent toxic effects on insects, he received the Nobel Prize for physiology or medicine. However, DDT proved to be a two-edged sword. With its chemical derivatives, DDT became the most widely used insecticide for more than 20 years and was a major factor in increased world food production and suppression of insect-borne diseases, but the widespread use of the chemical made it hazardous to wildlife and it was banned in 1970. Müller died on October 12, 1965, in Basel.

Mulliken, Robert S. (1896–1986) American *Chemist* Robert Sanderson Mulliken was born in Newburyport, Massachusetts, on June 7, 1896, to Samuel Parsons Mulliken, a professor of organic chemistry, and Katherine W. Mulliken. He received a B.Sc. degree in 1917 at the Massachusetts Institute of Technology, Cambridge, Massachusetts. He entered the chemical warfare service during the war but left due to illness and then became employed by New Jersey Zinc Company until he entered graduate school at the University of Chicago in the fall of 1919, where he received a Ph.D. degree in 1921.

While a graduate student in chemistry at Chicago, his research work on boundary layer or diffusion membrane played an integral role in the Manhattan Project. He also became interested in the interpretation of valence and chemical bonding from the work of IRVING LANGMUIR and G. N. Lewis. He taught at New York University (1926–28) and then joined the faculty of the University of Chicago (1928–85).

Mulliken worked on valence theory and molecular structure starting in the 1920s. In 1952 he developed a quantum-mechanical theory of the behavior of electron orbitals as different atoms merge to form molecules, and in 1966 he was awarded the Nobel Prize in chemistry "for his fundamental work concerning chemical bonds and the electronic structure of molecules by the molecular orbital method."

In 1929 he married Mary Helen von Noé, the daughter of a professor of paleobotany at the University of Chicago. They had two daughters.

Mulliken was a National Research Council fellow, University of Chicago, and Harvard University, 1921–25; a Guggenheim fellow, Germany and Europe, 1930 and 1932–33; and a Fulbright scholar, Oxford University, 1952–54. In 1975 the University of Chicago Press published his selected papers.

He died on October 31, 1986.

multicenter bond Representation of some MOLECULAR ENTITIES solely by localized two-electron two-center BONDS appears to be unsatisfactory. Instead, multicenter bonds have to be considered in which electron pairs occupy orbitals encompassing three or more atomic centers. Examples include the three-center bonds in diborane, the delocalized pi bonding of benzene, and BRIDGED CARBOCATIONS.

multicenter reaction A synonym for PERICYCLIC REACTION. The number of "centers" is the number of atoms not bonded initially, between which single bonds are breaking or new bonds are formed in the TRANSITION STATE. This number does not necessarily correspond to the ring size of the transition state for the pericyclic reaction. Thus, a Diels-Alder reaction is a "four-center" reaction. This terminology has largely been superseded by the more detailed one developed for the various pericyclic reactions.

See also CYCLOADDITION; SIGMATROPIC REARRANGEMENT.

multicopper oxidases A group of ENZYMEs that oxidize organic SUBSTRATEs and reduce dioxygen to water. These contain a combination of copper ions with different spectral features, called TYPE 1 centers, TYPE 2 centers, and TYPE 3 centers, where the type 2 and type 3 sites are clustered together as a triNUCLEAR unit. Well-known examples are LACCASE, ascorbate oxidase, and CERULOPLASMIN.

multident *See* AMBIDENT.

multienzyme A protein possessing more than one catalytic function contributed by distinct parts of a polypeptide chain (DOMAINs), or by distinct SUBUNITs, or both.

multiheme Refers to a protein containing two or more HEME groups.

multiple bond Some atoms can share multiple pairs of electrons, forming multiple covalent bonds. A single covalent bond is two atoms sharing a pair of electrons.

mu (μ) symbol Notation for a ligand (prefix) that bridges two or more metal centers. The symbol μ is used for dipole moments.

See also BRIDGING LIGAND.

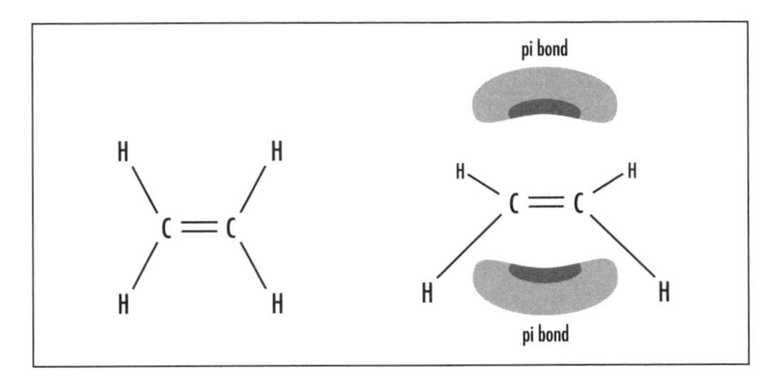

Multiple bonds. A bond between two atoms involving more than one pair of electrons (e.g., a double bond)

mutagen An agent that causes a permanent heritable change (i.e., a mutation) into the DNA (deoxyribonucleic acid) of an organism.

mutagenesis The introduction of permanent heritable changes, i.e., MUTATIONs, into the DNA of an organism. In the case of site-directed mutagenesis, the substitution or modification of a single amino acid at a defined location in a protein is performed by changing one or more base pairs in the DNA using recombinant DNA technology.

See also BASE PAIRING.

mutation A heritable change in the NUCLEOTIDE SEQUENCE of genomic DNA (or RNA in RNA viruses), or in the number of GENEs or chromosomes in a cell, that can occur spontaneously or be brought about by chemical mutagens or by radiation (induced mutation).

See also RIBONUCLEIC ACID.

mutual prodrug The association in a unique molecule of two, usually synergistic, DRUGs attached to each other, one drug being the carrier for the other and vice versa.

myocrysin *See* GOLD DRUGS.

myoglobin A monomeric dioxygen-binding heme-protein of muscle tissue, structurally similar to a SUBUNIT of HEMOGLOBIN.

NAD⁺ Oxidized form of nicotinamide adenine dinucleotide. Note that despite the plus sign in the symbol, the COENZYME is anionic under normal physiological conditions. NAD^+ is a coenzyme derived from the B vitamin niacin. It is transformed into NADH when it accepts a pair of high-energy electrons for transport in cells and is associated with catabolic and energy-yielding reactions.

NADH Reduced form of nicotinamide adenine dinucleotide (NAD). Called coenzyme I and is an electron donor essential for a variety of oxidation-reduction reactions.

NADP⁺ Oxidized form of nicotinamide adenine dinucleotide phosphate. Note that despite the plus sign in the symbol, the COENZYME is anionic under normal physiological conditions. An enzyme commonly associated with biosynthetic reactions. NADP is a hydrogen carrier in a wide range of redox reactions.

NADPH Reduced form of nicotinamide adenine dinucleotide phosphate. An energy-rich compound produced by the light-reaction of photosynthesis. It is used to synthesize carbohydrates in the dark-reaction.

nanoparticle A molecule or other particle measured in size on the order of tens of nanometers.

nanotube (buckytube) Any tube with nanoscale dimensions. Used mostly to refer to carbon nanotubes (sheets of graphite rolled up to make a tube).

narcissistic reaction A CHEMICAL REACTION that can be described as the conversion of a reactant into its mirror image, without rotation or translation of the product, so that the product ENANTIOMER actually coincides with the mirror image of the reactant molecule. Examples of such reactions are cited under the entries FLUXIONAL and DEGENERATE REARRANGEMENT.

native state The state when an element exists uncombined in nature and free of other elements.

natural gas A naturally occurring mixture of hydrocarbon and nonhydrocarbon gases that are found in porous/permeable geologic formations underneath the Earth's surface, and often found with petroleum.

natural product High-value chemical entities derived from plants or microbial sources.

natural radioactivity Spontaneous decomposition of an atom. Radioactivity associated with naturally occurring radioactive substances, e.g., C-14, K-40, U, and Th and some of their decay products.

NCE *See* NEW CHEMICAL ENTITY.

NDA (new drug application) The process of submitting a new drug for approval. After a new drug application (NDA) is received by the federal agency in charge, it undergoes a technical screening generally referred to as a completeness review and is evaluated to ensure that sufficient data and information have been submitted in each area to justify the filing.

neighboring-group participation The direct interaction of the reaction center (usually, but not necessarily, an incipient CARBENIUM CENTER) with a lone pair of electrons of an atom or with the electrons of a SIGMA or PI BOND contained within the parent molecule but not conjugated with the reaction center. A distinction is sometimes made between n, sigma, and pi participation.

A rate increase due to neighboring group participation is known as anchimeric assistance. "Synartetic acceleration" is the special case of anchimeric assistance, ascribed to participation by electrons binding a substituent to a carbon atom in a β-position relative to the leaving group attached to the α-carbon atom. According to the underlying model, these electrons then provide a three-center bond (or bridge) "fastening together" (as the word *synartetic* is intended to suggest) the α- and β-carbon atoms between which the charge is divided in the intermediate BRIDGED ION formed (and in the TRANSITION STATE preceding its formation). The term *synartetic acceleration* is not widely used.

See also INTRAMOLECULAR CATALYSIS; MULTICENTER BOND.

Nernst equation An equation that correlates chemical energy and the electric potential of a galvanic cell or battery. Links the actual reversible potential of an electrode (measured in volts), E, at nonstandard conditions of concentration or pressure, to the standard reversible potential of the electrode couple, E0, which is a thermodynamic value. The Nernst equation is named after the German physical chemist Walther Nernst.

net ionic equation A chemical equation used for a reaction that lists only those species participating in the reaction.

neuron The basic data processing unit of the nervous system; a specialized cell that carries information electrically from one part of the body to another by specialized processes or extensions called dendrites and axons. Widely branched dendrites carry nerve impulses toward the cell body, while axons carry them away and speed up transmitting nerve impulses (conduction) from one neuron to another. Each neuron has a nucleus within a cell body.

neurotransmitter A chemical made of amino acids and peptides that switch nerve impulses on or off across the synapse between neurons. Excitatory neurotransmitters stimulate the target cell, while inhibitory ones inhibit the target cells. Examples of neurotransmitters are acetylcholine, dopamine, noradrenaline, and serotonin.

Acetylcholine is the most abundant neurotransmitter in the body and the primary neurotransmitter between neurons and muscles and controls the stomach, spleen, bladder, liver, sweat glands, blood vessels, heart, and others. Dopamine is essential to the normal functioning of the central nervous system. Noradrenaline, or norepinephrine, acts in the sympathetic nervous system and produces powerful vasoconstriction. Serotonin is associated with the sleep cycle.

neutralization The resulting reaction when an acid reacts with a base to form salt and water.

neutron An atomic particle found in the nuclei of atoms that is similar to a PROTON in mass but has no electric charge.

See also ELECTRON.

new chemical entity A compound not previously described in the literature.

NHOMO *See* SUBJACENT ORBITAL.

nickel-cadmium cell (nicad battery) A dry cell in which the anode is cadmium (Cd), the cathode is NiO_2, and the electrolyte is basic. This "old" rechargeable battery technology is now being replaced by newer forms such as nickel-metal hydride.

nif A set of about 20 GENES required for the assembly of the NITROGENASE ENZYME complex.

NIH shift The INTRAMOLECULAR hydrogen MIGRATION that can be observed in enzymatic and chemical hydroxylations of aromatic rings. It is evidenced by appropriate deuterium labeling, for example:

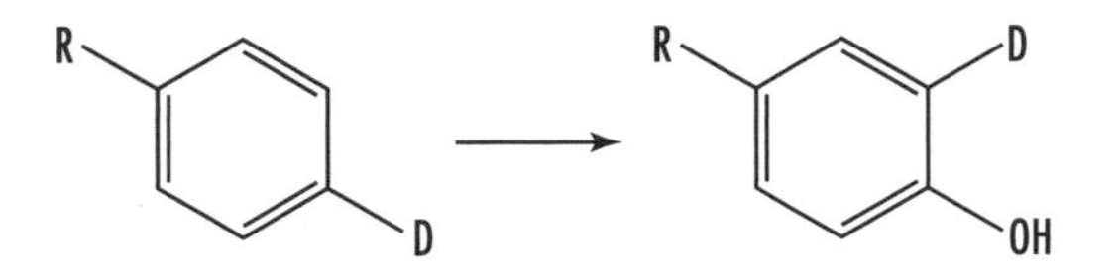

In enzymatic reactions, the NIH shift is generally thought to derive from the rearrangement of arene oxide intermediates, but other pathways have been suggested. NIH is the acronym of the National Institutes of Health, Bethesda, Maryland, where the shift was discovered.

nitrate reductase A METALLOENZYME containing molybdenum that reduces nitrate to nitrite.

nitrene Generic name for HN: and substitution derivatives thereof, containing an electrically neutral univalent nitrogen atom with four nonbonding electrons. Two of these are paired; the other two may have parallel spins (triplet state) or antiparallel spins (singlet state). The name is the strict analog of CARBENE and, as a generic name, it is preferred to a number of alternatives proposed (imene, imine radical, azene, azylene, azacarbene, imin, imidogen).

nitrenium ion The cation H_2N^+ and its N-hydrocarbyl derivatives R_2N^+, in which the nitrogen has a positive charge, and two unshared electrons. A synonymous term is aminylium ion.

nitrite reductase A METALLOENZYME that reduces nitrite. DISSIMILATORY nitrite reductases contain copper and reduce nitrite to nitrogen monoxide. ASSIMILATORY nitrite reductases contain SIROHEME and IRON-SULFUR CLUSTERS and reduce nitrite to ammonia.

nitrogenase An ENZYME complex from bacteria that catalyzes the reduction of dinitrogen to ammonia: N_2 + $8e^-$ $+10H^+$ $\rightarrow 2$ $^+NH_4$ + H_2 with the simultaneous HYDROLYSIS of at least 16 ATP molecules. The electron donor is reduced ferredoxin or flavodoxin. Dihydrogen is always a coproduct of the reaction. Ethyne (acetylene) can also be reduced to ethene (ethylene) and in some cases ethane. All nitrogenases are IRON-SULFUR PROTEINS. Three different types, which differ in the type of COFACTOR present, have been identified: molybdenum-nitrogenase (the most common, which contains the iron-molybdenum cofactor), vanadium-nitrogenase, and iron-only nitrogenase.

See also FEMO COFACTOR; REDUCTION.

nitrogen cycle A biochemical cycle in which occurs the transformation of nitrogen from an atmospheric gas to organic compounds in the soil, then to compounds in plants, and eventually back to the atmospheres as gas.

nitrogen fixation The natural process where atmospheric nitrogen, N_2, is converted to compounds that can be easily utilized by plants.

All organisms require nitrogen compounds, but few are able to utilize N_2, a relatively inert and unreactive form and, unfortunately, the most readily available. Most organisms require fixed forms such as NH_3, NO_3–, NO_2–, or organic-N. Bacteria perform nitrogen fixation by combining the nitrogen with hydrogen to form ammonia (NH_3) in the soil, which plants can then use. Cyanobacteria (blue-green algae) and bacteria (e.g., *Rhizobium* spp.; *Azotobacter* spp.) associated with legumes, like peas, can fix N_2 by reducing it to ammoniacal (ammonialike) N, mostly in the form of amino acids.

NMR *See* NUCLEAR MAGNETIC RESONANCE SPECTROSCOPY.

noble gases (rare gases) All the elements of the periodic Group 0; also called rare gases; formerly called inert gases: helium (He), neon (Ne), argon (Ar), krypton (Kr), xenon (Xe), and radon (Rn).

no-bond resonance *See* HYPERCONJUGATION.

nomenclature *See* BINOMIAL.

nonbonding orbital A MOLECULAR ORBITAL (a region in space within a molecule where electrons can be found) occupancy that does not significantly increase or decrease stability. Often, the main contribution to the molecular orbital comes only from an atomic orbital of one atom.

nonclassical carbocation A CARBOCATION the GROUND STATE of which has delocalized (bridged) bonding pi or sigma electrons. (Allylic and benzylic carbocations are not considered nonclassical.)

See also DELOCALIZATION.

nonclassical isostere Synonymous with BIOISOSTERE.

noncyclic electron flow The first stage of PHOTOSYNTHESIS; begins when light energy enters a cluster of pigment molecules called the PHOTOSYSTEM, located in the thylakoid; the light-induced flow of electrons from water to NADP in oxygen-evolving photosynthesis involving both photosystems I and II. Photosystems are a large complex of proteins and chlorophyll that capture energy from sunlight. Both systems I and II include special forms of chlorophyll A. Photosystem I, or P-700, includes chlorophyll A pigment with a specific absorbance of 700 nm (red light). Photosystem II, or P-680, contains the reaction center responsible for oxygen evolution and contains a special chlorophyll A that absorbs light at 680 nm (red light). If the photochemical reactions in photosystem II are inhibited, photosystem I is inhibited as well.

noncyclic photophosphorylation The formation of ATP by NONCYCLIC ELECTRON FLOW.

nonelectrolyte Any substance or material that does not conduct electricity when aqueous.

nonpolar covalent bond A covalent bond formed by the equal sharing of electrons between two atoms with the same electronegativity. Electronegativity is the tendency of an atom to attract electrons to itself in a COVALENT BOND.

normal kinetic isotope effect *See* ISOTOPE EFFECT.

normal mode (of vibration) In molecular vibrations, in a normal mode the atoms all move with the same frequency and phase; however, the amplitudes and directions of their motions differ. Generally, any stable mode or frequency at which the medium can vibrate independently.

n-σ delocalization (n-σ no bond resonance) DELOCALIZATION of a free electron pair (n) into an antibonding SIGMA ORBITAL (s).

See also HYPERCONJUGATION; RESONANCE.

N-terminal amino acid residue *See* AMINO ACID RESIDUE.

n-to-pi-star transition ($n \rightarrow \pi^*$) An electronic transition in which an electron is excited from a nonbonding orbital to an antibonding pi orbital, occurring in the UV-visible range.

n-type semiconductor A semiconductor where electrical conduction is mostly due to the movement of electrons.

nuclear binding energy Energy produced by the loss of mass from the formation of an atom from protons, electrons, and neutrons; energy released in the formation of an atom from the subatomic particles.

nuclear decay Disintegration of atomic nuclei that results in the emission of ALPHA or BETA PARTICLES (usually with gamma radiation).

nuclear fission The process of splitting nuclei with high mass number by a variety of processes (usually involving neutrons) into two nuclei of smaller mass (usually radioactive) and releasing energy and more neutrons.

nuclearity The number of CENTRAL ATOMS joined in a single COORDINATION entity by BRIDGING LIGANDS or metal-metal bonds is indicated by dinuclear, trinuclear, tetranuclear, polynuclear, etc.

nuclear magnetic resonance spectroscopy (NMR spectroscopy) NMR spectroscopy makes it possible to discriminate nuclei, typically protons, in different chemical environments. The electron distribution gives rise to a chemical shift of the resonance frequency. The chemical shift, δ, of a nucleus is expressed in parts per million (ppm) by its frequency, ν_n, relative to a standard, ν_{ref}, and is defined as $\delta = 10^6 (\nu_n - \nu_{ref})/\nu_o$, where ν_o is the operating frequency of the spectrometer. It is an indication of the chemical state of the group containing the nucleus. More information is derived from the SPIN-SPIN COUPLINGS between nuclei, which give rise to multiple patterns. Greater detail can be derived from two- or three-dimensional techniques. These use pulses of radiation at different nuclear frequencies, after which the response of the spin system is recorded as a free-induction decay (FID). Multidimensional techniques, such as COSY (correlated spectroscopy) and NOESY (nuclear overhauser effect [NOE] spectroscopy), make it possible to deduce the structure of a relatively complex molecule such as a small protein (molecular weight up to 25,000). In proteins containing PARAMAGNETIC centers, nuclear HYPERFINE interactions can give rise to relatively large shifts of resonant frequencies, known as contact and pseudo-contact (dipolar) shifts, and considerable increases in the nuclear spin relaxation rates. From this type of measurement, structural information can be obtained about the paramagnetic site.

nuclear radiation The radiation emitted during the spontaneous decay of an unstable atomic nucleus.

nuclear reaction Any reaction involving a change in the nucleus of an atom. For example, reaction between neutron, proton, or nucleus from a reactor or particle accelerator and a target nucleus resulting in the production of product nuclides, gamma rays, particles, and other radiations.

nuclear reactor A system where a fission chain reaction can be initiated, maintained, and controlled.

nucleation The process by which nuclei are formed; defined as the smallest solid-phase aggregate of atoms, molecules, or ions that is formed during a precipitation and that is capable of spontaneous growth.

nucleic acids Macromolecules composed of SEQUENCES of NUCLEOTIDES that perform several functions in living cells, e.g., the storage of genetic information and its transfer from one generation to the next (DNA), and the EXPRESSION of this information in protein synthesis (mRNA, tRNA). They may act as functional components of subcellular units such as RIBOSOMES (rRNA). RNA contains D-ribose; DNA contains 2-deoxy-D-ribose as the sugar component. Currently, synthetic nucleic acids can be made consisting of hundreds of NUCLEOTIDES.

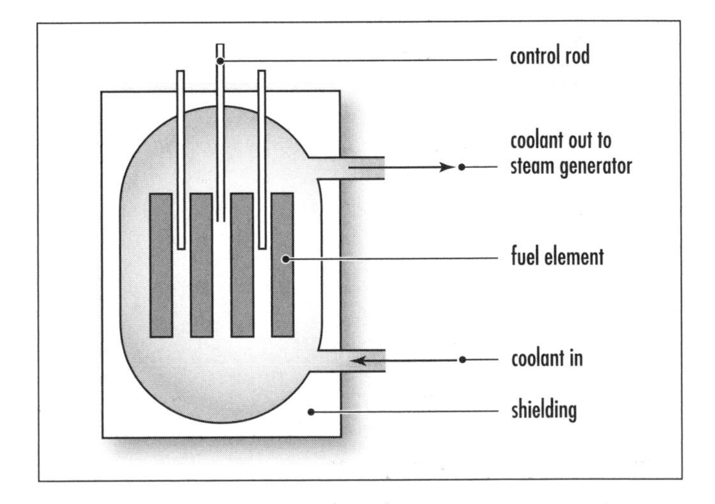

Nuclear reactor. A system where a fission chain reaction can be initiated, maintained, and controlled

Reactor hall of the Chernobyl nuclear power station, Ukraine. The floor squares in the foreground are located over the reactor's fuel channels. Four RBMK reactors were completed at Chernobyl during 1978–84; this is reactor 3. Each generates 925 megawatts of electricity. RBMK reactor cores use graphite to promote the nuclear chain reaction and water for cooling the vertical fuel channels. The RBMK design is flawed because its reactivity, and hence its power output, increases as coolant is lost from the fuel channels. This *positive void coefficient* was partly responsible for the major accident involving Chernobyl's Reactor 4 in 1986. *(Courtesy of Vaughan Melzer/JVZ/Science Photo Library)*

See also GENETIC CODE; MESSENGER RNA; OLIGONUCLEOTIDE; RIBONUCLEIC ACID; RIBOSOMAL RNA.

nucleobase *See* NUCLEOSIDE.

nucleofuge A LEAVING GROUP that carries away the bonding electron pair. For example, in the HYDROLYSIS of an alkyl chloride, Cl^- is the nucleofuge. The tendency of atoms or groups to depart with the bonding electron pair is called nucleofugality.

nucleons Particles making up the nucleus of an atom—protons and neutrons.

nucleophile (nucleophilic reagent) A nucleophile (or nucleophilic reagent) is a reagent that forms a bond to its reaction partner (the ELECTROPHILE) by donating both bonding electrons. Nucleophilic reagents are LEWIS BASES.

A nucleophilic substitution reaction is a HETEROLYTIC reaction in which the reagent supplying the entering group acts as a nucleophile. For example

$$MeO^- \text{ (nucleophile)} + Et\text{–}Cl \rightarrow MeOEt + Cl^- \text{ (nucleofuge)}$$

The term *nucleophilic* is also used to designate the apparent polar character of certain RADICALS, as inferred from their higher relative reactivity with reaction sites of lower electron density.

nucleophilic catalysis CATALYSIS by a LEWIS BASE, involving formation of a LEWIS ADDUCT as a REACTION INTERMEDIATE. For example, the hydrolysis of acetic anhydride in aqueous solution catalyzed by pyridine:

$$C_5H_5N + (CH_3CO)_2O \rightarrow [C_5H_5NCOCH_3]^+ + CH_3CO_2^-$$

$$[C_5H_5NCOCH_3]^+ + H_2O \rightarrow C_5H_5N + CH_3CO_2H + H^+_{aq}$$

See also ELECTROPHILIC; NUCLEOPHILICITY.

nucleophilicity (1) The property of being nucleophilic.

(2) The relative reactivity of a NUCLEOPHILIC REAGENT. (It is also sometimes referred to as nucleophilic power.) Qualitatively, the concept is related to Lewis basicity. However, whereas Lewis basicity is measured by relative equilibrium constants,

$$B: + A \overset{k}{\rightleftharpoons} B^+ - A^- \text{ (equilibrium constant K)}$$

nucleophilicity of a LEWIS BASE is measured by relative RATE CONSTANTS of different nucleophilic reagents toward a common SUBSTRATE, most commonly involving formation of a bond to carbon

$$B: + A\text{–}Z \rightarrow B^+\text{–}A^- + Z:^- \text{ (rate constant k)}$$

See also ELECTROPHILICITY; RITCHIE EQUATION; SWAIN-SCOTT EQUATION.

nucleoside Compound in which a purine or pyrimidine base is β-N-glycosidically bound to C-1 of either 2-deoxy-D-ribose or of D-ribose, but without any phosphate groups. The common nucleosides in biological systems are adenosine, guanosine, cytidine, and uridine (which contain ribose) and deoxyadenosine, deoxyguanosine, deoxycytidine, and thymidine (which contain deoxyribose).

See also NUCLEOTIDE.

nucleotide A NUCLEOSIDE with one or more phosphate groups esterified mainly to the 3′- or the 5′-position of the sugar moiety. Nucleotides found in cells are adenylic acid, guanylic acid, uridylic acid, cytidylic acid, deoxyadenylic acid, deoxyguanylic acid, deoxycytidylic acid, and thymidylic acid. A nucleotide is a nucleoside in which the primary hydroxy group of either 2-deoxy-D-ribose or of D-ribose is esterified by orthophosphoric acid.

See also ADENOSINE 5′-TRIPHOSPHATE; NAD^+; $NADP^+$.

nucleus The very small, dense core of the atom, where most of its mass and all of its positive charge is concentrated. Except for hydrogen, it consists of protons and neutrons and other subatomic particles. Also, the nucleus of a cell, i.e., the region containing the chromosomes.

nuclide The nucleus of a particular isotope.

nuclide symbol Symbol for an atom A/Z E, in which E is the symbol of an element, Z is its atomic number, and A is its mass number.

nylon Any of a group of high-strength and resilient synthetic polymers in which the molecules contain the recurring amide group –CONH.

O

obligate aerobe Any organism that must utilize atmospheric oxygen in its metabolic pathways and for cellular respiration, and cannot survive without it. The adjective *obligate* refers to an environmental factor.

See also AEROBE.

obligate anaerobe Any organism where atmospheric oxygen is toxic to its growth; grows only in an anaerobic environment. The adjective *obligate* refers to an environmental factor.

See also ANAEROBE.

oceanic zone The deep water of the oceans beyond the shelf break. The ocean is divided into zones. The whole mass of water is called the pelagic. This is divided into major subzones, including the neritic, which covers all water to a depth of 600 feet, and the oceanic zone, which covers all water below 600 feet. The oceanic zone is further divided into subzones. The mesopelagic, semidark waters, covers the depths from 650 feet to 3,200 feet, which is the middle layer between the upper (sunlit 650 feet) epipelagic and the lower cold and dark bathypelagic.

See also APHOTIC ZONE; LITTORAL ZONE.

ocean thermal vents A hydrothermal vent is an area where a major fissure occurs between plates making up the Earth's crust. As the plates gradually separate, underlying volcanic activity reaches the surface, and the molten volcanic rock meets ultracold seawater, causing physical and chemical reactions. These vents are the habitat of deep-sea animals formerly unknown to science.

See also APHOTIC ZONE.

octahedral Having symmetry of a regular octahedron. Molecules and polyatomic ions having one atom in the center and six atoms at the corners of an octahedron.

octahedral hole A cavity or space between six atoms or ions in a crystal in which these atoms represent the corners of an octahedron.

octahedron *See* COORDINATION.

octane number (octane rating) A number indicating the relative antiknock value of a gasoline. Octane numbers are based on a 100 scale on a comparison of reference fuels. Isooctane is 100 (minimal knock) and n-heptane is 0 (bad knock).

octet rule A principle that bonded atoms have eight outer electrons (including those shared with other atoms), although there are exceptions.

ODMR *See* OPTICALLY DETECTED MAGNETIC RESONANCE.

OEC *See* OXYGEN-EVOLVING COMPLEX.

oil A slippery, viscous, or liquefiable substance not miscible with water.

oil shale A dark-colored shale that contains a solid substance, kerogen, which is partially formed oil and can be extracted when crushed and heated to liberate oil.

olefin Unsaturated aliphatic hydrocarbons (C_nH_{2n}) characterized by relatively great chemical activity, e.g., ethylene, propylene, and butene.

olfaction The process of smell. In humans, chemoreceptors are located in a patch of tissue about the size of a postage stamp high in the nasal cavity, called the olfactory epithelium.

oligonucleotide Macromolecules composed of short SEQUENCES of NUCLEOTIDES that are usually synthetically prepared and used in SITE-DIRECTED MUTAGENESIS.

oligopeptide Four to 10 amino acids joined by peptide bonds.

oligotrophic lake A condition of a lake that has low concentrations of nutrients and algae, resulting in clear blue conditions. Contrast with mesotrophic lakes that have a moderate nutrient condition and EUTROPHIC LAKES that have excessive levels of nutrients.

onium ion (1) A cation (with its counterion) derived by addition of a hydron (hydrogen ion) to a mononuclear parent hydride of the nitrogen, chalcogen, and halogen family, e.g., H_4N^+ ammonium ion.

(2) Derivatives formed by substitution of the above parent ions by univalent groups, e.g., $(CH_3)_2S^+H$ dimethylsulfonium, $(CH_3CH_2)_4N^+$ tetraethylammonium.

(3) Derivatives formed by substitution of the above parent ions by groups having two or three free valencies on the same atom. Such derivatives are, whenever possible, designated by a specific class name, e.g., $R_2C{=}NH_2^+$ iminium ion.

See also CARBENIUM ION; CARBONIUM ION.

open sextet Species that have six electrons in the highest energy shell of the central element. Examples include many LEWIS ACIDS.

open system A system that can exchange both matter and energy with its surroundings.

operon A functional unit consisting of a PROMOTER, an operator, and a number of structural GENES, found mainly in PROKARYOTES. An example is the operon NIF. The structural genes commonly code for several functionally related ENZYMES, and although they are transcribed as one (polycistronic) mRNA, each has its separate TRANSLATION initiation site. In the typical operon, the operator region acts as a controlling element in switching on or off the synthesis of mRNA.

A group or sequence of closely linked genes that function as a unit in synthesizing enzymes needed for biosynthesis of a molecule and is controlled by operator and repressor genes; common in bacteria and phages. An operator gene is the region of the chromosome, next to the operon, where a repressor protein binds to prevent transcription of the operon. The repressor gene protein binds to an operator adjacent to the structural gene, preventing the transcription of the operon.

See also MESSENGER RNA.

opposing reaction *See* COMPOSITE REACTION.

optical activity A material that rotates the plane of polarization of any polarized light transmitted through it.

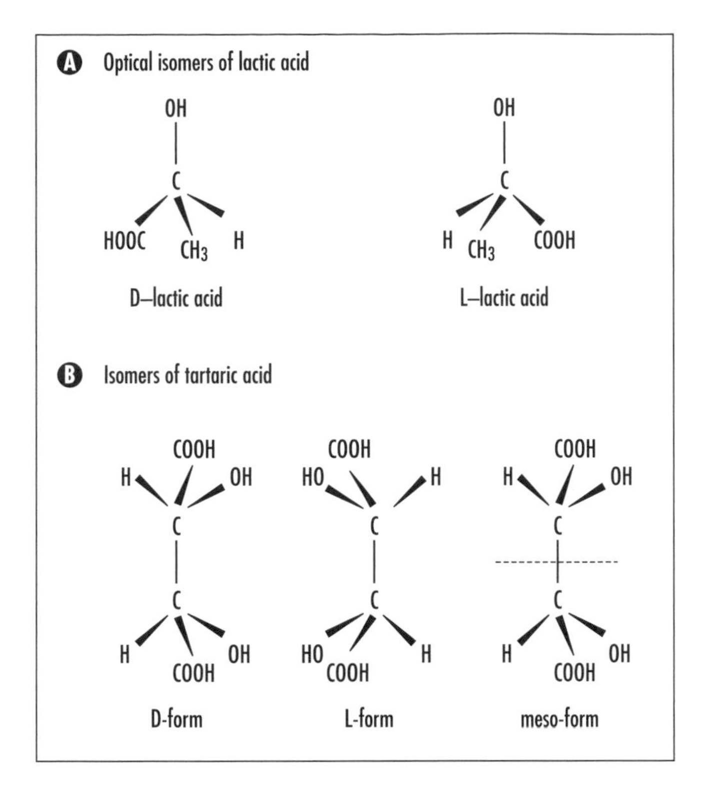

Optical isomerism occurs when a compound has no plane of symmetry and can exist in either left- or right-handed forms that are mirror images of each other.

optical isomers (enantiomers) Are nonsuperimposable mirror images of each other and are said to be chiral (not superimposable on their mirror image).

See also CHIRALITY.

optically detected magnetic resonance (ODMR) A double resonance technique in which transitions between spin sublevels are detected by optical means. Usually these are sublevels of a triplet, and the transitions are induced by microwaves.

optical yield In a CHEMICAL REACTION involving chiral reactants and products, the term *optical yield* refers to the ratio of the optical purity of the product to that of the precursor, reactant, or catalyst. This should not be confused with "enantiomeric excess." The optical yield is in no way related to the chemical yield of the reaction.

See also CHIRALITY; STEREOSELECTIVITY.

orbital *See* ATOMIC ORBITAL; MOLECULAR ORBITAL.

orbital steering A concept expressing that the stereochemistry of approach of two reacting species is governed by the most favorable overlap of their appropriate ORBITALS.

See also STEREOCHEMICAL.

orbital symmetry The behavior of an atomic or localized MOLECULAR ORBITAL under molecular symmetry operations characterizes its orbital symmetry. For example, under a reflection in an appropriate symmetry plane, the phase of the orbital may be unchanged (symmetric), or it may change sign (antisymmetric), i.e., the positive and negative lobes are interchanged.

A principal context for the use of orbital symmetry is the discussion of chemical changes that involve conservation of orbital symmetry. If a certain symmetry element (e.g., the reflection plane) is retained along a reaction pathway, that pathway is "allowed" by orbital symmetry conservation if each of the occupied orbitals of the reactant(s) is of the same symmetry type as a similarly (e.g., singly or doubly) occupied orbital of the product(s). This principle permits the qualitative construction of correlation diagrams to show how molecular orbitals transform (and how their energies change) during idealized chemical changes (e.g., CYCLOADDITIONS).

An idealized single bond is a SIGMA BOND—one that has cylindrical symmetry. In contrast, a *p*-orbital or PI-BOND orbital has pi symmetry—one that is antisymmetric with respect to reflection in a plane passing through the atomic centers with which it is associated. In ethene, the pi-bonding orbital is symmetric with respect to reflection in a plane perpendicular to and bisecting the C–C bond, whereas the pi-star-antibonding orbital is antisymmetric with respect to this operation.

Considerations of orbital symmetry are frequently grossly simplified in that, for example, the pi orbitals of a carbonyl group would be treated as having the same symmetry as those of ethene, and the fact that the carbonyl group in, for example, camphor (unlike that in formaldehyde) has no mirror planes would be ignored. These simplified considerations nevertheless afford the basis of one approach to the understanding of the rules that indicate whether PERICYCLIC

REACTIONS are likely to occur under thermal or photochemical conditions.

See also SIGMA, PI.

order of reaction *n* (SI unit: 1) If the macroscopic (observed, empirical, or phenomenological) rate of reaction (v) for any reaction can be expressed by an empirical differential rate equation (or rate law) that contains a factor of the form $k\,[A]^{\alpha}\,[B]^{\beta}$ ... (expressing in full the dependence of the rate of reaction on the concentrations [A], [B] ...) where α, β are constant exponents (independent of concentration and time) and k is independent of [A] and [B] etc. (rate constant, rate coefficient), then the reaction is said to be of order α with respect to A, of order β with respect to B, ..., and of (total or overall) order $v = \alpha + \beta + \ldots$. The exponents α, β... can be positive or negative integral or rational nonintegral numbers. They are the reaction orders with respect to A, B, ... and are sometimes called "partial orders of reaction." Orders of reaction deduced from the dependence of initial rates of reaction on concentration are called "orders of reaction with respect to concentration"; orders of reaction deduced from the dependence of the rate of reaction on time of reaction are called "orders of reaction with respect to time."

The concept of order of reaction is also applicable to chemical rate processes occurring in systems for which concentration changes (and hence the rate of reaction) are not themselves measurable, provided it is possible to measure a CHEMICAL FLUX. For example, if there is a dynamic equilibrium according to the equation

$$aA \rightleftharpoons pP$$

and if a chemical flux is experimentally found (e.g., by NMR line-shape analysis) to be related to concentrations by the equation

$$\varphi\text{–}A/a = k[A]^{\alpha}\,[L]^{\lambda}$$

then the corresponding reaction is of order α with respect to A ... and of total (or overall) order n (= $\alpha + \lambda + \ldots$).

The proportionality factor k above is called the (nth order) rate coefficient.

Rate coefficients referring to (or believed to refer to) ELEMENTARY REACTIONS are called "rate constants" or, more appropriately, "microscopic" (hypothetical, mechanistic) rate constants.

The (overall) order of a reaction cannot be deduced from measurements of a "rate of appearance" or "rate of disappearance" at a single value of the concentration of a species whose concentration is constant (or effectively constant) during the course of the reaction. If the overall rate of reaction is, for example, given by

$$v = k[A]^{\alpha}\,[B]^{\beta}$$

but [B] stays constant, then the order of the reaction (with respect to time), as observed from the concentration change of A with time, will be α, and the rate of disappearance of A can be expressed in the form

$$v_A = k_{obs}[A]^{\alpha}$$

The proportionality factor k_{obs} deduced from such an experiment is called the "observed rate coefficient," and it is related to the $(\alpha + \beta)$th-order rate coefficient k by the equation

$$k_{obs} = k[B]^{\beta}$$

For the common case when $\alpha = 1$, k_{obs} is often referred to as a "pseudo-first-order rate coefficient" (k_{Ψ}).

For simple (ELEMENTARY) REACTIONS, a partial order of reaction is the same as the stoichiometric number of the reactant concerned and must therefore be a positive integer (*see* RATE OF REACTION). The overall order is then the same as the MOLECULARITY. For STEPWISE REACTIONS, there is no general connection between stoichiometric numbers and partial orders. Such reactions may have more complex rate laws, so that an apparent order of reaction may vary with the concentrations of the CHEMICAL SPECIES involved and with the progress of the reaction. In such cases it is not useful to speak of orders of reaction, although apparent orders of reaction may be deducible from initial rates.

In a stepwise reaction, orders of reaction may in principle always be assigned to the elementary steps.

See also KINETIC EQUIVALENCE.

ore A natural mineral or elemental deposit that is extracted.

organic chemistry The study of carbon (organic) compounds; used to study the complex nature of living things. Organic compounds are composed mostly of carbon, hydrogen, oxygen, and nitrogen atoms bonded together. Organic chemistry is one of two main divisions of chemistry, the other being INORGANIC CHEMISTRY. Branches of these two include analytical, biochemical, and physical chemistry.

organism A living entity.

organochlorine compounds (chlorinated hydrocarbons) Organic pesticides that contain chlorine, carbon, and hydrogen, e.g., DDT and endrin. These pesticides affect the central nervous system.

organophosphorus compound A compound containing phosphorus and carbon, whose physiological effects include INHIBITION of acetylcholinesterase. Many pesticides and most nerve agents are organophosphorus compounds, e.g., malathion and parathion.

orphan drug A DRUG for the treatment of a rare disease for which reasonable recovery of the sponsoring firm's research and development expenditure is not expected within a reasonable time. The term is also used to describe substances intended for such uses.

osmoconformer Not actively changing internal osmolarity (total solute concentration) because an animal is isotonic (body fluids are of equal concentration with respect to osmotic pressure) with the environment.

osmolarity The concentration of a solution in terms of osmols per liter. An osmol is the amount of substance that dissociates in solution to form one mole of osmotically active particles. For example, a mole of nonelectrolyte forms one osmol of solute, but a substance that dissociates such as NaCl forms 2 osmols of solute. Mostly used in medicine.

osmoregulation A process to control water balance in a cell or organism with respect to the surrounding environment using osmosis; the ability by which organisms maintain a stable solute concentration by maintaining osmotic pressure on each side of a semipermeable membrane.

osmoregulator An organism that must take in or discharge excess water because its body fluids have a different osmolarity than the environment.

osmosis The diffusion or movement of water across a selectively permeable membrane from one aqueous system to another of different concentrations. Water moves from areas of high-water/low-solute concentration to areas of low-water/high-solute concentration.

osmotic pressure Pressure that is generated by a solution moving by osmosis into and out of a cell and caused by a concentration gradient.

Ostwald process An industrial process that produces oxide and nitric acid from ammonia and oxygen.

outer orbital complex Valence bond designation for a complex in which the metal ion utilizes *d* orbitals in the occupied or outermost shell in hybridization.

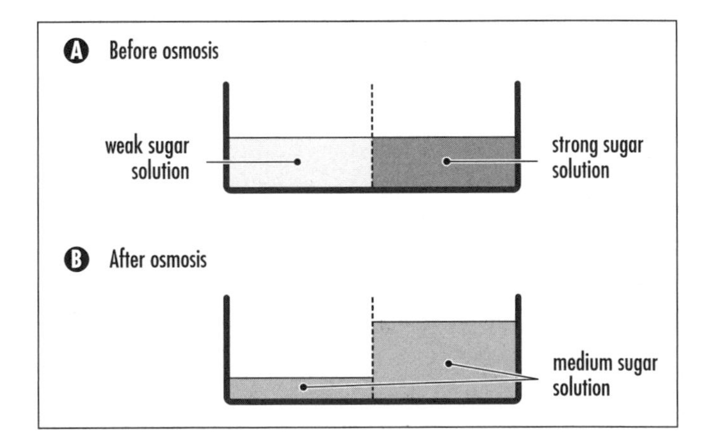

Osmosis. The diffusion or movement of water across a selectively permeable membrane from one aqueous system to another of different concentrations

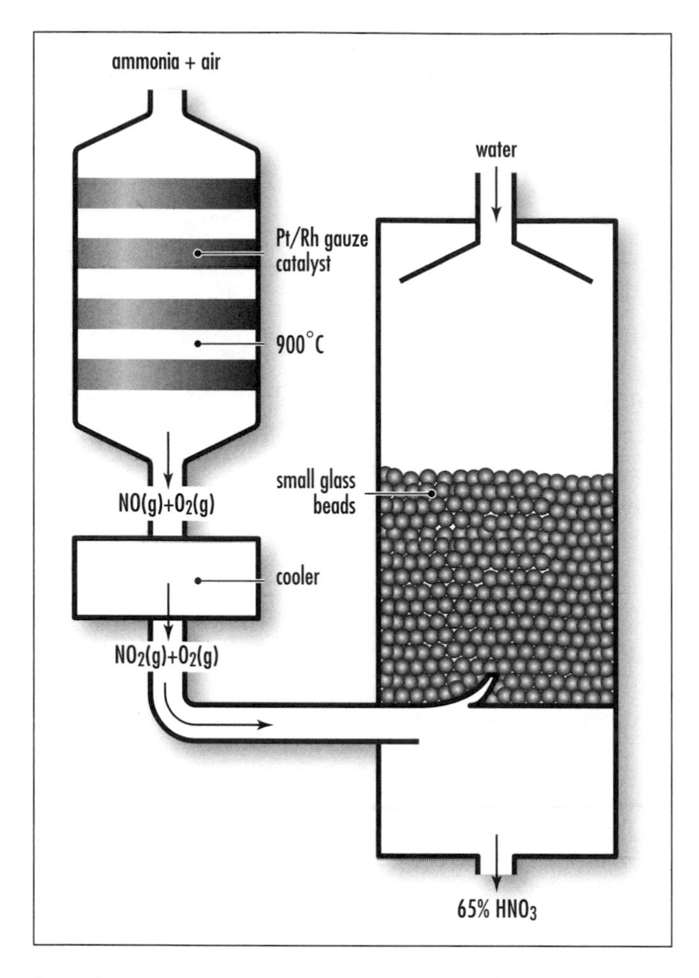

Ostwald process. An industrial process that produces oxide and nitric acid from ammonia and oxygen

outer-sphere electron transfer An outer-sphere electron transfer is a reaction in which the electron transfer takes place with no or very weak (4–16 kJ mol^{-1}) electronic interaction between the reactants in the transition state. If instead the donor and the acceptor exhibit a strong electronic coupling, the reaction is described as INNER-SPHERE ELECTRON TRANSFER. The two terms derive from studies concerning metal complexes, and it has been suggested that for organic reactions, the terms *nonbonded* and *bonded* electron transfer should be used.

overlap Interaction of orbitals on different atoms in the same region of space.

See also ATOMIC ORBITAL.

overpotential (**overvoltage**) Extra energy required in addition to the reduction potential in order for a reaction to proceed. In electrochemistry, it is the deviation of an electrode potential from its equilibrium value required to produce a net flow of current across an electrode/solution interface. It is the voltage in excess of the voltaic cell potential that is needed to cause electrolysis to occur.

ovotransferrin An iron-binding protein from eggs, structurally similar to the TRANSFERRINs.

oxidase An ENZYME that catalyzes the oxidation of SUBSTRATEs by O_2.

oxidation (1) The complete, net removal of one or more electrons from a MOLECULAR ENTITY (also called "de-electronation"); (2) an increase in the OXIDATION NUMBER of any atom within any substrate; (3) gain of oxygen and/or loss of hydrogen of an organic SUBSTRATE.

All oxidations meet criteria (1) and (2), and many meet criterion (3), but this is not always easy to demonstrate. Alternatively, an oxidation can be described as a TRANSFORMATION of an organic substrate that can be rationally dissected into steps or PRIMITIVE CHANGES. The latter consist in removal of one or several electrons from the substrate followed or preceded by gain or loss of water and/or HYDRONs or hydroxide ions, or by NUCLEOPHILIC substitution by water or its reverse and/or by an INTRAMOLECULAR MOLECULAR REARRANGEMENT.

This formal definition allows the original idea of oxidation (combination with oxygen), together with its extension to removal of hydrogen, as well as processes closely akin to this type of transformation (and generally regarded in current usage of the term in organic chemistry to be oxidations and to be effected by "oxidizing agents") to be descriptively related to definition (1). For example, the oxidation of methane to chloromethane can be considered as follows:

$$CH_4 - 2e^- - H^+ + OH^- = CH_3OH \text{ (oxidation)} \rightarrow CH_3Cl \text{ (reversal of hydrolysis)}$$

oxidation number The oxidation number of an element in any chemical entity is the number of charges that would remain on a given atom if the pairs of electrons in each bond to that atom were assigned to the more electronegative member of the bond pair. The oxidation (Stock) number of an element is indicated by a Roman numeral placed in parentheses immediately following the name (modified if necessary by an appropriate ending) of the element to which it refers. The oxidation number can be positive, negative, or zero. Zero, not a roman numeral, is represented by the usual cipher, 0. The positive sign is never used. An oxidation number is always positive unless the minus sign is explicitly used. Note that it cannot be nonintegral (*see* MIXED VALENCY). Nonintegral numbers may seem appropriate in some cases where a charge is spread over more than one atom, but such a use is not encouraged. In such ambiguous cases, the charge number, which designates ionic charge, can be used. A charge (EWENS-BASSETT) NUMBER is a number in parentheses written without a space immediately after the name of an ion, and whose magnitude is the ionic charge. Thus the number may refer to cations or anions, but never to neutral species. The charge is written in Arabic numerals and followed by the sign of the charge.

In a COORDINATION entity, the oxidation number of the CENTRAL ATOM is defined as the charge it would bear if all the LIGANDs were removed along with the electron pairs that were shared with the central atom. Neutral ligands are formally removed in their closed-shell configurations. Where it is not feasible or reasonable to define an oxidation state for each individual member of a group or CLUSTER, it is again recommended that the overall oxidation level of the group be defined by a formal ionic charge, the net charge on the coordination entity.

oxidation-reduction reactions (redox reactions) Reactions that involve oxidation of one reactant and reduction of another.

oxidation state Oxidation state shows the total number of electrons that have been removed from an element (giving a positive oxidation state) or added to an element (giving a negative oxidation state) to get to its present state. Oxidation involves an increase in oxidation state, while reduction involves a decrease in oxidation state.

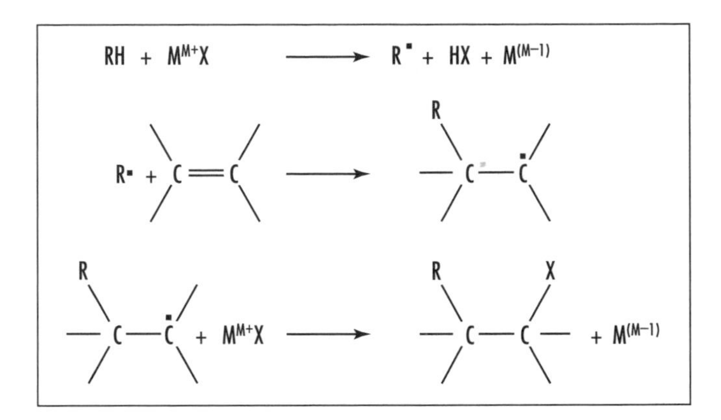

Oxidative addition

oxidative addition The INSERTION of a metal of a COORDINATION entity into a COVALENT BOND involving formally an overall two-electron loss on one metal or a one-electron loss on each of two metals.

oxidative coupling The coupling of two MOLECULAR ENTITIES through an oxidative process, usually catalyzed by a transition-metal compound and involving dioxygen as the oxidant; e.g.,

$$2\ CO + 2\ MeOH + 1/2\ O_2\ (+\ catalyst) \rightarrow MeOOC\text{-}COOMe + H_2O$$

oxidative phosphorylation An aerobic process of energy harnessing by the production of ATP (energy) in mitochondria by enzymatic phosphorylation of ADP coupled to an ELECTRON-TRANSPORT CHAIN (ETC). The ETC is a series of mitochondrial enzymes (protein carrier molecules) in the mitochondrial membranes. As high-energy electrons are shuttled down the chain via NADH and $FADH_2$ to oxygen molecules, they produce ATP and water.

oxide A compound of oxygen with another element; usually created by oxidation.

oxidizing agent An atom or ion that causes another to be oxidized and therefore it becomes reduced. It is a reactant that accepts electrons from another reactant.

Oxygen, chlorine, ozone, and peroxide compounds are examples of oxidizing agents.

oxidoreductase An ENZYME of EC class 1, which catalyzes an oxidation-reduction reaction.

oxygen One of the most important elements for biological systems and for other processes such as reacting with other substances to release energy. One tree can produce enough oxygen in one week to meet the demands of a person's daily oxygen need. Oxygen is needed in oxidation-reduction reactions within cells. Cellular respiration is the process that releases energy by breaking down food molecules in the presence of oxygen. Atomic symbol is O; atomic number is 8.

oxygen-evolving complex (OEC) The ENZYME that catalyzes the formation of O_2 in PHOTOSYNTHESIS. Contains a CLUSTER of probably four manganese ions.

ozone (O_3) A form of oxygen containing three atoms instead of the common two formed by high-energy ultraviolet radiation reacting with oxygen. Ozone accounts for the distinctive odor of the air after a thunderstorm or around electrical equipment, first reported as early as 1785; ozone's chemical constitution was established in 1872. The ozone layer in the upper atmosphere blocks harmful ultraviolet radiation that normally causes skin cancer. Ozone is an oxidizer and disinfectant, and it forms hydrogen peroxide when mixed with water. The Earth's ozone layer protects all life from the sun's harmful radiation, but human activities have damaged this shield. The United States, in cooperation with more than 140 other countries, is phasing out the production of ozone-depleting substances in an effort to safeguard the ozone layer.

See also OXYGEN.

ozonolysis The oxidation of an organic material by ozone. The process of treating an organic compound with ozone that splits a double bond in a hydrocarbon chain and forms an ozonide (an unstable intermediate); used to locate double bonds in molecules and to determine the structure of unsaturated fatty acids.

P

pairing energy Energy required to place two electrons in the same orbital.

See also ATOMIC ORBITAL; MOLECULAR ORBITAL.

paraffin A natural wax.

parallel reaction *See* COMPOSITE REACTION.

paramagnetic Substances having a positive MAGNETIC SUSCEPTIBILITY are paramagnetic. They are attracted by a magnetic field.

See also DIAMAGNETIC.

parent ion (precursor ion) An ion that dissociates to a smaller ion fragment, usually from a collision-induced dissociation in a MS/MS (mass spectrometry/mass spectrometry instrument) or tandem mass spectrometry experiment.

partial agonist An AGONIST that is unable to induce maximal activation of a RECEPTOR population, regardless of the amount of DRUG applied.

See also INTRINSIC ACTIVITY.

partial pressure Each gas in a mixture of gases exerts a pressure called the partial pressure; the pressure exerted by one gas in a mixture of gases.

partial rate factor The rate of substitution at one specific site in an AROMATIC compound relative to the rate of substitution at one position in benzene. For example, the partial rate factor fp^z for parasubstitution in a monosubstituted benzene C_6H_5Z is related to the rate constants $k(C_6H_5Z)$ and $k(C_6H_6)$ for the total reaction (i.e., at all positions) of C_6H_5Z and benzene, respectively, and %para (the percentage parasubstitution in the total product formed from C_6H_5Z) by the relation

$$f_p^Z = \frac{6k(C_6H_5Z)}{k(C_6H_6)} \frac{\%\text{para}}{100}$$

Similarly for metasubstitution:

$$f_m^Z = \frac{6k(C_6H_5Z)}{2k(C_6H_6)} \frac{\%\text{meta}}{100}$$

(The symbols p_f^z, m_f^z, o_f^z are also in use.) The term applies equally to the IPSO position, and it can be extended to other substituted SUBSTRATES undergoing parallel reactions at different sites with the same reagent according to the same RATE LAW.

See also SELECTIVITY.

particulate matter A generic term used for a type of air pollution that contains complex and varying mixtures of particles suspended in the air.

pascal The unit of pressure or stress under the International System of Units (SI). One pascal is equivalent to one newton (1 N) of force applied over an area of one meter squared (1 m^2).

passivation Describes the state where metals or other materials remain indefinitely unattacked because of modified or altered surface conditions, although it has conditions of thermodynamic instability. Also refers to a commercial process for protection of metals by creating an inert oxide layer, e.g., the dipping of steel fittings into a nitric acid solution to rapidly form a chromium oxide layer on the surface of the material, thus creating a passive film that protects the metal from further oxidation.

passive transport A molecule or ion that crosses a biological membrane by moving down a CONCENTRATION or ELECTROCHEMICAL GRADIENT with no expenditure of metabolic energy. Passive transport, in the same direction as a concentration gradient, can occur spontaneously, or proteins can mediate passive transport and provide the pathway for this movement across the lipid bilayer without the need to supply energy for the action. These proteins are called channels if they mediate ions and permeases for large molecules. This type of transport always operates from regions of greater concentration to regions of lesser concentration. Also called DIFFUSION.

See also ACTIVE TRANSPORT.

pattern recognition The identification of patterns in large data sets using appropriate mathematical methodologies.

Pauli exclusion principle States that no two electrons in the same atom may have identical sets of four quantum numbers. Introduced by Austrian-American physicist Wolfgang Pauli in 1925.

Pauling, Linus (1901–1994) American *Chemist* Linus Carl Pauling was born in Portland, Oregon, on February 28, 1901, to druggist Herman Henry William Pauling and Lucy Isabelle Darling. He attended the public elementary and high schools in the town of Condon and in Portland, Oregon, and entered the Oregon State College in 1917. He received a B.Sc. in chemical engineering in 1922. Pauling married Ava Helen Miller of Beaver Creek, Oregon, in 1923, and they had four children. Between 1919 and 1920 he taught quantitative analysis in the state college, later becoming appointed a teaching fellow in chemistry in the California Institute of Technology. He became a graduate student there from 1922 to 1925 and received his Ph.D. (summa cum laude) in chemistry, with minors in physics and mathematics.

Like Robert Mulliken, he was inspired by the works of IRVING LANGMUIR. In 1919 his interest lay in the field of molecular structure and the nature of the chemical bond.

Pauling was a controversial figure much of his life, but he made important discoveries in physical, structural, analytical, inorganic, and organic chemistry as well as biochemistry, combining theoretical physics, notably quantum theory and quantum mechanics, in his studies.

He wrote more than 1,000 articles and books for the field and for the general public—on science, peace, and health—with such popular books as *Vitamin C and the Common Cold; Cancer and Vitamin C* (with Ewan Cameron, M.D.); and *How to Live Longer and Feel Better.* His landmark book *The Nature of the Chemical Bond* has been called one of the most influential scientific books of the 20th century. His introductory textbook, *General Chemistry,* revised several times since its first printing in 1947 (translated into 13 languages), has been used by generations of undergraduates. He is also often considered the founding father of molecular biology.

In 1954 he was awarded the Nobel Prize in chemistry "for his research into the nature of the chemical bond and its application to the elucidation of the structure of complex substances," and in 1962 he received the Nobel Peace Prize, the only person to win two such prizes singularly. He received other awards too numerous to mention.

He died on August 19, 1994, at age 93, at his ranch near Big Sur, on the California coast.

***p*-block element** Elements in Group III to Group VIII , where the *p* orbitals are being filled.

- Group III elements are B, Al, Ga, In, and Tl. Their outer electron configuration is *s2p1*.
- Group IV contains the elements C, Si, Ge, Sn, and Pb. Their outer electron configuration is *s2p2*.
- Group V contains the elements N, P, As, Sb, and Bi. Their outer electron configuration is *s2p3*.
- Group VI (chalcogens) are the elements O, S, Se, Te, and Po. Their outer electron configuration is *s2p4*.
- Group VII (halogens) are the elements F, Cl, Br, I, and At. Their outer electron configuration is *s2p5*.
- Group VIII (inert gases or noble gases) are the elements He, Ne, Ar, Kr, Xe, and Rn. Their outer electron configuration is *s2p6*.

PCR *See* POLYMERASE CHAIN REACTION.

peptide bond The bond that links amino acids together. Created by a condensation reaction between the alpha-amino group of one amino acid and the alpha-carboxyl group of another amino acid; a covalent bond.

peptidoglycan A thick, rigid-layer, cross-linked polysaccharide-peptide complex that is found in the walls of bacteria. Composed of an overlapping lattice of two sugars, N-acetyl glucosamine (NAG) and N-acetyl muramic acid (NAM), that are cross-linked by amino-acid bridges found only in the cell walls of bacteria. This elaborate, covalently cross-linked structure provides great strength of the cell wall.

peptidomimetic A compound, containing nonpeptidic structural elements, that is capable of mimicking or antagonizing the biological action of a natural parent peptide. A peptidomimetic no longer has classical peptide characteristics such as enzymatically scissille peptidic bonds.

See also PEPTOID.

peptoid A PEPTIDOMIMETIC that results from the oligomeric assembly of N-substituted glycines.

percentage ionization The percentage of the weak electrolyte that ionizes in a solution of given concentration.

percent by mass The actual yield divided by theoretical yield multiplied by 100 percent. Number of grams of solute dissolved in 100 g of solution.

percent composition The relative measure of the mass of each different element present in the compound.

percent ionic character Ionic character refers to the amount of time that a bond exists in the ionic form, which is measured by the percent of ionic character. The numerical value of the percent of ionic character is related to the difference in electronegativity values for the two bonded atoms. If the difference is greater than 1.7, then the bond is ionic. If the difference is between 1.7 and 0.3, then the bond is polar covalent. If the difference is below 0.3, then the bond is nonpolar covalent.

percent purity The percent of a specified compound or element found in an impure sample.

perfect gas (ideal gas) A gas with molecules with no size that exhibits no intermolecular forces and that obeys the ideal gas equation relating pressure, volume, and temperature; $PV = nRT$, where R is a constant and n is the number of moles of gas. Real gases can approximate the behavior of an ideal gas at low pressure.

perfluorocarbon (PFC) A derivative of hydrocarbons in which all of the hydrogens have been replaced by fluorine. They are clear, colorless, odorless, nonconducting, and nonflammable liquids that are nearly twice as dense as water, and they are capable of dissolving large amounts of physiologically important gases (oxygen and carbon dioxide). PFCs are generally very chemically stable compounds that are not metabolized in body tissues. They are used in paints to make them spread easier, in textile manufacturing as a fabric protectant, as radiological imaging agents, and they have been explored as blood substitutes. Their vapors are also potent greenhouse gases.

pericyclic reaction A CHEMICAL REACTION in which concerted reorganization of bonding takes place throughout a cyclic array of continuously bonded atoms. It may be viewed as a reaction proceeding through a fully conjugated cyclic transition state. The number of atoms in the cyclic array is usually six, but other numbers are also possible. The term embraces a variety of processes, including CYCLOADDITION, CHELETROPIC REACTION, ELECTROCYCLIC REACTION, SIGMATROPIC REARRANGEMENT, etc. (provided they are CONCERTED PROCESSes).

See also MULTICENTER REACTION.

period Elements in a horizontal row of the periodic table.

periodicity With increasing atomic number, the electron configuration of the atoms displays a periodic variation.

periodic law When the elements are arranged by atomic number, their physical and chemical properties vary periodically. The properties of the elements are periodic functions of their atomic numbers.

periodic table of elements (**periodic chart of elements**) An arrangement of elements in an order of increasing atomic numbers that also emphasizes periodicity. *See* Appendix V, Periodic Table of Elements.

periselectivity The differentiation between two symmetry-allowed processes, for example the [2+4] vs. [4+6] CYCLOADDITION of cyclopentadiene to tropone.

peroxidase A HEME protein (donor:hydrogen-peroxide OXIDOREDUCTASE, EC class 1.11.1) that catalyzes the one-electron oxidation of a SUBSTRATE by dihydrogen peroxide. Substrates for different peroxidases include various organic compounds, CYTOCHROME c, halides, and Mn^{2+}.

See also EC NOMENCLATURE FOR ENZYMES.

peroxide A compound containing oxygen in the –1 oxidation state. An example is hydrogen peroxide H_2O_2.

petroleum A generic name for naturally occurring hydrocarbons, including crude oil, natural gas liquids, natural gases, and their products.

Pfeiffer's rule States that in a series of chiral compounds the EUDISMIC RATIO increases with increasing POTENCY of the EUTOMER.

pharmacokinetics The study of absorption, distribution, METABOLISM, and excretion (ADME) of bioactive compounds in a higher organism.

pharmacophore (**pharmacophoric pattern**) The ensemble of steric and electronic features that is necessary to ensure the optimal supramolecular interactions with a specific biological target structure and to trigger (or to block) its biological response.

A pharmacophore does not represent a real molecule or a real association of functional groups but is a purely abstract concept that accounts for the common molecular interaction capacities of a group of compounds toward their target structure. The pharmacophore can be considered to be the largest common denominator shared by a set of active molecules. This definition discards a misuse often found in the MEDICINAL CHEMISTRY literature, which consists of naming as pharmacophores simple chemical functionalities such as guanidines, sulfonamides, or dihydroimidazoles (formerly imidazolines), or typical structural skeletons such as flavones, phenothiazines, prostaglandins, or steroids.

pharmacophoric descriptors Are used to define a PHARMACOPHORE, including H-bonding hydrophobic and electrostatic interaction sites, defined by atoms, ring centers, and virtual points.

phase diagram A graphical representation of the effects of conditions on the equilibria among the various phases that may exist in a system. For a single substance, this usually represents the variation of melting point, boiling point, or the temperature at which one soilid phase can be converted to another with pressure.

For systems of more than one chemical component, the concentrations of each are additional variables. A typical two-component phase diagram displays the variation of vapor pressure or melting point with composition at a fixed pressure, for example.

phase rule **(Gibbs phase rule)** Describes the possible number of degrees of freedom in a (closed) system at equilibrium.

phase-transfer catalysis The phenomenon of rate enhancement of a reaction between chemical species located in different phases (immiscible liquids or solid and liquid) by addition of a small quantity of an agent (called the "phase-transfer CATALYST") that extracts one of the reactants, most commonly an anion, across the interface into the other phase so that reaction can proceed. These catalysts are salts of ONIUM IONs (e.g., tetraalkylammonium salts) or agents that complex inorganic cations (e.g., CROWN ethers). The catalyst cation is not consumed in the reaction, although an anion exchange does occur.

phenol Specifically, C_6H_5OH; generically, a hydrocarbon derivative containing an [OH] group bound to an aromatic ring. Found naturally in decaying dead organic matter like rotting vegetables and in coal.

phenonium ion *See* BRIDGED CARBOCATION.

pheromone A volatile chemical secreted and sent externally by an organism to send information to members of the same species via olfactory senses that induce a physiological or behavioral response, such as sexual attraction.

phosphatase An ENZYME that catalyzes the hydrolysis of orthophosphoric monoesters. Alkaline phosphatases (EC 3.1.3.1) have an optimum pH above 7 and are zinc-containing proteins. Acid phosphatases (EC 3.1.3.2) have an optimum pH below 7, and some of these contain a DINUCLEAR center of iron, or iron and zinc.

See also EC NOMENCLATURE FOR ENZYMES.

phosphate group Oxygenated phosphorus ($-PO_4$) that is attached to a carbon chain; important in energy transfer from ATP in cell signal transduction, the biochemical communication from one part of the cell to another; also part of a DNA NUCLEOTIDE.

phospholipase A **(phosphatide acylhydrolases)** Catalyzes the hydrolysis of one of the acyl groups of phosphoglycerides or glycerophosphatidates. Phospholipase A1 hydrolyzes the acyl group attached to the 1-position, while phospholipase A2 hydrolyzes the acyl group attached to the 2-position.

See also PHOSPHOLIPASES.

phospholipases A class of enzymes that catalyze the hydrolysis of phosphoglycerides or glycerophosphatidates.

See also PHOSPHOLIPASE A.

phospholipids The main component of cell walls; an amphiphilic molecule (lipid). A glycerol skeleton is attached to two fatty acids and a phosphate group, and onto the phosphate one of three nitrogen groups, so both phosphate and nitrogen groups make the "polar head" larger and more polar. The phosphate part of the molecule is water soluble, while the fatty-acid chains are fat soluble. The phospholipids have a polar hydrophilic head (phosphate) and nonpolar hydrophobic tail (fatty acids). When in water, phospholipids sort into spherical bilayers; the phosphate groups point to the cell exterior and interior, while the fatty acid groups point to the interior of the membrane.

Examples include lecithin, cephalin, sphingomyelin, phosphatidic acid, and plasmalogen. Two types of phospholipids exist: glycerophospholipid and sphingosyl phosphatide. A synthetic phospholipid, alkylphosphocholine, has been used in biological and therapeutic areas.

Structurally, phospholipids are similar to triglycerides except that a phosphate group replaces one of the fatty acids.

phosphorylation A process involving the transfer of a phosphate group (catalyzed by ENZYMEs) from a donor to a suitable acceptor; in general an ester linkage is formed, for example:

ATP + alcohol → ADP + phosphate ester

photic zone The upper layer within bodies of water reaching down to about 200 meters, where sunlight penetrates and promotes the production of photosynthesis; the richest and most diverse area of the ocean. A region where photosynthetic floating creatures (phytoplankton) are primary producers as well as a major food source. The LITTORAL ZONE and much of the sublittoral zone falls within the photic zone.

See also APHOTIC ZONE; OCEANIC ZONE.

photoautotroph An organism that uses sunlight to provide energy and carbon dioxide as the chief source of carbon, such as photosynthetic bacteria, cyanobacteria, and algae. Green plants are photoautotrophs.

See also PHOTOHETEROTROPH.

photochemical oxidants Photochemically produced oxidizing agents capable of causing damage to plants and animals. Formed when sunlight reacts to a mixture of chemicals in the atmosphere.

See also PHOTOCHEMICAL SMOG.

photochemical smog A brownish-colored smog in urban areas that receive large amounts of sunlight; caused by photochemical (light-induced) reactions from nitrogen oxides, hydrocarbons, and other components of polluted air that produce PHOTOCHEMICAL OXIDANTS.

photochemistry The branch of chemistry concerned with the chemical effects of light (far UV to IR).

photodissociation The splitting of a molecule by a photon.

photoelectric effect The emission of electrons from the surface of a metal caused by light hitting the surface.

photoheterotroph Like PHOTOAUTOTROPHs, any organism that uses light as a source of energy but must use organic compounds as a source of carbon, for example, green and purple nonsulfur bacteria.

photolysis A light-induced bond cleavage. The term is often used incorrectly to describe irradiation of a sample.

photon Name given to a quantum or packet of energy emitted in the form of electromagnetic radiation. A particle of light, gamma and X rays are examples.

photoperiodism The physiological response to length of day and night in a 24-hour period, such as flowering or budding in plants.

photophosphorylation The process of creating ADENOSINE-5′-TRIPHOSPHATE (ATP) from ADP and phosphate by using the energy of the sun. Takes place in the thylakoid membrane of chloroplasts.

photosynthesis A METABOLIC process in plants and certain bacteria, using light energy absorbed by CHLOROPHYLL and other photosynthetic pigments for the reduction of CO_2, followed by the formation of organic compounds.

See also PHOTOSYSTEM.

photosystem A membrane-bound protein complex in plants and photosynthetic bacteria, responsible for light harvesting and primary electron transfer. Com-

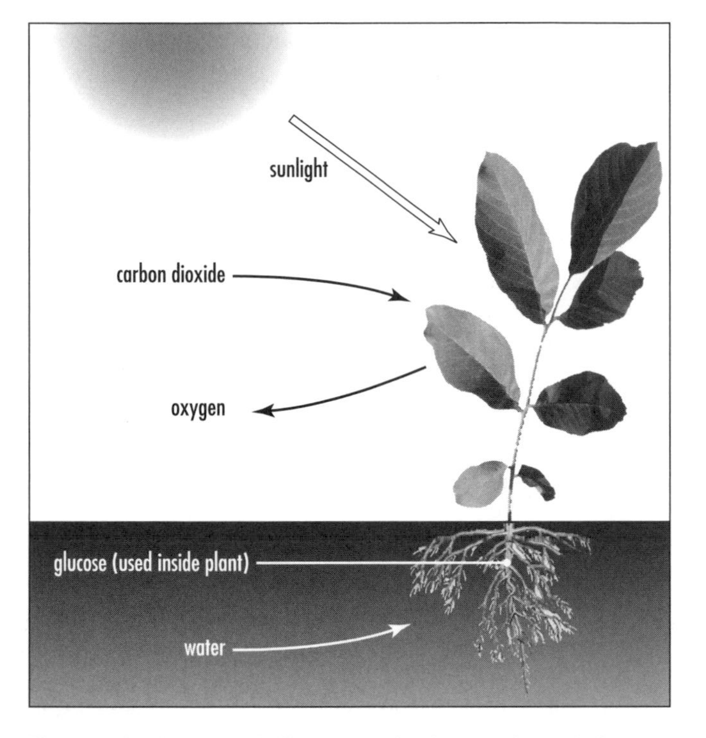

Photosynthesis. A metabolic process in plants and certain bacteria, using light energy absorbed by chlorophyll and other photosynthetic pigments for the reduction of CO_2, followed by the formation of organic compounds

prises light-harvesting pigments such as CHLOROPHYLL; a primary electron-transfer center, and secondary electron carriers. In green plant PHOTOSYNTHESIS, photosystem I transfers electrons from PLASTOCYANIN to a [2FE-2S] FERREDOXIN, and contains IRON-SULFUR PROTEINS. Photosystem II transfers electrons from the OXYGEN-EVOLVING COMPLEX to plastoquinone and contains an iron center.

photovoltaic cell (solar cell) A cell made of thin wafers of two slightly different types of silicon. One, the P-type (P for positive), is doped with tiny quantities of boron and contains positively charged "holes," that are missing electrons. The other type of silicon, called N-type (N for negative), is doped with small amounts of phosphorus and contains extra electrons. When these two thin P and N materials are put together, it produces a junction that, when exposed to light, produces a movement of electrons, thus producing an electric current. Photovoltaic cells convert light energy into electrical energy.

pH-rate profile A plot of observed rate coefficient, or more usually its decadic logarithm, against pH of solution, other variables being kept constant.

pH scale The concentration of hydrogen ions in a solution expressed as the negative logarithm of the concentration. The scale typically is regarded as running from 0 to 14, but in principle it can exceed these values if the hydrogen ion concentration is greater than 1 *M* or less than 10^{-4} *M*. Low pH corresponds to high hydrogen ion concentration, and high pH refers to low hydrogen concentration. A substance added to water that increases the concentration of hydrogen ions (i.e., lowers the pH) is called an acid, while a substance that reduces the concentration of hydrogen ions (i.e., raises the pH) is called a base. Acid in the stomach has a pH of 1, while a liquid drainer has a pH of 14. Pure water is neutral with a pH of 7. Compounds called BUFFERS can be added to a solution that will resist pH changes when an acid or base is added.

See also ACID; BASE.

physical change Refers to when a material changes from one physical state to another without formation of intermediate substances of different composition in the process, such as the change from gas to liquid.

phytoalexin A toxic substance that acts like an antibiotic that is produced by plants to inhibit or kill the growth of microorganisms such as certain fungi that would otherwise infect them; e.g., pisatin (produced by peas), phaseollin (produced by beans [*Phaseolus*]), camalexin (produced by *Arabidopsis thaliana*), resveratrol (grapes).

phytochelatin A peptide of higher plants, consisting of polymers of 2-11 glutathione (γ-glutamyl-cysteinyl-glycine) groups, that binds heavy metals.

phytochrome (red-light-sensitive system) Photoreceptor proteins that regulate light-dependent growth processes; absorb red and far-red light in a reversible system; one of the two light-sensing systems involved in PHOTOPERIODISM and photomorphogenesis. Plant responses regulated by phytochrome include photoperiodic induction of flowering, chloroplast development (minus chlorophyll synthesis), leaf senescence and leaf abscission, seed germination, and flower induction.

pi (π) adduct An ADDUCT formed by electron-pair donation from a PI (π) ORBITAL into a SIGMA (σ) ORBITAL, or from a sigma orbital into a pi orbital, or from a pi orbital into a pi orbital. For example:

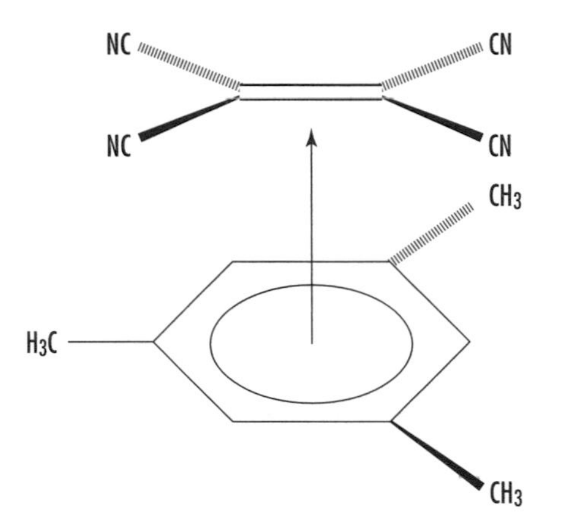

Such an adduct has commonly been known as a "pi complex" but, as the bonding is not necessarily weak, it is better to avoid the term COMPLEX, in accordance with the recommendations in this encyclopedia.

See also COORDINATION.

pi (π) bond A COVALENT BOND formed by the overlap between two *p* orbitals on different atoms. Pi bonds are superimposed on SIGMA (σ) BONDs, forming double or triple bonds.

See also SIGMA, PI.

picket-fence porphyrin A PORPHYRIN with a protective enclosure for binding oxygen at one side of the ring that is used to mimic the dioxygen-carrying properties of the HEME group.

See also BIOMIMETIC.

+pi (π) complex *See* PI (π) ADDUCT.

pi (π) electron acceptor, π-electron donor group A SUBSTITUENT capable of a +R (e.g., NO_2) or –R (e.g., OCH_3) effect, respectively.

See also ELECTRONIC EFFECT; POLAR EFFECT; SIGMA (σ) CONSTANT.

piezoelectric effect Crystals that acquire a charge when compressed, twisted, or distorted. The word *piezo* is Greek for "push." The effect was discovered by brothers Pierre and Jacques Curie in 1880.

pi (π) orbital Bonding orbitals with two lobes lying parallel to the bond axis.

See also SIGMA, PI.

placebo An inert substance or dosage form identical in appearance, flavor, and odor to the active substance or dosage form. It is used as a negative control in a BIOASSAY or in a clinical study.

Planck's constant *(h)* A physical constant used to describe the sizes of quanta (any quantity that can only take on integer multiples of some base value). Named after Max Planck, it plays a central role in the theory of quantum mechanics, in which Planck is one of the founders of quantum theory. Its value is expressed as:

$$h = 6.626 \times 10^{-34}\ \mathrm{J} \bullet \mathrm{s}$$

plasma In physics, a plasma is an ionized gas. In biology, this term has the following three meanings:

(1) Fluid component of blood in which the blood cells and platelets are suspended (blood plasma). Note the distinction between plasma, which describes a part of the blood (the fluid part of blood, outside the blood cells), and serum, which describes a fraction derived from blood by a manipulation (the fluid that separates when blood coagulates).

(2) Fluid component of semen produced by the accessory glands, the seminal vesicles, the prostate, and the bulbourethral glands.

(3) Cell substance outside the nucleus (CYTOPLASM).

plasmid An extrachromosomal GENETIC element consisting generally of circular double-stranded DNA, which can replicate independently of chromosomal DNA. R-plasmids are responsible for the mutual transfer of antibiotic resistance among microbes. Plasmids are used as vectors for CLONING DNA in bacteria or yeast host cells.

plastic A synthetic material made from long chains of molecules that has the capability of being molded or shaped, usually by the application of heat and pressure. A plastic is made up principally of a binder together with PLASTICIZERS, fillers, pigments, and other additives. There are two basic types of plastic: thermosetting, which cannot be resoftened after being subjected to heat and pressure, and thermoplastic, which can be repeatedly softened and remolded by heat and pressure.

plasticize To soften a material and make it plastic or moldable by the application of a PLASTICIZER or the application of heat.

plasticizer Plasticizers are added to a binder to increase flexibility and toughness. They are low-melting solids or high-boiling organic liquids. They have varying degrees of softening action and solvating ability resulting from a reduction of intermolecular forces in the polymer.

plastocyanin An ELECTRON-TRANSFER PROTEIN, containing a TYPE 1 COPPER site, involved in plant and cyanobacterial PHOTOSYNTHESIS, that transfers electrons to PHOTOSYSTEM I.

plutonium Element number 94, which was secretly discovered during World War II in 1940 but was publicized in 1946 by Glenn Seaborg and others. It is mostly used for nuclear weapons, as a fission energy source (a by-product of the fission process in nuclear reactors, due to neutron capture by uranium-238 in particular), and in deep-space probes. Extracted from uranium ore in 1947. Over one-third of the energy produced in most nuclear power plants comes from plutonium.

***p-n* junction** In a semiconductor, it is where a *p*-type material and an *n*-type material are in direct contact.

pOH The negative logarithm of the hydroxide ion concentration. In contrast, pH is the negative logarithm of the hydrogen ion concentration.

poikilotherm An organism (e.g., fish or reptile) whose body temperature varies or fluctuates with the temperature of its surroundings; an ectotherm.

See also ENDOTHERM.

polar aprotic solvent *See* DIPOLAR APROTIC SOLVENT.

polar covalent bond A covalent bond in which the electrons are not shared equally between the two atoms, but spend more time in the vicinity of the atom with higher ELECTRONEGATIVITY. For example, in the bond between hydrogen and oxygen, electrons shared by atoms spend a greater percentage of time closer to the oxygen nucleus than the hydrogen nucleus; bonds are polar, i.e., they have a partial electric charge (hydrogen is positive, oxygen is negative); in organisms, they can form weak HYDROGEN BONDS.

polar effect For a reactant molecule RY, the polar effect of the group R comprises all the processes whereby a substituent may modify the electrostatic forces operating at the reaction center Y, relative to the standard $R^{o}Y$. These forces can be governed by charge separations arising from differences in the ELECTRONEGATIVITY of atoms (leading to the presence of dipoles, the presence of unipoles, or electron DELOCALIZATION). It is synonymous with ELECTRONIC EFFECT or "electrical effect" of a substituent, as distinguished from other substituent effects, e.g., STERIC EFFECTS.

Sometimes, however, the term *polar effect* is taken to refer to the influence, other than steric, that nonconjugated substituents exert on reaction rates, i.e., effects connected with electron delocalization between a substituent and the molecular framework to which it is attached are excluded. Polar effect is then not synonymous with electronic effect.

See also FIELD EFFECT; INDUCTIVE EFFECT; MESOMERIC EFFECT.

polarimeter A device used to measure optical activity; determines the amount of polarization of light.

polarity When applied to solvents, this rather ill-defined term covers their overall SOLVATION capability (solvation power) for solutes (i.e., in chemical equilibria: reactants and products; in reaction rates: reactants and ACTIVATED COMPLEX; in light absorptions: ions or molecules in the GROUND and EXCITED STATE), which in turn depends on the action of all possible, nonspecific and specific, intermolecular interactions between solute ions or molecules and solvent molecules, excluding such interactions leading to definite chemical alterations of the ions or molecules of the solute. Occasionally, the term *solvent polarity* is restricted to nonspecific solute/solvent interactions only (i.e., to VAN DER WAALS FORCES).

See also DIMROTH-REICHARDT E_T PARAMETER; GRUNWALD-WINSTEIN EQUATION; IONIZING POWER; KAMLET-TAFT SOLVENT PARAMETERS; Z-VALUE.

polarizability The ease of distortion of the electron cloud of a MOLECULAR ENTITY by an electric field (such as that due to the proximity of a charged reagent). It is experimentally measured as the ratio of INDUCED DIPOLE MOMENT (μ_{ind}) to the field E that induces it:

$$\alpha = \mu_{ind}/E$$

The units of α are $C^2\ m^2\ V^{-1}$. In ordinary usage, the term refers to the "mean polarizability," the average over three rectilinear axes of the molecule. Polarizabilities in different directions (e.g., along the bond in Cl_2, called "longitudinal polarizability," and in the direction perpendicular to the bond, called "transverse polarizability") can be distinguished, at least in principle. Polarizability along the bond joining a substituent to the rest of the molecule is seen in certain modern theoretical approaches as a factor influencing chemical reactivity, etc., and parameterization thereof has been proposed.

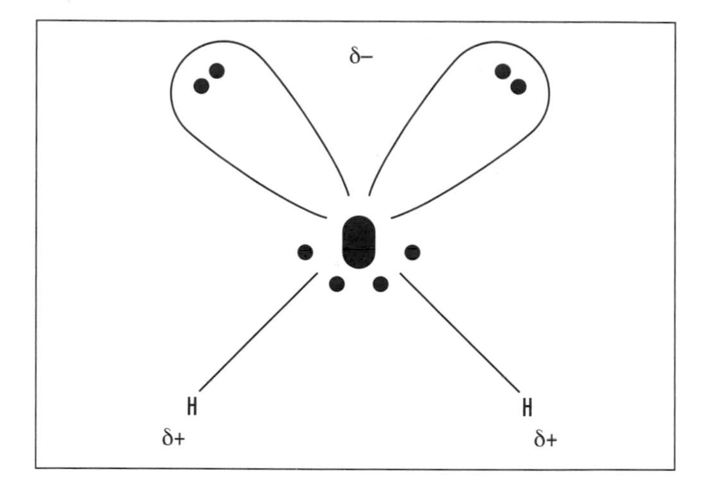

Polar molecule. A molecule that has both a positive and negative end, such as water

polar molecule A molecule that has both a positive and negative end, such as water.

polar solvent *See* POLARITY.

pollen Microscopic grains produced by plants to facilitate reproduction. Each plant has a pollinating period that varies, depending on the plant, climate, and region.

polychlorinated biphenyl (PCB) Polychlorinated biphenyls (PCBs) are a mixture of individual chemicals that are no longer produced in the United States but are still found in the environment. They are made up of two benzene rings attached by a C–C bond, with varying numbers of hydrogens replaced by chlorine. Health effects that have been associated with exposure to PCBs include acne-like skin conditions in adults and neurobehavioral and immunological changes in children, and PCBs are known to cause cancer in animals. PCBs have been found in at least 500 of the 1,598 National Priorities List sites identified by the Environmental Protection Agency (EPA).

Polychlorinated biphenyls are mixtures of up to 209 individual chlorinated compounds (known as CONGENERS). There are no known natural sources of PCBs, which are oily liquids or solids that are colorless to light yellow. Some PCBs can exist as a vapor in air. PCBs have no known smell or taste.

PCBs have been used as coolants and lubricants in transformers, capacitors, and other electrical equipment because they do not burn easily and are good insulators. The manufacture of PCBs was stopped in the United States in 1977 because of evidence that they build up in the environment and can cause harmful health effects. Cleanup of PCB-contaminated sites is an ongoing controversy.

polycyclic A cyclic structure containing many closed and fused rings of bonds, e.g., polycyclic AROMATIC HYDROCARBONS (PAH), hydrocarbons with multiple benzene rings such as naphthalene, the benzo pyrenes, fluoranthene, and chrysene. PAHs are typical components of asphalt, fuel, oils, and greases.

polydent A ligand that has two or more donor sites that can be used simultaneously.

See also AMBIDENT.

polydentate *See* CHELATION; DONOR ATOM SYMBOL.

polyene A compound containing more than one double bond per molecule.

polyester A group of polymers that consist basically of repeated units of an ester and are used especially in making fibers or plastics. A strong and durable human-made fabric that was used in many types of clothing made popular during the 1970s.

polyethylene A plastic made from ethylene. High-density forms are used in the manufacturing of bottles and other products that produce toxic fumes when burned and thus are often recycled.

polyhedral symbol The polyhedral symbol indicates the geometrical arrangements of the coordinating atoms about the CENTRAL ATOM. It consists of one or more capital italic letters derived from common geometric terms (tetrahedron, square plane, octahedron, etc.), which denote the idealized geometry of the LIGANDs around the COORDINATION center, and an Arabic numeral that is the coordination number of the central atom. The polyhedral symbol is used as an affix, enclosed in parentheses, and separated from the name by a hyphen. Examples are *T*-4, *SP*-4, *TBPY*-5, *SPY*-5, *OC*-6, and *CU*-8.

polymer A macromolecule of high relative molecular mass composed of many similar or identical MONOMERS linked together in chains. PLASTICS are polymers.

polymerase chain reaction (PCR) A laboratory technique used to rapidly amplify predetermined regions of double-stranded DNA. Generally involves the use of a heat-stable DNA polymerase.

polymerization The creation of large molecules by combining many small ones.

polymorphous Materials that crystallize in more than one crystalline arrangement.

polypeptide A polymer chain of amino acids linked by covalent peptide bonds. One or more polypeptides form proteins. Each polypeptide has two terminal ends: one called the amino terminal, or N-terminal, has a free amino group; the other end is called the carboxyl terminal, or C-terminal, with a free carboxyl group.

polyprotic acid An acid that can give up more than one proton per molecule in a reaction. Examples include sulfuric acid (H_2SO_4).

polysaccharide A carbohydrate (polymer) made by polymerizing over 1,000 monosaccharides; a complex sugar.

***p* orbital** An electron with one unit of angular momentum. Orbital letters are associated with the angular-momentum quantum number, which is assigned an integer value from 0 to 3 and where *s* cor-

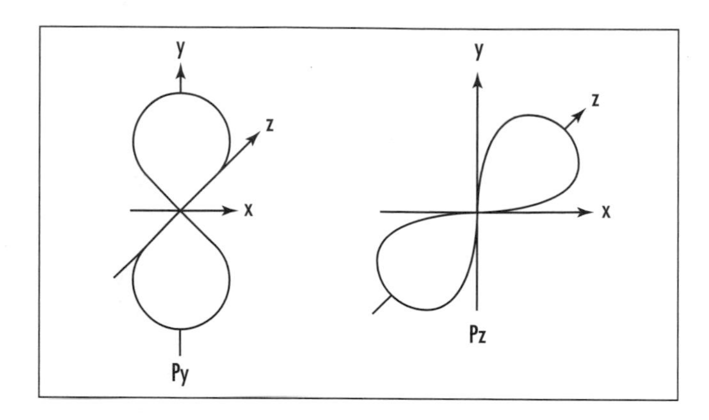

***P* orbital. An electron with one unit of angular momentum**

relates to 0, p = 1, d = 2, and f = 3. The p (principal) orbitals are said to be dumbbell shaped.

porins A class of proteins that create water-filled channels across cell membranes.

porphyrin A macrocyclic molecule that contains four pyrrole rings linked together by single carbon atom bridges between the alpha positions of the pyrrole rings. Porphyrins usually occur in their dianionic form coordinated to a metal ion.

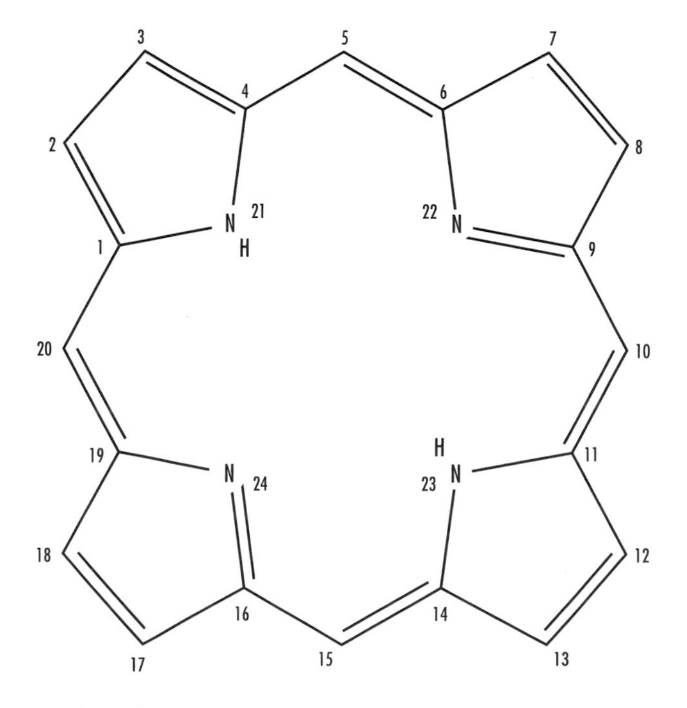

See also COORDINATION.

positron A particle equal in mass, but opposite in charge, to the electron (a positive electron). Also referred to as a β^+ particle when emitted in a nuclear reaction.

potency A comparative rather than an absolute expression of drug activity. Drug potency depends on both AFFINITY and EFFICACY. Thus, two AGONISTS can be equipotent but have different intrinsic efficacies with compensating differences in affinity. Potency is the dose of DRUG required to produce a specific effect of given intensity as compared with a standard reference.

potential energy Stored energy that can be released or harnessed to do work.

potential-energy profile A curve describing the variation of the potential energy of the system of atoms that make up the reactants and products of a reaction as a function of one geometric coordinate, and corresponding to the "energetically easiest passage" from reactants to products (i.e., along the line produced by joining the paths of steepest descent from the TRANSITION STATE to the reactants and to the products). For an ELEMENTARY REACTION, the relevant geometric coordinate is the REACTION COORDINATE; for a STEPWISE REACTION, it is the succession of reaction coordinates for the successive individual reaction steps. (The reaction coordinate is sometimes approximated by a quasichemical index of reaction progress, such as "degree of atom transfer" or BOND ORDER of some specified bond.)

See also POTENTIAL-ENERGY (REACTION) SURFACE; GIBBS ENERGY DIAGRAM.

potential-energy (reaction) surface A geometric hypersurface on which the potential energy of a set of reactants is plotted as a function of the coordinates representing the molecular geometries of the system.

For simple systems, two such coordinates (characterizing two variables that change during the progress from reactants to products) can be selected, and the potential energy plotted as a contour map.

For simple elementary reactions, e.g., A – B + C → A + B – C, the surface can show the potential energy for all values of the A, B, C geometry, providing that the ABC angle is fixed.

For more-complicated reactions, a different choice of two coordinates is sometimes preferred, e.g., the BOND ORDERS of two different BONDS. Such a diagram is often arranged so that reactants are located at the bottom left corner and products at the top right. If the trace of the representative point characterizing the route from reactants to products follows two adjacent edges of the diagram, the changes represented by the two coordinates take place in distinct succession; if the trace leaves the edges and crosses the interior of the diagram, the two changes are concerted. In many qualitative applications, it is convenient (although not strictly equivalent) for the third coordinate to represent the standard Gibbs energy rather than potential energy.

Using bond orders is, however, an oversimplification, since these are not well defined, even for the TRANSITION STATE. (Some reservations concerning the diagrammatic use of Gibbs energies are noted under GIBBS ENERGY DIAGRAM.)

The energetically easiest route from reactants to products on the potential-energy contour map defines the POTENTIAL-ENERGY PROFILE.

See also CONCERTED PROCESS; GIBBS ENERGY OF ACTIVATION; REACTION COORDINATE.

potentiometer (pot) A continuously variable resistor or instrument used to determine the precise measurement of electromotive forces; used to vary, or control, the amount of current that flows through an electronic circuit.

potentiometry The measurement of potential for quantitative electrochemical analysis.

Pourbaix diagram Pourbaix diagrams are often used to grab quickly an idea of which species predominates at specific pH and Eh (oxidation-reduction) conditions. The Pourbaix diagram is commonly used by corrosion engineers to determine the pH and potential where a metal will either be stable to corrosion, will corrode, or will form a passivating layer.

See also PREDOMINANCE AREA DIAGRAM.

power saturation A phenomenon used in ELECTRON PARAMAGNETIC RESONANCE SPECTROSCOPY to estimate the electron-spin relaxation times, providing information about distances between PARAMAGNETIC centers.

pre-association A step on the REACTION PATH of some STEPWISE REACTIONS in which the MOLECULAR ENTITY C is already present in an encounter pair or ENCOUNTER COMPLEX with A during the formation of B from A, e.g.,

$$A + C \underset{}{\overset{\text{Pre-association}}{\rightleftharpoons}} \underset{\text{encounter complex}}{(A\cdots\cdots C)} \longrightarrow \underset{\text{encounter complex}}{(B\cdots\cdots C)} \xrightarrow{\text{rapid}} \text{Product}$$

In this mechanism, the CHEMICAL SPECIES C may (but does not necessarily) assist in the formation of B from A, which may itself be a BIMOLECULAR reaction with some other reagent.

Pre-association is important when B is too short-lived to permit B and C to come together by diffusion.

See also MICROSCOPIC DIFFUSION CONTROL; SPECTATOR MECHANISM.

precipitate An insoluable solid that forms from a liquid suspension as a result of a chemical reaction, and usually settles out.

precursor complex *See* ENCOUNTER COMPLEX.

predominance area diagram Diagrams that delineate the relative stabilities of a chemical species in a given aqueous environment. Used to understand and predict equilibrium reactions in aqueous systems. The most well-known of such stability or predominance area diagrams is the Eh-pH diagram introduced by the work of Pourbaix. Others include the log {M}-pH and log {Ligand}-pH diagrams.

See also POURBAIX DIAGRAM.

pre-equilibrium (prior equilibrium) A rapidly reversible step preceding the RATE-LIMITING STEP in a STEPWISE REACTION. For example:

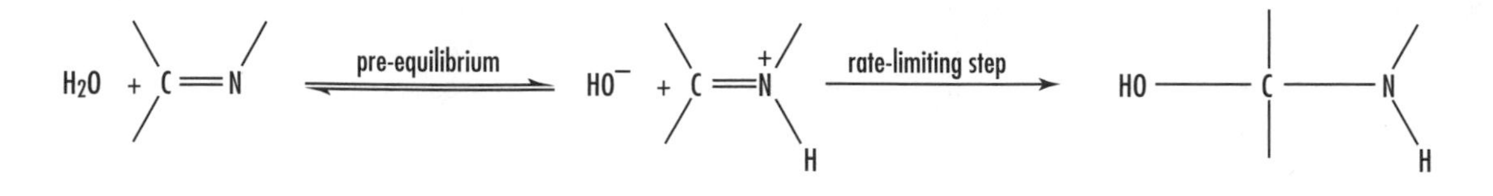

See also KINETIC EQUIVALENCE.

pre-exponential factor *See* ENERGY OF ACTIVATION; ENTROPY OF ACTIVATION.

Prigogine, Ilya (1917–2003) Belgian *Chemist* Ilya Prigogine was born in Moscow, Russia, on January 25, 1917. His scientific work dealt with the understanding of the role of time in the physical sciences and in biology and contributed significantly to the understanding of irreversible processes, particularly in systems far from equilibrium. He obtained both his undergraduate and graduate education in chemistry at the Universite Libre de Bruxelles.

He was awarded the Nobel Prize in chemistry in 1977 "for his contributions to non-equilibrium thermodynamics, particularly the theory of dissipative structures."

Before his death on May 28, 2003, he was regental professor and Ashbel Smith Professor of Physics and Chemical Engineering at the University of Texas at Austin. In 1967 he founded the Center for Statistical Mechanics, later renamed the Ilya Prigogine Center for Studies in Statistical Mechanics and Complex Systems.

In 1959 he became the director of the International Solvay Institutes in Brussels, Belgium. In 1989 Prigogine was awarded the title of viscount by the king of Belgium. He received 53 honorary degrees as well as numerous awards and was a member of 64 leading scientific organizations.

He authored several books, including *Modern Thermodynamics: From Heat Engines to Dissipative Structures* (with D. Kondepudi), 1998; *The End of Certainty, Time, Chaos and the New Laws of Nature* (with I. Stengers), 1997; and *Exploring Complexity* (with G. Nicolis), 1989.

primary kinetic electrolyte effect *See* KINETIC ELECTROLYTE EFFECT.

primary kinetic isotope effect *See* ISOTOPE EFFECT, KINETIC.

primary standard A standard that is designated or widely acknowledged as having the highest metrological qualities and whose value is accepted without reference to other standards of the same quantity.

primary structure The amino-acid SEQUENCE of a protein or NUCLEOTIDE sequence of DNA or RNA.

See also RIBONUCLEIC ACID.

primary voltaic cell A voltaic cell that cannot be recharged once the reactants are consumed.

primer A short preexisting polynucleotide chain to which new deoxyribonucleotides can be added by DNA polymerase.

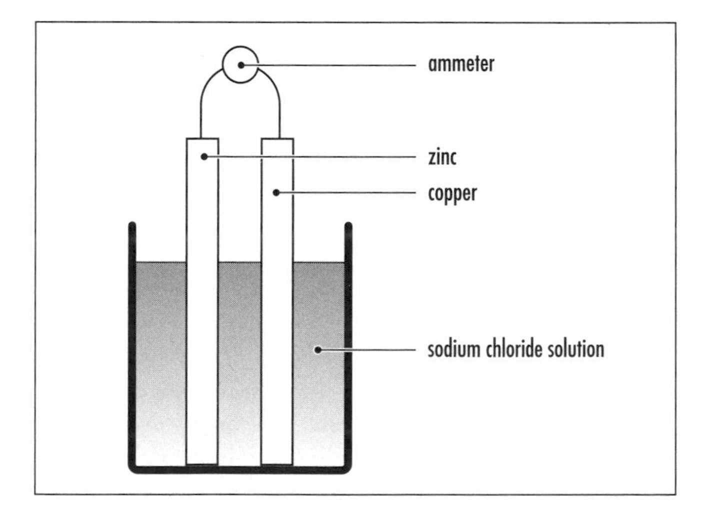

Primary voltaic cell. Produces electricity by chemical action and is irreversible

primitive change One of the conceptually simpler molecular changes into which an ELEMENTARY REACTION can be notionally dissected. Such changes include BOND rupture, bond formation, internal rotation, change of bond length or bond angle, bond MIGRATION, redistribution of charge, etc.

The concept of primitive changes is helpful in the detailed verbal description of elementary reactions, but a primitive change does not represent a process that is by itself necessarily observable as a component of an elementary reaction.

prior equilibrium *See* PRE-EQUILIBRIUM.

probability The statistical measure of likelihood.

prodrug Any compound that undergoes BIOTRANSFORMATION before exhibiting its pharmacological effects. Prodrugs can thus be viewed as DRUGS containing specialized nontoxic protective groups used in a transient manner to alter or to eliminate undesirable properties in the parent molecule.

See also DOUBLE PRODRUG.

product-determining step The step of a STEPWISE REACTION in which the product distribution is determined. The product-determining step may be identical to, or occur later than, the RATE-CONTROLLING STEP on the REACTION COORDINATE.

product-development control The term is used for reactions under KINETIC CONTROL where the SELECTIVITY parallels the relative (thermodynamic) stabilities of the products. Product-development control is usually associated with a TRANSITION STATE occurring late on the REACTION COORDINATE.

See also STERIC-APPROACH CONTROL; THERMODYNAMIC CONTROL.

promoter The DNA region, usually upstream to the coding SEQUENCE of a GENE or OPERON, that binds and directs RNA polymerase to the correct transcriptional start site and thus permits TRANSCRIPTION at a specific initiation site. (In catalysis, a promoter is used differently: a cocatalyst is usually present in much smaller amounts than the catalyst.)

promotion *See* PSEUDOCATALYSIS.

propagation *See* CHAIN REACTION.

prosthetic group A tightly bound, specific nonpolypeptide unit in a protein that is determining and that is involved in the protein's biological activity.

See also COFACTOR.

protein A molecule composed of many amino acids and with a complex structure, e.g., immunoglobulin, casein, etc.

See also AMINO ACID.

protein phosphatase (phosphoprotein phosphatase) An enzyme that removes a phosphate group from a protein by the use of hydrolysis; opposite effect of a protein kinase.

proteoglycan A type of glycoprotein with high carbohydrate content; component in the extracellular matrix of animal cells. Composed of one or more glycosaminoglycans; large, long polysaccharide chains covalently linked to protein cores.

proteomics The study of proteins—their location, structure, and function—through the combination of high-resolution protein separation techniques with mass spectrometry and modern sequence database mining tools.

protic *See* PROTOGENIC SOLVENT.

protogenic solvent Capable of acting as a PROTON (HYDRON) donor that is strongly or weakly acidic (as a BRONSTED ACID). The term is preferred to the synonym *protic* or the more ambiguous expression *acidic* by itself. Also called HBD (hydrogen-bond donor) solvent.

See also PROTOPHILIC SOLVENT.

protolysis This term has been used synonymously with PROTON (HYDRON)-TRANSFER REACTION. Because of its misleading similarity to HYDROLYSIS, PHOTOLYSIS, etc., its use is discouraged.

See also AUTOPROTOLYSIS.

proton A subatomic particle having a mass of 1.0073 amu and a charge of +1, found in the nuclei of atoms. The nucleus of normal hydrogen is made up of a proton; thus the ionized form of normal hydrogen is often called a proton.

proton affinity The negative of the enthalpy change in the gas-phase reaction (real or hypothetical) between a PROTON (more appropriately HYDRON) and the CHEMICAL SPECIES concerned, usually an electrically neutral species to give the CONJUGATE ACID of that species. Proton affinity is often, but unofficially, abbreviated as PA.

See also GAS-PHASE BASICITY.

protonation The addition of a proton to an atom, molecule, or ion.

proton motive force Energy or force created by the transfer of protons (hydrogen ions) on one side only of a cell membrane and across the membrane during chemiosmosis; an electrochemical gradient that has potential energy. This force can be channeled to operate rotating flagella, generate ATP, and other needed activities.

proton pump Proton pumps are a type of active transport and use the energy of ATP hydrolysis to force the transport of protons out of the cell and, in the process, create a membrane potential.

proton-transfer reaction A CHEMICAL REACTION, the main feature of which is the INTERMOLECULAR or INTRAMOLECULAR transfer of a proton (HYDRON) from one BINDING SITE to another. For example,

$$CH_3CO_2H + (CH_3)_2C{=}O \rightarrow CH_3CO_2^- + (CH_3)_2C{=}O^+H$$

In the detailed description of proton-transfer reactions, especially of rapid proton transfers between electronegative atoms, it should always be specified whether the term is used to refer to the overall process (including the more-or-less ENCOUNTER-CONTROLLED formation of a hydrogen-bonded complex and the separation of the products [*see* MICROSCOPIC DIFFUSION CONTROL]) or just to the proton-transfer event (including solvent rearrangement) by itself.

See also AUTOPROTOLYSIS; PROTOLYSIS; TAUTOMERISM.

protophilic solvent Capable of acting as proton acceptor, strongly or weakly basic (as a BRONSTED BASE). Also called HBA (hydrogen-bond acceptor) solvent.

See also PROTOGENIC SOLVENT.

protoplasm The living material within cells.

protoporphyrin IX The PORPHYRIN LIGAND of HEME b. Heme b is a Fe(II) porphyrin complex readily isolated from the hemoglobin of beef blood, but it is also found in other proteins, including other HEMOGLOBINs, MYOGLOBINs, CYTOCHROMEs P-450, CATALASEs, and PEROXIDASEs, as well as b-type cytochromes. Protoporphyrin IX contains four methyl groups in positions 2, 7, 12, and 18; two vinyl groups in positions 3 and 8; and two propionic acid groups in positions 13 and 17.

prototropic rearrangement **(prototropy)** *See* TAUTOMERISM.

pseudobinary ionic compounds Compounds containing more than two elements but that are named like binary compounds. Examples are sodium sulfate (Na_2SO_4), ammonium chloride (NH_4Cl), and ammonium phosphate ($(NH_4)_3PO_4$).

pseudocatalysis If an ACID or BASE is present in nearly constant concentration throughout a reaction in solution (owing to BUFFERing or the use of a large excess), it may be found to increase the rate of that reaction and also to be consumed during the process. The acid or base is then not a CATALYST, and the phenomenon cannot be called CATALYSIS according to the well-established meaning of these terms in chemical kinetics, although the MECHANISM of such a process is often intimately related to that of a catalyzed reaction. It is recommended that the term *pseudocatalysis* be used in these and analogous cases (not necessarily involving acids or bases). For example, if a BRONSTED ACID accelerates the hydrolysis of an ester to a carboxylic acid and an alcohol, this is properly called acid catalysis, whereas the acceleration, by the same acid, of hydrolysis of an amide should be described as pseudocatalysis by the acid: the "acid pseudocatalyst" is consumed during the reaction through formation of an ammonium ion. The terms *general acid pseudocatalysis* and *general base pseudocatalysis* can be used as the analogs of GENERAL ACID CATALYSIS and GENERAL BASE CATALYSIS.

The terms *base-promoted, base-accelerated,* or *base-induced* are sometimes used for reactions that are pseudocatalyzed by bases. However, the term *promotion* also has a different meaning in other chemical contexts.

pseudo-first-order rate coefficient *See* ORDER OF REACTION.

pseudomolecular rearrangement The use of this awkwardly formed term is discouraged. It is synonymous with "intermolecular rearrangement."

See also MOLECULAR REARRANGEMENT.

pseudopericyclic A concerted TRANSFORMATION is pseudopericyclic if the primary changes in bonding occur within a cyclic array of atoms at one (or more) sites where nonbonding and bonding ATOMIC ORBITALS interchange roles.

A formal example is the enol

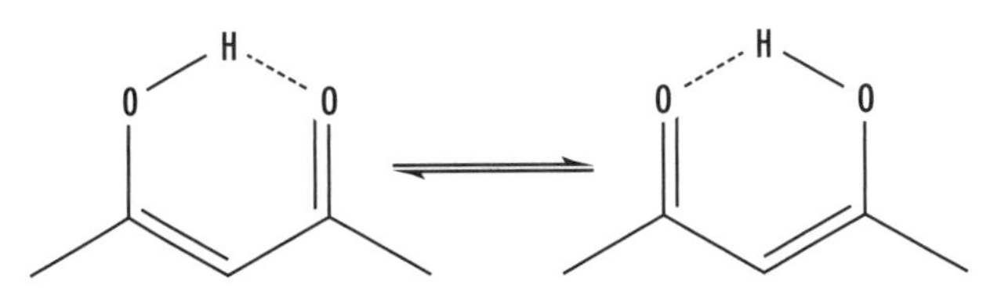

Because the pi and sigma atomic orbitals that interchange roles are orthogonal, such a reaction does not proceed through a fully CONJUGATED TRANSITION STATE and is thus not a pericyclic reaction and therefore is not governed by the rules that express ORBITAL SYMMETRY restrictions applicable to pericyclic reactions.

See also CONCERTED PROCESS.

pseudo-unimolecular A term sometimes used synonymously with PSEUDO-FIRST ORDER but is inherently meaningless.

See also MOLECULARITY; ORDER OF REACTION.

pterin 2-Amino-4-hydroxypteridine.

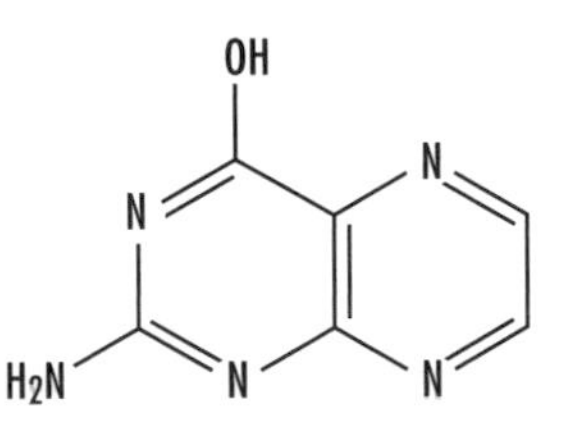

See also MOLYBDOPTERIN.

***p*-type semiconductor** A semiconductor that has an excess of conducting holes created by adding trace amounts of other elements to the original pure semiconductor crystal; a semiconductor doped with Group III elements.

pyrolysis THERMOLYSIS, usually associated with exposure to a high temperature.

See also FLASH-VACUUM PYROLYSIS.

QSAR *See* QUANTITATIVE STRUCTURE-ACTIVITY RELATIONSHIPS.

quantitative analysis The accurate determination of the quantity of a substance present in a specimen.

quantitative structure-activity relationships (QSAR) The building of structure-biological activity models by using regression analysis with physicochemical constants, indicator variables, or theoretical calculations. The term has been extended by some authors to include chemical reactivity, i.e., activity is regarded as synonymous with reactivity. This extension is, however, discouraged.

See also CORRELATION ANALYSIS.

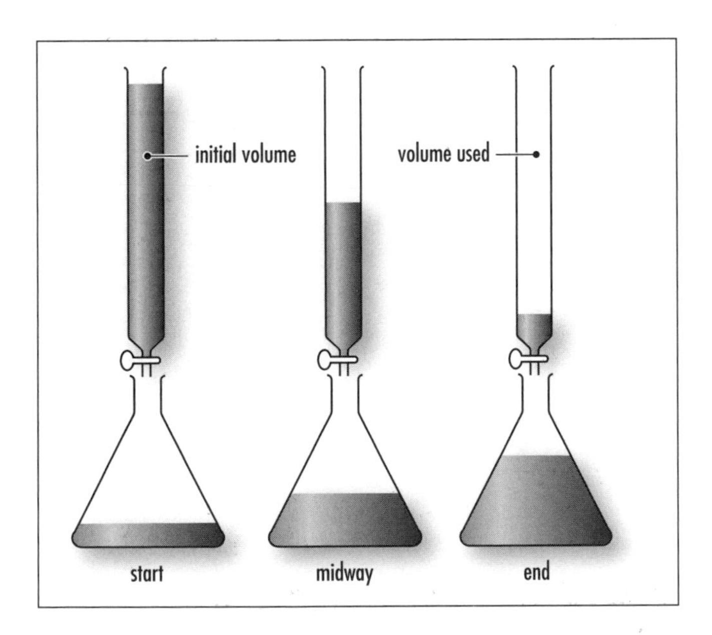

Quantitative analysis. The accurate determination of the quantity of a substance present in a specimen

quantum mechanics The fundamental physical theory of the universe; a mathematical method of treating particles on the basis of quantum theory, which states that the energy of small particles is not infinitely divisible.

quantum numbers Numbers that describe the energies of electrons in atoms; derived from quantum mechanical treatment. Consist of primary, secondary, azimuthal, and spin numbers.

quantum yield The number of defined events that occur per photon absorbed by the system. The integral quantum yield is

$$\phi = \frac{\text{number of events}}{\text{number of photons absorbed}}$$

For a photochemical reaction,

$$\phi = \frac{\text{moles of reactant consumed or product}}{\text{formed/moles of photons absorbed}}$$

The differential quantum yield is

$$\phi = \frac{d[X]/dt}{n}$$

where $d[X]/dt$ is the rate of change of the amount of (substance) concentration of a measurable quantity, and n is the amount of photons (mol or its equivalent einsteins) absorbed per unit time. φ can be used for photophysical processes or photochemical reactions.

quartz A crystalline mineral found worldwide in many forms, including amethyst, aventurine, citrin, opal, rock crystal, tiger's eye, rose quartz,and others. Quartz has a hardness of 7.0. It is a form of silica, SiO_2.

quaternary ammonium salts (quats) Substances that are used extensively as surfactants and antimicrobial agents. They contain at least one nitrogen atom linked to four *alkyl* or ARYL GROUPS.

quaternary nitrogen A quaternary nitrogen is one that is attached to four other carbon atoms.

quaternary structure There are four levels of structure found in POLYPEPTIDES and PROTEINS. The first or primary structure of a polypeptide of protein determines its secondary, tertiary, and quaternary structures. The primary structure is the AMINO ACID sequence. This is followed by the secondary structure, how the amino acids adjacent to each other are organized in the structure. The tertiary or third structure is the folded three-dimensional protein structure that allows it to perform its role, and the fourth or quaternary is the total protein structure that is made when all the SUBUNITS are in place. Quaternary structure is used to describe proteins composed of multiple subunits or multiple polypeptide molecules, each called a MONOMER. The arrangement of the monomers in the three-dimensional protein is the quaternary structure. A considerable range of quaternary structure is found in proteins.

R

racemic Pertaining to a racemate, an equimolar mixture of a pair of ENANTIOMERS. It does not exhibit optical activity.

radiation Energy released that travels through space or substance as particles or electromagnetic waves and includes visible and ultraviolet light, heat, X-rays, cosmic rays. Radiation can be nonionizing, such as infrared, visible light, ultraviolet, electromagnetic, or it can be ionizing, such as alpha, beta, gamma, and X-rays. Ionizing radiation can have severe effects on human health but is used in medical diagnostic equipment. Radiation has a host of other economic benefits ranging from electrical power generation to smoke detectors.

radical (free radical) A MOLECULAR ENTITY such as $\cdot CH_3$, $\cdot SnH_3$, $Cl\cdot$ possessing an unpaired electron. (In these formulae the dot, symbolizing the unpaired electron, should be placed so as to indicate the atom of highest spin density, if this is possible.) PARAMAGNETIC metal ions are not normally regarded as radicals. However, in the "isolobal analogy," the similarity between certain paramagnetic metal ions and radicals becomes apparent.

At least in the context of physical organic chemistry, it seems desirable to cease using the adjective *free* in the general name of this type of CHEMICAL SPECIES and MOLECULAR ENTITY, so that the term *free radical* may in future be restricted to those radicals that do not form parts of radical pairs.

Depending upon the core atom that possesses the unpaired electron, the radicals can be described as carbon-, oxygen-, nitrogen-, metal-centered radicals. If the unpaired electron occupies an orbital having considerable *s* or more or less pure *p* character, the respective radicals are termed sigma or pi radicals.

In the past, the term *radical* was used to designate a SUBSTITUENT group bound to a MOLECULAR ENTITY, as opposed to "free radical," which currently is called "radical." The bound entities may be called GROUPS or substituents, but should no longer be called radicals.

See also BIRADICAL.

radical anion *See* RADICAL ION.

radical center(s) The atom (or group of atoms) in a polyatomic radical on which an unpaired electron is largely localized. Attachment of a monovalent atom to a radical center gives a molecule for which it is possible to write a LEWIS FORMULA in which the normal stable valencies are assigned to all atoms.

radical combination *See* COLLIGATION.

radical ion A RADICAL that carries an electric charge. A positively charged radical is called a "radical cation" (e.g., the benzene radical cation $C_6H_6^+$); a negatively charged radical is called a "radical anion" (e.g., the benzene radical anion $C_6H_6^-$ or the benzophenone radical anion $Ph_2C–OO^-$). Commonly, but not necessarily, the odd electron and the charge are associated with the same atom.

Unless the positions of unpaired spin and charge can be associated with specific atoms, superscript dot and charge designations should be placed in the order $^{\cdot+}$ or $^{\cdot-}$ suggested by the name "radical ion" (e.g., $C_3H_6^{\cdot+}$).

radical pair (geminate pair) The term is used to identify two RADICALS in close proximity in solution, within a solvent CAGE. They may be formed simultaneously by some unimolecular process, e.g., peroxide decomposition, or they may have come together by DIFFUSION. While the radicals are together, correlation of the unpaired electron spins of the two species cannot be ignored: this correlation is responsible for the CIDNP phenomenon.

See also GEMINATE RECOMBINATION; MOLECULARITY.

radioactive isotope Atoms with the same number of protons but different numbers of neutrons are called isotopes. There are radioactive and nonradioactive isotopes, and some elements have both, such as carbon. Each radioactive isotope has its own unique HALF-LIFE, which is the time it takes for half of the parent radioactive element to decay to a daughter product. Some examples of radioactive elements, their stable daughters, and half-lives are: potassium 40–argon 40 (1.25 billion years); rubidium 87–strontium 87 (48.8 billion years), thorium 232–lead 208 (14 billion years); uranium 235–lead 207 (704 million years); uranium 238–lead 206 (4.47 billion years); carbon 14–nitrogen 14 (5,730 years).

See also ELEMENT.

radioactive tracer (radioactive label) A small amount of RADIOACTIVE ISOTOPE used to replace a nonradioactive isotope of the element in a compound whose path is to be traced or monitored by detection of RADIOACTIVITY.

radioactivity The spontaneous disintegration of atomic nuclei, which liberates particles and/or energy (e.g., alpha or beta particles, neutrons, and gamma rays).

radiocarbon dating A dating method used to determine the age of samples containing carbon, particularly useful to archaeologists. The method measures the disintegration of the ^{14}C atom, which is produced in the atmosphere by cosmic ray bombardment and has a HALF-LIFE of 5,570 years, making it useful for dating samples in the range of 0–40,000 years. The carbon-14 method was developed by the American physicist WILLARD FRANK LIBBY in 1947.

radiolysis The cleavage of one or several bonds resulting from exposure to high-energy RADIATION. The term is also often used loosely to specify the method of irradiation (pulse radiolysis) used in any radiochemical reaction, not necessarily one involving bond cleavage.

radiometric dating The use of RADIOACTIVE ISOTOPES and their HALF-LIVES to give absolute dates to rock formations, artifacts, and fossils. Radioactive elements tend to accumulate in human-made artifacts, igneous rocks, and the continental crust, and so they are not very useful for sedimentary rocks, although in some cases when certain elements are found, it is possible to date them using this technique. Other radiometric dating techniques used are:

Electron Dating Spin Resonance

Electrons become trapped in the crystal lattice of minerals from adjacent radioactive material and alter the magnetic field of the mineral at a known rate. This technique is used for dating bone and shell, since it does not destroy the material, such as carbonates (calcium) in limestone, coral, egg shells, and teeth, by exposing it to different magnetic fields.

Fission Track Dating

This technique is used for dating glassy material like obsidian or any artifacts that contain uranium-bearing material such as natural or human-made glass, ceramics, or stones that were used in hearths for food prepa-

Nuclear waste storage. Steel drums containing nuclear waste at a dump at a nuclear weapons testing site. The drums contain solid waste that is made up of transuranics, radioactive chemical elements with heavy atomic nuclei that are made in nuclear reactions. These drums will be packed, 50 at a time, into steel cargo containers for storage aboveground. Corrosion of these containers will be minimized because of the dry desert climate of the site. Photographed at the Area 5 Radioactive Waste Management Site at Frenchman Flat in Nevada. ***(Courtesy of U.S. Department of Energy/Science Photo Library)***

ration. Narrow fission tracks from the release of high-energy charged alpha particles burn into the material as a result of the decay of uranium 238 to lead 206 (half-life of 4.51 billion years) or induced by the irradiation of uranium 235 to lead 207 (704 million years). The number of tracks is proportional to the time passed since the material cooled from its original molten condition, i.e., fission tracks are created at a constant rate throughout time, so it is possible to determine the amount of time that has passed since the track accumulation began from the number of tracks present. This technique is good for dates ranging from 20 million to 1 billion years ago. U-238 fission track techniques are from spontaneous fission, and induced fission track from U-235 is a technique developed by controlled irradiation of the artifact with thermal neutrons of the U-235. Both techniques give a thermal age for the material in question. The spontaneous fission of uranium 238 was first discovered by the Russian scientists K.A. Petrzhak and G.N. Flerov in 1940.

Potassium-Argon Dating

This method has been used to date rocks as old as 4 billion years and is a popular dating technique for archaeological material. Potassium 40, with a half-life of 1.3 billion years in volcanic rock, decays into argon 40 and calcium 40 at a known rate. Dates are determined by measuring the amount of argon 40 in a sample. Argon 40 and argon 39 ratios can also be used for dating the same way. Potassium-argon dating is accurate from 4.3 billion years (the age of the Earth) to about 100,000 years before the present.

Radiocarbon Dating
See RADIOCARBON DATING.

Thermoluminescence Dating
A technique used for dating ceramics, bricks, sediment layers, burnt flint, lava, and even cave structures like stalactites and stalagmites, based on the fact that some materials, when heated, give off a flash of light. The intensity of the light is used to date the specimen and is proportional to the quantity of radiation to which it has been exposed and the time span since it was heated. The technique is similar to electron spin resonance (ESR). Good for dates between 10,000 and 230,000 years.

radionuclide A radioactive NUCLIDE. The term *nuclide* implies an atom of specified atomic number and mass number. In the study of biochemical processes, RADIOACTIVE ISOTOPEs are used for labeling compounds that subsequently are used to investigate various aspects of the reactivity or METABOLISM of proteins, carbohydrates, and lipids, or as sources of radiation in IMAGING. The fate of the radionuclide in reactive products or metabolites is determined by following (counting) the emitted radiation. Prominent among the radionuclides used in biochemical research are : ^{3}H, ^{14}C, ^{32}P, ^{35}Ca, ^{99m}Tc, ^{125}I, and ^{131}I.

radius ratio The radius of the positive ion, r^+, divided by the radius of the negative ion, r^-, in an ionic compound.

radon **(Rn)** A radioactive element with atomic number 86. It has several ISOTOPEs, the most important of which has atomic weight 222 (radon 222). Radon is a colorless, tasteless, odorless, naturally occurring inert gas derived from the natural radioactive decay of three radioactive isotopes: uranium 238 (Rn-222), uranium 235 (Rn-219), and thorium 232 (Rn-220). The chemically inert gas enters homes through soil, water, and building materials and is the second leading cause of lung cancer.

Raman spectroscopy A spectroscopic method of analysis that utilizes the Raman effect to extract information about the vibrational structure of a molecule. The Raman effect is a collision between an incident photon and a molecule, resulting in the vibrational or rotational energy of the molecule being changed, and the energy of the scattered photon differing from the energy of the incident photon.

Raoult's law Physical law relating the change in VAPOR PRESSURE of a liquid to the amount of solute dissolved in it. The vapor pressure of a solvent in an ideal solution decreases as its mole fraction decreases.

rate coefficient *See* ORDER OF REACTION; KINETIC EQUIVALENCE.

rate constant ***k*** (SI unit: s^{-1} $[dm^3\ mol^{-1}]^n-1$) *See* ORDER OF REACTION.

rate-controlling step (rate-determining step, rate-limiting step) A rate-controlling (rate-determining or rate-limiting) step in a reaction occurring by a composite reaction sequence is an ELEMENTARY REACTION, the rate constant for which exerts a strong effect—stronger than that of any other rate constant—on the overall rate. It is recommended that the expressions *rate-controlling*, RATE-DETERMINING, and *rate-limiting* be regarded as synonymous, but some special meanings sometimes given to the last two expressions are considered under a separate heading.

A rate-controlling step can be formally defined on the basis of a control function (or control factor), CF, identified for an elementary reaction having a rate constant k_i by

$$CF = (\partial \ln v / \partial \ln k_i)_{K_j,k_j}$$

where v is the overall rate of reaction. In performing the partial differentiation, all equilibrium constants K_j and all rate constants except k_i are held constant. The elementary reaction having the largest control factor exerts the strongest influence on the rate v, and a step having a CF much larger than any other step may be said to be rate-controlling.

A rate-controlling step defined in the way recommended here has the advantage that it is directly related to the interpretation of kinetic isotope effects (*see* ISOTOPE EFFECT, KINETIC).

As formulated, this implies that all rate constants are of the same dimensionality. Consider, however, the reaction of A and B to give an intermediate C, which then reacts further with D to give products:

$$A + B \underset{k_{-1}}{\overset{k_1}{\rightleftharpoons}} C \qquad (1)$$

$$C + D \xrightarrow{k_2} \text{Products} \qquad (2)$$

Assuming that C reaches a STEADY STATE, then the observed rate is given by

$$v = \frac{k_1 k_2 \,[A][B][D]}{k_{-1} + k_2\,[D]}$$

Considering k_2[D], a pseudo-first-order rate constant, then $k_2[D] >> k_{-1}$, and the observed rate $v = k_1[A][B]$ and $k_{obs} = k_1$

Step (1) is said to be the rate-controlling step.

If $k_2[D] << k_{-1}$, then the observed rate

$$v = \frac{k_1 k_2}{k_{-1}}\,[A][B][D]$$

$$= Kk_2\,[A][B][D]$$

where K is the equilibrium constant for the pre-equilibrium (1) and is equal to k_1/k_{-1}, and $k_{obs} = Kk_2$.

Step (2) is said to be the rate-controlling step.

See also GIBBS ENERGY DIAGRAM; MICROSCOPIC DIFFUSION CONTROL; MIXING CONTROL; ORDER OF REACTION; RATE-DETERMINING STEP.

rate-determining step (rate-limiting step) These terms are best regarded as synonymous with RATE-CONTROLLING STEP. However, other meanings that have been given to them should be mentioned, as it is necessary to be aware of them in order to avoid confusion.

Sometimes the term *rate-determining* is used as a special case of *rate-controlling,* being assigned only to an initial slow step that is followed by rapid steps. Such a step imposes an upper limit on the rate and has also been called rate-limiting.

In view of the considerable danger of confusion when special meanings are applied to the terms *rate-determining* and *rate-limiting,* it is recommended that they be regarded as synonymous with the meaning explained under the entry for rate-controlling step.

See also MICHAELIS-MENTEN KINETICS.

rate law (empirical differential rate equation) An expression for the RATE OF REACTION of a particular reaction in terms of concentrations of CHEMICAL SPECIES and constant parameters (normally RATE COEFFICIENTS and partial ORDERS OF REACTION) only. For examples of rate laws, see equations (1) to (3) under KINETIC EQUIVALENCE, and (1) under STEADY STATE.

rate of appearance *See* RATE OF REACTION.

rate of reaction For the general CHEMICAL REACTION

$$aA + bB \rightarrow pP + qQ\ldots$$

occurring under constant-volume conditions, without an appreciable buildup of reaction INTERMEDIATES, the rate of reaction v is defined as

$$\upsilon = -\frac{1}{a}\frac{d[A]}{dt} = -\frac{1}{b}\frac{d[B]}{dt} = +\frac{1}{p}\frac{d[P]}{dt} = +\frac{1}{q}\frac{d[Q]}{dt}$$

where symbols placed inside square brackets denote amount (or amount of substance) concentrations (conventionally expressed in units of mol dm^{-3}). The symbols R and r are also commonly used in place of v. It is recommended that the unit of time should always be the second.

In such a case, the rate of reaction differs from the rate of increase of concentration of a product P by a constant factor (the reciprocal of its coefficient in the stoichiometric equation, p) and from the rate of decrease of concentration of the reactant A by α^{-1}.

The quantity:

$$\dot{\xi} = -\frac{d\xi}{dt}$$

defined by the equation

$$\dot{\xi} = -\frac{1}{a}\frac{dn_A}{dt} = -\frac{1}{b}\frac{dn_B}{dt} = +\frac{1}{p}\frac{dn_P}{dt} = +\frac{1}{q}\frac{dn_Q}{dt}$$

(where n_A designates the amount of substance A, conventionally expressed in units of mole) may be called the "rate of conversion" and is appropriate when the use of concentrations is inconvenient, e.g., under conditions of varying volume. In a system of constant volume, the rate of reaction is equal to the rate of conversion per unit volume throughout the reaction.

For a STEPWISE REACTION this definition of "rate of reaction" (and "extent of reaction," π) will apply only if there is no accumulation of intermediate or formation of side products. It is therefore recommended that the term *rate of reaction* be used only in cases where it is experimentally established that these conditions apply. More generally, it is recommended that, instead, the terms *rate of disappearance* or *rate of consumption* of A (i.e., $-d[A]/dt$, the rate of decrease of concentration of A) or "rate of appearance" of P (i.e., $d[P]/dt$, the rate of increase of concentration of product P) be used, depending on the concentration change of the particular CHEMICAL SPECIES that is actually observed. In some cases, reference to the CHEMICAL FLUX observed may be more appropriate.

The symbol v (without lettered subscript) should be used only for rate of reaction; v with a lettered subscript (e.g., v_A) refers to a rate of appearance or rate of disappearance (e.g., of the chemical species A).

See also CHEMICAL RELAXATION; LIFETIME; ORDER OF REACTION.

reactants The materials consumed in a chemical reaction.

reacting bond rules (1) For an internal motion of a MOLECULAR ENTITY corresponding to progress over a TRANSITION STATE (energy maximum), any change that makes the motion more difficult will lead to a new molecular geometry at the energy maximum, in which the motion has proceeded further. Changes that make the motion less difficult will have the opposite effect. (This rule corresponds to the HAMMOND PRINCIPLE).

(2) For an internal motion of a molecular entity that corresponds to a vibration, any change that tends to modify the equilibrium point of the vibration in a particular direction will actually shift the equilibrium in that direction.

(3) Effects on reacting bonds (bonds made or broken in the reaction) are the most significant. The bonds nearest the site of structural change are those most strongly affected.

reaction *See* CHEMICAL REACTION.

reaction coordinate A geometric parameter that changes during the conversion of one (or more) reactant MOLECULAR ENTITIES into one (or more) product molecular entities and whose value can be taken for a measure of the progress of an ELEMENTARY REACTION (for example, a bond length or bond angle or a combination of bond lengths and bond angles; it is sometimes approximated by a nongeometric parameter, such as the BOND ORDER of some specified bond). In the formalism of TRANSITION-STATE theory, the reaction coordinate is that coordinate—in a set of curvilinear coordinates obtained from the conventional ones for the reactant—which, for each reaction step, leads smoothly from the configuration of the reactants through that of the transition state to the configuration of the products. The reaction coordinate is typically chosen to follow the path along the gradient (path of shallowest ascent/deepest descent) of potential energy from reactants to products.

The term has also been used interchangeably with the term TRANSITION COORDINATE, applicable to the coordinate in the immediate vicinity of the potential energy maximum. Being more specific, the name *transition coordinate* is to be preferred in that context.

See also POTENTIAL-ENERGY PROFILE; POTENTIAL-ENERGY (REACTION) SURFACE.

reaction mechanism *See* MECHANISM.

reaction path (1) A synonym for MECHANISM.

(2) A trajectory on the POTENTIAL-ENERGY SURFACE.

(3) A sequence of synthetic steps.

reaction ratio The relative amounts of reactants and products involved in a reaction.

reaction stage A set of one or more (possibly experimentally inseparable) REACTION STEPs leading to or from a detectable or presumed reaction INTERMEDIATE.

reaction step An ELEMENTARY REACTION, constituting one of the stages of a STEPWISE REACTION in which a reaction INTERMEDIATE (or, for the first step, the reactants) is converted into the next reaction intermediate (or, for the last step, the products) in the sequence of intermediates between reactants and products.

reactive, reactivity As applied to a CHEMICAL SPECIES, the term expresses a kinetic property. A species is said to be more reactive or to have a higher reactivity in some given context than some other (reference) species if it has a larger rate constant for a specified ELEMENTARY REACTION. The term has meaning only by reference to some explicitly stated or implicitly assumed set of conditions. It is not to be used for reactions or reaction patterns of compounds in general. The term is also more loosely used as a phenomenological description not restricted to elementary reactions. When applied in this sense, the property under consideration may reflect not only rate but also equilibrium constants.

See also STABLE; UNREACTIVE; UNSTABLE.

reactivity index Any numerical index derived from quantum mechanical model calculations that permits the prediction of relative reactivities of different molecular sites. Many indices are in use, based on a variety of theories and relating to various types of reaction. The more successful applications have been to the SUBSTITUTION REACTIONs of CONJUGATED SYSTEMs, where relative reactivities are determined largely by changes of pi-electron energy.

reactivity-selectivity principle (RSP) This idea can be expressed loosely as: the more REACTIVE a reagent is, the less selective it is.

Consider two substrates S^1 and S^2 undergoing the same type of reaction with two reagents R^1 and R^2, S^2 being more reactive than S^1, and R^2 more reactive than R^1 in the given type of reaction. The relative reactivities (in log units; *see* SELECTIVITY) for the four possible reactions can notionally be represented as follows:

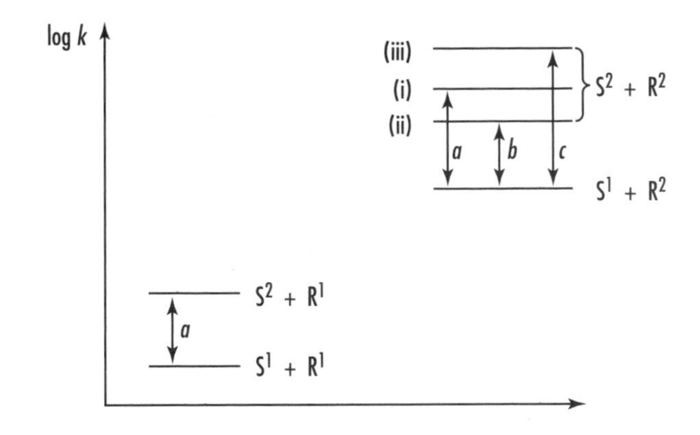

With the positions of ($S^1 + R^1$), ($S^2 + R^1$), and ($S^1 + R^2$) fixed, there are three types of positions for ($S^2 + R^2$):

In position (i) the selectivity of R^2 for the two substrates, measured by *a,* is the same as the selectivity of R^1 for the two substrates, also *a.*

In position (ii) the selectivity of R^2 for the two substrates, measured by *b,* is less than the selectivity of R^1 for the two substrates, i.e., $b < a$. It is this situation that is in accord with the RSP.

In position (iii) the selectivity of R^2 for the two substrates, measured by *c,* is greater than the selectivity of R^1 for the two substrates, i.e., $c > a$. This situation can be described as anti-RSP.

There are many examples in which the RSP is followed, but there are also many examples corresponding to situations (i) and (iii). The RSP is in accord with intuitive feeling and certainly holds in the limiting case when reactivity is controlled by DIFFUSION. However, the validity of the RSP is a matter of great controversy, and diverse opinions have been expressed, from declaring the reactivity-selectivity principle as a universal law up to virtually useless in practice as a general rule.

real gas Gases under physical conditions that give nonideal behavior. The repulsion between molecules at very close distances is due to the repulsion between the nuclei of the two molecules. These forces give rise to relationships between the pressure, temperature, volume, and quantity of a substance that do not exactly obey the ideal gas law and therefore give nonideal behavior.

rearrangement *See* DEGENERATE REARRANGEMENT; MOLECULAR REARRANGEMENT; SIGMATROPIC REARRANGEMENT.

rearrangement stage The ELEMENTARY REACTION or REACTION STAGE (of a MOLECULAR REARRANGEMENT) in which there is both making and breaking of bonds between atoms common to a reactant and a reaction product or a reaction INTERMEDIATE. If the rearrangement stage consists of a single elementary reaction, this is a "rearrangement step."

Réaumur, René-Antoine Ferchault de (1683–1757) French *Philosopher, naturalist* René-Antoine Réaumur was born in La Rochelle, France, in 1683. After studying mathematics in Bourges, he moved to Paris in 1703 at age 20 under the care of a relative. Like most scientists of the time, he made contributions in a number of areas, including meteorology. His work in mathematics allowed him entrance to the Academy of Sciences in 1708. Two years later, he was put in charge of compiling a description of the industrial and natural resources in France, and as a result developed a broad-based view of the sciences. It also inspired him into the annals of weather and climate with the invention of a thermometer and temperature scale.

In 1713 he made spun glass fibers, today the building blocks of Ethernet networking and fiber optics, which are still made of the same material. A few years later, in 1719, after observing wasps building nests, he suggested that paper could be made from wood in response to a critical shortage of papermaking materials (rags) at the time. He was also impressed by the geometrical perfection of the beehive's hexagonal cells and proposed that they be used as a unit of measurement.

He turned his interests from steel to temperature, and in 1730 he presented to the Paris Academy his study "A Guide for the Production of Thermometers with Comparable Scales." He wanted to improve the reliability of thermometers based on the work of Guillaume Amontons, though he appears not to be familiar with DANIEL GABRIEL FAHRENHEIT's earlier work.

His thermometer of 1731 used a mixture of alcohol (wine) and water instead of mercury, perhaps creating the first alcohol thermometer, and it was calibrated with a scale he created called the Réaumur scale. This scale had 0° for freezing and 80° for boiling points of water. The scale is no longer used today; however, at the time most of Europe, with the exception of the British Isles and Scandinavia, adopted his thermometer and scale.

Unfortunately, errors in the way he fixed his points were criticized by many in the scientific community at the time, and even with modifications in the scale, instrument makers favored mercury-based thermometers. Réaumur's scale, however, lasted over a century, and in some places well into the late 20th century.

Between 1734 and 1742, Réaumur wrote six volumes of *Mémoires pour servir à l'histoire des insectes* (Memoirs serving as a natural history of insects). Although unfinished, this work was an important contribution to entomology. He also noticed that crayfish have the ability to regenerate lost limbs and demonstrated that corals were animals, not plants. In 1735 he introduced the concept of growing degree-days, later known as Réaumur's thermal constant of phenology. This idea led to the heat-unit system used today to study plant-temperature relationships.

In 1737 he became an honorary member of the Russian Academy of Sciences, and the following year he became a fellow of the Royal Society.

After studying the chemical composition of Chinese porcelain in 1740, he formulated his own Réaumur porcelain. In 1750, while investigating the animal world, he designed an egg incubator. Two years later, in 1752, he discovered that digestion is a chemical process by isolating gastric juice and studied its role in food digestion by observing hawks and dogs.

He died in La Bermondiere on October 18, 1757, and bequeathed to the Academy of Science his cabinet of natural history with his collections of minerals and plants.

receptor A molecule or a polymeric structure in or on a cell that specifically recognizes and binds a compound acting as a molecular messenger (neurotransmitter, HORMONE, lymphokine, lectin, DRUG, etc.).

receptor mapping The technique used to describe the geometric or electronic features of a binding site when insufficient structural data for this RECEPTOR or

ENZYME are available. Generally the active site cavity is defined by comparing the superposition of active to that of inactive molecules.

receptor-mediated endocytosis Cells use receptor-mediated endocytosis for ingestion of nutrients, hormones, and growth factors, a method where specific molecules are ingested into the cell; the specificity results from a RECEPTOR-LIGAND (a molecule or ion that can bind another molecule) interaction. Other ligands that can be ingested include toxins and lectins, viruses, and serum transport proteins and antibodies. Receptors, a specific binding protein such as clathrin on the plasma membrane of the target tissue, will specifically bind to ligands on the outside of the cell. An endocytotic process results: the cell folds inward with a portion of the plasma membrane, and the resulting clathrin-coated pit is pinched off, forming a membrane-enclosed bubble or vesicle, called an endosome. After entering the cytoplasm, the endocytotic vesicle loses its clathrin coat and the ligand (multiple ligands can enter the cell in the same coated pit) is ingested. The receptor can be recycled to the surface by vesicles that bud from the endosome targeting the plasma membrane. After these recycling vesicles fuse with the plasma membrane, the receptor is returned to the cell surface for binding and activity once more.

receptor potential A change in a neuron's membrane potential (a change in voltage across the receptor membrane) caused by redistribution of ions responding to the strength of the stimulus. If the potential is high enough, an action potential will be fired in an afferent neuron. The more action potentials fired, the more neurotransmitters released, and the stronger the signals reaching the brain. Also called an end-plate potential.

redox potential Any OXIDATION-REDUCTION (redox) REACTION can be divided into two half reactions: one in which a chemical species undergoes oxidation and one in which another chemical species undergoes reduction. If a half-reaction is written as a reduction, the driving force is the reduction potential. If the half-reaction is written as oxidation, the driving force is the oxidation potential related to the reduction potential by a sign change. So the redox potential is the reduction/oxidation potential of a compound measured under standard conditions against a standard reference HALF-CELL. In biological systems, the standard redox potential is defined at pH = 7.0 versus the hydrogen electrode and partial pressure of hydrogen = 1 bar.

See also ELECTRODE POTENTIAL.

reducing agent The reactant that donates its electrons and in turn becomes oxidized when another substance is reduced.

See also OXIDATION-REDUCTION REACTION.

reductase *See* OXIDOREDUCTASE.

reduction The part of a redox reaction where the reactant has a net gain of electrons and in which a different reactant must oxidize (lose electrons).

See also OXIDATION.

reductive elimination The reverse of OXIDATIVE ADDITION.

refractory material/metal Any material that withstands high heat or a metal with a very high melting point, e.g., clay, tungsten, molybdenum, tantalum, niobium, chromium, vanadium, and rhenium.

regioselectivity, regioselective A regioselective reaction is one in which one direction of bond making or breaking occurs preferentially over all other possible directions. Reactions are termed completely (100 percent) regioselective if the discrimination is complete, or partially (x percent) if the product of reaction at one site predominates over the product of reaction at other sites. The discrimination may also semiquantitatively be referred to as high or low regioselectivity. (Originally the term was restricted to ADDITION REACTIONS of unsymmetrical reagents to unsymmetrical alkenes.)

In the past, the term *regiospecificity* was proposed for 100 percent regioselectivity. This terminology is not recommended owing to inconsistency with the terms STEREOSELECTIVITY and STEREOSPECIFICITY.

See also CHEMOSELECTIVITY.

regulation Refers to control of activity of an ENZYME (system) or GENE EXPRESSION.

Reichstein, Tadeus (1897–1996) Swiss *Chemist* Tadeus Reichstein was born on July 20, 1897, at Wloclawek, Poland, to Isidor Reichstein and Gastava Brockmann. He was educated at a boarding school at Jena before his family moved to Zurich in 1906 (where he was naturalized). He had a private tutor and then attended the Oberrealschule (technical school of junior college grade) and the Eidgenössische Technische Hochschule (E.T.H.) (state technical college).

In 1916 he began to study chemistry at the E.T.H. at Zurich and graduated in 1920. In 1922 he began a nine-year research project on the composition of the flavoring substances in roasted coffee.

In 1931 he turned to other scientific research, and by 1938 he was professor in pharmaceutical chemistry and director of the pharmaceutical institute in the University of Basel. From 1946 to 1967 he was professor of organic chemistry at the University of Basel.

In 1933 he synthesized vitamin C (ascorbic acid) and worked on plant glycosides. From 1953 to 1954 he worked with several other scientists and was the first to isolate and explain the constitution of aldosterone, a hormone of the adrenal cortex. He also collaborated with E. C. KENDALL and P. S. Hench in their work on the hormones of the adrenal cortex. For this work, Reichstein, Kendall, and Hench were jointly awarded the Nobel Prize for physiology or medicine in 1950.

After 1967 he worked on the study of ferns and published many papers on the subject. He died on August 1, 1996, in Basel.

relative configuration The CONFIGURATION of any stereogenic (asymmetric) center with respect to any other stereogenic center contained within the same MOLECULAR ENTITY. A stereogenic unit is a grouping within a molecular entity that can be considered a focus of STEREOISOMERISM.

relaxation If a system is disturbed from its state of equilibrium, it returns to that state, and the process is referred to as relaxation.

releaser A signal stimulus that functions as a communication signal between individuals of the same species and initiates a fixed action pattern (FAP), a stereotyped species-to-species behavior.

reorganization energy In a one-electron transfer reaction

$$A + D \rightleftharpoons A^{\cdot -} + D^{\cdot +}$$

the reorganization energy λ is the energy required for all structural adjustments (in the reactants and in the surrounding solvent molecules) that are needed in order that A and D assume the configuration required for the transfer of the electron.

See also INTRINSIC BARRIER; MARCUS EQUATION.

residence time The duration of time a specific molecule of water remains in a particular flow system. In geochemistry, it is the average time spent by an atom or molecule within the reservoir (the mass of an element or a compound within a defined "container") between the time it entered and exited it. Residence time is calculated by dividing the reservoir size by the input (or the output).

resolving power A property of instruments like microscopes and telescopes that distinguish objects that are close to each other; the smaller the minimum distance at which two objects can be distinguished, the greater the resolving power.

resonance In the context of chemistry, the term refers to the representation of the electronic structure of a MOLECULAR ENTITY in terms of CONTRIBUTING STRUCTURES. Resonance among contributing structures means that the wave function is represented by "mixing" the wave functions of the contributing structures. The concept is the basis of the quantum mechanical valence bond methods. The resulting stabilization is linked to the quantum mechanical concept of "resonance energy." The term *resonance* is also used to refer to the DELOCALIZATION phenomenon itself.

See also MESOMERISM.

resonance effect This is the term most commonly used to describe the influence (on reactivity, spectra, etc.) of a SUBSTITUENT through electron DELOCALIZATION into or from the substituent. The use of the term obviates the need to attempt to distinguish between the operation of the MESOMERIC EFFECT and the ELECTROMERIC EFFECT. (An alternative term with essentially the same meaning is *conjugative effect.* At one time "tautomeric effect" was also used, but it was abandoned because TAUTOMERISM implies reorganization of the atomic nuclei.) The effect is symbolized by R.

resonance energy The difference in potential energy between the actual MOLECULAR ENTITY and the CONTRIBUTING STRUCTURE of lowest potential energy. The resonance energy cannot be measured, but only estimated, since contributing structures are not observable MOLECULAR ENTITIES.

See also RESONANCE.

resonance Raman spectroscopy A spectroscopic technique increasingly used in bioinorganic chemistry for characterization and assignment of vibrations directly connected with a CHROMOPHORE, as well as for the assignment of the chromophore. The excitation frequency is applied close to the absorption maximum of the chromophore. Particularly useful for deeply colored species.

See also RAMAN SPECTROSCOPY.

respiration The process where mitochondria in the cells of plants and animals release chemical energy from sugar and other organic molecules through chemical oxidation.

See also KREBS CYCLE.

resting potential The state of a neuron's charge—the gradient of electric potential across the membrane—when it is in a resting state and ready to receive a nerve impulse (the action potential); usually consists of a negative charge on the inside of the cell relative to the outside. At rest, the cell membrane electrical gradient maintains a negative interior charge of –70 mV.

restriction enzyme (restriction endonuclease) A DNA-cutting protein that recognizes a specific nucleotide sequence in a DNA molecule and excises the DNA; found in bacteria. Some sites are common and occur every several hundred base pairs, while others are less common. Bacteria possess several hundred restriction enzymes that cut more than 100 different DNA sequences. Each restriction enzyme has a single, specific recognition sequence and cuts the DNA molecule at a specific site. Some restriction enzymes have been used in recombinant DNA technology.

retroaddition *See* CYCLOELIMINATION.

retrocycloaddition *See* CYCLOELIMINATION.

retro-ene reaction *See* ENE REACTION.

reverse micelle (reversed micelle) *See* INVERTED MICELLE.

reverse osmosis During osmosis, water flows from a less-concentrated solution through a semipermeable membrane to a more-concentrated saline solution until concentrations on both sides of the membrane are equal. Reverse osmosis forces molecules to flow through a semipermeable membrane from the concentrated solution into a diluted solution using greater hydrostatic pressure on the concentrated-solution side.

See also OSMOSIS.

reversible reaction A CHEMICAL REACTION that can occur in both directions, either OXIDATION or REDUCTION.

Rh blood types A blood group that involves 45 different antigens on the surface of red cells that are controlled by two closely linked genes on chromosome 1. Can lead to serious medical complications between a mother and her developing fetus if they are different (+ or –)—mother is Rh– and the father is Rh+.

See also KARL LANDSTEINER.

rhodopsin The red photosensitive pigment found in the retina's rod cells; contains the protein opsin linked to retinene (or retinal), a light-absorbing chemical derived from vitamin A and formed only in the dark. Light striking the rhodopsin molecule forces it to change shape and generate signals while the retinene splits from the opsin, which is then reattached in the dark, and completes the visual cycle.

rho-sigma (ρσ) equation *See* HAMMETT EQUATION; RHO (ρ) VALUE; SIGMA (σ) CONSTANT; TAFT EQUATION.

rho (ρ) value A measure of the susceptibility to the influence of SUBSTITUENT groups on the RATE CONSTANT or equilibrium constant of a particular organic reaction involving a family of related SUBSTRATES. Defined by Hammett for the effect of ring substituents in meta- and para-positions of aromatic side-chain reactions by the empirical "ρσ-equation" of the general form

$$\lg(kx/k_H) = \rho\sigma X$$

in which σx is a constant characteristic of the substituent X and of its position in the reactant molecule.

More generally (and not only for aromatic series), ρ-values (modified with appropriate subscripts and superscripts) are used to designate the susceptibility of reaction series for families of various organic compounds to any substituent effects, as given by the modified set of σ-constants in an empirical ρσ-correlation.

Reactions with a positive ρ-value are accelerated (or the equilibrium constants of analogous equilibria are increased) by substituents with positive σ-constants. Since the sign of σ was defined so that substituents with a positive σ increase the acidity of benzoic acid, such substituents are generally described as attracting electrons away from the aromatic ring. It follows that reactions with a positive ρ-value are considered to involve a TRANSITION STATE (or reaction product) so that the difference in energy between this state and the reactants is decreased by a reduction in electron density at the reactive site of the substrate.

See also HAMMETT EQUATION; SIGMA (σ) CONSTANT; STABILITY CONSTANT; TAFT EQUATION.

ribonucleic acid (RNA) Linear polymer molecules composed of a chain of ribose units linked between positions 3 and 5 by phosphodiester groups. The bases adenine or guanine (via atom N-9) or uracil or cytosine (via atom N-1), respectively, are attached to ribose at its atom C-1 by β-N-glycosidic bonds (*see* NUCLEOTIDE). The three most important types of RNA in the cell are: MESSENGER RNA (mRNA), transfer RNA (tRNA), and RIBOSOMAL RNA (rRNA).

ribonucleotide reductases ENZYMES that catalyze the reduction of ribonucleotide diphosphates or triphosphates to the corresponding deoxyribonucleotides by a RADICAL-dependent reaction. The enzyme of animal, yeast, and AEROBIC *E. coli* cells contains an oxo-bridged DINUCLEAR iron center and a tyrosyl radical cation, and it uses thioredoxin, a thiol-containing protein, as a reductant. At least three other ribonucleotide reductases are known from bacteria, containing, respectively, an IRON-SULFUR CLUSTER with a glycyl radical, adenosyl COBALAMIN, and a dinuclear manganese CLUSTER.

ribose The sugar component of RNA; a five-carbon (pentose) aldose.

ribosomal RNA (rRNA) The most common form of RNA. When combined with certain proteins, it forms ribosomes that are responsible for translation of MESSENGER RNA (mRNA) into protein chains.

ribosomes A subcellular unit composed of specific RIBOSOMAL RNA (rRNA) and proteins that are responsible for the TRANSLATION of MESSENGER RNA (mRNA) into protein synthesis.

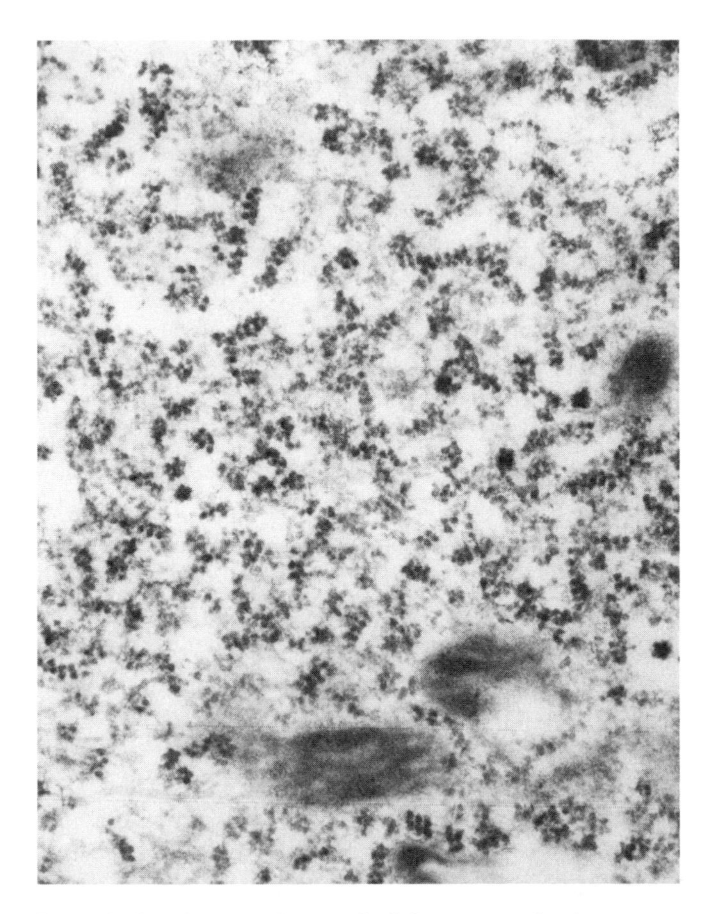

Transmission electron micrograph of ribosomes and polysomes from chick embryo fibroblast cells. Ribosomes are the site of protein synthesis in the cell. Each consists of a large and small subunit made of roughly equal ratios of ribosomal RNA (ribonucleic acid) and 40 or more different types of protein. During protein synthesis, a molecule of messenger RNA attaches to a groove in the ribosome and slides along. As it passes, amino acids are bound together in an order spelt out by the base code on the messenger RNA. Groups of ribosomes attached to the same messenger RNA molecule are called polysomes. Magnification ×135,000 at 10 × 8-inch size. ***(Courtesy of Dr. Gopal Murti/Science Photo Library)***

ribozyme RNA with enzymatic or catalytic ability specifically to cleave (break down) or bind RNA molecules. Also known as autocatalytic or catalytic RNA.

ribulose-1,5-bisphosphate carboxylase/oxygenase (rubisco) A magnesium-dependent ENZYME. The primary enzyme of carbon dioxide fixation in plants and AUTOTROPHIC bacteria. It catalyzes the synthesis of 3-phospho-D-glycerate from ribulose bisphosphate and also the oxidation of ribulose bisphosphate by O_2 to 3-phospho-D-glycerate and 2-phosphoglycolate.

Rieske iron-sulfur protein An IRON-SULFUR PROTEIN of the MITOCHONDRIAL respiratory chain, in which the [2FE-2S] CLUSTER is coordinated to two sulfur LIGANDS from cysteine and two imidazole ligands from histidine. The term is also applied to similar proteins isolated from photosynthetic organisms and microorganisms and other proteins containing [2Fe-2S] clusters with similar COORDINATION.

See also PHOTOSYNTHESIS.

Ritchie equation The LINEAR FREE-ENERGY RELATION

$$\log k_N = \log k_0 + N_+$$

applied to the reactions between nucleophiles and certain large and relatively stable organic cations, e.g., arenediazonium, triarylmethyl, and aryltropylium cations in various solvents. k_N is the rate constant for reaction of a given cation with a given nucleophilic system (i.e., given nucleophile in a given solvent); k_0 is the rate constant for the same cation with water in water; and N_+ is a parameter that is characteristic of the nucleophilic system and independent of the cation. A surprising feature of the equation is the absence of a coefficient of N_+, characteristic of the substrate (this is the s in the SWAIN-SCOTT EQUATION), even though values of N_+ vary over 13 log units. The equation thus involves a gigantic breakdown of the REACTIVITY-SELECTIVITY PRINCIPLE. The equation has been extended both in form and in range of application.

RNA *See* RIBONUCLEIC ACID.

roentgen A measure of ionizing radiation named after Wilhelm Roentgen, a German scientist who discovered X rays in 1895. One roentgen is the amount of gamma or X rays needed to produce ions resulting in a charge of 0.000258 coulombs/kilogram of air under standard conditions.

rotamer One of a set of conformers (one of a set of stereoisomers, each of which is characterized by a CON-

FORMATION corresponding to a distinct POTENTIAL-ENERGY minimum), arising from restricted rotation about one single bond.

rotational energy Energy due to the tumbling motion of the molecule.

rotational spectrum Molecular spectrum resulting from transitions between rotational levels of a molecule.

rubisco *See* RIBULOSE-1,5-BISPHOSPHATE CARBOXYLASE/OXYGENASE.

rubredoxin An IRON-SULFUR PROTEIN without ACID-LABILE SULFUR, in which an iron center is COORDINATED by four sulfur-containing LIGANDs, usually cysteine. The function, where known, is as an electron carrier.

rubrerythrin A protein assumed to contain both a RUBREDOXIN-like iron center and a HEMERYTHRIN-like DINUCLEAR iron center.

ruby A red gemstone composed of corundum (Al_2O_3) containing a small amount of chromium. Because of its radiative properties, ruby crystals were used in some of the first LASERs (ruby lasers, whose emissions were in the red spectrum).

rusticyanin An ELECTRON-TRANSFER PROTEIN, containing a TYPE 1 COPPER site, from the periplasm of the iron-oxidizing bacterium *Thiobacillus ferrooxidans.*

Rutherford, Ernest (1871–1937) British *Physicist* Ernest Rutherford was born on August 30, 1871, in Nelson, New Zealand, to James Rutherford, a Scottish wheelwright who emigrated to New Zealand in 1842. His mother, Martha Thompson, an English schoolteacher, followed in 1855.

Educated in government schools, Rutherford entered Nelson College at the age of 15, and in 1889 he was awarded a university scholarship to the University of New Zealand, Wellington, entering Canterbury College. He received a M.A. in 1893 with a double first in mathematics and physical science, and a bachelors of science the following year. In 1895 after receiving a scholarship, he entered Trinity College, Cambridge, as a research student at the Cavendish Laboratory under J. J. Thomson.

In 1892 he was awarded a B.A. research degree and the Coutts-Trotter Studentship of Trinity College, and the following year he accepted the Macdonald chair of physics at McGill University, Montreal.

In 1900 Rutherford married Mary Newton, and they had one daughter.

In 1907 he returned to England to become the Langworthy professor of physics at the University of Manchester, and in 1919 became the Cavendish professor of physics at Cambridge and chairman of the advisory council, H. M. Government, Department of Scientific and Industrial Research; professor of natural philosophy, Royal Institution, London; and director of the Royal Society Mond Laboratory, Cambridge.

Rutherford's work has made him known as the father of nuclear physics with his research on radioactivity (alpha and beta particles and protons, which he named), and he was the first to describe the concepts of half-life and decay constant. He showed that elements such as uranium transmute (become different elements) through radioactive decay, and he was the first to observe nuclear reactions (split the atom in 1917). In 1908 he received the Nobel Prize in chemistry "for his investigations into the disintegration of the elements, and the chemistry of radioactive substances." He was president of the Royal Society (1926–30) and of the Institute of Physics (1931–33) and was decorated with the Order of Merit (1925). He became Lord Rutherford in 1931.

He died in Cambridge on October 19, 1937, and his ashes were buried in Westminster Abbey, next to Lord Kelvin and just west of Sir Isaac Newton's tomb.

Rydberg series A Rydberg state is a state of an atom or molecule in which one of the electrons has been excited to a high principal quantum number orbital. A Rydberg series is the set of bound states of the excited electron for a given set of excited electron angular momentum quantum numbers and ion core state.

sacrificial anode The ANODE in a CATHODIC PROTECTION system. For cathodic protection, a more active metal is placed next to a less active metal. The more active metal will then serve as an anode and will become corroded instead of the less active metal. The anode has an oxidation potential that is greater than that of the other metal, so it is preferentially oxidized and is called a sacrificial anode.

sacrificial hyperconjugation *See* HYPERCONJUGATION.

saline Salty. Describes the condition of water that has a high concentration of salt, making it unfit for human consumption.

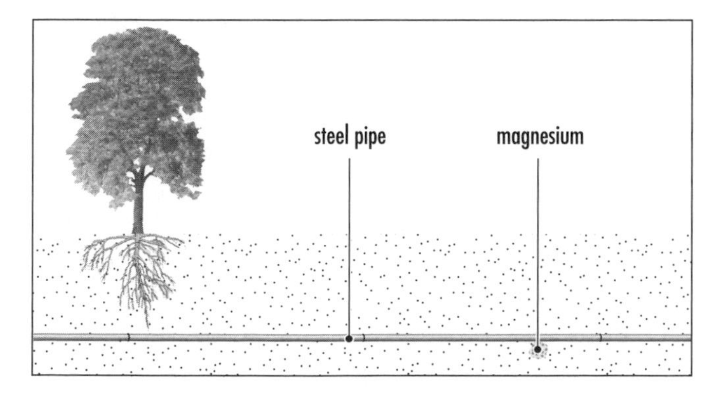

Sacrificial anode. The anode in a cathodic protection system

salinity A measure of the salt content of water, soil, etc.

salt An ionic compound made up of a cation and an anion and produced in neutralization reactions, e.g., an acid combined with a base produces a salt.

salt bridge A U-shaped tube containing an electrolyte that connects two HALF-CELLS of a VOLTAIC CELL allowing ion flow without extensive mixing of the different solutions. Also a permeable material soaked in a salt solution that allows ions to be transferred from one container to another.

sandwich compound A class of organometallic compounds (organic compound with one or more metal atoms in its molecules) whose molecules have a metal atom or ion bound between two plane parallel organic rings, e.g., metallocenes [Metal$(C_5H_5)_2$]; ferrocene [$Fe(C_5H_5)_2$].

saponification The chemical reaction that makes soap. The reaction between triglycerides and sodium or potassium hydroxide (lye) that produces glycerol and soap.

saprobe Any organism that obtains its nutrients and energy from dead or decaying matter. Formerly called saprophyte.

saturated fatty acid FATTY ACID with a hydrocarbon chain that contains no double bonds but has the maximum possible number of hydrogen atoms attached to every carbon atom; it is "saturated" with hydrogen atoms. A fatty acid is the building block of fat, which is composed of one or more fatty acids, and exists mostly solid at room temperature. It raises the blood LDL-cholesterol ("bad" cholesterol) level.

saturated hydrocarbons Hydrocarbons that contain only single carbon-carbon and carbon-hydrogen bonds. Also referred to as alkanes or paraffin hydrocarbons.

saturated solution A solution in which no more solute will dissolve; a solution in equilibrium with the dissolved material. A SUPERSATURATED SOLUTION contains more dissolved substance than a saturated solution.

scanning electron microscope A special kind of microscope in which a finely focused beam of electrons is scanned across a specimen, and the electron intensity variations are used to construct an image of the specimen.

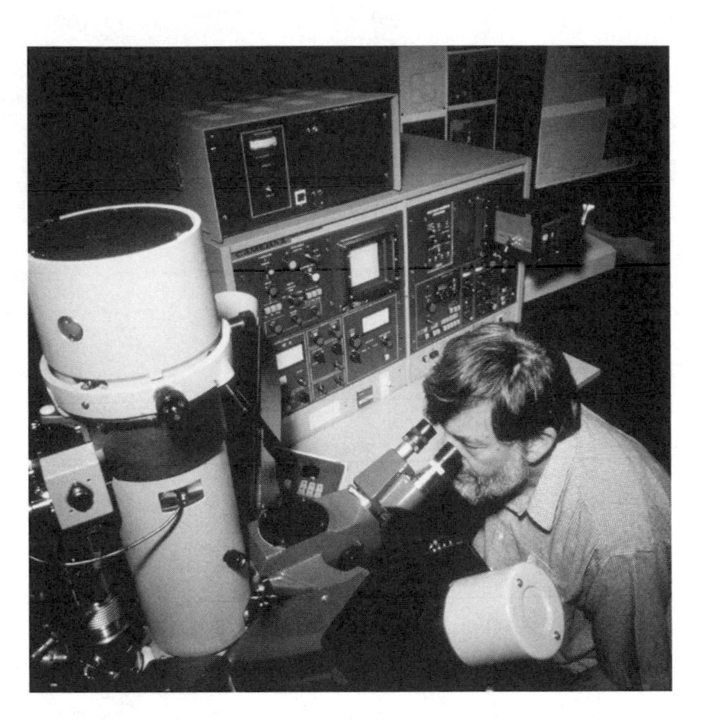

A scientist using a scanning electron microscope (SEM). The SEM uses a beam of electrons, rather than light, to illuminate the subject. The wavelength of electrons is considerably shorter than optical wavelengths, allowing a much larger magnification—up to about 40,000×. The beam is scanned across the subject that emits secondary electrons from its surface. These can be focused onto a phosphor screen or detected electronically and shown on a monitor. Nonmetallic subjects are usually coated with a thin film of gold to reflect the beam current. ***(Courtesy of Chris Taylor/CSIRO/Science Photo Library)***

scavenger A substance that reacts with (or otherwise removes) a trace component or traps a reactive reaction INTERMEDIATE (as in the scavenging of RADICALs or free electrons in radiation chemistry). Also an animal that feeds on dead animals.

Schiff base An imine bearing a hydrocarbyl group on the nitrogen atom: $R_2C=NR'$ ($R' \neq H$).

Schroedinger equation (wavefunction) Original equation in quantum mechanics used to write wave functions for particles; governs the evolution of probability waves in quantum mechanics.

scientific method The steps taken by a scientist after he or she develops a hypothesis, tests its predictions through experimentation, and then changes or discards the hypothesis if the predictions are not supported by the results.

scientific notation The system used when studying very small or very large quantities. A way to represent real numbers as a number between one and 10 multiplied by 10 to some power, rather than in full written form.

scrambling *See* ISOTOPIC SCRAMBLING.

secondary alcohol An alcohol in which the OH group is attached to a secondary carbon atom. Also called a *sec*-alcohol.

secondary compound Compounds that are not directly a function in the process of growth and development in a plant but are normal parts of a plant's METABOLISM. They often serve to discourage consumption by making it taste bad or having a toxic effect on the consumer. Examples of secondary compounds are nicotine and caffeine.

secondary kinetic electrolyte effect *See* KINETIC ELECTROLYTE EFFECT.

secondary kinetic isotope effect *See* ISOTOPE EFFECT, KINETIC.

secondary productivity The rate of new biomass production that is nutrient material synthesized by consumers over a specific time frame in an ecosystem.

secondary standard Standard prepared by dilution of an aliquot of a PRIMARY STANDARD solution with a known volume of solvent, or by subsequent serial dilutions; or a standard solution measured by reference to a primary standard solution.

secondary structure Level of structural organization in proteins described by the folding of the polypeptide chain into structural MOTIFs such as ALPHA HELICES and BETA SHEETs, which involve hydrogen bonding of backbone atoms. Secondary structure is also formed in NUCLEIC ACIDs, especially in single-stranded RNAs by internal BASE PAIRING.

See also PRIMARY STRUCTURE.

secondary voltaic cell VOLTAIC CELL that has the ability to be recharged.

See also PRIMARY VOLTAIC CELL.

second law of thermodynamics States that ENTROPY, a measure of disorder, increases in the universe and is spontaneous. Elements in a closed system will tend to seek their most probable distribution, and entropy always increases; it is a measure of unusable energy and a gauge of randomness or chaos within a closed system. Also called the law of increased entropy. Along with the FIRST LAW OF THERMODYNAMICS, these two laws serve as the fundamental principles of physics.

second messenger An intracellular METABOLITE or ion increasing or decreasing as a response to the stimulation of RECEPTORs by AGONISTs, which are considered the "first messenger." This generic term usually does not prejudge the rank order of intracellular biochemical events.

second-order reaction A reaction where two reactant molecules come together for the reaction to occur. The reaction rate is proportional to the square of the substrate concentration or to the first power of each of two reactants.

See also RATE OF REACTION.

selective permeability Refers to the control a cell membrane has over what can cross, deciding which specific molecules may enter or leave the cell by using either PASSIVE or ACTIVE TRANSPORT, or by way of a vesicle.

selectivity The discrimination shown by a reagent in competitive attack on two or more SUBSTRATEs or on two or more positions in the same substrate. It is quantitatively expressed by ratios of RATE CONSTANTs of the competing reactions or by the decadic logarithms of such ratios.

See also ISOSELECTIVE RELATIONSHIP; PARTIAL RATE FACTOR; REGIOSELECTIVITY; SELECTIVITY FACTOR; STEREOSELECTIVITY.

selectivity factor A quantitative representation of SELECTIVITY in AROMATIC SUBSTITUTION REACTIONs (usually ELECTROPHILIC for monosubstituted benzene derivatives). If the PARTIAL RATE FACTOR, *f*, expresses the reactivity of a specified position in the aromatic compound PhX relative to that of a single position in benzene, then the selectivity factor S_f (expressing discrimination between *p*- and *m*-positions in PhX) is defined as

$$S_f = \lg (f_p^X / f_m^X).$$

s electron An electron that occupies spherical space about the nucleus of an atom.

semiconductor A material that conducts electricity at high temperatures but not at low temperature. A material having a small number of mobile (free) electrons (*n*-type) or a small number of electron vacancies (holes; *p*-type) in an otherwise filled band, and which can conduct electricity as a result of the electrons moving under an applied field. In the *n*-type, electrons carry the charge (negative charge carriers). In the *p*-type, electrons move into a "hole," thereby creating a new hole in the previous place, so that it is the holes that seem to move relative to all the other negative electrons. Hence the holes are considered the charge carriers and are taken to be positive relative to the surrounding electrons. A characteristic of semiconductor conductivity behavior is that conductivity increases with increasing temperature rather than decreasing, as is characteristic of regular conductors or metallic conductivity (which also is much greater because of the larger number of electrons free to move).

semimetal (metalloid) Any element with properties between those of a metal and nonmetal, e.g., boron (B), silicon (Si), germanium (Ge), arsenic (As), and tellurium (Te); they are electrical SEMICONDUCTORS.

semipermeable A membrane where some substances will pass through but others will not.

sensory neuron A specialized neuron that sends messages that it receives from external or internal stimuli such as light, sound, smell, and chemicals to the central nervous system.

sensory receptor A cell or organ that converts a stimulus from a form of sound, light, smell or from thermal, chemical, or mechanical stimuli into a signal, or action potential, that can be transmitted through the organism. Sensory receptors are specific to the stimulus to which they are responding and fall into specific types such as chemoreceptors, photoreceptors, thermoreceptors, mechanoreceptors, and pain receptors, to name a few. Each changes the polarization of the cell that may eventually cause an action potential. Phasic receptors send action potentials quickly when first stimulated and then soon reduce the frequency of action potentials even if the stimulus continues, e.g., odor or pressure. They are useful for signaling sudden changes in the environment. Receptors that respond to light or mechanics are tonic receptors and produce a constant signal, after an initial amount of high-frequency action potentials, while the stimulus is being applied.

sequence The order of neighboring amino acids in a protein or the purine and pyrimidine bases in RNA or DNA.

See also PRIMARY STRUCTURE.

sequence-directed mutagenesis *See* MUTAGENESIS.

serotonin (5-Hydroxy tryptamine (5-HT)) A biogenic monoamine (like histamine, epinephrine, norepinephrine), a neurotransmitter, and an important vasoactive substance. A mediator of immediate hypersensitivity. Serotonin is one of many mediators released by circulating basophils and tissue mast cells. Found in blood platelets, in the gastrointestinal tract, and in certain regions of the brain.

Serotonin plays a major role in blood clotting, stimulating strong heartbeats, initiating sleep, fighting depression, and causing migraine headaches.

serum *See* PLASMA.

shell (of electrons) The layers in which electrons orbit the nucleus of an atom. Shell(s) of electrons are present in all atoms. The specific radius of each electron shell is determined by the energy level of the electrons in that shell. The inner shell of any atom has a capacity of two electrons, and any further shells have a capacity of eight electrons. A shell is designated by the value of the principal quantum number *n*.

shielding In the context of NMR spectroscopy, shielding is the effect of the electron shells of the observed and the neighboring nuclei on the external magnetic field. The external field induces circulations in the electron cloud. The resulting magnetic moment is oriented in the opposite direction to the external field, so that the local field at the central nucleus is weakened, although it may be strengthened at other nuclei (deshielding).

The phenomenon is the origin of the structural dependence of the resonance frequencies of the nuclei.

See also CHEMICAL SHIFT.

shielding effect The shielding of the effect of protons in the nucleus on the outer-shell electrons by the electrons in the filled sets of the *s* and *p* orbitals.

side chain A part of an amino acid that confers its identity and can range from a single hydrogen atom to groups of 15 or more atoms. Each of the 20 amino acids differs from the others by having a unique side chain, and the different identity and arrangement of the atoms in the side chains give different physical and chemical properties to each amino acid. May also refer to groups attached to the main backbone of linear polymers.

siderophore Generic term for Fe(III)-complexing compounds released into the medium by bacteria for the purpose of scavenging iron.

See also SCAVENGER.

sigma, pi (σ, π) The terms are symmetry designations, with pi molecular orbitals being antisymmetric with respect to a defining plane containing at least one atom (e.g., the molecular plane of ethene), and sigma molecular orbitals being symmetric with respect to the same plane. In practice, the terms are used both in this rigorous sense (for orbitals encompassing the entire molecule) and also for localized two-center orbitals or bonds, and it is necessary to make a clear distinction between the two usages.

In the case of two-center bonds, a pi bond has a nodal plane that includes the internuclear bond axis, whereas a sigma bond has no such nodal plane. (A delta bond in organometallic or inorganic molecular species has two nodes.) RADICALS are classified by analogy into sigma and pi radicals.

Such two-center orbitals may take part in molecular orbitals of sigma or pi symmetry. For example, the methyl group in propene contains three C–H bonds, each of which is of local sigma symmetry (i.e., without a nodal plane including the internuclear axis), but these three "sigma bonds" can in turn be combined to form a set of group orbitals, one of which has pi symmetry with respect to the principal molecular plane and can accordingly interact with the two-center orbital of pi symmetry (pi bond) of the double-bonded carbon atoms to form a molecular orbital of pi symmetry.

Such an interaction between the CH_3 group and the double bond is an example of what is called HYPERCONJUGATION. This cannot rigorously be described as "sigma-pi conjugation," since sigma and pi here refer to different defining planes, and interaction between orbitals of different symmetries (with respect to the same defining plane) is forbidden.

sigma (σ) adduct The product formed by the ATTACHMENT of an ELECTROPHILIC or NUCLEOPHILIC entering group or of a RADICAL to a ring carbon of an aromatic species so that a new sigma bond is formed and the original CONJUGATION is disrupted. (This has generally been called a σ-complex, but "adduct" is more appropriate than "complex" according to the definitions given.) The term may also be used for analogous adducts to unsaturated (and conjugated) systems in general.

See also MEISENHEIMER ADDUCT.

sigma (σ) bond A covalent bond where most of the electrons are located in between the nuclei. (The distribution is cylindrically symmetrical about the internuclear axis.)

See also SIGMA, PI.

sigma (σ) constant Specifically, the substituent constant for meta- and for para-substituents in benzene derivatives as defined by Hammett on the basis of the ionization constant of a substituted benzoic acid in water at 25°C, i.e., $\lg(K_a/K_a^o)$, where K_a is the ionization constant of an m- or p-substituted benzoic acid and K_a^o is that of benzoic acid itself.

The term is also used as a collective description for related electronic substituent constants based on other standard reaction series, of which σ^+, σ^-, and σ^o are typical; it is also used for constants that represent dissected electronic effects, such as σ_I and σ_R. For this purpose it might be better always to spell out the term in full, i.e., as "Hammett sigma constant," and restrict σ-constants to the scale of substituent constants, based on benzoic acid.

A large positive σ-value implies high electron-withdrawing power by inductive and/or resonance effect relative to H; a large negative σ-value implies high electron-releasing power relative to H.

See also HAMMETT EQUATION; RHO (ρ) VALUE; TAFT EQUATION.

sigma (σ) orbital Molecular orbital bonding with cylindrical symmetry about the internuclear axis.

See also SIGMA, PI.

sigmatropic rearrangement A MOLECULAR REARRANGEMENT that involves both the creation of a new sigma bond between atoms previously not directly linked and the breaking of an existing sigma bond. There is normally a concurrent relocation of pi bonds in the molecule concerned, but the total number of pi and sigma bonds does not change. The term was originally restricted to INTRAMOLECULAR PERICYCLIC REACTIONs, and many authors use it with this connotation. It is, however, also applied in a more general, purely structural, sense.

If such reactions are INTRAMOLECULAR, their TRANSITION STATE may be visualized as an ASSOCIATION of two fragments connected at their termini by two partial sigma bonds, one being broken and the other being formed as, for example, the two allyl fragments in (a′). Considering only atoms within the (real or hypothetical) cyclic array undergoing reorganization, if the numbers of these in the two fragments are designated i and j, then the rearrangement is said to be a sigmatropic change of order [i,j] (conventionally [i] ≤ [j]). Thus the rearrangement (a) is of order [3,3], while reaction (b) is a [1,5]sigmatropic shift of hydrogen. (Note that by convention, the square brackets [...] here refer to numbers of atoms, in contrast with current usage in the context of cycloaddition.)

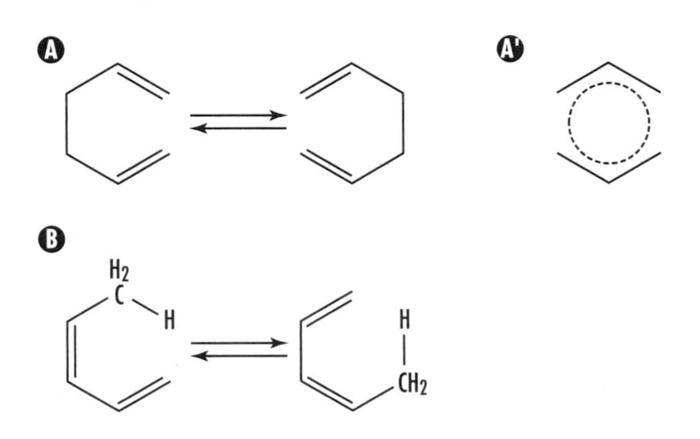

The descriptors a and s (ANTARAFACIAL and SUPRAFACIAL) may also be annexed to the numbers i and j; (b) is then described as a [1s,5s] sigmatropic rearrangement, since it is suprafacial with respect both to the hydrogen atom and to the pentadienyl system:

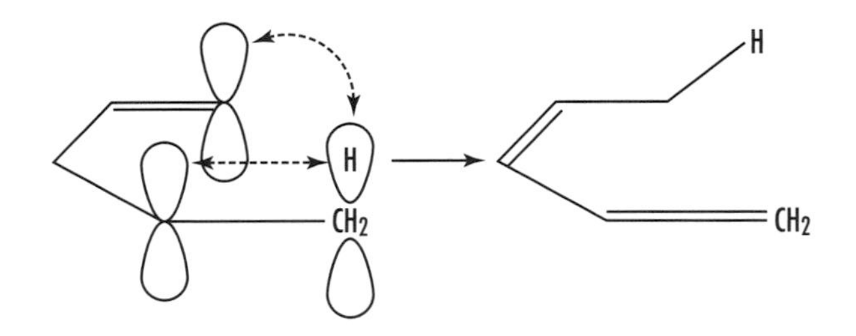

The prefix *homo* (meaning one extra atom, interrupting CONJUGATION, cf. HOMOAROMATICITY) has frequently been applied to sigmatropic rearrangements, but it is misleading.

See also CYCLOADDITION; TAUTOMERISM.

signal peptide A sequence of amino acids that determines the location of a protein in a eukaryotic cell.

See also EUKARYOTE.

signal transduction pathway Signal transduction refers to the movement of signals from outside the cell to the inside and is a mechanism connecting the stimulus to a cellular response.

silica, silicon dioxide (SiO_2) A crystalline or amorphous material that occurs naturally in impure forms such as quartz and sand.

silicate A mineral composed of silicon and oxygen. Silicates comprise the most abundant mineral group in the Earth's crust.

silicones Organic polymers with a backbone of alternating silicon and oxygen atoms.

silylene (1) Generic name for H_2Si: and substitution derivatives thereof, containing an electrically neutral bivalent silicon atom with two nonbonding electrons. (The definition is analogous to that given for CARBENE.)

(2) The silanediyl group (H_2Si), analogous to the methylene group (H_2C).

single covalent bond The sharing of an electron pair by two atoms.

single-electron transfer mechanism (SET) A reaction MECHANISM characterized by the transfer of a single electron between the species occurring on the REACTION COORDINATE of one of the elementary steps.

single-step reaction A reaction that proceeds through a single TRANSITION STATE.

siroheme A HEME-like PROSTHETIC GROUP found in a class of ENZYMEs that catalyze the six-electron reduction of sulfite and nitrite to sulfide and ammonia, respectively.

See also NITRITE REDUCTASE; SULFITE REDUCTASE.

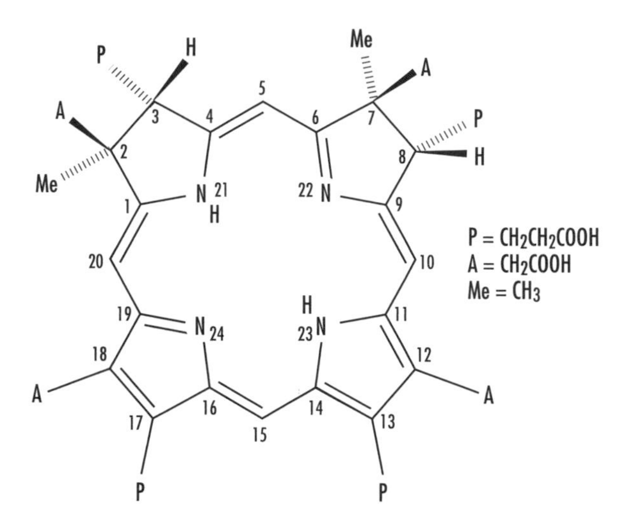

site-directed mutagenesis *See* MUTAGENESIS.

site-specific delivery An approach to target a DRUG to a specific tissue, using PRODRUGs or antibody recognition systems.

Slater-type orbital An approximate atomic orbital that attempts to allow for electron-electron repulsion by scaling the nuclear charge for each orbital.

smectic phase A phase type found in LIQUID CRYSTALs, a phase of matter whose order is intermediate between a liquid and a crystal. The molecules are typically rod-shaped. The nematic phase is characterized by the control of the orientational order and optical properties of the constituent molecules with applied electric fields. Nematics are the most commonly used phase in liquid crystal displays (LCDs). The smectic phase, found at lower temperatures than the nematic, forms well-defined layers that can slide over one another like soap. Smectics are positionally ordered along one direction and are characterized by a layered structure. As many as 12 smectic phases have been identified. Also used to describe clays in which water can be present between the aluminosilicate layers and which show a large change in volume upon wetting or drying.

smelting A process using high temperatures that separates a pure metal, usually in a molten form, from an ore.

smog A noxious mixture of air pollutants including gases and fine particles that can often be seen as a brownish-yellow or greyish-white haze.

See also PHOTOCHEMICAL SMOG.

SN1/SN2 reactions Substitution reactions (SN1/SN2). There are two basic kinds of substitution reactions:

SN2 = substitution nucleophilic bimolecular
SN1 = substitution nucleophilic unimolecular

A NUCLEOPHILE is an electron-rich species that will react with an electron-poor species. A substitution implies that one group replaces another, with the formation of a new bond to the nucleophile and the breaking of a bond to the leaving group.

soap A material used for cleaning made from the salts of vegetable or animal fats. The chemical reaction that makes soap and glycerin from fatty acids and sodium or potassium hydroxide is called SAPONIFICATION.

SOD *See* SUPEROXIDE DISMUTASE.

Soddy, Frederick (1877–1956) British *Physicist* Frederick Soddy was born at Eastbourne, Sussex, England, on September 2, 1877, to Benjamin Soddy, a London merchant. He was educated at Eastbourne College and the University College of Wales, Aberystwyth.

After receiving a scholarship at Merton College, Oxford, in 1895, he graduated three years later with first-class honors in chemistry. After spending two years doing research at Oxford, he moved to Canada, and during 1900–02 he was demonstrator in the chemistry department of McGill University, Montreal, working with ERNEST RUTHERFORD on the problems of radioactivity. He married Winifred Beilby in 1908. From 1904 to 1914 he was a lecturer in physical chemistry and radioactivity at the University of Glasgow, and after teaching in Scotland, he was appointed professor of chemistry at Oxford (1919–36).

He was one of the first to conclude in 1912 that some elements can exist in forms that "are chemically identical, and save only as regards the relatively few physical properties which depend on atomic mass directly, physically identical also." He called them isotopes. Later he promoted their use in determining geologic age. He is credited (with others) with the discovery of the element protactinium in 1917.

In 1921 he was awarded the Nobel Prize in chemistry "for his contributions to our knowledge of the chemistry of radioactive substances, and his investigations into the origin and nature of isotopes."

He published several books, including: *Radioactivity* (1904), *The Interpretation of Radium* (1909), *The*

(continued on p. 250)

The Role of Chemistry

by Karl F. Moschner, Ph.D.

Chemistry—the science of the properties, composition, structure, and reactions of substances—has always been an integral part of mankind's everyday life. Long before chemistry was a recognized field of science and even before recorded history, chemistry yielded numerous useful discoveries to inquisitive minds. Even common materials and everyday activities offer opportunities for new discoveries and a deeper understanding of the chemical processes that impact our lives and the world around us.

Nothing dramatizes this point more than the use of fire, one of the four ancient elements, the others being air, earth, and water. In Greek mythology, Prometheus stole fire from the gods and gave it to mankind. While fire presented many dangers, over thousands of years it provided mankind with light, warmth, and numerous other benefits. Even the by-products of incomplete combustion such as charcoal and carbon black found uses. Yet it was not until the 1980s that modern chemists discovered that soot, from combustion of organic matter, contained a symmetrical, soccer-ball-shaped molecule composed of 60 carbon atoms, named buckminsterfullerene in honor of the architect Buckminster Fuller, a leading proponent of the geodesic dome that it resembled. The serendipitous discovery of this ancient and ubiquitous third allotrope of the element carbon led to an explosion in new research with wide-ranging applications from anticancer drugs to nanotechnology.

Fire was also essential for mankind's discovery and exploitation of the elements of copper, gold, silver, tin, lead, iron, and mercury, and, by trial and error, their alloys such as bronze (copper and tin), pewter (tin and lead), and steel (iron and carbon). Since the 18th century, scientists have discovered more exotic metals and created novel alloys, some of which are important industrial catalysts—accelerators of chemical reactions. Two of the largest scale uses of catalysts are petroleum cracking, optimization of fuel production, and hydrogenation of vegetable oils (hardening) for oleomargarines. Other alloys act as semiconductors used in integrated circuits and superconductors such as those used in magnets for medical magnetic resonance imaging (MRI).

Oils and fats have been important throughout human history not only for food, but also as lubricants, polishes, ointments, and fuel. The reaction of oils and fats with alkali (saponification) produces soap (salts of fatty acids) and glycerin. This chemical process was known to the Romans and continues to be of significant commercial importance. Today, tens of thousands of tons of soap are produced annually from tallow and plant oils. Tallow is a by-product of the meat industry, while the principal plant oils are dependent on extensive plantations—palm and palm kernel oils from Indonesia, Malaysia, and India, and coconut oil from the Philippines and Brazil. Twentieth-century chemists designed more effective synthetic, crude-oil-based surface-active agents (surfactants, e.g., sodium linearalkylbenzenesulfonate or LAS) for fabric, household, and industrial cleaning applications, and specialty surfactants to meet the needs of consumer products industry such as milder skin and hair cleansers.

The surface-active properties of dissolved soaps and surfactants are attributed to their amphipathic structure, having both hydrophilic (water liking) and hydrophobic (water disliking) parts. In solution, surfactants condense at interphases, with the hydrophilic end in solution and with their hydrophobic tails aligned away from the solution. As a result, surfactants change the solution surface properties, lowering the surface tension and improving wetting and spreading. Early studies of surfactant monolayers provided insights into the surfactant molecule size and shape and the intermolecular forces that influence molecular packing. In the bulk solution, surfactants aggregate to form microscopic micelles when the so-called critical micelle concentration (CMC) is exceeded. Typically, the micelles initially assume consistent spherical aggregates. As the concentration of surfactants increases, the micelles may also assume the shape of rods and plates or disks. Under the proper conditions, some surfactants may form vesicles, cell-like spherical structures consisting of a surfactant bilayer "skin" separating an internal phase from the surrounding bulk solution. Chemists design special surfactants and control formulation compositions in order to exploit these phenomena in a broad range of novel products and applications.

Scientists have long recognized that specific molecular interactions and aggregation phenomena are crucial in biological systems. Lipid bilayers are essential for forming the complex cell membrane, while hydrogen-bonding interactions between nucleic-acid base pairs give rise to the double helical or twisted ladder structure of DNA, arguably the most important discovery of the 20th century. More importantly, the specific hydrogen bonding in DNA is also responsible for the ability of multiplying cells to precisely reproduce their genetic code. Modern scientists have learned how to mimic the cell's ability to replicate DNA, a technique now commonly used in forensics for DNA matches. Molecular interactions also determine the three-dimensional structure of enzymes and other biologically important proteins and thereby their function. Pharmaceutical chemists have long designed drugs that target proteins in disease-causing organisms to cure infections. Recently, scientists discovered that short-chain proteins, called prions, can catalyze the protein misfolding that causes BSE or "mad cow disease." They now hope that further research may lead to a cure for BSE and other ailments caused by misfolded proteins such as Alzheimer's and Parkinson's.

Chemists are now venturing to exploit their knowledge about molecular interactions to design molecules that can be controlled to self-assemble into novel structures. Some scientists hope to create a set of molecular building blocks to construct "molecular machines," while others are designing nanostructures with electrical properties useful for creating the next generation of computer chips. A few scientists even dare to attempt to create "artificial life."

The world of chemistry is all around us. Modern science has deciphered many of nature's chemical mysteries, but there are still many more to be discovered, sometimes in the most common but overlooked places. These few offer only a glimpse at the important role chemistry plays in science, its contributions to mankind, and the opportunities available to the curious.

— **Karl F. Moschner, Ph.D.**, is an organic chemistry and scientific computing consultant in Troy, New York.

(continued from p. 248)

Chemistry of the Radioactive Elements (1912–14), *Matter and Energy* (1912), *Science and Life* (1920), *The Interpretation of the Atom* (1932), *The Story of Atomic Energy* (1949), and *Atomic Transmutation* (1953).

Soddy died on September 22, 1956, at Brighton, Sussex.

sodium-potassium pump An ACTIVE TRANSPORT mechanism of cell membranes to regulate pressure between the inside and outside of the cell and to pump potassium ions into the cell and keep sodium out, thereby preventing water retention and swelling within the cell. The sodium-potassium pump also maintains the electrical charge within each cell. ATP is used as the energy source for the pump.

soft acid *See* HARD ACID.

soft base *See* HARD BASE.

soft drug A compound that is degraded in vivo to predictable nontoxic and inactive METABOLITES after having achieved its therapeutic role.

sol A colloid (HETEROGENEOUS MIXTURE of tiny particles suspended in another material) with solid particles suspended in a liquid, e.g., protoplasm.

solar energy Heat and light radiated from the Sun.

solar radiation Electromagnetic radiation (light energy in the form of photons) emitted by the Sun that hits the Earth.

sol-gel process Sol-gel is a colloidal suspension of silica particles that is gelled to form a solid. The sol-gel process, created by Bell Labs, allows silica to be modified with a variety of dopants to produce novel kinds of glass. The technique has been extended to the production of other types of materials such as ceramics.

solid solution A HOMOGENEOUS MIXTURE of two or more components. A main component and (normally) a small amount of another component exist within the crystal structure of the main component.

solubility product constant Equilibrium constant that applies to the solution of a slightly soluble ionic compound added to water.

solubility product principle The expression of the solubility product constant for a slightly soluble compound is the product of the concentrations of the constituent ions, each raised to the power that corresponds to the number of ions in one formula unit.

solute Any dissolved substance in a solution.

solution Any liquid mixture of two or more substances that is homogeneous.

solvation Any stabilizing interaction of a solute (or solute MOIETY) with the solvent, or a similar interaction of solvent with groups of an insoluble material (i.e., the ionic groups of an ion-exchange resin). Such interactions generally involve electrostatic forces and VAN DER WAALS FORCES, as well as chemically more specific effects such as HYDROGEN BOND formation.

See also CYBOTACTIC REGION.

solvatochromic relationship A LINEAR FREE-ENERGY RELATIONSHIP based on SOLVATOCHROMISM.

See also KAMLET-TAFT SOLVENT PARAMETERS.

solvatochromism The (pronounced) change in position and sometimes intensity of an electronic absorption or emission band, accompanying a change in the polarity of the medium. Negative (positive) solvatochromism corresponds to a HYPSOCHROMIC (BATHOCHROMIC) SHIFT with increasing solvent POLARITY.

See also DIMROTH-REICHARDT E_T PARAMETER; Z-VALUE.

Array of solar panels (photovoltaic cells) on the roof of terraced houses in Heerhugowaard in the Netherlands. This is part of a Dutch government pilot project to investigate clean, renewable energy forms such as solar power. The 10 houses in this terrace (only two seen here) have a total of 200 square meters of solar panels on their roofs. The total power output of these solar panels is 24.6 kilowatts. The photovoltaic cells contain a semiconducting material that converts sunlight directly into electricity. Solar power is relatively cheap, as once the panels are installed, they continue to produce electricity indefinitely. ***(Courtesy of Martin Bond/Science Photo Library)***

solvent Any liquid that dissolves another solute and forms a homogeneous solution. Several types of solvents exist, such as organic solvents (acetone, ethanol) or hydrocarbon solvents such as mineral spirits. Water is the most common solvent.

solvent extraction A method of removing a substance (solute) from solution by allowing the solution to contact a second liquid that is immiscible with the first solvent and in which the solute is more soluble.

solvent isotope effect *See* ISOTOPE EFFECT, SOLVENT.

solvent parameter Quantitative measures of the capability of solvents for interaction with solutes.

Such parameters have been based on numerous different physicochemical quantities, e.g., RATE CONSTANTS, solvatochromic shifts in ultraviolet/visible spectra, solvent-induced shifts in infrared frequencies, etc. Some solvent parameters are purely empirical in nature, i.e., they are based directly on some experimental measurement. It may be possible to interpret such a parameter as measuring some particular aspect of solvent-solute interaction, or it may be regarded simply as a measure of solvent POLARITY. Other solvent parameters are based on analyzing experimental results. Such a parameter is considered to quantify some particular aspect of solvent capability for interaction with solutes.

See also DIMROTH-REICHARDT E_T PARAMETER; GRUNWALD-WINSTEIN EQUATION; KAMLET-TAFT SOLVENT PARAMETERS; KOPPEL-PALM SOLVENT PARAMETERS; SOLVOPHOBICITY PARAMETER; Z-VALUE.

solvent polarity *See* POLARITY.

solvent-separated ion pair *See* ION PAIR.

solvolysis Generally, reaction with a solvent or with a LYONIUM ION or LYATE ION, involving the rupture of one or more bonds in the reacting solute. More specifically, the term is used for ELIMINATION, FRAGMENTATION, and SUBSTITUTION REACTIONs in which a solvent species is the NUCLEOPHILE ("alcoholysis" if the solvent is an alcohol, etc.).

solvophobicity parameter A SOLVENT PARAMETER defined by

$$S_p = 1 - M/M(\text{hexadecane})$$

derived from the GIBBS ENERGY of transfer ($\Delta_t G^o$) of a series of solutes from water to numerous aqueous-organic mixtures and to pure solvents:

$$\Delta_t G^o \text{ (to solvent)} = MR_T + D$$

where R_T is a solute parameter, and M and D characterize the solvent. The M values are used to define a solvent solvophobic effect so that S_p values are scaled from unity (water) to zero (hexadecane).

SOMO A singly occupied molecular orbital (such as the half-filled HOMO of a radical).

See also FRONTIER ORBITALS; HOMO.

sonication Irradiation with (often ultra-) sound waves, e.g., to increase the rate of a reaction or to prepare vesicles in mixtures of surfactants and water.

***s* orbital** An orbital is the area in space about an atom or molecule in which the probability of finding an electron is greatest. There is one *s*-orbital for each shell (orbital quantum number $l = 0$), three *p*-orbitals, and five *d*-orbitals. A spherically symmetrical atomic orbital.

soret band A very strong absorption band in the blue region of the optical absorption spectrum of a HEME protein.

SPC *See* STRUCTURE-PROPERTY CORRELATIONS.

special salt effect The initial steep rate increase observed in the KINETIC ELECTROLYTE EFFECT on certain SOLVOLYSIS reactions upon addition of some noncommon ion salts, especially $LiClO_4$.

speciation Refers to the chemical form or compound in which an element occurs in both nonliving and living systems. It may also refer to the quantitative distribution of an element. In biology, it refers to the origination of a new species.

See also BIOAVAILABILITY.

species *See* CHEMICAL SPECIES.

specific catalysis The acceleration of a reaction by a unique CATALYST, rather than by a family of related substances. The term is most commonly used in connection with specific hydrogen-ion or hydroxide-ion (LYONIUM ION or LYATE ION) catalysis.

See also GENERAL ACID CATALYSIS; GENERAL BASE CATALYSIS; PSEUDOCATALYSIS.

specific gravity The ratio of the density of a substance to the density of water.

specific heat The amount of heat per unit mass required to move the temperature by 1°C. Every substance has its own specific heat. Measured in joules per gram-degree Celsius (J/g°C), the specific heat of water is 2.02 J/g°C in gas phase, 4.184 in liquid phase, and 2.06 in solid phase.

spectator ion Ions in a solution that do not take part in a chemical reaction.

spectator mechanism A PRE-ASSOCIATION mechanism in which one of the MOLECULAR ENTITIES, C, is already present in an ENCOUNTER PAIR with A during formation of B from A, but does not assist the formation of B, for example:

$$A + C \underset{}{\overset{\text{Pre-association}}{\rightleftharpoons}} \underset{\text{encounter complex}}{(A \cdots\cdots C)} \longrightarrow \underset{\text{encounter complex}}{(B \cdots\cdots C)} \xrightarrow{\text{rapid}} \text{Product}$$

The formation of B from A may itself be a bimolecular reaction with some other reagent. Since C does not assist the formation of A, it is described as being present as a spectator, and hence such a mechanism is sometimes referred to as a spectator mechanism.

See also MICROSCOPIC DIFFUSION CONTROL.

spectral line Any of the lines in a spectrum due to the emission or absorption of ELECTROMAGNETIC RADIATION at a discrete wavelength; represents the energy difference between two energy levels. Different atoms or molecules can be identified by the unique sequence of spectral lines associated with them.

spectrochemical series A sequence of LIGANDs in order of increasing ligand field strength.

spectrograph (spectrometer, spectroscope) Any instrument that breaks down ELECTROMAGNETIC RADIATION into its component frequencies and wavelengths for study; a photograph or image of a spectrum.

spectrophotometer An instrument that measures the intensity of light versus its wavelength.

spectrum The display of component wavelengths of ELECTROMAGNETIC RADIATION.

spin adduct *See* SPIN TRAPPING.

spin counting *See* SPIN TRAPPING.

spin density The unpaired ELECTRON DENSITY at a position of interest, usually at carbon, in a RADICAL. It is often measured experimentally by ELECTRON PARAMAGNETIC RESONANCE (EPR, ESR [electron spin resonance]) SPECTROSCOPY through HYPERFINE coupling constants of the atom or an attached hydrogen.

See also RADICAL CENTER.

spin label A STABLE PARAMAGNETIC group that is attached to a part of another MOLECULAR ENTITY whose microscopic environment is of interest and may be revealed by the ELECTRON PARAMAGNETIC RESONANCE spectrum of the spin label. When a simple paramagnetic molecular entity is used in this way without covalent attachment to the molecular entity of interest, it is frequently referred to as a "spin probe."

spin-orbit coupling The interaction of the electron spin MAGNETIC MOMENT with the magnetic moment due to the orbital motion of the electron.

spin probe *See* SPIN LABEL.

spin-spin coupling The interaction between the spin MAGNETIC MOMENTs of different electrons and/or nuclei. In NMR SPECTROSCOPY it gives rise to multiple patterns and cross-peaks in two-dimensional NMR

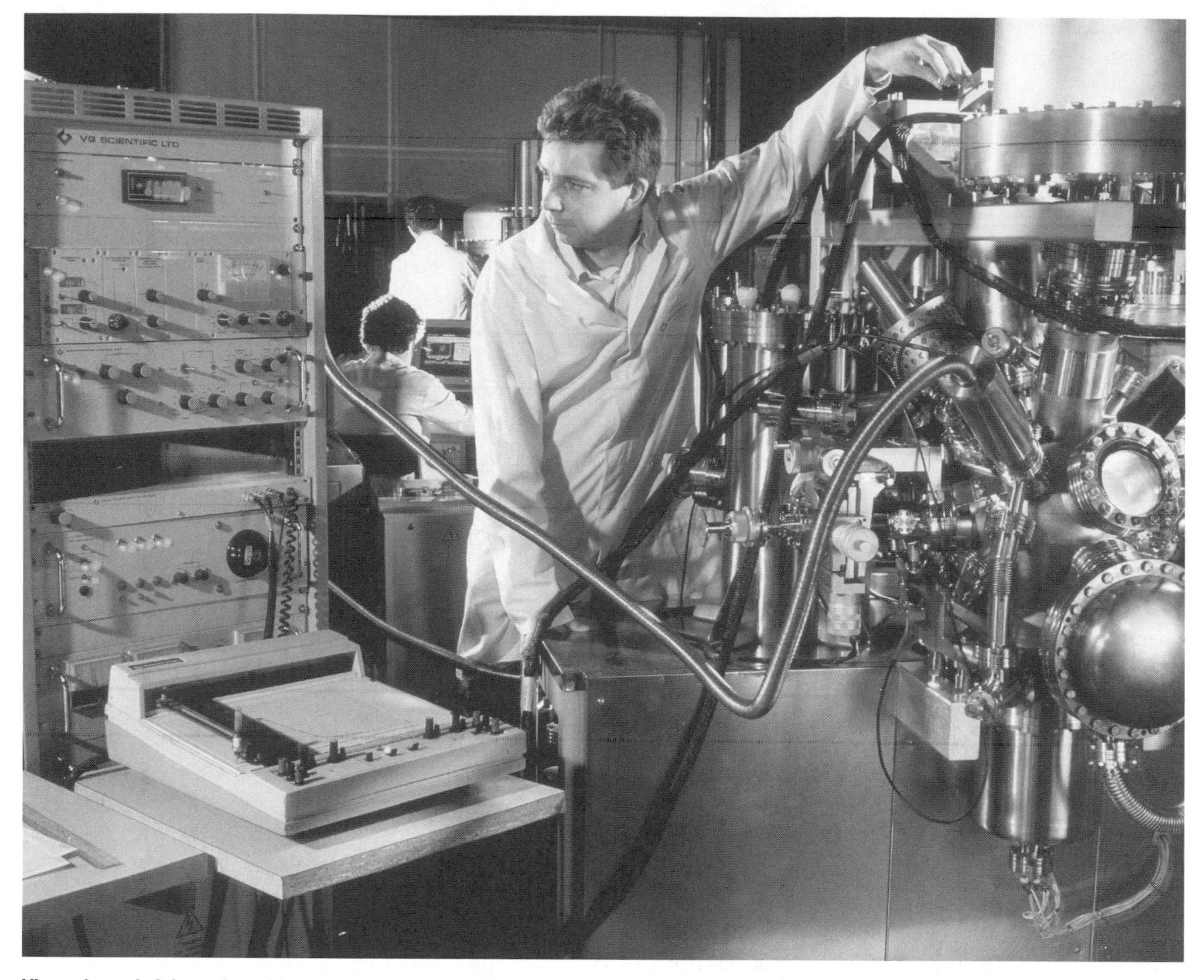

View of a technician using a high-resolution spectrometer in a research laboratory. The instrument is being used to analyze minute samples of particular chemicals or substances. Spectrometers work by producing a spectrum of light that can vary from X rays to infrared radiation (spectrophotometers). The effect of a particular substance on a spectrum is analyzed and interpreted by the spectrometer. The results, displayed graphically on a screen or printout, provide information on chemical composition or molecular organization of the substance. ***(Courtesy of Geoff Tompkinson/Science Photo Library)***

spectra. Between electron and nuclear spins, this is termed the nuclear HYPERFINE interaction. Between electron spins, it gives rise to relaxation effects and splitting of the EPR spectrum.

spin trapping In certain solution reactions, a transient RADICAL will interact with a DIAMAGNETIC reagent to form a more persistent radical. The product radical accumulates to a concentration where detection and, frequently, identification are possible by ELECTRON PARAMAGNETIC RESONANCE SPECTROSCOPY. the key reaction is usually one of attachment; the diamagnetic reagent is said to be a "spin trap," and the persistent product radical is then the "spin adduct."

spontaneous fission Fissioning that occurs without any outside cause, such as bombardment with a neutron.

spontaneous reaction A reaction that will continue to proceed without any outside energy. Spontaneous reactions are those that will take place by themselves, e.g., precipitation of calcium carbonate in the stalactites in a cave over millions of years.

square planar Molecules and polyatomic ions with one atom in the center and four atoms at the corners of a square.

square planar complex When metal is in the center of a square plane, and the LIGAND donor atoms are at each of the four corners.

square plane *See* COORDINATION.

stability constant An equilibrium constant that expresses the propensity of a species to form from its component parts. The larger the stability constant, the more STABLE is the species. The stability constant (formation constant) is the reciprocal of the instability constant (dissociation constant).

stable A term describing a system in a state of equilibrium corresponding to a local minimum of the appropriate thermodynamic potential for the specified constraints on the system. Stability cannot be defined in an absolute sense, but if several states are in principle accessible to the system under given conditions, that with the lowest potential is called the stable state, while the other states are described as metastable. Unstable states are not at a local minimum. Transitions between metastable and stable states occur at rates that depend on the magnitude of the appropriate activation energy barriers that separate them.

standard electrode potential The potential recorded by an electrode in which all reacting species are present at unit activity (approximately 1 *M* and 1 atm).

standard molar enthalpy of formation The enthalpy change when a mole of substance is formed from its elements in their most stable form under standard conditions.

standard molar volume The volume occupied by 1 mole of an ideal gas under standard conditions; 22.4 liters.

starch A polysaccharide containing glucose (long-chain polymer of amylose and amylopectin) that is the energy storage reserve in plants.

state function A function whose value depends only on the state of a substance and not on the path by which the state was reached, e.g., energy, enthalpy, temperature, volume, pressure, and temperature.

stationary state *See* STEADY STATE.

steady state (stationary state) (1) In a kinetic analysis of a complex reaction involving UNSTABLE intermediates in low concentration, the rate of change of each such INTERMEDIATE is set equal to zero, so that the rate equation can be expressed as a function of the concentrations of CHEMICAL SPECIES present in macroscopic amounts. For example, assume that X is an unstable intermediate in the reaction sequence:

$$\mathrm{A} \underset{k_{-1}}{\overset{k_1}{\rightleftharpoons}} \mathrm{X}$$

$$\mathrm{X} + \mathrm{C} \xrightarrow{k_2} \mathrm{D}$$

Conservation of mass requires that:

$$[\mathrm{A}] + [\mathrm{X}] + [\mathrm{D}] = [\mathrm{A}]_0$$

which, since $[\mathrm{A}]_0$ is constant, implies:

$$-d[\mathrm{X}]/dt = d[\mathrm{A}]/dt + d[\mathrm{D}]/dt.$$

Since [X] is negligibly small, the rate of formation of D is essentially equal to the rate of disappearance of A, and the rate of change of [X] can be set equal to zero. Applying the steady-state approximation, ($d[\mathrm{X}]/dt = 0$) allows the elimination of [X] from the kinetic equations, whereupon the rate of reaction is expressed:

$$d[D]/dt = -d[A]/dt = \frac{k_1k_2[A][C]}{k_{-1} + k_2[C]}$$

Note that the steady-state approximation does not imply that [X] is even approximately constant, only that its absolute rate of change is very much smaller than that of [A] and [D]. Since according to the reaction scheme d[D]/dt = k_2[X][C], the assumption that [X] is constant would lead—for the case in which C is in large excess—to the absurd conclusion that formation of the product D will continue at a constant rate even after the reactant A has been consumed. (2) In a stirred-flow reactor, a steady state implies a regime so that all concentrations are independent of time.

stellacyanin An ELECTRON-TRANSFER PROTEIN, containing a TYPE 1 COPPER site, isolated from exudates of the Japanese lacquer tree.

stepwise reaction A CHEMICAL REACTION with at least one reaction INTERMEDIATE and involving at least two consecutive ELEMENTARY REACTIONS.

See also COMPOSITE REACTION.

stereochemical Refers to the three-dimensional view of a molecule either as such or in a projection.

stereoelectronic Pertaining to the dependence of the properties (especially the energy) of a MOLECULAR ENTITY in a particular electronic state (or of a TRANSITION STATE) on relative nuclear geometry. The electronic GROUND STATE is usually considered, but the term can apply to excited states as well. Stereoelectronic effects are ascribed to the different alignment of electronic orbitals in different arrangements of nuclear geometry.

stereoelectronic control Control of the nature of the products of a CHEMICAL REACTION (or of its rate) by STEREOELECTRONIC factors. The term is usually applied in the framework of an orbital approximation. The variations of MOLECULAR ORBITAL energies with relative nuclear geometry (along a REACTION COORDINATE) are then seen as consequences of variations in basis-orbital overlaps.

stereoisomeric Isomerism due to the difference in the spatial arrangement of atoms without any difference in connectivity or bond multiplicity between the isomers.

stereoisomers Compounds with the same chemical formula but having different three-dimensional structures, differing only in the way that atoms are oriented in space.

stereoselectivity, stereoselective Stereoselectivity is the preferential formation in a CHEMICAL REACTION of one stereoisomer over another. When the stereoisomers are ENANTIOMERS, the phenomenon is called enantioselectivity and is quantitatively expressed by the enantiomer excess; when they are DIASTEREOISOMERS, it is called diastereoselectivity and is quantitatively expressed by the diastereomer excess. Reactions are termed (100 percent) stereoselective if the discrimination is complete or partially (x percent) stereoselective if one product predominates. The discrimination may also be referred to semiquantitatively as high or low stereoselectivity.

stereospecificity, stereospecific (1) A reaction is termed stereospecific if starting materials differing only in their CONFIGURATION are converted into STEREOISOMERIC products. According to this definition, a stereospecific process is necessarily STEREOSELECTIVE, but not all stereoselective processes are stereospecific. Stereospecificity may be total (100 percent) or partial. The term is also applied to situations where a reaction can be performed with only one stereoisomer. For example, the exclusive formation of *trans*-1,2-dibromocyclohexane upon bromination of cyclohexene is a stereospecific process, although the analogous reaction with (E)-cyclohexene has not been performed.

(2) The term has also been applied to describe a reaction of very high stereoselectivity, but this usage is unnecessary and is discouraged.

steric-approach control Control of STEREOSELECTIVITY of a reaction by steric hindrance toward attack of the reagent, which is directed to the less hindered face of the molecule. Partial bond making is strong enough at the TRANSITION STATE for steric control to take place. This suggests that the transition state should not be close to products.

See also PRODUCT-DEVELOPMENT CONTROL.

steric effect The effect on a chemical or physical property (structure, rate, or equilibrium constant) upon introduction of substituents having different steric requirements. The steric effect in a reaction is ascribed to the difference in steric energy between, on the one hand, reactants and, on the other hand, a TRANSITION STATE (or products). A steric effect on a rate process may result in a rate increase (steric acceleration) or a decrease (steric retardation). (The adjective *steric* is not to be confused with *stereochemical.*)

Steric effects arise from contributions ascribed to STRAIN as the sum of (1) nonbonded repulsions, (2) bond angle strain, and (3) bond stretches or compressions.

For the purpose of CORRELATION ANALYSIS or LINEAR FREE-ENERGY RELATIONs, various scales of steric parameters have been proposed, notably A VALUES, Taft's E_s, and Charton's υ scales.

In a reactant molecule RY and an appropriate reference molecule R°Y, the "primary steric effect" of R is the direct result of differences in compressions, which occur because R differs from R° in the vicinity of the reaction center Y. A "secondary steric effect" involves the differential moderation of electron delocalization by nonbonded compressions.

Some authors make a distinction between "steric" effects attributed to van der Waals repulsions alone, and "strain" effects, attributed to deviations of bond angles from "ideal" values.

See also STABILITY CONSTANT; TAFT EQUATION; VAN DER WAALS FORCES.

steric hindrance The original term for a STERIC EFFECT arising from crowding of substituents.

steric isotope effect *See* ISOTOPE EFFECT, STERIC.

steroid hormone (steroid drug) A large family of structurally similar chemicals. Various steroids have sex-determining, anti-inflammatory, and growth-regulatory roles. Examples include corticosteroid and glucocorticoid.

stimulus The cause for making a nerve cell respond; can be in the form of light, sounds, taste, touch, smell, temperature, etc.

See also SENSORY RECEPTOR.

stochastic Mathematical probability; a stochastic model takes into account variations in outcome that occur because of chance alone.

stock number *See* OXIDATION NUMBER.

stoichiometric number *See* RATE OF REACTION.

stoichiometry The calculation of the quantities of reactants and products (among elements and compounds) involved in a chemical reaction.

stopped flow A technique for following the kinetics of reactions in solution (usually in the millisecond time range) in which two reactant solutions are rapidly mixed by being forced through a mixing chamber. The flow of the mixed solution along a uniform tube is then suddenly arrested. At a fixed position along the tube, the solution is monitored (as a function of time following the stoppage of the flow) by some method with a rapid response (e.g., photometry).

See also MIXING CONTROL.

strain (strain energy) Strain is present in a MOLECULAR ENTITY or TRANSITION STRUCTURE if the energy is enhanced because of unfavorable bond lengths, bond angles, or dihedral angles (torsional strain) relative to a standard.

It is quantitatively defined as the standard enthalpy of a structure relative to a strainless structure (real or hypothetical) made up from the same atoms with the same types of bonding. (The enthalpy of formation of cyclopropane is 53.6 kJ mol^{-1}, whereas the enthalpy of formation based on three "normal" methylene groups, from acyclic models, is –62 kJ mol^{-1}. On this basis, cyclopropane is destabilized by ca. 115 kJ mol^{-1} of strain energy.)

See also MOLECULAR MECHANICS CALCULATION.

stratosphere The section of the atmosphere between 8 and 16 km to 50 km above the surface of the Earth. Temperature generally increases with increasing height.

strict aerobe An organism that utilizes aerobic respiration in that it can only survive in an atmosphere of oxygen.

strong acid An acid that is completely dissociated (ionized) in an aqueous (water) solution, e.g., hydrochloric acid (HCl), nitric acid (HNO_3).

strong base A metal hydroxide that completely dissociates (ionizes) into water, e.g., potassium hydroxide (KOH).

strong electrolyte A substance that conducts electricity well in a dilute aqueous solution and gives a 100 percent yield of ions when dissolved in water. A substance that is completely converted to ions when dissolved in water.

strong-field ligand LIGAND (a molecule or ion that can bind another molecule) that exerts a strong crystal or ligand electrical field and generally forms low-spin complexes with metal ions when possible; causes a large splitting of *d*-orbital energy of the metal ion in a COORDINATION COMPLEX.

structural formula The structural formula of a compound shows how the atoms and bonds are arranged in the molecule; indicates the arrangement of the atoms in space.

structural isomers Compounds that contain identical numbers of identical kinds of atoms in different geometric arrangements differing in the sequence in which the atoms or groups of atoms are bonded together.

structure-activity relationship (SAR) The relationship between chemical structure and pharmacological activity for a series of compounds.

structure-based design A DRUG design strategy based on the three-dimensional structure of the target obtained by X ray or NMR.

structure-property correlations (SPC) All statistical mathematical methods used to correlate any structural property with any other property (intrinsic, chemical, or biological) using statistical regression and PATTERN RECOGNITION techniques.

subjacent orbital The next-to-highest occupied molecular orbital (NHOMO, also called HOMO) and the second lowest unoccupied molecular orbital (SLUMO). Subjacent orbitals are sometimes found to play an important role in the interpretation of molecular interactions according to the FRONTIER ORBITAL approach.

sublimation The direct vaporization or transition of a solid directly to a vapor without passing through the liquid state.

substance All forms of matter that have the same chemical and physical properties.

substituent An atom or GROUP of bonded atoms that can be considered to have replaced a hydrogen atom (or two hydrogen atoms in the special case of bivalent groups) in a parent MOLECULAR ENTITY (real or hypothetical).

substituent electronegativity *See* ELECTRONEGATIVITY.

substitution reaction A reaction in which atoms are replaced by other atoms.

substrate A substrate can be a chemical species of particular interest, of which the reaction with some other chemical reagent is under observation (e.g., a compound that is transformed under the influence of a catalyst). It can also be the chemical entity whose conversion to a product or products is catalyzed by an ENZYME, or it could be a solution or dry mixture containing all ingredients necessary for the growth of a microbial culture or for product formation. Finally, it can be a component in the nutrient medium, supplying the organisms with carbon (C-substrate), nitrogen (N-substrate), etc.

substrate-level phosphorylation The formation or synthesis of the energy source ATP (adenosine triphosphate) by transferring an inorganic phosphate group to ADP (adenosine diphosphate).

subunit An individual polypeptide chain in a protein containing more than one polypeptide chain. Different types of subunits are frequently designated by α, β, γ, etc.

successor complex The radical ion pair that forms by the transfer of an electron from the donor D to the acceptor A after these species have diffused together to form the PRECURSOR or ENCOUNTER COMPLEX:

$$A + D \rightarrow (A\ D) \rightarrow (A^{\cdot -} D^{\cdot +})$$

sugar An organic compound that has the general chemical formula $(CH_2O)n$, also called saccharides. All sugars are carbohydrates, but not all carbohydrates are called sugars, e.g., starch, cellulose.

suicide inhibition *See* MECHANISM-BASED INHIBITION.

sulfite reductase ENZYMES that catalyze the reduction of sulfite to sulfide. All known enzymes of this type contain SIROHEME and IRON-SULFUR CLUSTERS.

superacid A medium having a high ACIDITY, generally greater than that of 100 wt.% sulfuric acid. The common superacids are made by dissolving a powerful LEWIS ACID (e.g., SbF_5) in a suitable BRONSTED ACID such as HF or HSO_3F. (An equimolar mixture of HSO_3F and SbF_5 is known by the trade name "magic acid.")

In a biochemical context, superacid catalysis is sometimes used to denote catalysis by metal ions, analogous to catalysis by hydrogen ions.

Oppositely, a compound having a very high basicity, such as lithium diisopropylamide, is called a "superbase."

superconductivity The ability of certain kinds of materials to conduct an electric current with virtually no resistance and under low temperature.

superconductor An intermetallic alloy, element, or compound that will conduct electricity without resistance below a certain temperature. Once the current begins to flow in a closed loop of superconducting material, it will flow forever. There are two types of superconductors: Type 1 and Type 2. Type 1 or "soft" superconductors are mainly composed of metals and metalloids that show *some* conductivity at room temperature but require incredibly cold temperatures to slow down molecular vibrations enough to facilitate unimpeded electron flow. Type 2 or "hard" superconductors are composed of metallic compounds and alloys (with the exception of elements vanadium, technetium, and niobium). Superconductors are used in maglev trains, medical imaging, and electricity generation, among other applications.

supercooled liquids Cooled substances existing as liquids at a temperature below their freezing point. An amorphous solid that will continue to flow very slowly under the influence of gravity.

supercritical fluid Highly compressed gases that combine properties of gases and liquids; a substance at a temperature above its CRITICAL TEMPERATURE and pressure (CRITICAL POINT).

superhyperfine *See* ELECTRON PARAMAGNETIC RESONANCE SPECTROSCOPY.

superoxide dismutases (SOD) ENZYMES that catalyze the DISMUTATION reaction of superoxide anion to dihydrogen peroxide and dioxygen. The enzymes have ACTIVE SITES containing either copper and zinc (Cu/Zn-superoxide dismutase), iron (Fe-superoxide dismutase), or manganese (Mn-superoxide dismutase).

supersaturated solution One that contains a higher-than-saturation concentration of solute. It is an unstable system, and excess solute will come out of solution under the proper nucleation conditions.

suprafacial *See* ANTARAFACIAL.

supramolecular chemistry This is defined as the chemistry of molecular assemblies and of the intermolecular bond as "chemistry beyond the molecule," bearing on the organized entities of higher complexity that result from the association of two or more chemical species held together by intermolecular forces. Thus, supramolecular chemistry may be considered to represent a generalized COORDINATION chemistry extending beyond the coordination of TRANSITION ELEMENTs by organic and inorganic LIGANDs to the bonding of all kinds of SUBSTRATES: cationic, anionic, and neutral species of either inorganic, organic, or biological nature.

supramolecule A system of two or more MOLECULAR ENTITIES held together and organized by means of intermolecular (noncovalent) binding interactions.

surface tension The cohesive forces between liquid molecules. Surface tension is typically measured in dynes/cm, the force in dynes required to break a film of length 1 cm. Water at 20°C has a surface tension of 72.8 dynes/cm. SURFACTANTs act to reduce the surface tension of a liquid.

surfactant Any material that lowers the SURFACE TENSION of a liquid, e.g., soap, dish detergent.

suspension Particles dispersed or undissolved in a fluid or solid.

sustainable A process that can continue indefinitely without overusing resources and causing damage to the environment.

sustainable agriculture Agricultural techniques and systems that, while economically viable to meet the need to provide safe and nutritious foods, utilize nonenvironmentally destructive methods such as the use of organic fertilizers, biological control of pests instead of pesticides, and minimal use of nonrenewable fuel.

sustainable development Development and economic growth that meets the requirements of the present generation without compromising the ability of future generations to meet their needs; a strategy seeking a balance between development and conservation of natural resources.

Sutherland, Earl W., Jr. (1915–1974) American *Pharmacologist* Earl Sutherland was born on November 19, 1915, in Burlingame, Kansas. He received a B.S. from Washburn College in 1937 and an M.D. from the Washington University in St. Louis School of Medicine in 1942.

He joined the faculty of Washington University and in 1953 became director of the department of medicine at Case Western Reserve University in Cleveland, Ohio. Here he discovered cyclic AMP (adenosine monophosphate) in 1956.

In 1963 he became professor of physiology at Vanderbilt University in Nashville, Tennessee. He was awarded the Nobel Prize for physiology or medicine in 1971 for isolation of cyclic adenosine monophosphate (cyclic AMP) and for demonstrating its involvement in numerous animal metabolic processes. From 1973 until his death he was a member of the faculty of the University of Miami Medical School.

Sutherland died on March 9, 1974, in Miami, Florida. The Earl Sutherland Prize award is presented annually by the chancellor of Vanderbilt University to a Vanderbilt faculty member who has made a nationally recognized impact in a particular discipline.

Swain-Lupton equation A dual-parameter approach to the correlation analysis of substituent effects, which involves a field constant (F) and a resonance constant (R). The original treatment was modified later. The procedure has been considerably applied, but also much criticized.

Swain-Scott equation The LINEAR FREE-ENERGY RELATION of the form

$$\lg(k/k_0) = sn$$

applied to the variation of REACTIVITY of a given ELECTROPHILIC substrate toward a series of nucleophilic reagents, where n is characteristic of the reagent (i.e., a measure of its NUCLEOPHILICITY) and s is characteristic of the substrate (i.e., a measure of its sensitivity to the nucleophilicity of the reagent). A scale of n values is based on the rate coefficients k for the reaction of methyl bromide with nucleophiles in water at 25°C, s being defined as 1.00 for these reactions and n being defined as 0.00 for the hydrolysis of methyl bromide. (Other scales have been devised.)

symbiosis The term was originally applied to describe the maximum flocking of either hard or soft LIGANDs in the same complexes. For hydrocarbon molecules, symbiosis implies that those containing a maximum number of C–H bonds (e.g., CH_4) or C–C bonds (e.g., Me_4C) are the most stable.

symproportionation Synonymous with COMPROPORTIONATION.

syn- *See* ANTI.

synapse A gap or junction between the ends of two neurons in a neural pathway where nerve impulses pass from one to the other. An impulse causes the release of a neurotransmitter at the synapse that diffuses across the gap and triggers the next neuron's electrical impulse.

synaptic terminal (synaptic vesicle) A membrane or bulb at the end of an axon terminal that stores and releases neurotransmitters.

synartetic acceleration *See* NEIGHBORING-GROUP PARTICIPATION.

synchronization (principle of nonperfect synchronization) This principle applies to reactions in which there is a lack of synchronization between bond formation or bond rupture and other PRIMITIVE CHANGES that affect the stability of products and reactants, such as RESONANCE, SOLVATION, electrostatic, HYDROGEN BONDING, and POLARIZABILITY effects. The principle states that a product-stabilizing factor whose development lags behind bond changes at the TRANSITION STATE, or a reactant-stabilizing factor whose loss is ahead of bond changes at the transition state, increases the INTRINSIC BARRIER and decreases the "intrinsic rate constant" of a reaction. For a product-stabilizing factor whose development is ahead of bond changes, or for reactant factors whose loss lags behind bond changes, the opposite relations hold. The reverse effects are observable for factors that destabilize a reactant or product.

See also IMBALANCE; SYNCHRONOUS.

synchronous A CONCERTED PROCESS in which the PRIMITIVE CHANGEs concerned (generally bond rupture and bond formation) have progressed to the same extent as the TRANSITION STATE is said to be synchronous. The term figuratively implies a more-or-less synchronized progress of the changes. However, the progress of the bonding change (or other primitive change) has not been defined quantitatively in terms of a single parameter applicable to different bonds or different bonding changes. The concept is therefore in general only qualitatively descriptive and does not admit an exact definition except in the case of concerted processes involving changes in two identical bonds.

See also IMBALANCE.

synchrotron An instrument that accelerates electrically charged particles close to the speed of light and maintains them in circular orbits; a source of very intense X rays.

synchrotron radiation Electromagnetic radiation from very-high-energy electrons moving in a magnetic field. Synchrotron radiation is highly polarized and continuous, and its intensity and frequency are directly related to the strength of the magnetic field and the energy of the charged particles affected by the field. The stronger the magnetic field and the higher the energy of the particles, the greater are the intensity and frequency of the emitted radiation.

synergism The combination of two or more substances in which the sum result of this combination is greater than the effects that would be exhibited from each of the individual substances alone.

synthase An ENZYME that catalyzes a reaction in which a particular molecule is synthesized, not necessarily by formation of a bond between two molecules.

See also LIGASE.

synthetase *See* LIGASE.

Szent-Györgyi, Albert von (1893–1986) Hungarian *Biochemist* Albert von Szent-Györgyi was born in Budapest on September 16, 1893, to Nicolaus von Szent-Györgyi and Josefine, whose father, Joseph Lenhossék, and brother Michael were both professors of anatomy in the University of Budapest.

He took his medical degree at the University of Sciences in Budapest in 1917, and in 1920 he became an assistant at the University Institute of Pharmacology in Leiden. From 1922 to 1926 he worked with H. J. Hamburger at the Physiology Institute, Groningen, Netherlands. In 1927 he went to Cambridge as a Rockefeller fellow, working under F. G. HOPKINS, and spent one year at the Mayo Foundation, Rochester, Minnesota, before returning to Cambridge. In 1930 he became chair of medical chemistry at the University of Szeged, and five years later he was also chair in organic chemistry. At the end of World War II, he was chair of medical chemistry at Budapest.

Szent-Györgyi's early researches concerned the chemistry of cell respiration. He pioneered the study of biological oxidation mechanisms and proved that hexuronic acid, which he isolated and renamed ascorbic acid, was identical to vitamin C and that it could be extracted from paprika. He won the 1937 Nobel Prize in physiology or medicine for his discoveries, especially of vitamin C.

In the late 1930s his work on muscle research quickly led him to discover the proteins actin and myosin and their complexes. This led to the foundation of muscle research in the following decades. He also worked on cancer research in his later years. In 1947 he moved to the United States, where he became director of research, Institute of Muscle Research, Woods Hole, Massachusetts.

His publications include *Oxidation, Fermentation, Vitamins, Health and Disease* (1939); *Muscular Contraction* (1947); *The Nature of Life* (1947); *Contraction in Body and Heart Muscle* (1953); *Bioenergetics* (1957); and *The Crazy Ape* (1970), a commentary on science and the future of the human race. He died on October 22, 1986.

T

Taft equation Various equations are associated with R.W. Taft, but the term is most often used to designate the family of equations that emerged from Taft's analysis of the reactivities of aliphatic esters and that involved the polar substituent constant σ^* and the steric substituent constant E_s

$$\lg k = \lg k_o + \rho^*\sigma^* + \delta E_s$$

or the one-parameter forms applicable when the role of either the polar term or the steric term can be neglected. Nowadays σ^* is usually replaced by the related constant σ_I.

See also HAMMETT EQUATION; RHO (ρ) VALUE; SIGMA (σ) CONSTANT.

tautomeric effect *See* ELECTROMERIC EFFECT.

tautomerism Isomerism of the general form

$$G\text{–}X\text{–}Y\text{=}Z \rightleftharpoons X\text{=}Y\text{–}Z\text{–}G$$

where the isomers (called tautomers) are readily interconvertible; the atoms connecting the groups X,Y,Z are typically any of C, H, O, or S, and G is a group that becomes an ELECTROFUGE or NUCLEOFUGE during isomerization. The most common case, when the electrofuge is H^+, is also known as "prototropy."

Examples, written so as to illustrate the general pattern given above, include: Keto-enol tautomerism, such as

$$H\text{–}O\text{–}C(CH_3)\text{=}CH\text{–}CO_2Et \text{ (enol)} \rightleftharpoons (CH_3)C(\text{=}O)\text{–}CH_2\text{–}CO_2Et \text{ (keto)}$$

$$(G = H, X = O, Y = CCH_3, Z = CHCO_2Et)$$

$$ArCH_2\text{–}N\text{=}CHAr' \rightleftharpoons ArCH\text{=}N\text{–}CH_2Ar'$$

$$(G = H, X = CHAr, Y = N, Z = CHAr')$$

The grouping Y may itself be a three-atom (or five-atom) chain extending the conjugation, as in

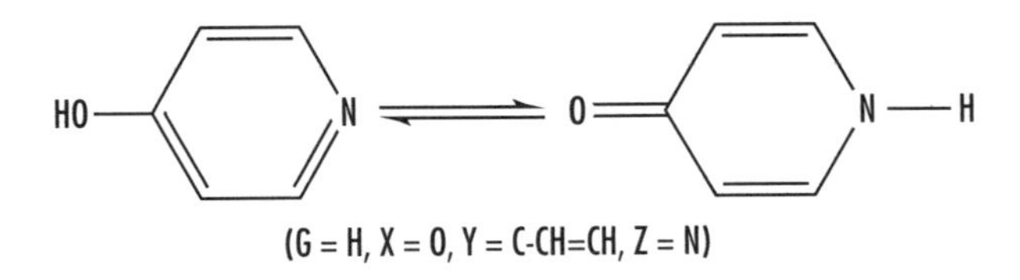

The double bond between Y and Z may be replaced by a ring, when the phenomenon is called ring-chain tautomerism, as in

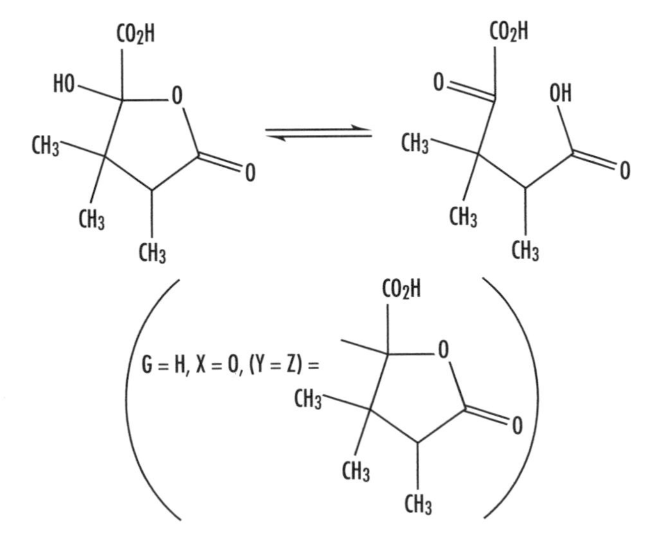

See also AMBIDENT; SIGMATROPIC REARRANGEMENT; TAUTOMERIZATION; VALENCE TAUTOMERIZATION.

tautomerization The ISOMERIZATION by which tautomers are interconverted. It is a HETEROLYTIC MOLECULAR REARRANGEMENT and is frequently very rapid.

See also TAUTOMERISM.

taxon A grouping of organisms by taxonomic rank, such as order, family, genus, etc.; also a group of organisms or other taxa sharing a single common ancestor.

tele-substitution A SUBSTITUTION REACTION in which the ENTERING GROUP takes up a position more than one atom away from the atom to which the LEAVING GROUP was attached.

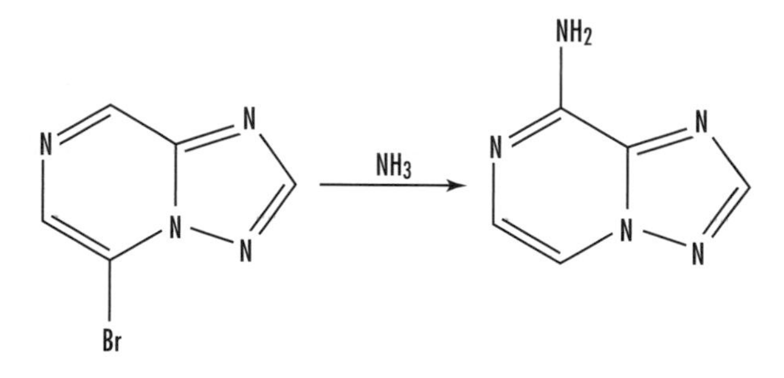

See also CINE-SUBSTITUTION.

telomerization The formation of an ADDITION oligomer, having uniform end groups X′...X″, by a CHAIN REACTION in which a CHAIN TRANSFER limits the length of the polymer ("telomer") produced. An example is the polymerization of styrene in bromotrichloromethane solution (X′ = CCl_3, X″ = Br), where Cl_3C radicals are formed in the initiation step to produce $Cl_3C[CH_2CHPh]_nBr$, with n greater than 1 and often less than 10.

temperature A measure of the energy in a substance. The more heat energy in the substance, the higher is the temperature. There are a number of temperature scales that have evolved over time, but three are used presently: Fahrenheit, Celsius, and Kelvin. The Fahrenheit temperature scale is a scale based on 32 for the freezing point of water and 212 for the boiling point of water, the interval between the two being divided into 180 parts. This scale is named after its inventor, the 18th-century German physicist DANIEL GABRIEL FAHRENHEIT. The Celsius or centigrade temperature scale is based on 0 for the freezing point of water and 100 for the boiling point of water. This scale was invented in 1742 by the Swedish astronomer ANDERS CELSIUS, and was once called the centigrade scale because of the 100-degree interval between the defined points. The Kelvin temperature scale is the base unit of thermodynamic temperature measurement and is defined as 1/273.16 of the TRIPLE POINT (equilibrium among the solid, liquid, and gaseous phases) of pure water. The Kelvin is also the base unit of the Kelvin scale, an absolute-temperature scale named for the British physicist William Thomson, Baron Kelvin. Such a scale has as its zero point absolute zero, the theoretical temperature at which the molecules of a substance have the lowest energy.

termination The steps in a CHAIN REACTION in which reactive INTERMEDIATES are destroyed or rendered inactive, thus ending the chain.

ternary acid An acid containing H, O, and another element, often a nonmetal. Examples are HNO_3

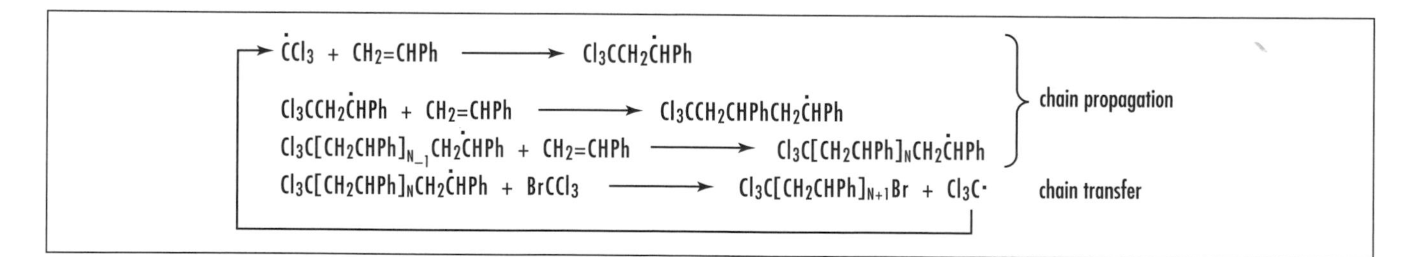

The formation of an addition oligomer

(nitric acid), $HClO_3$ (chloric acid), and $HC_2H_3O_2$ (acetic acid). Acetic is a ternary acid by this definition, but in general organic acids are not included in this class.

Also called oxyacids (or oxoacids).

terpenes Hydrocarbon solvents, compounds composed of molecules of hydrogen and carbon. They form the primary constituents in the aromatic fractions of scented plants, e.g., pine oil. Others include turpentine and camphor oil.

tertiary structure The overall three-dimensional structure of a BIOPOLYMER. For proteins, this involves the side-chain interactions and packing of SECONDARY STRUCTURE motifs. For NUCLEIC ACIDs this may be the packing of stem loops or supercoiling of DOUBLE HELICES.

tetrahedral Refers to a solid with four plane faces. Describes a configuration with one atom in the center and four atoms at the corners of a tetrahedron.

tetrahedral hole A hole or space in a crystal surrounded by four atoms or ions arranged toward the corners of a tetrahedron.

tetrahedral intermediate A reaction INTERMEDIATE in which the bond arrangement around an initially double-bonded carbon atom (typically a carbonyl carbon atom) has been transformed from trigonal to tetrahedral. For example, aldol in the CONDENSATION REACTION of acetaldehyde (but most tetrahedral intermediates have a more fleeting existence).

tetrahedron *See* COORDINATION.

tetrahydrofolate A reduced FOLATE derivative that contains additional hydrogen atoms in positions 5, 6, 7, and 8. Tetrahydrofolates are the carriers of activated one-carbon units and are important in the biosynthesis of amino acids and precursors needed for DNA synthesis.

See also FOLATE COENZYME.

thalassemia A chronic inherited disease characterized by defective synthesis of HEMOGLOBIN. Defective synthesis of the α chain of hemoglobin is called α-thalassemia, and defective synthesis of the β-chain of hemoglobin is called β-thalassemia. Thalassemias result in anemia that can be severe and are found more frequently in areas where malaria is endemic.

Theorell, Axel Hugo Theodor (1903–1982) Swedish *Biochemist* Axel Hugo Theodor Theorell was born at Linköping, Sweden, on July 6, 1903, to Thure Theorell, surgeon major to the First Life Grenadiers practicing medicine in Linköping, and his wife, Armida Bill.

Theorell was educated at a state secondary school in Linköping and started studying medicine in 1921 at the Karolinska Institute. In 1924 he graduated with a Bachelor of Medicine and spent three months studying bacteriology at the Pasteur Institute in Paris. He received an M.D. in 1930 and became lecturer in physiological chemistry at the Karolinska Institute.

In 1924 he became part of the staff of the Medico-Chemical Institution as an associate assistant and temporary associate professor working on the influence of the lipids on the sedimentation of the blood corpuscles. In 1931 at Uppsala University, he studied the molecular weight of myoglobin. The following year he was appointed associate professor in medical and physiological chemistry at Uppsala University and continued and extended his work on myoglobin.

From 1933 until 1935 he held a Rockefeller fellowship and became interested in oxidation enzymes. He produced, for the first time, the oxidation enzyme called "the yellow ferment" and he succeeded in splitting it reversibly into a coenzyme part, which was found to be flavinmononucleotide, and a colorless protein part.

Returning to Sweden in 1935, he worked at the Karolinska Institute, and in 1936 he was appointed head of the newly established biochemical department of the Nobel Medical Institute, opened in 1937.

He carried out research on various oxidation enzymes, contributing to the knowledge of cytochrome c, peroxidases, catalases, flavoproteins, and pyridinproteins, particularly the alcohol dehydrogenases. For his work on the nature and effects of oxidation enzymes he was awarded the Nobel Prize for physiology or medicine for 1955.

He was a member of numerous scientific organizations and in 1954 was chief editor of the journal *Nordisk Medicin*. He died on August 15, 1982.

theoretical chemistry A branch of chemistry that deals with structure theory, molecular dynamics, and statistical mechanics.

theoretical yield The maximum quantity of a product that can be formed in a chemical reaction if all the LIMITING REACTANT reacted to form products.

therapeutic index For a substance used to alleviate disease, pain, or injury, the therapeutic index is the ratio between toxic and therapeutic doses (the higher the ratio, the greater the safety of the therapeutic dose).

thermal analysis Measurement of the relationship between the physical and/or chemical properties of a sample and its temperature. Some methods that are commonly used are differential thermal analysis (DTA): temperature difference between the sample and a standard; differential scanning calorimetry (DSC): heat-flow difference between the sample and a standard; thermogravimetric analysis (TGA): mass vs. temperature; and thermomechanical analysis (TMA): dimension vs. temperature.

thermal cracking The degrading of heavy oil molecules into lighter fractions using high temperature and without using a catalyst. Used in the petroleum refining industry and used to convert gasoline into naphtha.

thermal neutron A neutron in thermal equilibrium within the medium in which it exists. A neutron with about the same energy as the surrounding matter (less than 0.4 eV).

thermal pollution Excessive raising or lowering of water temperature above or below normal seasonal ranges in streams, lakes, estuaries, oceans, or other bodies of water as the result of discharge effluents (hot/cold).

thermodynamic control (of product composition) The term characterizes conditions that lead to reaction products in a proportion governed by the equilibrium constant for their interconversion and/or for the interconversion of reaction INTERMEDIATES formed in or after the RATE-LIMITING STEP. (Some workers prefer to describe this phenomenon as "equilibrium control.")

See also KINETIC CONTROL.

thermodynamics The study of the energy transfers or conversion of energy in physical and chemical processes.

thermogravimetric analysis (TGA) Precise measurement of the weight change of a solid as it is heated at a controlled rate.

thermoluminescence A phenomenon exhibited in certain minerals in which they give off light, as if glowing, when heated. When heated to high temperatures, the trapped electrons are released and they give off energy in the form of light. Used by archaeologists as a dating technique in pottery and rock that has been previously fired.

thermolysin A calcium- and zinc-containing neutral protease isolated from certain bacteria.

thermolysis An uncatalyzed bond cleavage resulting from exposure of a compound to a raised temperature.

thermonuclear energy Energy created from nuclear fusion reactions.

third law of thermodynamics If all the thermal motion of molecules (kinetic energy) could be removed, absolute zero would occur (a temperature of 0 Kelvin or –273.15°C).

See also FIRST LAW OF THERMODYNAMICS; SECOND LAW OF THERMODYNAMICS.

thorium-lead dating A method to measure the age of rocks and other materials that contain thorium and lead through the use of the natural radioactive decay of ^{232}Th (HALF-LIFE about 4.7 billion years) as it decays to ^{208}Pb. Because of the very long half-life, this dating technique is usually restricted to ages greater than 10 million years.

three-center bond Bonding of three atoms by a pair of electrons in a molecular orbital formed from the overlap of three atomic orbitals.

three-dimensional quantitative structure-activity relationship (3D-QSAR) The analysis of the quantitative relationship between the biological activity of a set of compounds and their spatial properties using statistical methods.

thylakoid Enclosed membrane structure inside CHLOROPLASTS and PHOTOSYNTHETIC bacteria.

tight ion pair *See* ION PAIR.

titration A method in which one solution is added to another until the reaction between the two is complete.

T-jump *See* CHEMICAL RELAXATION.

topliss tree A topliss tree is an operational scheme for ANALOG design.

torquoselectivity The term refers to the preference for "inward" or "outward" rotation of substituents in conrotatory or disrotatory electrocyclic ring-opening reactions.

torr A unit of measure of pressure. At sea level, 760 torr is standard pressure.

torsion angle, dihedral angle The relative position, or angle, between the A–X bonds and the B–Y bonds when considering four atoms connected in the order A–X–Y–B; also is the angle between two planes defined as A–X–Y and X–Y–B.

total energy The sum of both kinetic and potential energy.

total ionic equation Equation for a chemical reaction to show the predominant form of all species in aqueous solution or in contact with water. For example:

$$CaCO_3(s) + 2H^+(aq) + 2Cl^-(aq) \rightleftharpoons Ca_2^+(aq) + 2Cl^-(aq) + H_2O(l) + CO_2(g)$$

toxicity The action of poisons (including XENOBIOTICS) on biochemical reactions or processes in living organisms or ecological systems. A study of this action is the subject matter of toxicology.

toxin A poisonous material that can cause damage to living tissues.

trace elements Elements required for physiological functions in very small amounts that vary for different organisms. Included among the trace elements are Co, Cu, F, Fe, I, Mn, Mo, Ni, Se, V, W, and Zn. Excess mineral intake may produce toxic symptoms.

tracer A radioactively tagged compound used to produce a nuclear image, often used in medicine.

trans- In inorganic nomenclature, a structural prefix designating two groups directly across a CENTRAL

ATOM from each other (not generally recommended for precise nomenclature purposes of complicated systems). In organic systems, the prefix designates groups on opposite sides of a double bond.

See also CIS.

transcription The process by which the GENETIC information encoded in a linear SEQUENCE of NUCLEOTIDES in one strand of DNA is copied into an exactly complementary sequence of RNA.

transduction (1) The transfer of GENETIC information from one bacterium to another by means of a transducing bacteriophage. When the phage is grown on the first host, a fragment of the host DNA can be incorporated into the phage particles. This foreign DNA can be transferred to the second host upon infection with progeny phage from the first experiment.

(2) In cell biology, the transduction of a signal (mechanical signal, hormone, etc.) to cells or tissues summarizes the chain of events between the primary reception of the signal and the final response (change in growth and/or METABOLISM) of the target cells or tissues. Inorganic substances (e.g., calcium ions) are frequently involved in the transduction of signals.

transferability Transferability assumes invariance of properties that are associated conceptually with an atom or a fragment present in a variety of molecules. The property, such as ELECTRONEGATIVITY, nucleophilicity, NMR CHEMICAL SHIFT, etc., is held as retaining a similar value in all these occurrences.

transferase An ENZYME of EC class 2 that catalyzes the transfer of a group from one SUBSTRATE to another.

See also EC NOMENCLATURE FOR ENZYMES.

transferrin An iron-transport protein of blood PLASMA that comprises two similar iron-binding DOMAINs with high affinity for Fe(III). Similar proteins are found in milk (lactoferrin) and eggs (ovotransferrin).

transformation The conversion of a SUBSTRATE into a particular product, irrespective of reagents or MECHANISMs involved. For example, the transformation of aniline ($C_6H_5NH_2$) into N-phenylacetamide ($C_6H_5NHCOCH_3$) can be effected by the use of acetyl chloride or acetic anhydride or ketene. A transformation is distinct from a reaction, the full description of which would state or imply all the reactants and all the products.

See also CHEMICAL REACTION.

transient (chemical) species Relating to a short-lived reaction INTERMEDIATE. It can be defined only in relation to a time scale fixed by the experimental conditions and the limitations of the technique employed in the detection of the intermediate. The term is a relative one. Transient species are sometimes also said to be "metastable." However, this latter term should be avoided because it relates a thermodynamic term to a kinetic property, although most transients are also thermodynamically UNSTABLE with respect to reactants and products.

transition coordinate The reaction coordinate at the transition state corresponding to a vibration with an imaginary frequency. Motion along it in the two opposite senses leads toward the reactants or toward the products.

See also REACTION COORDINATE; TRANSITION STATE.

transition element A transition element is an element whose atom has an incomplete d-subshell, or which gives rise to a cation or cations with an incomplete d-subshell. The first transition series of elements in which the 3d subshell is incomplete is Sc, Ti, V, Cr, Mn, Fe, Co, Ni, and Cu. The second and third transition series are similarly derived from the 4d and 5d subshells.

transition state In theories describing ELEMENTARY REACTIONs, it is usually assumed that there is a transition state of more positive molar Gibbs energy between the reactants and the products through which an assembly of atoms (initially composing the MOLECULAR ENTITIES of the reactants) must pass upon going from reactants to products in either direction. In the formalism of "transition state theory" the transition state of an elementary reaction is that set of states (each characterized by its own geometry and energy) in which an assembly of atoms, when randomly placed there, would have an equal probability of forming the reactants or of forming the products of that elementary reaction. The transition state is characterized by one and only one imaginary frequency. The assembly of atoms at the transition state has been called an ACTIVATED COMPLEX. (It is not a COMPLEX according to the definition in this encyclopedia.)

It may be noted that the calculations of reaction rates by the transition-state method and based on calculated POTENTIAL-ENERGY SURFACES refer to the potential-energy maximum at the saddle point, as this is the only point for which the requisite separability of transition-state coordinates may be assumed. The ratio of the number of assemblies of atoms that pass through to the products to the number of those that reach the saddle point from the reactants can be less than unity, and this fraction is the "transmission coefficient" κ. (There are also reactions, such as the gas-phase COLLIGATION of simple RADICALs, that do not require "activation" and which therefore do not involve a transition state.)

See also GIBBS ENERGY OF ACTIVATION; HAMMOND PRINCIPLE; POTENTIAL-ENERGY PROFILE; TRANSITION STRUCTURE.

transition-state analog A compound that mimics the transition state of a substrate bound to an ENZYME.

transition structure A saddle point on a POTENTIAL-ENERGY SURFACE. It has one negative force constant in the harmonic force-constant matrix.

See also ACTIVATED COMPLEX; TRANSITION STATE.

transmission coefficient *See* TRANSITION STATE.

transpiration The loss of water from a plant or tree by way of the stomata.

transport control *See* MICROSCOPIC DIFFUSION CONTROL.

transuranium element A radioactive element with atomic number greater than that of uranium (at. no. 92). Up to and including fermium (at. no. 100), the transuranium elements are produced by the capture of neutrons. These include neptunium (at. no. 93), plutonium (at. no. 94), americium (at. no. 95), curium (at. no. 96), berkelium (at. no. 97), californium (at. no. 98), einsteinium (at. no. 99), fermium (at. no. 100), mendelevium (at. no. 101), nobelium (at. no. 102), and lawrencium (at. no. 103), rutherfordium, dubnium, and seaborgium (at. no. 104, 105, and 106, respectively), bohrium (at. no. 107), hassium, meitnerium, ununnilium, unununium, and ununbium (at. no. 108 through 112, respectively), ununquadium (at. no. 114), and ununhexium (at. no. 116).

trapping The interception of a reactive molecule or reaction intermediate so that it is removed from the system or converted into a more STABLE form for study or identification.

triboluminescence Optical phenomenon in which light is generated via friction, e.g., scratching a peppermint in the dark. The term comes from the Greek *tribein* (to rub) and the Latin *lumin* (light). Some minerals glow when you scratch them.

triple bond A chemical bond consisting of three distinct covalent bonds linking two atoms in a molecule.

triple point Temperature and pressure at which the three phases of a substance are in equilibrium.

tritium An isotope of hydrogen that has one proton and two neutrons.

trophic structure The distribution of the energy flow and its relationships through the various trophic levels.

tropism The movement of a plant toward (positive tropism) or away (negative tropism) from an environmental stimulus by elongating cells at different rates. Phototropism is induced by light; gravitropism is

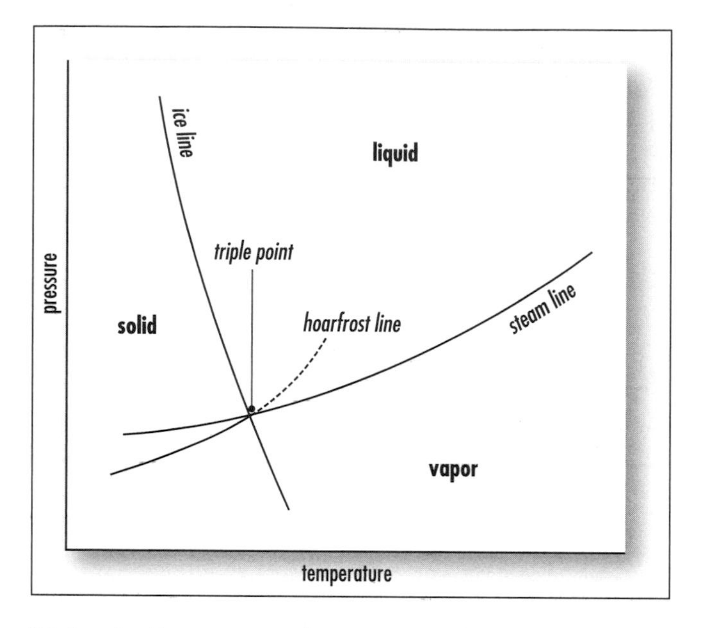

Triple point. Temperature and pressure at which the three phases of a substance are in equilibrium

induced by gravitational pull; hydrotropism is a response to water gradients; and thigmotropism is a response to touch.

troposphere The lowest layer of the atmosphere. It ranges in thickness from 8 km at the poles to 16 km over the equator and is bounded above by the tropopause, a boundary marked by stable temperatures. The troposphere is the layer where most of the world's weather takes place.

Trouton's rule At the normal boiling temperature, the entropy of vaporization is constant.

tunneling The process by which a particle or a set of particles crosses a barrier on its POTENTIAL-ENERGY SURFACE without having the energy required to surmount this barrier. Since the rate of tunneling decreases with increasing reduced mass, it is significant in the context of ISOTOPE EFFECTS of hydrogen isotopes.

turgor pressure Like air pressure in a car tire, it is the outward pressure that is exerted against the inside surface of a plant cell wall under the conditions of water flowing into the cell by osmosis, and the resulting resistance by the cell wall to further expansion.

Tyndall effect Light scattering by colloidal particles.

type 1,2,3 copper Different classes of copper-BINDING SITES in proteins, classified by their spectroscopic properties as Cu(II). In type 1, or BLUE COPPER centers, the copper is COORDINATED to at least two imidazole nitrogens from histidine and one sulfur from cysteine. They are characterized by small copper HYPERFINE couplings and a strong visible absorption in the Cu(II) state. In type 2, or non-blue copper sites, the copper is mainly bound to imidazole nitrogens from histidine. Type 3 copper centers comprise two spin-coupled copper ions bound to imidazole nitrogens.

tyrosinase A copper protein containing an antiferromagnetically coupled DINUCLEAR copper unit (TYPE 3–like site) that oxygenates the tyrosine group to catechol and further oxidizes this to the quinone.

tyrosine kinase (PTK) Protein enzymes that modulate a wide variety of cellular events, including differentiation, growth, metabolism, and apoptosis. Protein kinases add phosphate groups to proteins. Enzymes that add phosphate groups to tyrosine residues are called protein tyrosine kinases. These enzymes have important roles in signal transduction and regulation of cell growth, and their activity is regulated by a set of molecules called protein tyrosine phosphatases that remove the phosphate from the tyrosine residues. A tyrosine kinase is an enzyme that specifically phosphorylates (attaches phosphate groups to) tyrosine residues in proteins and is critical in T- and B-cell activation.

tyrosine kinase receptor Proteins found in the plasma membrane of the cell that can phosphorylate (attach phosphate groups to) a tyrosine residue in a protein. Insulin is an example of a hormone whose receptor is a TYROSINE KINASE. Following binding of the hormone, the receptor undergoes a conformational change, phosphorylates itself, then phosphorylates a variety of intracellular targets.

U

ultrasound The use of sound with a frequency higher than 20,000 Hz. Used to obtain images for medical diagnostic purposes, especially during pregnancy.

ultraviolet radiation The energy range just beyond the violet end of the visible spectrum. Although ultraviolet radiation constitutes only about 5 percent of the total energy emitted from the sun, it is the major energy source for the stratosphere and mesosphere, playing a dominant role in both energy balance and chemical composition.

Most ultraviolet radiation is blocked by Earth's atmosphere, but some solar ultraviolet radiation penetrates the ozone and aids in plant photosynthesis and the production of vitamin D in humans. Too much ultraviolet radiation can burn the skin, cause skin cancer and cataracts, and damage vegetation.

ultraviolet spectrum (UV) The electromagnetic spectrum beyond the violet end of the visible spectrum. Wavelengths in the 400-nm or less range.

umpolung Any process by which the normal alternating donor and acceptor reactivity pattern of a chain, which is due to the presence of O or N heteroatoms, is interchanged. Reactivity umpolung is most often achieved by temporary exchange of heteroatoms (N, O) by others, such as P, S, and Se.

The original meaning of the term has been extended to the reversal of any commonly accepted reactivity pattern. For example, reaction of R–C≡CX (X = halide) as a synthon for "R–C≡C$^+$ (i.e., ELECTROPHILIC acetylene) is an umpolung of the normal, more common acetylide, R–C≡C$^-$ (i.e., NUCLEOPHILIC) reactivity.

unified atomic mass unit (**u or m_u**) A unit of mass defined as the mass of one atom of ^{12}C divided by 12. Its approximate value is u = (1.660 565±0.000 008 6) Å E10^{-27} kg.

unimolecular *See* MOLECULARITY.

unit cell The smallest repeating unit of a crystalline solid that can be used to describe the entire structure. Can be used as a template and copied to produce an entire crystal.

unreactive Failing to react with a specified CHEMICAL SPECIES under specified conditions. The term should not be used in place of STABLE, since a relatively more stable species may nevertheless be more REACTIVE than some reference species toward a given reaction partner.

unsaturated fatty acids Fatty acids are the essential building blocks of all fats in our food supply and body, but not all of them have beneficial results. There are five major fatty acid types: saturated (SAFA), unsaturated (UFA), monounsaturated (MUFA), polyunsaturated (PUFA), and essential (EFA).

An unsaturated fatty acid is a long-chain CARBOXYLIC ACID that contains one or more carbon C=C double bonds. They occur when all of the carbons in a chemical chain are not saturated with hydrogen so that the fat molecule contains one or more double bonds and produces a fat that is fluid at room temperature.

There are three types of unsaturated fatty acids: monounsaturated, such as oleic acid found in olive and sesame oils that contain one double bond; polyunsaturated, such as corn, soybean, and sunflower oils that contain more than one double bond; and essential fatty acids (EFA) that are important, although they cannot be created in the body. They are linoleic acid (LA) and alpha-linolenic acid (LNA).

The double bonds in a molecule of an unsaturated fatty acid can be found in two forms known as CIS and TRANS.

Cis double bonds produce a kink, or a bend, of about 30 degrees for each double bond into the backbone, and these can flip over to the *trans* form under high temperatures. *Trans* double bonds allow the molecule to lie in a straight line; however, the human body cannot convert the *trans* form into nutrients and so prevents the metabolic activities from converting it to the active *cis* forms. This can lead to a deficiency in essential fatty acids. The more double bonds, and therefore more kinks, the more beneficial it is to human health. By completely changing the physical and chemical properties, the kinks allow essential protein associations to form more easily, thus permitting more saturated fatty acids to disperse and interact with water or blood.

See also SATURATED FATTY ACID.

unsaturated hydrocarbon A hydrocarbon (organic compound) that contains double (alkenes) or triple (alkynes) carbon-carbon bonds.

unsaturated solution A solution that contains less than the maximum possible equilibrium concentration of a solute.

unstable The opposite of STABLE, i.e., the CHEMICAL SPECIES concerned has a higher molar GIBBS FREE ENERGY than some assumed standard. The term should not be used in place of REACTIVE or TRANSIENT, although species that are more reactive or transient are frequently also more unstable. (Very unstable chemical species tend to undergo exothermic UNIMOLECULAR decompositions. Variations in the structure of the related chemical species of this kind generally affect the energy of the TRANSITION STATES for these decompositions less than they affect the stability of the decomposing chemical species. Low stability may therefore parallel a relatively high rate of unimolecular decomposition.)

upfield *See* CHEMICAL SHIFT.

urea A nitrogen-containing waste product of metabolism. A result of the normal breakdown of protein in the liver in mammals. It is created in liver cells from ammonia and carbon dioxide and carried via the bloodstream to the kidneys, where it is excreted in the urine. Urea accumulates in the body of people with renal failure. Urea is also a synthetic source of nitrogen made from natural gas.

urease A nickel ENZYME, urea amidohydrolase catalyzes the HYDROLYSIS of UREA to ammonia and carbon dioxide. The ACTIVE SITE comprises two Ni(II) ions, bridged by a carbamate.

Urey, Harold (1893–1981) American *Physical chemist, geophysicist* Harold Clayton Urey was born in Walkerton, Indiana, on April 29, 1893, to Rev. Samuel Clayton Urey and Cora Rebecca Reinsehl. His early education was in rural schools, and he graduated from high school in 1911 and taught for three years in country schools.

He entered the University of Montana in 1914 and received a B.S. degree in zoology in 1917. In 1921 he entered the University of California and received a Ph.D. in chemistry in 1923.

In 1924 he was a fellow in Copenhagen at Professor Niels Bohr's Institute for Theoretical Physics as the American-Scandinavian Foundation Fellow to Denmark. He then became an associate in chemistry at Johns Hopkins University, where he championed the application of quantum mechanics to chemistry. He married Frieda Daum in 1926 and they had four children. In 1929 he became associate professor in chemistry at Columbia University, finally reaching full professor in 1934 (to 1945).

In 1931, with F. Brickwedde and G. Murphy, he discovered deuterium, or heavy hydrogen, for which Urey was awarded the 1934 Nobel Prize in chemistry. He and his team also discovered isotopes of oxygen, carbon, nitrogen, and sulfur.

During the war period (1940–45), though a pacifist, he was director of war research on the Manhattan Project at Columbia University, and after the war he moved to the Institute for Nuclear Studies, University of Chicago, in 1945 as distinguished service professor of chemistry and as Martin A. Ryerson professor in 1952. He became an advocate of nuclear arms control, working with other scientists to promote global cooperation and to prevent nuclear proliferation and conflict. His other interests were research on the origin and evolution of life on Earth and the solar system. He also calculated the temperature of ancient oceans by measuring the isotopic quantities in fossils.

Urey authored *Atoms, Molecules and Quanta* (1930, with A. E. Ruark) and *The Planets* (1952), and he was editor of the *Journal of Chemical Physics* (1933–40). He was awarded numerous honors, prizes, and awards throughout his lifetime.

He died on January 5, 1981.

uric acid The end result of urine breakdown, a product of protein metabolism, that is the major pathway for excreting metabolic nitrogen out of the body. Too much uric acid in the blood and its salts in joints lead to gout, which causes pain and swelling in the joints. When urine contains too much uric acid, "kidney" or uric-acid stones can develop.

valence The maximum number of univalent atoms (originally hydrogen or chlorine atoms) that can combine with an atom of the element under consideration, or with a fragment, or for which an atom of this element can be substituted.

valence bond theory Covalent bonds are formed when atomic orbitals on different atoms overlap and the electrons are shared.

valence electrons Outermost electrons of atoms; often involved in bonding.

valence isomer A constitutional isomer interrelated with another by PERICYCLIC REACTIONs. For example, Dewar benzene, prismane, and benzvalene are valence isomers of benzene.

See also TAUTOMERISM.

valence shell Valence electrons are the electrons located in the outermost, highest-energy orbits or SHELL of an atom. The shell is more of a field density and indicates the region where the electrons are located. The valence electrons determine the chemical properties of an element, since it is these valence electrons that are gained or lost during a chemical reaction.

valence-shell electron-pair repulsion theory The repulsion between pairs of valence electrons is used to predict the shape or geometry of a molecule.

valence tautomerization The term describes simple reversible and generally rapid ISOMERIZATIONs or DEGENERATE REARRANGEMENTs involving the formation and rupture of single and/or double BONDs, without MIGRATION of atoms or GROUPs; for example,

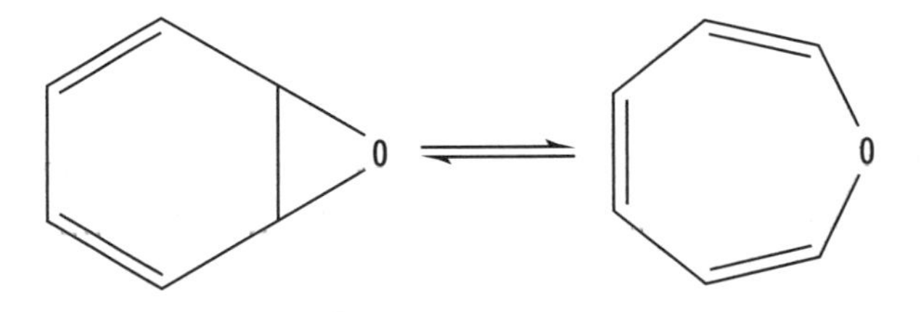

See also FLUXIONAL; TAUTOMERISM.

van der Waals equation Extends the IDEAL GAS LAW to real gases by including two empirically determined parameters, which are different for different gases.

van der Waals forces The attractive or repulsive forces between MOLECULAR ENTITIES (or between groups within the same molecular entity) other than those due to BOND formation or to the electrostatic interaction of ions or of ionic GROUPs with one another or with neutral molecules. The term includes: DIPOLE-

DIPOLE, DIPOLE-INDUCED DIPOLE, and LONDON FORCES (instantaneous induced dipole-induced dipole forces).

The term is sometimes used loosely for the totality of nonspecific attractive or repulsive intermolecular forces.

van der Waals radius Radius of an imaginary hard sphere used to model the atom, determined from measurements of atomic spacing between pairs of unbonded atoms in crystals. Named for Johannes Diderik van der Waals, winner of the 1910 Nobel Prize in physics.

vapor A gas created by boiling or evaporating a liquid.

vapor pressure Pressure of a vapor at the surface of its parent liquid.

ventilation Ventilation is the passage of air into and out of the respiratory tract. Ventilation exchange (VE) is the exchange of oxygen and carbon dioxide and other gases during the passage of air into and out of the respiratory passages.

visible light The portion of the electromagnetic spectrum that we perceive. This excludes radio waves, microwaves, infrared light, ultraviolet light, X rays, and gamma rays. Each of the visible light wavelengths are detected as various colors (red, orange, yellow, green, blue, indigo, and violet, along with various combinations and shades of these colors) by the human eye based on their wavelength, ranging in wavelength from about 400 nm to about 700 nm. It travels at the same speed as all other radiation (186,000 miles per second), and its wavelength is longer than ultraviolet light but shorter than X rays. Violet has the shortest wavelength, while red has the longest.

See also ENERGY.

vitamin An essential organic nutrient that is needed in small amounts by an organism for metabolism and other processes. Organisms either synthesize them or obtain them in other ways. Examples of vitamins are vitamin C and vitamin E, both antioxidants. A vitamin usually functions as a coenzyme or a component of a coenzyme and is soluble in either water or organic solvent. The lack of certain vitamins can lead to disease such as in rickets (vitamin D), tooth decay (vitamin K), bone softening (vitamin D), or night blindness (vitamin A). Other vitamins include vitamin B1 (thiamin), vitamin B2 (riboflavin), niacinamide (niacin-vitamin B3), vitamin B6 (pyridoxine), vitamin B12 (COBALAMIN), pantothenic acid (vitamin B5), pyridoxal (vitamin B6), phylloquinone (vitamin K), biotin, folic acid, inositol, choline, and PABA (para amino benzoic acid). Vitamin supplements are a billion-dollar-a-year industry.

Scanning electron micrograph of crystals of vitamin C (ascorbic acid) showing monoclinic crystal structure (characterized by three unequal axes, one pair of which are not at right angles). Magnification ×140. ***(Courtesy of Dr. Jeremy Burgess/Science Photo Library)***

vitamin B12 *See* COBALAMIN.

volt The electromotive force that causes current to flow. One volt equals one joule of energy per coloumb of charge.

voltage Potential difference between two electrodes.

voltage-gated channel Ion channels are pores in cell membranes that allow the passage of ions in and out of cells. There are two types, voltage-gated and chemically gated channels. The opening and closing of the voltage-dependent ion channels is regulated by voltage, the electrical charge or potential difference between the inside and outside of the membrane, while chemical stimuli are responsible for opening and closing the chemically gated channels. Neurons use these channels to pass sodium and potassium ions through them.

voltaic cells Electrochemical cells in which spontaneous chemical reactions produce electricity. Common household batteries are of this type. Also called galvanic cells.

volume of activation ($\Delta^{‡}V$) A quantity derived from the pressure dependence of the RATE CONSTANT of a reaction (mainly used for reactions in solution), defined by the equation

$$\Delta^{‡}V = -RT(\partial \ln k/\partial p)T$$

providing that the rate constants of all reactions (except first-order reactions) are expressed in pressure-independent concentration units, such as mol dm^{-3} at a fixed temperature and pressure. The volume of activation is interpreted, according to TRANSITION STATE theory, as the difference between the partial molar volumes of the transition state ($^{‡}V$) and the sums of the partial volumes of the reactants at the same temperature and pressure, i.e.,

$$\Delta^{‡}V = {}^{‡}V - \Sigma\,(\mathrm{r}V_{\mathrm{R}})$$

where r is the order in the reactant R and V_R its partial molar volume.

VSEPR (valence-shell electron-pair repulsion) A theory that allows the prediction of shapes of simple polyatomic molecules by applying a set of rules.

W

Waksman, Selman Abraham (1888–1973) American *Biochemist* Selman Abraham Waksman was born in Priluka, near Kiev, Russia, on July 22, 1888, to Jacob Waksman and Fradia London. He received his early education from private tutors and took school training in Odessa in an evening school, also with private tutors.

In 1911 he entered Rutgers College, having won a state scholarship the previous spring, and received a B.S. in agriculture in 1915. He was appointed research assistant in soil bacteriology at the New Jersey Agricultural Experiment Station and continued graduate work at Rutgers, obtaining an M.S. in 1916, the year he became a naturalized U.S. citizen. In 1918 he was appointed a research fellow at the University of California, where he received his Ph.D. in biochemistry the same year.

He was invited back to Rutgers and by 1930 was a professor. When the department of microbiology was organized in 1940, he became professor of microbiology and head of the department, and nine years later was appointed director of the Institute of Microbiology. He retired in 1958.

Waksman brought medicine from the soil. By studying soil-based acintomycetes, he was able to extract a number of antibiotics such as actinomycin (1940), clavacin and streptothricin (1942), streptomycin (1943), grisein (1946), neomycin (1948), fradicin, candicidin, candidin, and more. His discovery of streptomycin, the first effective treatment against tuberculosis, brought him the 1952 Nobel Prize for physiology or medicine.

He published more than 400 scientific papers and has written, alone or with others, 18 books, including *Principles of Soil Microbiology* (1927) and *My Life with the Microbes* (1954), an autobiography. He was a member of numerous scientific organizations. In 1950 he was made commander of the French Légion d'Honneur. He died on August 16, 1973, in Hyannis, Massachusetts.

Warburg, Otto Heinrich (1883–1970) German *Biochemist* Otto Heinrich Warburg was born on October 8, 1883, in Freiburg, Baden, to physicist Emil Warburg. He studied chemistry under EMIL FISCHER and received his doctor of chemistry from the University of Berlin in 1906, and a doctor of medicine from the University of Heidelberg in 1911.

In 1918 he was appointed professor at the Kaiser Wilhelm Institute for Biology, Berlin-Dahlem, and from 1931 to 1953 he was director of the Kaiser Wilhelm Institute for Cell Physiology (now Max Planck Institute) in Berlin.

He specialized in the investigation of metabolism in tumors and respiration of cells. He discovered that flavins and the nicotinamides were the active groups of the hydrogen-transferring enzymes, and his early discovery of iron-oxygenase provided details of oxidation and reduction (redox reactions) in living organisms. For his discovery of the nature and mode of action of the respiratory enzymes that enable cells to process oxygen, he was awarded the Nobel Prize in 1931. He was offered a second Nobel Prize in 1944 for his enzyme work but was not allowed to accept it, since he was living under the Hitler regime. He later discovered how the conversion of light energy to chemical energy is activated in photosynthesis. He even showed the carcinogenic nature of food additives and cigarette smoke and demonstrated how cancer cells are destroyed by radiation during the 1930s.

Warburg is the author of *New Methods of Cell Physiology* (1962). He died on August 1, 1970.

water Two atoms of hydrogen and one atom of oxygen.

water cycle The process by which water is transpired and evaporated from the land and water, condensed in the clouds, and precipitated out onto the earth once again to replenish the bodies of water on the Earth.

water equivalent The amount of water that would absorb the same amount of heat as the calorimeter per degree of temperature increase.

water of crystallization Water that is in chemical combination with a crystal. While it is necessary for the maintenance of crystalline properties, it is capable of being removed by sufficient heat; water that is present in hydrated compounds.

water potential Direction of water flow based on solute concentration and pressure. OSMOSIS is an example, which is the DIFFUSION of water across a semipermeable barrier such as a cell membrane, from high water potential to lower water potential.

It is also a measure of the moisture stress in plants or soil, measured in megapascals. A more negative value indicates greater moisture stress. Soils with no moisture stress have a water potential of 0 to –1 mPa. Two methods of measuring soil water potential are the heat-dissipation method or the electrical-resistance method.

water table The level below the Earth's surface at which the ground becomes saturated with water; usually mimics the surface contour and is set where hydrostatic pressure equals atmospheric pressure.

water vapor Water present in the atmosphere in gaseous form. Water vapor is an important part of the natural GREENHOUSE EFFECT. While humans are not significantly increasing its concentration, it contributes to the enhanced greenhouse effect because the warming influence of greenhouse gases leads to a positive water vapor feedback. In addition to its role as a natural greenhouse gas, water vapor plays an important role in regulating the temperature of the planet because clouds form when excess water vapor in the atmosphere condenses to form ice and water droplets and precipitation.

Watson-Crick model A DNA molecule consisting of two polynucleotide strands coiled around each other in a helical "twisted ladder" structure.

wave function A function of the coordinates of an electron's position in three-dimensional space that describes the properties of the electron.

wavelength The physical distance between points of a corresponding phase of two consecutive cycles of a wave.

weak acid/base A BRONSTED ACID that only partially dissociates into hydrogen ions and anions in solution. A weak BASE only partially reacts to form ions in solution, e.g., ammonia.

weak electrolyte Any substance that poorly conducts electricity in an aqueous solution; gives a low-percentage yield of ions when dissolved in water.

wetting agent SURFACTANT for use in spray formulations to assist dispersion of a powder in the diluent or spreading of spray droplets on surfaces. A surface-active agent that promotes wetting by decreasing the cohesion within a liquid (i.e., reduces the SURFACE TEN-

SION), and therefore increases its adhesion to a solid surface.

Wheland intermediate *See* MEISENHEIMER ADDUCT; SIGMA (σ) ADDUCT.

Whipple, George Hoyt (1878–1976) American *Pathologist* George Hoyt Whipple was born on August 28, 1878, in Ashland, New Hampshire, to Dr. Ashley Cooper Whipple and Frances Hoyt. Whipple was educated at Phillips Academy in Andover and received a B.A. at Yale University in 1900. He then completed Johns Hopkins University and received his M.D. degree in 1905 and was appointed an assistant in pathology at the Johns Hopkins Medical School. In 1914 he was appointed professor of research medicine at the University of California Medical School and director of the Hooper Foundation for Medical Research at that university, serving as dean of the medical school during the years 1920 and 1921. In 1921 he was appointed professor of pathology and dean of the School of Medicine and Dentistry at the University of Rochester and became the founding dean of the university's School of Medicine (1921–53) and chair of the pathology department.

Whipple's main researches were concerned with anemia and the physiology and pathology of the liver. In 1908 he began a study of bile pigments that led to his interest in the body's manufacture of the oxygen-carrying hemoglobin, an important element in the production of bile pigments. His studies dealt with the effect of foods on the regeneration of blood cells and hemoglobin in 1918. Between 1923 and 1925, his experiments in artificial anemia were instrumental in determining that iron is the most potent inorganic factor to form red blood cells.

For his work on liver research and treatment of anemia he was awarded, together with George R. Minot and William P. Murphy, the Nobel Prize for physiology or medicine in 1934. Whipple published many scientific papers in physiological journals.

He died on February 2, 1976, at Rochester, New York. His birthplace home on Pleasant Street in Ashland was listed on the National Register in 1978.

Wilson's disease An inherited condition in which copper fails to be excreted in the bile. Copper accumulates progressively in the liver, brain, kidney, and red blood cells. As the amount of copper accumulates, hemolytic anemia, chronic liver disease, and a neurological syndrome develop.

See also CHELATION THERAPY.

Woodward-Hoffmann rules *See* ORBITAL SYMMETRY.

work The movement of an object against some force over a distance.

work function Energy needed to remove an electron from the Fermi level (energy of the highest occupied state at zero temperature) in a metal to a point at infinite distance away outside the surface. The minimum energy that must be supplied to extract an electron from a solid.

XANES (X-ray absorption near-edge structure) *See* EXTENDED X-RAY ABSORPTION FINE STRUCTURE.

xenobiotic A xenobiotic (Greek *xenos* "foreign" and *bios* "life") is a compound that is foreign to a living organism. Principal xenobiotics include: drugs, carcinogens, and various compounds that have been introduced into the environment by artificial means.

X ray Electromagnetic radiation of shorter wavelength than ultraviolet radiation (10^{-11} m to 10^{-9} m or 0.01 nm to 1 nm [0.00001 to 3,000 angstroms]) produced by bombardment of atoms with high-quantum-energy particles. Can pass through many forms of matter and is used medically and industrially to examine internal structures of organs, bodies, and materials.

X-ray absorption near-edge structure (XANES) *See* EXTENDED X-RAY ABSORPTION FINE STRUCTURE.

X-ray diffraction An analytical technique used to determine the structures of crystalline solids.

Y

yeast Mothers and brewers have been familiar with yeast for centuries. It is considered to be the oldest "plant" cultivated by humans. Yeast is a unicellular fungus that belongs to the family Saccharomycetaceae. It lives in the soil, on plants, and in the air, and it has been used in the production of bread, beer, and wine because it is responsible for the process of fermentation. It produces carbon dioxide and alcohol when in the presence of sugar. There are actually many species of yeasts.

yield The quantity of product obtained from a chemical reaction.

ylide A CHEMICAL SPECIES produced (actually or notationally) by loss of a HYDRON from an atom directly attached to the central atom of an ONIUM ION, e.g.,

$$Ph_3P^+\text{–}CHRR' \rightarrow (Ph_3P^+\text{–}C^-RR' \underset{\text{ylide}}{\longleftrightarrow} Ph_3P{=}CRR')$$

Yukawa-Tsuno equation A multiparameter extension of the HAMMETT EQUATION to quantify the role of enhanced RESONANCE effects on the reactivity of meta- and para-substituted benzene derivatives, for example,

$$\lg k = \lg k_o + \rho[\sigma + r(\sigma^+ - \sigma)]$$

The parameter r gives the enhanced resonance effect on the scale $(\sigma^+ – \sigma)$ or $(\sigma^- – \sigma)$, respectively.

See also RHO (ρ) VALUE; SIGMA (σ) CONSTANT.

Z

zeolite A group of hydrated aluminosilicate minerals that occur in nature or are manufactured. They are crystalline compounds composed of silicon, oxygen, and some aluminum atoms. Zeolite ion-exchange units are extensively used in water conditioning to remove calcium and magnesium ions by replacing them with sodium ions. Zeolites are used for separating mixtures by selective absorption.

zero-order reaction A reaction whose rate is independent of the concentration of reactants.

zero-point energy The energy of the lowest state of a quantum system.

Ziegler-Natta catalyst (ZNC) A catalyst involved in creating highly structured (stereoregular) polyolefin polymer chains. Used to mass-produce polyethylene and polypropylene.

zinc finger A DOMAIN, found in certain DNA-binding proteins, comprising a HELIX-loop structure in which a zinc ion is COORDINATED to 2–4 cysteine sulfurs, the remaining LIGANDs being histidines. In many proteins of this type, the domain is repeated several times.

zone refining A technique for producing solids of extreme purity. By moving a heater slowly along a bar of the material to be refined, a molten region is formed that carries impurities with it along the bar.

Zucker-Hammett hypothesis This hypothesis states that, if in an acid-catalyzed reaction, lg k_1 (first-order RATE CONSTANT of the reaction) is linear in H_o (Hammett ACIDITY FUNCTION), water is not involved in the TRANSITION STATE of the RATE-CONTROLLING STEP. However, if lg k_1 is linear in lg$[H^+]$, then water is involved. This has been shown to be incorrect by Hammett.

Z-value An index of the IONIZING POWER of a solvent based on the frequency of the longest-wavelength electronic absorption maximum of 1-ethyl-4-methoxycarbonylpyridinium iodide in the solvent. The Z-value is defined by

$$Z = 2.859 \times 10^4/\lambda$$

where Z is in kcal mol^{-1} and λ is in nm.

See also DIMROTH-REICHARDT E_T PARAMETER; GRUNWALD-WINSTEIN EQUATION.

zwitterionic compound A neutral compound having electrical charges of opposite sign, delocalized or not, on adjacent or nonadjacent atoms. Zwitterionic compounds have no uncharged canonical representations. Sometimes referred to as inner salts, ampholytes, dipolar ions (a misnomer). For example: $H_3N^+CH_2C(=O)O^-$, glycine.

See also YLIDE.

APPENDIX I

BIBLIOGRAPHY

Aboul-Enein, Hassan Y., Raluca-Ioana Stefan, and George Baiulescu. *Quality and Reliability in Analytical Chemistry.* Analytical Chemistry Series. Boca Raton, Fla.: CRC Press, 2001.

Adams, Dave, Paul J. Dyson, and Stewart Tavener. *Chemistry in Alternative Reaction Media.* Hoboken, N.J.: Wiley, 2004.

Adams, Donald D., and Walter P. Page. *Acid Deposition: Environmental, Economic, and Policy Issues.* New York: Plenum Press, 1985.

Adams, D. M. *Inorganic Solids; an Introduction to Concepts in Solid-State Structural Chemistry.* New York: Wiley, 1974.

Agarwala, R. P. *Soft Chemistry Leading to Novel Materials, Diffusion and Defect Data.* Defect and Diffusion Forum, vol. 191, part A. Enfield, N.H.: Scitech Publications, 2001.

Aggeli, Amalia, Neville Boden, and Shuguang Zhang. *Self-Assembling Peptide Systems in Biology, Medicine, and Engineering.* Boston: Kluwer Academic Publishers, 2001.

Agosta, William C. *Thieves, Deceivers, and Killers: Tales of Chemistry in Nature.* Princeton, N.J.: Princeton University Press, 2001.

Alder, John F., John G. Baker, and Royal Society of Chemistry (Great Britain). *Quantitative Millimetre Wavelength Spectrometry.* RSC Analytical Spectroscopy Monographs. Cambridge, U.K.: Royal Society of Chemistry, 2002.

Alder, Roger W., Ray Baker, and John Michael Brown. *Mechanism in Organic Chemistry.* New York: Wiley-Interscience, 1971.

Allcock, H. R., James E. Mark, and Frederick Walter Lampe. *Contemporary Polymer Chemistry.* 3rd ed. Upper Saddle River, N.J.: Prentice Hall, 2003.

Allen, D. W., John C. Tebby, and ebrary Inc. *Organophosphorus Chemistry.* Specialist Periodical Reports. Cambridge, U.K.: Royal Society of Chemistry, 2002.

———. *Organophosphorus Chemistry.* Specialist Periodical Reports. Cambridge, U.K.: Royal Society of Chemistry, 2001.

———. *Organophosphorus Chemistry.* Specialist Periodical Reports. Cambridge, U.K.: Royal Society of Chemistry, 2000.

———. *Organophosphorus Chemistry.* Specialist Periodical Reports. Cambridge, U.K.: Royal Society of Chemistry, 1999.

Allen, G. *Molecular Structure and Properties.* Vol. 2 of *Physical Chemistry.* Series 1. Baltimore, University Park Press, 1972.

Alpers, Charles N., J. L. Jambor, Darrell Kirk Nordstrom, Paul H. Ribbe, and Mineralogical Society of America. *Sulfate Minerals: Crystallography, Geochemistry, and Environmental Significance.* Vol. 40 of *Reviews in Mineralogy and Geochemistry.* Washington, D.C.: Mineralogical Society of America, 2000.

Amjad, Zahid. *Advances in Crystal Growth Inhibition Technologies.* New York: Kluwer Academic/Plenum Publishers, 2000.

Anastas, Paul T., Paul H. Bickart, and Mary M. Kirchhoff. *Designing Safer Polymers.* New York: Wiley-Interscience, 2000.

Anastas, Paul T., Lauren G. Heine, and Tracy C. Williamson. *Green Chemical Syntheses and Processes.* ACS Symposium Series, no. 767. Washington, D.C.: American Chemical Society, 2000.

Anastas, Paul T., and John Charles Warner. *Green Chemistry: Theory and Practice.* New York: Oxford University Press, 1998.

Anastas, Paul T., and Tracy C. Williamson. *Green Chemistry: Frontiers in Benign Chemical Syntheses and Processes.* New York: Oxford University Press, 1998.

Antonietti, M. *Colloid Chemistry.* Topics in Current Chemistry, no. 226–227. New York: Springer, 2003.

Archer, Ronald D. *Inorganic and Organometallic Polymers.* Special Topics in Inorganic Chemistry. New York: Wiley-VCH, 2001.

Askadskii, Audrey A. *Lectures on the Physico-Chemistry of Polymers.* New York: Nova Science, 2003.

Asperger, Smiljko. *Chemical Kinetics and Inorganic Reaction Mechanisms.* 2nd ed. New York: Kluwer Academic/Plenum Publishers, 2003.

Aspinall, Helen C. *Chemistry of the F-Block Elements.* Vol. 5 of *Advanced Chemistry Texts.* Australia: Gordon & Breach, 2001.

Auner, Norbert, and Johann Weis. *Organosilicon Chemistry IV: From Molecules to Materials.* New York: Wiley-VCH, 2000.

———. *Organosilicon Chemistry III: From Molecules to Materials.* New York: Wiley-VCH, 1998.

Bagdasarian, Khristofor Stepanovich. *Theory of Free-Radical Polymerization.* Jerusalem, Israel: Program for Scientific Translations, 1968.

Bagdassarov, N., Didier Laporte, and Alan Bruce Thompson. *Physics and Chemistry of Partially Molten Rocks.* Vol. 11 of *Petrology and Structural Geology.* Boston: Kluwer Academic Publisers, 2000.

Baggott, J. E. *The Meaning of Quantum Theory: A Guide for Students of Chemistry and Physics.* Oxford Science Publications. New York: Oxford University Press, 1992.

Bai, Chunli. *Scanning Tunneling Microscopy and Its Applications.* 2nd rev. ed. Springer Series in Surface Sciences, no. 32. New York: Shanghai Scientific and Technical Publishers, 2000.

Bailey, Ronald Albert. *Chemistry of the Environment.* 2nd ed. San Diego, Calif.: Academic Press, 2002.

Bandrauk, André D., and North Atlantic Treaty Organization, Scientific Affairs Division. *Atomic and Molecular Processes with Short Intense Laser Pulses.* Vol. 171 of *Physics.* NATO ASI Series, ser. B. New York: Plenum Press, 1988.

Baraton, Marie-Isabelle, and Irina Uvarova. *Functional Gradient Materials and Surface Layers Prepared by Fine Particles Technology.* Vol. 16 of *Mathematics, Physics, and Chemistry.* NATO Science Series, ser. II. Boston: Kluwer Academic Publishers, 2001.

Bar-Cohen, Yoseph. *Electroactive Polymer (EAP) Actuators as Artificial Muscles: Reality, Potential, and Challenges.* SPIE Press Monograph, PM 98. Bellingham, Wash.: SPIE Press, 2001.

Bard, Allen J., and Michael V. Mirkin. *Scanning Electrochemical Microscopy.* Monographs in Electroanalytical Chemistry and Electrochemistry. New York: Marcel Dekker, 2001.

Barrante, James R. *Applied Mathematics for Physical Chemistry.* 3rd ed. Upper Saddle River, N.J.: Pearson Prentice Hall, 2004.

Bartle, Keith D., Peter Myers, and Royal Society of Chemistry (Great Britain). *Capillary Electrochromatography.* RSC Chromatography Monographs. Cambridge, U.K.: Royal Society of Chemistry, 2001.

Baszkin, Adam, and Willem Norde. *Physical Chemistry of Biological Interfaces.* New York: Marcel Dekker, 2000.

Beaucage, Serge L. *Current Protocols in Nucleic Acid Chemistry.* New York: Wiley, 2001.

Beer, Tom, Alik Ismail-Zadeh, and North Atlantic Treaty Organization Scientific Affairs Division. *Risk Science and Sustainability: Science for Reduction of Risk and Sustainable Development of Society.* Vol. 112 of *Mathematics, Physics, and Chemistry.* NATO Science Series, ser. II. Boston: Kluwer, 2003.

Beesley, Thomas E., Benjamin Buglio, and Raymond P. W. Scott. *Quantitative Chromatographic Analysis.* Chromatographic Science Series, vol. 85. New York: Marcel Dekker, 2001.

Benedek, G., P. Milani, Victor G. Ralchenko, and North Atlantic Treaty Organization Scientific Affairs Division. *Nanostructured Carbon for Advanced Applications.* Vol. 24 of *Mathematics, Physics, and Chemistry.* NATO Science Series, ser. II. Boston: Kluwer Academic Publishers, 2001.

Benedetti-Pichler, A. A. *Introduction to the Microtechnique of Inorganic Analysis.* New York: Wiley, 1942.

Benedikt, Rudolf, and Edmund Knecht. *The Chemistry of the Coal-Tar Colours.* Technological Handbooks. London: G. Bell and sons, 1886.

Berry, R. Stephen, Stuart Alan Rice, and John Ross. *Physical Chemistry.* New York: Wiley, 1980.

Berson, Jerome A. *Chemical Discovery and the Logicians' Program: A Problematic Pairing.* Weinheim, Germany: Wiley-VCH, 2003.

Bertini, Ivano, Russell S. Drago, C. Luchinat, and North Atlantic Treaty Organization Scientific Affairs Division. *The Coordination Chemistry of Metalloenzymes: The Role of Metals in Reactions Involving Water, Dioxygen, and Related Species.* Proceedings of the NATO Advanced Study Institute, held at San Miniato, Pisa, Italy, May 28–June 8, 1982. Boston: D. Reidel, 1983.

Bertrand, G. *Carbene Chemistry: From Fleeting Intermediates to Powerful Reagents*. New York: Marcel Dekker, 2002.

Bertucco, A., and Gerhard Vetter. *High Pressure Process Technology: Fundamentals and Applications*. Vol. 9 of *Industrial Chemistry Library*. New York: Elsevier, 2001.

Best, M. D., United States Environmental Protection Agency Office of Research and Development., Lockheed Engineering and Management Services Co., Radian Corp., Northrop Services Inc., and United States Dept. of Energy. *Eastern Lake Survey, Phase 1: Quality Assurance Report*. Washington, D.C.: U.S. Environmental Protection Agency, Office of Research and Development, 1987.

Bhushan, Bharat. *Fundamentals of Tribology and Bridging the Gap between the Macro- and Micro/Nanoscales*. Vol. 10 of *Mathematics, Physics, and Chemistry*. NATO Science Series, ser. II. Boston: Kluwer Academic Publishers, 2001.

Bhushan, Nalini, and Stuart M. Rosenfeld. *Of Minds and Molecules: New Philosophical Perspectives on Chemistry*. New York: Oxford University Press, 2000.

Billing, Gert D. *Dynamics of Molecule Surface Interactions*. New York: Wiley, 2000.

Billmeyer, Fred W. *Textbook of Polymer Chemistry*. New York: Interscience Publishers, 1957.

Billo, E. Joseph. *Excel for Chemists: A Comprehensive Guide*. 2nd ed. New York: Wiley-VCH, 2001.

Birdi, K. S. *Handbook of Surface and Colloid Chemistry*. 2nd ed. Boca Raton, Fla.: CRC Press, 2003.

Birikh, Rudolph V. *Liquid Interfacial Systems: Oscillations and Instability*. Surfactant Science Series, vol. 113. New York: Marcel Dekker, 2003.

Blümich, Bernhard. *NMR Imaging of Materials*. Monographs on the Physics and Chemistry of Materials, no. 57. New York: Oxford University Press, 2000.

Böhm, Hans-Joachim, and Gisbert Schneider. *Protein-Ligand Interactions from Molecular Recognition to Drug Design*. Vol. 19 of *Methods and Principles in Medicinal Chemistry*. Weinheim, Germany: Wiley-VCH, 2003.

———. *Virtual Screening for Bioactive Molecules*. Vol. 10 of *Methods and Principles in Medicinal Chemistry*. New York: Wiley-VCH, 2000.

Böhm, M. C. *One-Dimensional Organometallic Materials: An Analysis of Electronic Structure Effects*. Lecture Notes in Chemistry, no. 45. New York: Springer-Verlag, 1987.

Bohme, Diethard K. *Chemistry and Spectroscopy of Interstellar Molecules*. Tokyo: University of Tokyo Press, 1992.

Böhme, Gernot, and Wolf Schäfer. *Finalization in Science: The Social Orientation of Scientific Progress*. Vol. 77 of *Boston Studies in the Philosophy of Science*. Boston: D. Reidel, 1983.

Bohn, Hinrich L., Brian Lester McNeal, and George A. O'Connor. *Soil Chemistry*. New York: Wiley, 1979.

Borissov, Anatoly A., L. De Luca, and Aleksandr Grigorevich Merzhanov. *Self-Propagating High-Temperature Synthesis of Materials*. Combustion Science and Technology Series, vol. 5. New York: Taylor & Francis, 2002.

Borówko, Magorzata. *Computational Methods in Surface and Colloid Science*. Surfactant Science Series, vol. 89. New York: Marcel Dekker, 2000.

Breitmaier, E. *Structure Elucidation by NMR in Organic Chemistry: A Practical Guide*. 3rd rev. ed. Hoboken, N.J.: Wiley, 2002.

Brereton, Richard G. *Chemometrics: Data Analysis for the Laboratory and Chemical Plant*. Hoboken, N.J.: Wiley, 2003.

Brock, W. H. *The Chemical Tree: A History of Chemistry*. Norton History of Science. New York: Norton, 2000.

———. *The Norton History of Chemistry*. 1st American ed. Norton History of Science. New York: W.W. Norton, 1993.

Brockington, John. *Students' Guide to Inorganic Chemistry*. London: Butterworths, 1966.

Brook, Michael A. *Silicon in Organic, Organometallic, and Polymer Chemistry*. New York: Wiley, 2000.

Brown, I. David. *The Chemical Bond in Inorganic Chemistry: The Bond Valence Model*. International Union of Crystallography Monographs on Crystallography, no. 12. New York: Oxford University Press, 2002.

Bruno, Thomas J., and Paris D. N. Svoronos. *Handbook of Basic Tables for Chemical Analysis*. 2nd ed. Boca Raton, Fla.: CRC Press, 2003.

Buchanan, Bob B., Wilhelm Gruissem, and Russell L. Jones. *Biochemistry and Molecular Biology of Plants*. Rockville, Md.: American Society of Plant Physiologists, 2000.

Bultinck, Patrick. *Computational Medicinal Chemistry for Drug Discovery*. New York: Marcel Dekker, 2004.

Buncel, E., and T. Durst. *Comprehensive Carbanion Chemistry*. Studies in Organic Chemistry, no. 5. New York: Elsevier, 1980.

Buncel, E., C. C. Lee, and N. C. Deno. *Isotopes in Molecular Rearrangements*. Vol. 1 of *Isotopes in Organic Chemistry*. New York: Elsevier, 1975.

Buncel, E., and William Hundley Saunders. *Heavy Atom Isotope Effects*. Vol. 8 of *Isotopes in Organic Chemistry*. New York: Elsevier, 1992.

Buncel, E., R. A. Stairs, and H. Wilson. *The Role of the Solvent in Chemical Reactions*. Oxford Chemistry Masters, no. 6. New York: Oxford University Press, 2003.

Burger, Alfred, and Manfred E. Wolff. *Burger's Medicinal Chemistry and Drug Discovery*. 6th ed. Edited by Donald J. Abraham. New York: Wiley, 2003.

Burger, K. *Organic Reagents in Metal Analysis.* 1st ed. International Series of Monographs in Analytical Chemistry, vol. 54. New York: Pergamon, 1973.

Burke, Robert. *Hazardous Materials Chemistry for Emergency Responders.* 2nd ed. Boca Raton, Fla.: Lewis Publishers, 2003.

Bustoz, Joaquin, Mourad Ismail, and S. K. Suslov. *Special Functions 2000: Current Perspective and Future Directions.* Vol. 30 of *Mathematics, Physics, and Chemistry.* NATO Science Series, ser. II. Boston: Kluwer Academic Publishers, 2001.

Butt, Hans-Jürgen, Karlheinz Graf, and Michael Kappl. *Physics and Chemistry of Interfaces.* Hoboken, N.J.: Wiley, 2003.

Cai, Yong, O. C. Braids, and American Chemical Society. *Biogeochemistry of Environmentally Important Trace Elements.* ACS Symposium Series, no. 835. Washington, D.C.: American Chemical Society, 2003.

Calvert, Jack G. *The Mechanisms of Atmospheric Oxidation of Aromatic Hydrocarbons.* New York: Oxford University Press, 2002.

Calvert, Jack G., and International Union of Pure and Applied Chemistry. *The Chemistry of the Atmosphere: Its Impact on Global Change.* Chemistry for the 21st Century monograph. Boston: Blackwell Scientific, 1994.

Carbó, Ramón. *Molecular Quantum Similarity in QSAR and Drug Design.* Lecture Notes in Chemistry, no. 73. New York: Springer, 2000.

Carey, Francis A., and Richard J. Sundberg. *Advanced Organic Chemistry.* 3rd ed. New York: Plenum Press, 1990.

Carley, A. F., and M. W. Roberts. *Surface Chemistry and Catalysis, Fundamental and Applied Catalysis.* New York: Kluwer Academic/Plenum Publishers, 2002.

Caroli, Sergio. *Element Speciation in Bioinorganic Chemistry.* Vol. 135 of *Chemical Analysis.* New York: John Wiley & Sons, 1996.

Caroli, Sergio, Paolo Cescon, and D. W. H. Walton. *Environmental Contamination in Antarctica: A Challenge to Analytical Chemistry.* New York: Elsevier, 2001.

Carraher, Charles E., Graham Swift, American Chemical Society Division of Polymeric Materials: Science and Engineering, and American Chemical Society. *Functional Condensation Polymers.* New York: Kluwer Academic/Plenum Publishers, 2002.

Carraher, Charles E., Minoru Tsuda, and American Chemical Society Division of Organic Coatings and Plastics Chemistry. *Modification of Polymers.* ACS Symposium Series, no. 121. Washington, D.C.: American Chemical Society, 1980.

Chalmers, John M. *Spectroscopy in Process Analysis, Sheffield Analytical Chemistry.* Boca Raton, Fla.: CRC Press, 2000.

Chang, Laura. *Scientists at Work: Profiles of Today's Groundbreaking Scientists from* Science Times. New York: McGraw-Hill, 2000.

Chien, James C. W. *Polyacetylene: Chemistry, Physics, and Material Science.* Orlando, Fla.: Academic Press, 1984.

Chikishev, Andrey Yu, Natsyianalnaia akademiia navuk Belarusi., Belaruski respublikanski fond fundamentalnykh dasledavanniau, and Society of Photo-optical Instrumentation Engineers. *Icono 2001: Novel Trends in Nonlinear Laser Spectroscopy and Optical Diagnostics and Lasers in Chemistry, Biophysics, and Biomedicine.* SPIE Proceedings Series, vol. 4749. Bellingham, Wash.: SPIE, 2002.

Choppin, Gregory R. *Nuclei and Radioactivity.* The General Chemistry Monograph Series. New York: Benjamin, 1964.

———. *Experimental Nuclear Chemistry.* Prentice-Hall Chemistry Series. Englewood Cliffs, N.J.; Prentice-Hall, 1961.

Choppin, Gregory R., Jan-Olov Liljenzin, and Jan Rydberg. *Radiochemistry and Nuclear Chemistry.* 3rd ed. Boston: Butterworth-Heinemann, 2002.

Chu, D. C. K. *Recent Advances in Nucleosides: Chemistry and Chemotherapy.* Boston: Elsevier, 2002.

Ciferri, A. *Supramolecular Polymers.* New York: Marcel Dekker, 2000.

Ciliberto, C. *Applications of Algebraic Geometry to Coding Theory, Physics, and Computation.* Vol. 36 of *Mathematics, Physics, and Chemistry.* NATO Science Series, ser. II. Boston: Kluwer Academic Publishers, 2001.

Ciliberto, E., and G. Spoto. *Modern Analytical Methods in Art and Archaeology.* Vol. 155 of *Chemical Analysis.* New York: Wiley, 2000.

Cioslowski, Jerzy. *Quantum-Mechanical Prediction of Thermochemical Data.* Vol. 22 of *Understanding Chemical Reactivity.* Boston: Kluwer Academic Publishers, 2001.

Clark, James H., and Duncan J. Macquarrie. *Handbook of Green Chemistry and Technology.* Malden, Mass.: Blackwell Science, 2002.

Considine, Douglas M. *Chemical and Process Technology Encyclopedia.* New York: McGraw-Hill, 1974.

Considine, Glenn D., and Peter H. Kulik. *Van Nostrand's Scientific Encyclopedia.* 9th ed. New York: Wiley-Interscience, 2002.

Consonni, Viviana, and Roberto Todeschini. *Handbook of Molecular Descriptors.* Vol. 11 of *Methods and Principles in Medicinal Chemistry.* New York: Wiley-VCH, 2000.

Cork, David G., and Tohru Sugawara. *Laboratory Automation in the Chemical Industries.* New York: Marcel Dekker, 2002.

Cotton, F. Albert. *Advanced Inorganic Chemistry.* 6th ed. New York: Wiley, 1999.

———. *Progress in Inorganic Chemistry.* Vol. 1. New York: Wiley, 1959.

Crabtree, Robert H. *The Organometallic Chemistry of the Transition Metals.* 3rd ed. New York: Wiley, 2001.

Cramer, Christopher J. *Essentials of Computational Chemistry: Theories and Models.* New York: Wiley, 2002.

Cramer, W. A., and D. B. Knaff. *Energy Transduction in Biological Membranes: A Textbook of Bioenergetics.* Springer Advanced Texts in Chemistry. New York: Springer-Verlag, 1990.

Cundari, Thomas R. *Computational Organometallic Chemistry.* New York: Marcel Dekker, 2001.

Cuyper, Marcel de, and Jeff W. M. Bulte. *Physics and Chemistry Basis of Biotechnology.* Vol. 7 of *Focus on Biotechnology.* Boston: Kluwer Academic Publishers, 2001.

Czanderna, Alvin Warren. *Methods of Surface Analysis.* Vol. 1 of *Methods and Phenomena, Their Applications in Science and Technology.* New York: Elsevier, 1975.

Czepulkowski, B. H. *Analyzing Chromosomes, The Basics.* New York: Springer-Verlag, 2001.

Dabrowski, Jarek, and Hans-Joachim Müssig. *Silicon Surfaces and Formation of Interfaces: Basic Science in the Industrial World.* River Edge, N.J.: World Scientific, 2000.

Daintith, John. *A Dictionary of Chemistry.* 4th ed. Oxford: Oxford University Press, 2000.

Davidson, George, and ebrary Inc. *Spectroscopic Properties of Inorganic and Organometallic Compounds: A Review of the Literature Published up to Late 2000.* Cambridge, U.K.: Royal Society of Chemistry, 2001.

———. *Spectroscopic Properties of Inorganic and Organometallic Compounds.* Specialist Periodical Reports. Cambridge, U.K.: Royal Society of Chemistry, 2000.

———. *Spectroscopic Properties of Inorganic and Organometallic Compounds: A Review of the Literature Published up to Late 1998.* Specialist Periodical Reports. Cambridge, U.K.: Royal Society of Chemistry, 1999.

———. *Spectroscopic Properties of Inorganic and Organometallic Compounds: A Review of the Literature Published up to Late 1996.* Specialist Periodical Report, no. 31. Cambridge, U.K.: The Royal Society of Chemistry, 1998.

Davies, J. S., and ebrary Inc. *Amino Acids, Peptides and Proteins.* Cambridge, U.K.: Royal Society of Chemistry, 2001.

———. *Amino Acids, Peptides and Proteins.* Chapman and Hall Chemistry Sourcebooks. New York: Chapman and Hall, 2000.

———. *Amino Acids, Peptides and Proteins.* Cambridge, U.K.: Royal Society of Chemistry, 1999.

———. *Amino Acids, Peptides and Proteins.* Chapman and Hall Chemistry Sourcebooks. New York: Chapman and Hall, 1998.

Dean, P. M., Richard A. Lewis, and ebrary Inc. *Molecular Diversity in Drug Design.* London: Kluwer Academic, 2002.

De Bièvre, Paul, and Helmut Günzler. *Measurement Uncertainty in Chemical Analysis.* New York: Springer, 2003.

DeLaat, John C., and NASA Glenn Research Center. *Active Combustion Control for Aircraft Gas Turbine Engines.* NASA/TM 2000-210346. Hanover, Md.: NASA Center for Aerospace Information, 2000.

Demaison, J., Kamil Sarka, Edward A. Cohen, and North Atlantic Treaty Organization, Scientific Affairs Division. *Spectroscopy from Space.* Vol. 20 of *Mathematics, Physics, and Chemistry.* NATO Science Series, ser. II. Boston: Kluwer Academic Publishers, 2001.

Denton, M. Bonner, and Royal Society of Chemistry (Great Britain). *Further Developments in Scientific Optical Imaging.* Special publication no. 254. Cambridge, U.K.: Royal Society of Chemistry, 2000.

Derouane, E. G. *Combinatorial Catalysis and High Throughput Catalyst Design and Testing.* Vol. 560 of *Mathematical and Physical Sciences.* NATO Science Series, ser. C. Boston: Kluwer Academic Publishers, 2000.

DeSimone, Joseph M., and William Tumas. *Green Chemistry Using Liquid and Supercritical Carbon Dioxide.* New York: Oxford University Press, 2003.

DeVoe, Howard. *Thermodynamics and Chemistry.* Upper Saddle River, N.J.: Prentice Hall, 2001.

Dewar, Michael James Steuart. *The Molecular Orbital Theory of Organic Chemistry.* New York: McGraw Hill, 1968.

Dill, Ken A., and Sarina Bromberg. *Molecular Driving Forces: Statistical Thermodynamics in Chemistry and Biology.* New York: Garland Science, 2003.

Di Toro, Dominic M. *Sediment Flux Modeling.* New York: Wiley-Interscience, 2001.

Dodziuk, Helena. *Introduction to Supramolecular Chemistry.* Boston: Kluwer Academic Publishers, 2002.

Doremus, R. H. *Diffusion of Reactive Molecules in Solids and Melts.* New York: Wiley, 2002.

Dressler, Hans. *Resorcinol: Its Uses and Derivatives, Topics in Applied Chemistry.* New York: Plenum Press, 1994.

Dressler, Rainer A. *Chemical Dynamics in Extreme Environments.* Advanced Series in Physical Chemistry, vol. 11. River Edge, N.J.: World Scientific, 2001.

Duffey, George H. *Modern Physical Chemistry: A Molecular Approach.* New York: Kluwer Academic/Plenum Publishers, 2000.

Duffy, J. A. *General Inorganic Chemistry.* London: Longmans, 1966.

Dupuy, François, and ebrary Inc. *The Chemistry of Change Problems, Phases, and Strategy.* New York: Palgrave, 2002.

Dybkov, V. I. *Reaction Diffusion and Solid State Chemical Kinetics*. Kyiv: IPMS Publications, 2002.

ebrary Inc. *The Carbon Dioxide Dilemma: Promising Technologies and Policies*. Proceedings of a symposium, April 23–24, 2002. Washington, D.C.: National Academies Press, 2003.

———. *McGraw-Hill Dictionary of Chemistry*. 2nd ed. New York: McGraw-Hill, 2003.

———. *Minorities in the Chemical Workforce: Diversity Models That Work*. Workshop report to the Chemical Sciences Roundtable. Washington, D.C.: National Academies Press, 2003.

———. *Graduate Education in the Chemical Sciences: Issues for the 21st Century*. Report of a Workshop, The Compass Series. Washington, D.C.: National Academy Press, 2000.

———. *Women in the Chemical Workforce*. Workshop Report to the Chemical Sciences Roundtable, The Compass Series. Washington, D.C.: National Academy Press, 2000.

———. *Improving Research Capabilities in Chemical and Biomedical Sciences*. Proceedings of a multi-site electronic workshop, The Compass Series. Washington, D.C.: National Academy Press, 1999.

———. *Research Teams and Partnerships: Trends in the Chemical Sciences*. Report of a workshop, The Compass Series. Washington, D.C.: National Academy Press, 1999.

———. *Environmental Impact of Power Generation*. Issues in Environmental Science and Technology, no. 11. Cambridge, U.K.: Royal Society of Chemistry, 1999.

———. *Assessing the Value of Research in the Chemical Sciences*. Report of a workshop, The Compass Series. Washington, D.C.: National Academy Press, 1998.

———. *The Atmospheric Sciences Entering the Twenty-First Century*. Washington, D.C.: National Academy Press, 1998.

———. *Risk Assessment and Risk Management*. Issues in Environmental Science and Technology, no. 9. Cambridge, U.K.: Royal Society of Chemistry, 1998.

Ehrfeld, Wolfgang, Holger Löwe, and Volker Hessel. *Microreactors: New Technology for Modern Chemistry*. New York: Wiley-VCH, 2000.

Eisler, Ronald. *Handbook of Chemical Risk Assessment: Health Hazards to Humans, Plants, and Animals*. Boca Raton, Fla.: Lewis Publishers, 2000.

Enke, Christie G. *The Art and Science of Chemical Analysis*. New York: Wiley, 2001.

Eriksson, Leif A. *Theoretical Biochemistry: Processes and Properties of Biological Systems*. 1st ed. Theoretical and Computational Chemistry, no. 9. New York: Elsevier, 2001.

Esumi, Kunio, and Minoru Ueno. *Structure-Performance Relationships in Surfactants*. 2nd rev. ed. Surfactant Science Series, vol. 112. New York: Marcel Dekker, 2003.

Feringa, Ben L. *Molecular Switches*. New York: Wiley-VCH, 2001.

Ferrier, Robert J., R. Blattner, and ebrary Inc. *Carbohydrate Chemistry: Monosaccharides, Disaccharides, and Specific Oligosaccharides*. Specialist Periodical Reports. London: Royal Society of Chemistry, 2002.

———. *Carbohydrate Chemistry: Monosaccharides, Disaccharides, and Specific Oligosaccharides*. Specialist Periodical Reports. London: Royal Society of Chemistry, 2001.

———. *Carbohydrate Chemistry: Monosaccharides, Disaccharides, and Specific Oligosaccharides*. Specialist Periodical Reports. London: Royal Society of Chemistry, 2000.

———. *Carbohydrate Chemistry: Monosaccharides, Disaccharides, and Specific Oligosaccharides*. Specialist Periodical Reports. London: Royal Society of Chemistry, 1998.

Fifield, F. W., and P. J. Haines. *Environmental Analytical Chemistry*. 2nd ed. Malden, Mass.: Blackwell Science, 2000.

Fifield, F. W., D. Kealey, and NetLibrary Inc. *Principles and Practice of Analytical Chemistry*. 5th ed. Malden, Mass.: Blackwell Science, 2000.

Fingerman, Milton, and Rachakonda Nagabhushanam. *Bio-Organic Compounds: Chemistry and Biomedical Applications*. Vol. 6 of *Recent Advances in Marine Biotechnology*. Enfield, N.H.: Science Pub., 2001.

Finkelstein, Alexei V., and Oleg B. Ptitsyn. *Protein Physics: A Course of Lectures*. San Diego, Calif.: Academic Press, 2002.

Floudas, Christodoulos A., and P. M. Pardalos. *Optimization in Computational Chemistry and Molecular Biology: Local and Global Approaches*. Vol. 40 of *Nonconvex Optimization and Its Applications*. Boston: Kluwer Academic Publishers, 2000.

Franks, Felix. *Water, a Comprehensive Treatise*. New York: Plenum Press, 1972.

Franks, Felix, and Royal Society of Chemistry. *Water: A Matrix of Life*. 2nd ed. Cambridge, U.K.: Royal Society of Chemistry, 2000.

Freund, Jan A., and Thorsten Pöschel. *Stochastic Processes in Physics, Chemistry, and Biology*. Lecture Notes in Physics, no. 557. New York: Springer, 2000.

Frommer, Jane, René M. Overney, American Chemical Society Division of Colloid and Surface Chemistry, and American Chemical Society. *Interfacial Properties on the Submicrometer Scale*. ACS Symposium Series, no. 781. Washington, D.C.: American Chemical Society, 2001.

Furukawa, Junji. *Physical Chemistry of Polymer Rheology.* Springer Series in Chemical Physics, vol. 72. New York: Springer, 2003.

Gallup, Gordon A., and ebrary Inc. *Valence Bond Methods: Theory and Applications.* New York: Cambridge University Press, 2002.

García-Campaña, Ana M., and Willy R. G. Baeyens. *Chemiluminescence in Analytical Chemistry.* New York: Marcel Dekker, 2001.

Gardziella, A., L. Pilato, and A. Knop. *Phenolic Resins: Chemistry, Applications, Standardization, Safety, and Ecology.* 2nd rev. ed. New York: Springer, 2000.

Garnovskii, A. D., and Boris I. Kharisov. *Synthetic Coordination and Organometallic Chemistry.* New York: Marcel Dekker, 2003.

Gibb, Thomas R. P. *Optical Methods of Chemical Analysis.* International Chemical Series. New York: McGraw-Hill, 1942.

Gibb, Thomas R. P., and American Chemical Society Division of Analytical Chemistry. *Analytical Methods in Oceanography: A Symposium Sponsored by the Division of Analytical Chemistry at the 168th Meeting of the American Chemical Society.* Advances in Chemistry Series, no. 147. Washington, D.C.: American Chemical Society, 1975.

Gilbert, A., and ebrary Inc. *Photochemistry.* Specialist Periodical Reports. Cambridge, U.K.: Royal Society of Chemistry, 2001.

———. *Photochemistry.* Specialist Periodical Reports. Cambridge, U.K.: Royal Society of Chemistry, 2000.

———. *Photochemistry.* Specialist Periodical Reports. Cambridge, U.K.: Royal Society of Chemistry, 1999.

———. *Photochemistry.* Specialist Periodical Reports. Cambridge, U.K.: Royal Society of Chemistry, 1998.

Gilbert, B. C., and ebrary Inc. *Electron Paramagnetic Resonance.* Specialist Periodical Reports. Cambridge, U.K.: Royal Society of Chemistry, 2000.

———. *Electron Paramagnetic Resonance.* Specialist Periodical Reports. Cambridge, U.K.: Royal Society of Chemistry, 1998.

Gilbert, Everett E. *Sulfonation and Related Reactions.* Interscience Monographs on Chemistry. New York: Interscience Publishers, 1965.

Gilbert, Robert G., and Sean C. Smith. *Theory of Unimolecular and Recombination Reactions.* Physical Chemistry Texts. Boston: Blackwell Scientific Publications, 1990.

Gillespie, Ronald J. *Chemistry.* 2nd ed. Englewood Cliffs, N.J.: Prentice Hall, 1989.

———. *Chemistry.* Boston: Allyn and Bacon, 1986.

Gillespie, Ronald J., and Paul L. A. Popelier. *Chemical Bonding and Molecular Geometry: From Lewis to Electron Densities.* Topics in Inorganic Chemistry. New York: Oxford University Press, 2001.

Goldberg, David E., and ebrary Inc. *How to Solve Word Problems in Chemistry.* New York: McGraw-Hill, 2001.

———. *Schaum's Outline of Theory and Problems of Beginning Chemistry.* 2nd ed. Schaum's Outline Series. New York: McGraw-Hill, 1999.

Goldberg, Herbert S. *Antibiotics: Their Chemistry and Non-Medical Uses.* Princeton, N.J., Van Nostrand, 1959.

Goldberg, Marvin C., and American Chemical Society Division of Environmental Chemistry. *Luminescence Applications in Biological, Chemical, Environmental, and Hydrological Sciences.* ACS Symposium Series, no. 383. Washington, D.C.: American Chemical Society, 1989.

Goodrich, F. C. *A Primer of Quantum Chemistry.* New York: Wiley-Interscience, 1972.

Goodwin, A. R. H., K. N. Marsh, W. A. Wakeham, and International Union of Pure and Applied Chemistry, Commission on Thermodynamics. *Measurement of the Thermodynamic Properties of Single Phases.* 1st ed. IUPAC Chemical Data Series, no. 40. Boston: Elsevier, 2003.

Goodwin, Brian L. *Handbook of Intermediary Metabolism of Aromatic Compounds.* New York: Wiley, 1976; distributed in U.S. by Halsted Press.

Goodwin, James W., Roy W. Hughes, and Royal Society of Chemistry (Great Britain). *Rheology for Chemists: An Introduction.* Cambridge, U.K.: Royal Society of Chemistry, 2000.

Goodwin, J. W., and Royal Society of Chemistry (Great Britain), Continuing Education Committee. *Colloidal Dispersions.* Special Publication / Royal Society of Chemistry, no. 43. London: Royal Society of Chemistry (Great Britain), 1982.

Goodwin, T. W., and E. I. Mercer. *Introduction to Plant Biochemistry.* 2nd ed. Pergamon International Library of Science, Technology, Engineering, and Social Studies. New York: Pergamon Press, 1983.

———. *Introduction to Plant Biochemistry.* 1st ed. *The Commonwealth and International Library.* New York: Pergamon Press, 1972.

Graja, Andrzej, B. R. Buka, and F. Kajzar. *Molecular Low Dimensional and Nanostructured Materials for Advanced Applications.* Vol. 59 of *Mathematics, Physics, and Chemistry.* NATO Science Series, ser. II. London: Kluwer Academic, 2002.

Green, M. *Solid State Surface Science.* Vol. 1. New York: Marcel Dekker, 1969.

Green, M., and ebrary Inc. *Organometallic Chemistry.* Specialist Periodical Reports. Cambridge, U.K.: Royal Society of Chemistry, 2001.

———. *Organometallic Chemistry.* Specialist Periodical Reports. Cambridge, U.K.: Royal Society of Chemistry, 2000.

———. *Organometallic Chemistry.* Specialist Periodical Reports. Cambridge, U.K.: Royal Society of Chemistry, 1999.

———. *Organometallic Chemistry.* Specialist Periodical Reports. Cambridge, U.K.: Royal Society of Chemistry, 1998.

Greenberg, Arthur. *The Art of Chemistry: Myths, Medicines, and Materials.* Hoboken, N.J.: Wiley-Interscience, 2003.

Greenberg, Arthur, and Joel F. Liebman. *Strained Organic Molecules.* Organic Chemistry Series, vol. 38. New York: Academic Press, 1978.

Greenberg, J. Mayo, V. Pirronello, and North Atlantic Treaty Organization, Scientific Affairs Division. *Chemistry in Space.* Boston: Kluwer Academic Publishers, 1991.

Griffen, Dana T. *Silicate Crystal Chemistry.* New York: Oxford University Press, 1992.

Griffin, Anselm Clyde, Julian Frank Johnson, and American Chemical Society, Division of Colloid and Surface Chemistry. *Liquid Crystals and Ordered Fluids.* Vol. 4. New York: Plenum Press, 1984.

Griffin, G. J. L. *Chemistry and Technology of Biodegradable Polymers.* 1st ed. New York: Blackie Academic & Professional, 1994.

Griffin, J. P., John O'Grady, and ebrary Inc. *The Textbook of Pharmaceutical Medicine.* 4th ed. London: BMJ Books, 2002.

Griffin, Rodger W. *Modern Organic Chemistry.* New York: McGraw-Hill, 1969.

Griffith, Edward J. *Phosphate Fibers.* Topics in Applied Chemistry. New York: Plenum Press, 1995.

Griffith, J. S. *The Theory of Transition-Metal Ions.* Cambridge, U.K.: Cambridge University Press, 1961.

Griffith, R. O., and A. McKeown. *Photo-Processes in Gaseous and Liquid Systems.* Textbooks of Physical Chemistry. New York: Longmans Green, 1929.

Griffith, William Pettit. *The Chemistry of the Rarer Platinum Metals (Os, Ru, Ir, and Rh).* Interscience Monographs on Chemistry, Inorganic Chemistry Section. New York: Interscience Publishers, 1967.

Griffiths, Peter R. *Transform Techniques in Chemistry.* Modern Analytical Chemistry. New York: Plenum Press, 1978.

Grimes, Russell N. *Carboranes, Organometallic Chemistry.* New York: Academic Press, 1970.

Grimshaw, James. *Electrochemical Reactions and Mechanisms in Organic Chemistry.* New York: Elsevier, 2000.

Grimshaw, Rex W., and Alfred B. Searle. *The Chemistry and Physics of Clays and Allied Ceramic Materials.* 4th rev. ed. New York: Wiley-Interscience, 1971.

Gross, Erhard, and Johannes Meienhofer. *The Peptides: Analysis, Synthesis, Biology.* New York: Academic Press, 1979.

Gross, Helmut, and Wolfgang Borsdorf. *Dictionary of Chemistry and Chemical Technology, English-German.* New York: Elsevier, 1984.

Gross, Mihal E., Joseph M. Jasinski, and John T. Yates. *Chemical Perspectives of Microelectronic Materials.* Materials Research Society Symposium Proceedings, vol. 131. Pittsburgh, Pa.: Materials Research Society, 1989.

Gross, Richard A., H. N. Cheng, American Chemical Society, Division of Polymer Chemistry, and American Chemical Society. *Biocatalysis in Polymer Science.* ACS Symposium Series, no. 840. Washington, D.C.: American Chemical Society, 2003.

Grove, Evelina. *Applied Atomic Spectroscopy.* Modern Analytical Chemistry. New York: Plenum Press, 1978.

Grubbs, Robert H. *Handbook of Metathesis.* Hoboken, N.J.: Wiley, 2003.

Hadziioannou, Georges, and Paul F. van Hutten. *Semiconducting Polymers: Chemistry, Physics, and Engineering.* New York: Wiley-VCH, 2000.

Hall, Nina. *The New Chemistry.* New York: Cambridge University Press, 2000.

Handley, Alan J. *Extraction Methods in Organic Analysis, Sheffield Analytical Chemistry.* Boca Raton, Fla.: CRC Press, 1999.

Handley, Alan J., and E. R. Adlard. *Gas Chromatographic Techniques and Applications.* Sheffield Analytical Chemistry. Boca Raton, Fla.: CRC Press, 2001.

Harrison, P. *Computational Methods in Physics, Chemistry, and Biology: An Introduction.* New York: Wiley, 2001.

Harrison, Roy M. *Understanding Our Environment: An Introduction to Environmental Chemistry and Pollution.* 2nd ed. Cambridge, U.K.: Royal Society of Chemistry, 1992.

Harrison, Roy M., R. E. Hester, and ebrary Inc. *Chemistry in the Marine Environment.* Issues in Environmental Science and Technology, no. 13. Cambridge, U.K.: Royal Society of Chemistry, 2000.

Harrison, Roy M., R. E. Hester, and Royal Society of Chemistry (Great Britain). *Chemistry in the Marine Environment.* Issues in Environmental Science and Technology, no. 13. Cambridge, U.K.: Royal Society of Chemistry, 2000.

Harrison, Roy M., and Roger Perry. *Handbook of Air Pollution Analysis.* 2nd ed. New York: Chapman and Hall, 1986.

Harrison, Roy M., and Royal Society of Chemistry (Great Britain). *Understanding Our Environment: An Introduction to Environmental Chemistry and Pollution.* 3rd ed. Cambridge, U.K.: Royal Society of Chemistry, 1999.

Harrison, Walter A. *Electronic Structure and the Properties of Solids: The Physics of the Chemical Bond.* New York: Dover Publications, 1989.

———. *Electronic Structure and the Properties of Solids: The Physics of the Chemical Bond.* San Francisco: Freeman, 1980.

Hassner, Alfred. *Small Ring Heterocycles.* Vol. 42 of *The Chemistry of Heterocyclic Compounds,* pt. 1. New York: Wiley, 1983.

Hassner, Alfred, and C. Stumer. *Organic Syntheses Based on Name Reactions.* 2nd ed. Tetrahedron Organic Chemistry Series, vol. 22. Boston: Pergamon, 2002.

Hastie, John W., D. W. Bonnell, P. K. Schenck, and National Institute of Standards and Technology (U.S.). *Laser-Assisted Vaporization Mass Spectrometry Application to Thermochemistry at Very High Temperatures.* NISTIR, 6793. Gaithersburg, Md.: U.S. Dept. of Commerce, Technology Administration, National Institute of Standards and Technology, 2001.

Hastie, John W., and National Institute of Standards and Technology (U.S.). *A Predictive Ionization Cross Section Model for Inorganic Molecules.* NISTIR, 6768. Gaithersburg, Md.: U.S. Dept. of Commerce, Technology Administration, National Institute of Standards and Technology, 2001.

Hawley, Gessner Goodrich. *The Condensed Chemical Dictionary.* 10th ed. New York: Van Nostrand Reinhold, 1981.

Hawley, Gessner Goodrich, and Richard J. Lewis. *Hawley's Condensed Chemical Dictionary.* 14th ed. New York: Wiley, 2001.

Hazen, Robert M. *The New Alchemists: Breaking through the Barriers of High Pressure.* New York: Times Books, 1993.

Hazen, Robert M., Robert T. Downs, and Mineralogical Society of America. *High-Temperature and High-Pressure Crystal Chemistry.* Vol. 41 of *Reviews in Mineralogy and Geochemistry.* Washington, D.C.: Mineralogical Society of America and the Geochemical Society, 2000.

Hazen, Robert M., and Larry W. Finger. *Comparative Crystal Chemistry: Temperature, Pressure, Composition, and the Variation of Crystal Structure.* New York: Wiley, 1982.

Hemond, Harold F., and Elizabeth J. Fechner-Levy. *Chemical Fate and Transport in the Environment.* 2nd ed. San Diego, Calif.: Academic, 2000.

Herzberg, Gerhard. *The Spectra and Structures of Simple Free Radicals: An Introduction to Molecular Spectroscopy.* Dover Phoenix Editions. Mineola, N.Y.: Dover Publications, 2003.

———. *The Spectra and Structures of Simple Free Radicals: An Introduction to Molecular Spectroscopy.* The George Fisher Baker Non-Resident Lectureship in Chemistry at Cornell University. Ithaca, N.Y.: Cornell University Press, 1971.

Heslop, R. B., and P. L. Robinson. *Inorganic Chemistry: A Guide to Advanced Study.* 3rd ed. New York: Elsevier, 1967.

Hess, Bernd A. *Relativistic Effects in Heavy-Element Chemistry and Physics.* Wiley Series in Theoretical Chemistry. Hoboken, N.J.: Wiley, 2003.

Hess, Dennis W., and Klavs F. Jensen. *Microelectronics Processing: Chemical Engineering Aspects.* Advances in Chemistry Series, no. 221. Washington, D.C.: American Chemical Society, 1989.

Hesse, Manfred. *Ring Enlargement in Organic Chemistry.* New York: Wiley, 1991.

———. *Alkaloid Chemistry.* New York: Wiley, 1981.

Hesse, Peter Ralston. *A Textbook of Soil Chemical Analysis.* New York: Chemical Pub. Co., 1972.

Hester, R. E., Roy M. Harrison, and ebrary Inc. *Environmental and Health Impact of Solid Waste Management Activities.* Issues in Environmental Science and Technology, no. 18. Cambridge, U.K.: Royal Society of Chemistry, 2002.

———. *Global Environmental Change.* Issues in Environmental Science and Technology, no. 17. Cambridge, U.K.: Royal Society of Chemistry, 2002.

———. *Food Safety and Food Quality.* Issues in Environmental Science and Technology, no. 15. Cambridge, U.K.: Royal Society of Chemistry, 2001.

———. *Assessment and Reclamation of Contaminated Land.* Issues in Environmental Science and Technology, no. 16. Cambridge, U.K.: Royal Society of Chemistry, 2001.

———. *Causes and Environmental Implications of Increased UV-B Radiation.* Issues in Environmental Science and Technology, no. 14. Cambridge, U.K.: Royal Society of Chemistry, 2000.

———. *Endocrine Disrupting Chemicals.* Issues in Environmental Science and Technology, no. 12. Cambridge, U.K.: Royal Society of Chemistry, 1999.

———. *Air Pollution and Health.* Issues in Environmental Science and Technology, no. 10. Cambridge, U.K.: Royal Society of Chemistry, 1998.

———. *Contaminated Land and Its Reclamation.* Issues in Environmental Science and Technology, no. 7. Cambridge, U.K.: Royal Society of Chemistry, 1997.

———. *Air Quality Management.* Issues in Environmental Science and Technology, no. 8. Cambridge, U.K.: Royal Society of Chemistry Information Services, 1997.

———. *Chlorinated Organic Micropollutants.* Issues in Environmental Science and Technology, no. 6. Cambridge, U.K.: Royal Society of Chemistry, 1996.

———. *Agricultural Chemicals and the Environment.* Issues in Environmental Science and Technology, no. 5. Cambridge, U.K.: Royal Society of Chemistry Information Services, 1996.

———. *Waste Treatment and Disposal.* Issues in Environmental Science and Technology, no. 3. Cambridge, U.K.: Royal Society of Chemistry, 1995.

———. *Volatile Organic Compounds in the Atmosphere.* Issues in Environmental Science and Technology, no. 4. Cambridge, U.K.: Royal Society of Chemistry, 1995.

———. *Mining and Its Environmental Impact.* Issues in Environmental Science and Technology, no. 1. Cambridge, U.K.: Royal Society of Chemistry, 1994.

———. *Waste Incineration and the Environment.* Issues in Environmental Science and Technology, no. 2. Cambridge, U.K.: Royal Society of Chemistry, 1994.

Hester, R. E., Roy M. Harrison, and Royal Society of Chemistry (Great Britain). *Mining and Its Environmental Impact.* Issues in Environmental Science and Technology, no. 1. Cambridge, U.K.: Royal Society of Chemistry, 1994.

Hillis, P., Royal Society of Chemistry, European Desalination Society, and Society of Chemical Industry. *Membrane Technology in Water and Wastewater Treatment.* Special Publication, no. 249. Cambridge, U.K.: Royal Society of Chemistry, 2000.

Hillisch, A., and R. Hilgenfeld. *Modern Methods of Drug Discovery.* Boston: Birkhäuser Verlag, 2003.

Hinchliffe, Alan. *Modelling Molecular Structures.* 2nd ed. *Wiley Series in Theoretical Chemistry.* New York: Wiley, 2000.

———. *Chemical Modeling: From Atoms to Liquids.* New York: Wiley, 1999.

———. *Computational Quantum Chemistry.* Chichester; New York: Wiley, 1988.

———. *Ab Initio Determination of Molecular Properties.* Bristol: Hilger, 1987.

Hinchliffe, Alan, and ebrary Inc. *Chemical Modelling Applications and Theory.* Specialist Periodical Reports. Cambridge, U.K.: Royal Society of Chemistry, 2002.

———. *Chemical Modelling Applications and Theory.* A Specialist Periodical Report. Cambridge, U.K.: Royal Society of Chemistry, 2000.

Hoenig, Steven L., and ebrary Inc. *Basic Training in Chemistry.* New York: Kluwer Academic/Plenum Publishers, 2001.

Holmberg, Krister. *Surfactants and Polymers in Aqueous Solution.* 2nd ed. Hoboken, N.J.: Wiley, 2003.

Hopf, Henning. *Classics in Hydrocarbon Chemistry: Syntheses, Concepts, Perspectives.* New York: Wiley-VCH, 2000.

Horton, Derek. *Advances in Carbohydrate Chemistry and Biochemistry.* Oxford, U.K.: Elsevier/Academic, 2003.

Howard, Gary C., and William E. Brown. *Modern Protein Chemistry: Practical Aspects.* Boca Raton, Fla.: CRC Press, 2002.

Hubbard, Arthur T. *Encyclopedia of Surface and Colloid Science.* New York: Marcel Dekker, 2002.

———. *The Handbook of Surface Imaging and Visualization.* Boca Raton, Fla.: CRC Press, 1995.

Hubbard, Colin D., and Rudi van Eldik. *Chemistry under Extreme or Non-Classical Conditions.* New York: Wiley, 1997.

Iozzo, Renato V. *Proteoglycan Protocols.* Methods in Molecular Biology, no. 171. Totowa, N.J.: Humana Press, 2001.

Irwin, John S., and Air Resources Laboratory (U.S.). *A Historical Look at the Development of Regulatory Air Quality Models for the United States Environmental Protection Agency.* NOAA technical memorandum OAR ARL, no. 244. Silver Spring, Md.: U.S. Dept. of Commerce, National Oceanic and Atmospheric Administration, Oceanic and Atmospheric Research Laboratories, Air Resources Laboratory, 2002.

Iwamoto, Mitsumasa, and Chen-Xu Wu. *The Physical Properties of Organic Monolayers.* River Edge, N.J.: World Scientific, 2001.

Jensen, Frank. *Introduction to Computational Chemistry.* New York: Wiley, 1999.

Jensen, Frederick R., and Bruce Rickborn. *Electrophilic Substitution of Organomercurials.* McGraw-Hill Series in Advanced Chemistry. New York: McGraw-Hill, 1968.

Jensen, Howard Barnett, Siggis Sidney, American Chemical Society Division of Petroleum Chemistry, and American Chemical Society Division of Analytical Chemistry. *Analytical Chemistry of Liquid Fuel Sources: Tar Sands, Oil Shale, Coal, and Petroleum.* ACS Advances in Chemistry Series, no. 170. Washington, D.C.: American Chemical Society, 1978.

Jensen, James N. *A Problem-Solving Approach to Aquatic Chemistry.* New York: Wiley, 2003.

Jensen, Klavs F., Donald G. Truhlar, American Chemical Society Division of Industrial and Engineering Chemistry. *Supercomputer Research in Chemistry and Chemical Engineering.* ACS Symposium Series, no. 353. Washington, D.C.: American Chemical Society, 1987.

Jensen, William A. *Botanical Histochemistry: Principles and Practice.* A Series of Biology Books. San Francisco: W. H. Freeman, 1962.

Jollès, Pierre, and Hans Jörnvall. *Proteomics in Functional Genomics: Protein Structure Analysis.* EXS, no. 88. Boston: Birkhäuser Verlag, 2000.

———. *Interface between Chemistry and Biochemistry.* EXS, no. 73. Boston: Birkhäuser Verlag, 1995.

Jones, William. *Organic Molecular Solids: Properties and Applications.* Boca Raton, Fla.: CRC Press, 1997.

Jutzi, Peter, and U. Schubert. *Silicon Chemistry: From the Atom to Extended Systems.* Cambridge, U.K.: Wiley-VCH, 2003.

Kallay, Nikola. *Interfacial Dynamics.* Surfactant Science Series, vol. 88. New York: Marcel Dekker, 2000.

Kamide, Kenji, and Toshiaki Dobashi. *Physical Chemistry of Polymer Solutions: Theoretical Background.* New York: Elsevier, 2000.

Karim, Alamgir, and Kum ara Sanata. *Polymer Surfaces, Interfaces and Thin Films.* River Edge, N.J.: World Scientific, 2000.

Kehew, Alan E. *Applied Chemical Hydrogeology.* Upper Saddle River, N.J.: Prentice Hall, 2001.

Kendall, Kevin. *Molecular Adhesion and Its Applications: The Sticky Universe.* New York: Kluwer Academic/Plenum Publishers, 2001.

Khopkar, Shripad Moreshwar. *Analytical Chemistry of Macrocyclic and Supramolecular Compounds.* London: Springer-Verlag, 2002.

Khosravi, E., and T. Szymanska-Buzar. *Ring Opening Metathesis Polymerisation and Related Chemistry: State of the Art and Visions for the New Century.* Vol. 56 of *Mathematics, Physics, and Chemistry.* NATO Science Series, ser. II. Boston: Kluwer Academic Publishers, 2002.

Kiihne, S. R., and H. J. M. De Groot. *Perspectives on Solid State NMR in Biology.* Vol. 1 of *Focus on Structural Biology.* Boston: Kluwer Academic Publishers, 2001.

Kirchner, Karl, and Walter Weissensteiner. *Organometallic Chemistry and Catalysis.* New York: Springer, 2001.

Kirk, Raymond E., and Donald F. Othmer. *Kirk-Othmer Encyclopedia of Chemical Technology.* 4th ed. New York: Wiley, 2000.

Kirk, Raymond E., Donald F. Othmer, Martin Grayson, and David Eckroth. *Kirk-Othmer Concise Encyclopedia of Chemical Technology.* New York: Wiley, 1985.

Kirk, Raymond E., Donald F. Othmer, Jacqueline I. Kroschwitz, and Mary Howe-Grant. *Encyclopedia of Chemical Technology.* 4th ed. New York: Wiley, 1991.

Kirk, T. Kent, Takayoshi Higuchi, Hou-min Chang, and United States–Japan Cooperative Science Program. *Lignin Biodegradation: Microbiology, Chemistry, and Potential Applications.* Boca Raton, Fla.: CRC Press,

Klabunde, Kenneth J. *Nanoscale Materials in Chemistry.* New York: Wiley-Interscience, 2001.

———. *Chemistry of Free Atoms and Particles.* New York: Academic Press, 1980.

Kleiböhmer, Wolfgang. *Environmental Analysis.* Vol. 3 of *Handbook of Analytical Separations.* New York: Elsevier, 2001.

Knovel (firm). *Knovel Critical Tables.* Norwich, N.Y.: Knovel, 2003.

———. *Lange's Handbook of Chemistry.* 15th ed. Vol. coverage as of Mar. 19, 2001. Binghamton, N.Y.: Knovel, 1999.

Kobayashi, S., and Karl Anker Jørgensen. *Cycloaddition Reactions in Organic Synthesis.* Weinheim, Germany: Wiley-VCH, 2002.

Kohn, Matthew J., John F. Rakovan, and John M. Hughes. *Phosphates: Geochemical, Geobiological, and Materials Importance.* Vol. 48 of *Reviews in Mineralogy and Geochemistry.* Washington, D.C.: Mineralogical Society of America, 2002.

Kokorin, A. I., and D. Bahnemann. *Chemical Physics of Nanostructured Semiconductors.* Boston: VSP, 2003.

Kolasinski, Kurt W. *Surface Science: Foundations of Catalysis and Nanoscience.* New York: Wiley, 2002.

Kordal, Richard Joseph, Arthur M. Usmani, Wai Tak Law, American Chemical Society, Division of Industrial and Engineering Chemistry, and American Chemical Society. *Microfabricated Sensors: Application of Optical Technology for DNA Analysis.* ACS Symposium Series, no. 815. Washington, D.C.: American Chemical Society, 2002.

Kozma, David. *CRC Handbook of Optical Resolutions via Diastereomeric Salt Formation.* Boca Raton, Fla.: CRC Press, 2002.

Kralchevsky, Peter A., and Kuniaki Nagayama. *Particles at Fluids Interfaces and Membranes: Attachment of Colloid Particles and Proteins to Interfaces and Formation of Two-Dimensional Arrays.* Vol. 10 of *Studies in Interface Science.* New York: Elsevier, 2001.

Krane, Dan E., and Michael L. Raymer. *Fundamental Concepts of Bioinformatics.* San Francisco: Benjamin Cummings, 2003.

Krause, Norbert. *Modern Organocopper Chemistry.* Weinheim: Wiley-VCH, 2002.

Krebs, Heinz. *Fundamentals of Inorganic Crystal Chemistry.* European Chemistry Series. New York: McGraw-Hill, 1968.

Krebs, Robert E. *The History and Use of Our Earth's Chemical Elements: A Reference Guide.* Westport, Conn.: Greenwood Press, 1998.

Kuran, Witold. *Principles of Coordination Polymerisation: Heterogeneous and Homogeneous Catalysis in Polymer Chemistry—Polymerisation of Hydrocarbon, Heterocyclic, and Heterounsaturated Monomers.* New York: Wiley, 2001.

Kurian, George Thomas. *The Nobel Scientists: A Biographical Encyclopedia.* Amherst, N.Y.: Prometheus Books, 2002.

Lagally, Max G., and North Atlantic Treaty Organization, Scientific Affairs Division. *Kinetics of Ordering and Growth at Surfaces.* Vol. 239 of *Physics.* NATO ASI Series, ser. B. New York: Plenum Press, 1990.

Lagaly, Gerhard, and Kolloid-Gesellschaft, Hauptversammlung. *Molecular Organisation on Interfaces,* Vol. 121 of *Progress in Colloid and Polymer Science.* New York: Springer, 2002.

Laue, Carola A., Kenneth L. Nash, and American Chemical Society, Division of Nuclear Chemistry and Technology. *Radioanalytical Methods in Interdisciplinary Research: Fundamentals in Cutting-Edge Applications.* ACS Symposium Series, no. 868. Washington, D.C.: American Chemical Society, 2004.

Laue, Thomas, and Andreas Plagens. *Named Organic Reactions.* New York: Wiley, 1998.

Le Bris, Claude. *Computational Chemistry: Special Volume.* Vol. 10 of *Handbook of Numerical Analysis.* Boston: North Holland, 2003.

Lee, Mike S. *LC/MS Applications in Drug Development.* Wiley-Interscience Series on Mass Spectrometry. New York: Wiley, 2002.

Lehr, Jay H., and Janet K. Lehr. *Standard Handbook of Environmental Science, Health, and Technology.* McGraw-Hill Standard Handbooks. New York: McGraw-Hill, 2000.

Leroy, Francis. *A Century of Nobel Prize Recipients: Chemistry, Physics, and Medicine.* New York: Marcel Dekker, 2003.

Leskovac, Vladimir. *Comprehensive Enzyme Kinetics.* New York: Kluwer Academic/Plenum, 2003.

Levere, Trevor Harvey. *Affinity and Matter: Elements of Chemical Philosophy, 1800–1865.* Oxford, U.K.: Clarendon Press, 1971.

Levere, Trevor Harvey, and ebrary Inc. *Transforming Matter: A History of Chemistry from Alchemy to the Buckyball.* Johns Hopkins Introductory Studies in the History of Science. Baltimore: Johns Hopkins University Press, 2001.

Lewars, Errol. *Computational Chemistry: Introduction to the Theory and Applications of Molecular and Quantum Mechanics.* Boston: Kluwer Academic, 2003.

Li, Jie Jack. *Name Reactions: A Collection of Detailed Reaction Mechanisms.* 2nd ed. New York: Springer, 2003.

Li, Jie Jack, and Gordon W. Gribble. *Palladium in Heterocyclic Chemistry: A Guide for the Synthetic Chemist.* Tetrahedron Organic Chemistry Series, vol. 20. New York: Pergamon, 2000.

Lieser, Karl Heinrich. *Nuclear and Radiochemistry: Fundamentals and Applications.* 2nd ed. New York: Wiley-VCH, 2001.

Lindhorst, Thisbe K. *Essentials of Carbohydrate Chemistry and Biochemistry.* 2nd ed. Weinheim, Germany: Wiley-VCH, 2003.

Lindoy, Leonard F. *The Chemistry of Macrocyclic Ligand Complexes.* New York: Cambridge University Press, 1989.

Lindoy, Leonard F., Ian M. Atkinson, and Royal Society of Chemistry (Great Britain). *Self-Assembly in Supramolecular Systems.* Monographs in Supramolecular Chemistry, no. 7. Cambridge, U.K.: Royal Society of Chemistry, 2000.

Loconto, Paul R. *Trace Environmental Quantitative Analysis: Principles, Techniques and Applications.* New York: Marcel Dekker, 2001.

Lund, A., and M. Shiotani. *EPR of Free Radicals in Solids: Trends in Methods and Applications.* Vol. 10 of *Progress in Theoretical Chemistry and Physics.* Boston: Kluwer Academic, 2003.

Ma, T. S., and V. Horák. *Microscale Manipulations in Chemistry.* Vol. 44 of *Chemical Analysis.* New York: Wiley, 1976.

Ma, T. S., and Athanasios S. Ladas. *Organic Functional Group Analysis by Gas Chromatography.* The Analysis of Organic Materials, no. 10. New York: Academic Press, 1976.

Ma, T. S., and Robert E. Lang. *Quantitative Analysis of Organic Mixtures.* New York: Wiley, 1979.

Machamer, Peter K., Marcello Pera, and Aristeid es Baltas. *Scientific Controversies: Philosophical and Historical Perspectives.* New York: Oxford University Press, 2000.

Majoral, J. P. *New Aspects in Phosphorous Chemistry I.* Topics in Current Chemistry, no. 220. New York: Springer, 2002.

Makriyannis, Alexandros, and Diane Biegel. *Drug Discovery Strategies and Methods.* New York: Marcel Dekker, 2004.

Malinowski, Edmund R. *Factor Analysis in Chemistry.* 3rd ed. New York: Wiley, 2002.

Malmsten, Martin. *Surfactants and Polymers in Drug Delivery.* Vol. 122 of *Drugs and the Pharmaceutical Sciences.* New York: Marcel Dekker, 2002.

Manahan, Stanley E. *Environmental Chemistry.* 6th ed. Boca Raton, Fla.: Lewis Publications, 1994.

———. *Industrial Ecology: Environmental Chemistry and Hazardous Waste.* Boca Raton, Fla.: Lewis Publications, 1999.

———. *Toxicological Chemistry and Biochemistry.* 3rd ed. Boca Raton, Fla.: Lewis Publications, 2003.

Marcus, P. M., and F. Jona. *Determination of Surface Structure by Leed.* The IBM Research Symposia Series. New York: Plenum Press, 1984.

Marcus, Y. *Solvent Mixtures: Properties and Selective Solvation.* New York: Marcel Dekker, 2002.

Marcus, Y., and A. S. Kertes. *Ion Exchange and Solvent Extraction of Metal Complexes.* New York: Wiley-Interscience, 1969.

Martin, Dean Frederick. *Marine Chemistry.* 2nd ed. New York: Marcel Dekker, 1972.

Maruani, Jean, and ebrary Inc. *New Trends in Quantum Systems in Chemistry and Physics.* Vol. 7 of *Progress in Theoretical Chemistry and Physics.* London: Kluwer Academic Publishers, 2002.

———. *New Trends in Quantum Systems in Chemistry and Physics.* Vol. 6 of *Progress in Theoretical Chemistry and Physics.* London: Kluwer Academic, 2001.

Mason, T. J., and J. Phillip Lorimer. *Applied Sonochemistry: The Uses of Power Ultrasound in Chemistry and Processing.* Weinheim, Germany: Wiley-VCH, 2002.

May, Clayton A. *Epoxy Resins: Chemistry and Technology.* 2nd ed. New York: Marcel Dekker, 1988.

May, Volkhard, and Oliver Kühn. *Charge and Energy Transfer Dynamics in Molecular Systems.* 2nd ed. Weinheim, Germany: Wiley-VCH, 2003.

Mazor, Imanuel. *Chemical and Isotopic Groundwater Hydrology.* 3rd ed. Books in Soils, Plants and the Environment, no. 98. New York: Marcel Dekker, 2004.

Mázor, László. *Analytical Chemistry of Organic Halogen Compounds.* International Series in Analytical Chemistry, vol. 58. New York: Pergamon Press, 1975.

McCreery, Richard L. *Raman Spectroscopy for Chemical Analysis.* Vol. 157 of *Chemical Analysis.* New York: Wiley, 2000.

McGrayne, Sharon Bertsch, and ebrary Inc. *Prometheans in the Lab Chemistry and the Making of the Modern World.* New York: McGraw-Hill, 2001.

Mei, Houng-Yau, Anthony W. Czarnik, and ebrary Inc. *Integrated Drug Discovery Technologies.* New York: Marcel Dekker, 2002.

Meier, Hans. *Organic Semiconductors: Dark- and Photoconductivity of Organic Solids.* Vol. 2 of *Monographs in Modern Chemistry.* Weinheim, Germany: Verlag Chemie, 1974.

Meier, Mark S., Jennifer L. Muzyka, Herbert Meislich, Howard Nechamkin, Jacob Sharefkin, and ebrary Inc. *Organic Chemistry: Based on Schaum's Outline of Organic Chemistry.* Schaum's Easy Outlines. New York: McGraw-Hill, 2000.

Meier, Mark S., Jennifer L. Muzyka, Herbert Meislich, Howard Nechamkin, Jacob Sharefkin, and NetLibrary Inc. *Schaum's Easy Outlines.* Schaum's Outline Series. New York: McGraw-Hill, 2000.

Meier, Peter C., and Richard E. Zünd. *Statistical Methods in Analytical Chemistry.* Vol. 123 of *Chemical Analysis.* New York: Wiley, 1993.

Meier, Walter, J. B. Uytterhoeven, Eidgenössische Technische Hochschule., and Schweizerische Chemische Gesellschaft. *Molecular Sieves.* Advances in Chemistry Series, no. 121. Washington, D.C.: American Chemical Society, 1973.

Meiwes-Broer, Karl-Heinz. *Metal Clusters at Surfaces: Structure, Quantum Properties, Physical Chemistry.* New York: Springer, 2000.

Meyer, Gerd, Dieter Naumann, and Lars Wesemann. *Inorganic Chemistry Highlights.* Weinheim, Germany: Wiley-VCH, 2002.

Meyers, Robert A. *Encyclopedia of Analytical Chemistry: Applications, Theory, and Instrumentation.* New York: Wiley, 2000.

———. *Encyclopedia of Lasers and Optical Technology.* San Diego, Calif.: Academic Press, 1991.

Mezey, Paul G. *Shape in Chemistry: An Introduction to Molecular Shape and Topology.* New York: VCH, 1993.

———. *Potential Energy Hypersurfaces.* New York: Elsevier, 1987.

Mezey, Paul G., Beverly E. Robertson, and ebrary Inc. *Electron, Spin and Momentum Densities and Chemical Reactivity.* Vol. 21 of *Understanding Chemical Reactivity.* Boston: Kluwer Academic Publishers, 2000.

Miller, Audrey, and Philippa H. Solomon. *Writing Reaction Mechanisms in Organic Chemistry.* 2nd ed. Advanced Organic Chemistry Series. San Diego, Calif.: Harcourt/Academic Press, 2000.

Miller, A. R. *The Theory of Solutions of High Polymers.* Oxford: Clarendon Press, 1948.

Miller, Bernard. *Advanced Organic Chemistry: Reactions and Mechanisms.* 2nd ed. Upper Saddle River, N.J.: Pearson Education, 2004.

Miller, Lawrence P. *Phytochemistry.* New York: Van Nostrand Reinhold, 1973.

Miller, L. S., and J. B. Mullin. *Electronic Materials: From Silicon to Organics.* New York: Plenum Press, 1991.

Millero, Frank J. *The Physical Chemistry of Natural Waters.* Wiley-Interscience Series in Geochemistry. New York: Wiley-Interscience, 2001.

Mitra, S. *Sample Preparation Techniques in Analytical Chemistry.* Vol. 162 of *Chemical Analysis.* Hoboken, N.J.: Wiley-Interscience, 2003.

Mittal, K. L., and D. O. Shah. *Adsorption and Aggregation of Surfactants in Solution.* Surfactant Science Series, vol. 109. New York: Marcel Dekker, 2003.

Morra, Marco. *Water in Biomaterials Surface Science.* New York: Wiley, 2001.

Morris, Richard, and ebrary Inc. *The Last Sorcerers: The Path from Alchemy to the Periodic Table.* Washington, D.C.: Joseph Henry Press, 2003.

Moss, Robert A., Matthew Platz, and Maitland Jones. *Reactive Intermediate Chemistry.* Hoboken, N.J.: Wiley-Interscience, 2004.

Moss, Tom. *DNA-Protein Interactions: Principles and Protocols.* 2nd ed. Vol. 148 of *Methods in Molecular Biology.* Totowa, N.J.: Humana Press, 2001.

Moss, Walker H., and ebrary Inc. *Annual Reports in Combinatorial Chemistry and Molecular Diversity.* New York: Kluwer Academic Publishers, 1997.

Mottana, Annibale, Mineralogical Society of America, and Accademia nazionale dei Lincei. *Micas: Crystal Chemistry and Metamorphic Petrology.* Vol. 46 of *Reviews in Mineralogy and Geochemistry.* Washington, D.C.: Mineralogical Society of America, 2002.

Mueller, Michael R. *Fundamentals of Quantum Chemistry: Molecular Spectroscopy and Modern Electronic Structure Computations.* New York: Kluwer Academic/Plenum Publishers, 2001.

Mulchandani, Ashok, Omowunmi A. Sadik, American Chemical Society Division of Environmental Chemistry, and American Chemical Society. *Chemical and Biological Sensors for Environmental Monitoring.* ACS Symposium Series, no. 762. Washington, D.C.: American Chemical Society, 2000.

Mulliken, Robert Sanderson, D. A. Ramsay, and Jürgen Hinze. *Selected Papers of Robert S. Mulliken.* Chicago: University of Chicago Press, 1975.

Mulliken, Samuel Parsons. *A Method for the Identification of Pure Organic Compounds by a Systematic Analytical Procedure Based on Physical Properties and Chemical Reactions.* New York: Wiley, Chapman & Hall, 1904.

Mullins, Eamonn, and Royal Society of Chemistry. *Statistics for the Quality Control Chemistry Laboratory.* Cambridge, U.K.: Royal Society of Chemistry, 2003.

Myasoedova, Vera V. *Physical Chemistry of Non-Aqueous Solutions of Cellulose and Its Derivatives.* Wiley Series in Solution Chemistry, vol. 5. New York: Wiley, 2000.

Myers, Anne B., and Thomas R. Rizzo. *Laser Techniques in Chemistry.* Vol. 23 of *Techniques of Chemistry.* New York: Wiley, 1995.

Myers, Drew. *Surfaces, Interfaces, and Colloids: Principles and Applications.* 2nd ed. New York: Wiley-VCH, 1999.

———. *Surfactant Science and Technology.* 2nd ed. New York: VCH, 1992.

———. *Surfaces, Interfaces, and Colloids: Principles and Applications.* New York: VCH Publishers, 1991.

Nail, Steven L., and Michael J. Akers. *Development and Manufacture of Protein Pharmaceuticals.* Vol. 14 of *Pharmaceutical Biotechnology.* New York: Kluwer Academic/Plenum Publishers, 2002.

Nair, Padmanabhan P., and David Kritchevsky. *The Bile Acids; Chemistry, Physiology, and Metabolism.* New York: Plenum Press, 1971.

Nakamura, Akira, Norikazu Ueyama, and Kizashi Yamaguchi. *Organometallic Conjugation: Structures, Reactions, and Functions of D-D and D-[Pi] Conjugated Systems.* Springer Series in Chemical Physics, vol. 73. New York: Springer, 2002.

Nakamura, T. *Chemistry of Nanomolecular Systems: Towards the Realization of Molecular Devices.* Springer Series in Chemical Physics, vol. 70. New York: Springer, 2002.

Nalwa, Hari Singh. *Advances in Surface Science.* Vol. 38 of *Experimental Methods in the Physical Sciences.* San Diego, Calif.: Academic Press, 2001.

———. *Handbook of Surfaces and Interfaces of Materials.* San Diego, Calif.: Academic Press, 2001.

———. *Ferroelectric Polymers: Chemistry, Physics, and Applications.* Plastics Engineering, no. 28. New York: Marcel Dekker, 1995.

Nascimento, Marco Antonio Chaer. *Theoretical Aspects of Heterogeneous Catalysis.* Vol. 8 of *Progress in Theoretical Chemistry and Physics.* Boston: Kluwer Academic Publishers, 2001.

National Research Council (U.S.) Committee on Challenges for the Chemical Sciences in the 21st Century, and ebrary Inc. *Beyond the Molecular Frontier Challenges for Chemistry and Chemical Engineering.* Washington, D.C.: National Academies Press, 2003.

Nicholson, John W., and Royal Society of Chemistry (Great Britain). *The Chemistry of Medical and Dental Materials.* RSC Materials Monographs. Cambridge, U.K.: Royal Society of Chemistry, 2002.

———. *The Chemistry of Polymers.* Royal Society of Chemistry Paperbacks. Cambridge, U.K.: Royal Society of Chemistry, 1991.

Nyquist, Richard A. *Interpreting Infrared, Raman, and Nuclear Magnetic Resonance Spectra.* San Diego, Calif.: Academic Press, 2001.

Nyquist, Richard A., and Ronald O. Kagel. *Infrared Spectra of Inorganic Compounds (3800–45 cm).* Sup. 1. New York: Academic Press, 1971.

Odegard, Gregory M., and Langley Research Center. *Equivalent-Continuum Modeling of Nano-Structured Materials.* NASA Technical Memorandum, NASA/TM-2001-210863. Hampton, Va.: National Aeronautics and Space Administration Langley Research Center; available from NASA Center for AeroSpace Information (CASI), 2001.

Ogden, J. S. *Introduction to Molecular Symmetry.* Oxford Chemistry Primers, no. 97. New York: Oxford University Press, 2001.

Ojima, Iwao, James R. McCarthy, John T. Welch, American Chemical Society Division of Fluorine Chemistry, and American Chemical Society Division of Medicinal Chemistry. *Biomedical Frontiers of Fluorine Chemistry.* ACS Symposium Series, no. 639. Washington, D.C.: American Chemical Society, 1996.

Ojima, Iwao, Gregory D. Vite, Karl-Heinz Altmann, American Chemical Society Division of Organic Chemistry, American Chemical Society Division of Medicinal Chemistry, and American Chemical Society. *Anticancer Agents: Frontiers in Cancer Chemotherapy.* ACS Symposium Series, no. 796. Washington, D.C.: American Chemical Society, 2001.

Olah, George A. *A Life of Magic Chemistry: Autobiographical Reflections of a Nobel Prize Winner.* New York: Wiley-Interscience, 2001.

———. *Friedel-Crafts Chemistry.* Interscience Monographs on Organic Chemistry. New York: Wiley, 1973.

Olah, George A., Ripudaman Malhotra, and Subhash C. Narang. *Nitration: Methods and Mechanisms.* Organic

Nitro Chemistry Series. New York: VCH Publishers, 1989.

Olah, George A., and Árpád Molnár. *Hydrocarbon Chemistry.* 2nd ed. Hoboken, N.J.: Wiley-Interscience, 2003.

Olah, George A., and Paul von R. Schleyer. *Cage Hydrocarbons.* New York: Wiley, 1990.

Olah, George A., and David R. Squire. *Chemistry of Energetic Materials.* San Diego: Academic Press, 1991.

Ono, Noboru. *The Nitro Group in Organic Synthesis.* Organic Nitro Chemistry Series. New York: Wiley-VCH, 2001.

Ott, J. Bevan, and Juliana Boerio-Goates. *Chemical Thermodynamics: Principles and Applications.* San Diego, Calif.: Academic Press, 2000.

Parcher, J. F., and T. L. Chester. *Unified Chromatography.* ACS Symposium Series, no. 748. Washington, D.C.: American Chemical Society, 2000.

Parsons, A. F. *An Introduction to Free Radical Chemistry.* Malden, Mass.: Blackwell Science, 2000.

Parsons, Cyril Russell Hamilton, and Clare Dover. *The Elements and Their Order.* Foundations of Inorganic Chemistry, Foundations of Science Library, The Chemical Sciences. London: S. Low Marston, 1966.

Pasto, Daniel J., and Carl R. Johnson. *Laboratory Text for Organic Chemistry: A Source Book of Chemical and Physical Techniques.* Englewood Cliffs, N.J.: Prentice-Hall, 1979.

———. *Organic Structure Determination.* Prentice-Hall International Series in Chemistry. Englewood Cliffs, N.J.; Prentice-Hall, 1969.

Patai, Saul. *The Chemistry of Organic Germanium, Tin, and Lead Compounds.* The Chemistry of Functional Groups. New York: Wiley, 1995.

———. *The Chemistry of Organic Arsenic, Antimony, and Bismuth Compounds.* The Chemistry of Functional Groups. New York: Wiley, 1994.

———. *The Chemistry of Hydroxyl, Ether and Peroxide Groups.* The Chemistry of Functional Groups, sup. E2. New York: Wiley, 1993.

———. *The Chemistry of Sulphenic Acids and Their Derivatives.* The Chemistry of Functional Groups. New York: Wiley, 1990.

———. *The Chemistry of Sulphinic Acids, Esters and Their Derivatives.* The Chemistry of Functional Groups. New York: Wiley, 1990.

———. *The Chemistry of Double-Bonded Functional Groups.* The Chemistry of Functional Groups, sup. A. New York: Wiley, 1989.

———. *Patai's Guide to the Chemistry of Functional Groups.* The Chemistry of Functional Groups. New York: Wiley, 1989.

———. *The Chemistry of Peroxides.* The Chemistry of Functional Groups. New York: Wiley, 1983.

———. *The Chemistry of Ketenes, Allenes, and Related Compounds.* The Chemistry of Functional Groups. Chichester; New York: Wiley, 1980.

———. *Chemistry of Ethers, Crown Ethers, Hydroxyl Groups and Their Sulphur Analogues.* The Chemistry of Functional Groups, sup. E, pt. 1–2. New York: J. Wiley, 1980.

———. *The Chemistry of Acid Derivatives.* The Chemistry of Functional Groups, sup. B, pt. 1–2. New York: Wiley, 1979.

———. *The Chemistry of the Carbon-Carbon Triple Bond.* The Chemistry of Functional Groups. New York: Wiley, 1978.

———. *The Chemistry of Diazonium and Diazo Groups.* The Chemistry of Functional Groups. New York: Wiley, 1978.

———. *The Chemistry of Cyanates and Their Thio Derivatives.* The Chemistry of Functional Groups. New York: Wiley, 1977.

———. *The Chemistry of Double-Bonded Functional Groups.* The Chemistry of Functional Groups, sup. A, pt. 1–2. New York: Wiley, 1977.

———. *The Chemistry of Amidines and Imidates.* The Chemistry of Functional Groups. New York: Wiley, 1975.

———. *The Chemistry of the Thiol Group.* The Chemistry of Functional Groups. New York: Wiley, 1974.

———. *The Chemistry of the Quinonoid Compounds.* The Chemistry of Functional Groups. New York: Interscience, 1974.

———. *The Chemistry of the Azido Group.* The Chemistry of Functional Groups. New York: Interscience Publishers, 1971.

———. *The Chemistry of the Carbon-Nitrogen Double Bond.* The Chemistry of Functional Groups. New York: Interscience Publishers, 1970.

———. *The Chemistry of Carboxylic Acids and Esters.* The Chemistry of Functional Groups. New York: Interscience Publishers, 1969.

———. *The Chemistry of the Carbonyl Group.* The Chemistry of Functional Groups. New York: Interscience Publishers, 1966.

———. *Glossary of Organic Chemistry, Including Physical Organic Chemistry.* New York: Interscience Publishers, 1962.

Patai, Saul, and Zvi Rappoport. *The Chemistry of Halides, Pseudo-Halides, and Azides.* rev. ed. The Chemistry of Functional Groups, sup. D2. New York: Wiley, 1995.

———. *The Chemistry of Sulphur-Containing Functional Groups.* The Chemistry of Functional Groups, sup. New York: Wiley, 1993.

———. *The Chemistry of Amidines and Imidates.* The Chemistry of Functional Groups. New York: Wiley, 1991.

———. *The Chemistry of Sulphonic Acids, Esters, and Their Derivatives.* The Chemistry of Functional Groups. New York: Wiley, 1991.

———. *The Chemistry of Organic Silicon Compounds.* The Chemistry of Functional Groups. New York: Wiley, 1989.

———. *The Chemistry of Enones.* The Chemistry of Functional Groups. New York: Wiley, 1989.

———. *The Chemistry of Organic Selenium and Tellurium Compounds.* The Chemistry of Functional Groups. New York: Wiley, 1986.

———. *The Chemistry of Halides, Pseudo-Halides, and Azides.* The Chemistry of Functional Groups, sup. D. New York: Wiley, 1983.

———. *The Chemistry of Triple-Bonded Functional Groups.* The Chemistry of Functional Groups, sup. C. New York: Wiley, 1983.

Patil, S. H., and K. T. Tang. *Asymptotic Methods in Quantum Mechanics: Application to Atoms, Molecules, and Nuclei.* Springer Series in Chemical Physics, vol. 64. New York: Springer, 2000.

Pauleau, Y. *Chemical Physics of Thin Film Deposition Processes for Micro- and Nano-Technologies.* Vol. 5 of *Mathematics, Physics, and Chemistry.* NATO Science Series, ser. II. Boston: Kluwer Academic Publishers, 2002.

Pauling, Linus. *College Chemistry: An Introductory Textbook of General Chemistry.* 3rd ed. A Series of Books in Chemistry. San Francisco: W. H. Freeman, 1964.

———. *The Nature of the Chemical Bond and the Structure of Molecules and Crystals: An Introduction to Modern Structural Chemistry.* 3rd ed. The George Fisher Baker Non-Resident Lectureship in Chemistry at Cornell University. Vol. 18. Ithaca, N.Y.: Cornell University Press, 1960.

———. *General Chemistry: An Introduction to Descriptive Chemistry and Modern Chemical Theory.* San Francisco: W. H. Freeman, 1947.

Pauling, Linus, and Peter Pauling. *Chemistry.* A Series of Books in Chemistry. San Francisco: W. H. Freeman, 1975.

Pauling, Linus, and E. Bright Wilson. *Introduction to Quantum Mechanics, with Applications to Chemistry.* New York: McGraw-Hill, 1935.

Pauson, Peter L. *Organometallic Chemistry.* London: Edward Arnold, 1967.

Pavesi, Lorenzo, and E. V. Buzaneva. *Frontiers of Nano-Optoelectronic Systems.* Boston: Kluwer Academic Publishers, 2000.

Pawliszyn, Janusz. *Sampling and Sample Preparation for Field and Laboratory: Fundamental and New Directions in Sample Preparation.* Vol. 37 of *Comprehensive Analytical Chemistry.* Boston: Elsevier Science, 2002.

———. *Solid Phase Microextraction: Theory and Practice.* New York: Wiley-VCH, 1997.

Penadés, Soledad, and Jürgen-Hinrich Fuhrhop. *Host-Guest Chemistry: Mimetic Approaches To Study Carbohydrate Recognition.* Topics in Current Chemistry, no. 218. New York: Springer, 2002.

Pennington, S. R., and Michael J. Dunn. *Proteomics: From Protein Sequence to Function.* New York: Springer-Verlag, 2001.

Poss, Andrew J. *Library Handbook for Organic Chemists.* New York: Chemical Pub. Co., 2000.

Quinkert, Gerhard, and M. Volkan Kisakürek. *Essays in Contemporary Chemistry: From Molecular Structure Towards Biology.* New York: Wiley-VCH, 2001.

Rahman, Atta-ur. *New Advances in Analytical Chemistry.* Amsterdam: Harwood Academic, 2000.

Rahman, Atta-ur, and Philip W. Le Quesne. *New Trends in Natural Products Chemistry 1986.* Proceedings of the Second International Symposium and Pakistan-U.S. Binational Workshop on Natural Products Chemistry, Karachi, Pakistan, 18–25 January 1986, Studies in Organic Chemistry, no. 26. Oxford: Elsevier, 1986.

Rahman, Atta-ur, and Zahir Shah. *Stereoselective Synthesis in Organic Chemistry.* New York: Springer-Verlag, 1993.

Rauk, Arvi. *Orbital Interaction Theory of Organic Chemistry.* 2nd ed. New York: Wiley, 2001.

Ravve, A. *Principles of Polymer Chemistry.* 2nd ed. New York: Kluwer Academic/Plenum Publishers, 2000.

Razumas, V. I., Björn Lindman, and T. Nylander. *Surface and Colloid Science.* Vol. 116 of *Progress in Colloid and Polymer Science.* New York: Springer-Verlag, 2001.

Reeve, Roger N. *Introduction to Environmental Analysis.* Analytical Techniques in the Sciences. New York: Wiley, 2002.

Reeve, Roger N., and John D. Barnes. *Environmental Analysis (Analytical Chemistry by Open Learning).* New York: Wiley, 1994.

Reichardt, C. *Solvents and Solvent Effects in Organic Chemistry.* 3rd ed. Weinheim, Germany: Wiley-VCH, 2003.

Reis, Rui L., Daniel Cohn, and North Atlantic Treaty Organization Scientific Affairs Division. *Polymer Based Systems on Tissue Engineering, Replacement and Regeneration.* Vol. 86 of *Mathematics, Physics, and Chemistry.* NATO Science Series, ser. II. Boston: Kluwer Academic Publishers, 2002.

Reynolds, Charles A. *Principles of Analytical Chemistry.* Boston: Allyn and Bacon, 1966.

Ricca, Renzo L. *An Introduction to the Geometry and Topology of Fluid Flows.* Vol. 47 of *Mathematics, Physics, and*

Chemistry. NATO Science Series, ser. II. Boston: Kluwer Academic, 2001.

Richet, Pascal. *The Physical Basis of Thermodynamics: With Applications to Chemistry.* New York: Kluwer Academic/Plenum Publishers, 2001.

Riegel, Emil Raymond, and James Albert Kent. *Riegel's Handbook of Industrial Chemistry.* 9th ed. New York: Chapman & Hall, 1992.

Rieger, Philip Henri, Jerome Laib Rosenberg, Lawrence M. Epstein, and NetLibrary Inc. *College Chemistry Crash Course: Based on Schaum's Outline of College Chemistry by Jerome L. Rosenberg and Lawrence M. Epstein.* Schaum's Easy Outlines. New York: McGraw-Hill, 2000.

Ritchie, Calvin D. *Physical Organic Chemistry: The Fundamental Concepts.* 2nd ed. New York: Marcel Dekker, 1990.

———. *Physical Organic Chemistry: The Fundamental Concepts.* Vol. 4 of *Studies in Organic Chemistry.* New York: Marcel Dekker, 1975.

Ritchie, G. A. D., and D. S. Sivia. *Foundations of Physics for Chemists.* Oxford Chemistry Primers, no. 93. New York: Oxford University Press, 2000.

Rogers, Allen, and C. C. Furnas. *Rogers' Industrial Chemistry: A Manual for the Student and Manufacturer.* New York: D. Van Nostrand, 1942.

Rogers, Donald. *Computational Chemistry Using the PC.* 3rd ed. Hoboken, N.J.: Wiley-Interscience, 2003.

Rogers, J. E., and William Barnaby Whitman. *Microbial Production and Consumption of Greenhouse Gases: Methane, Nitrogen Oxides, and Halomethanes.* Washington, D.C.: American Society for Microbiology, 1991.

Rogers, L. J., and J. R. Gallon. *Biochemistry of the Algae and Cyanobacteria.* New York: Clarendon Press, 1988.

Ropp, R. C. *Solid State Chemistry.* Boston: Elsevier, 2003.

———. *The Chemistry of Artificial Lighting Devices: Lamps, Phosphors and Cathode Ray Tubes.* Studies in Inorganic Chemistry, no. 17. New York: Elsevier, 1993.

———. *Inorganic Polymeric Glasses.* Studies in Inorganic Chemistry, no. 15. New York: Elsevier, 1992.

Rosoff, Morton. *Nano-Surface Chemistry.* New York: Marcel Dekker, 2002.

Rosoff, Morton, and NetLibrary Inc. *Nano-Surface Chemistry.* New York: Marcel Dekker, 2001.

Rossi, David T., and Michael W. Sinz. *Mass Spectrometry in Drug Discovery.* New York: Marcel Dekker, 2002.

Rowlinson, J. S. "The Perfect Gas," In *The International Encyclopedia of Physical Chemistry and Chemical Physics (Oxford).* Topic 10: The Fluid State, vol. 5. Oxford, U.K.: Pergamon, 1963.

Royal Society of Chemistry (Great Britain), and Chemical Society (Great Britain). *Organic Compounds of Sulphur, Selenium, and Tellurium.* Vol. 1 of *A Specialist Periodical Report.* London: Royal Society of Chemistry, 1969–1970.

Sacksteder, Kurt R., and NASA Glenn Research Center. *Sixth International Microgravity Combustion Workshop.* NASA/CP-2001-210826. Hanover, Md.: NASA Center for Aerospace Information, 2001.

Saifutdinov, Rafik Galimzyanovich. *Electron Paramagnetic Resonance in Biochemistry and Medicine.* Topics in Applied Chemistry. New York: Kluwer Academic/Plenum, 2001.

Savignac, Philippe, and Bogdan Iorga. *Modern Phosphonate Chemistry.* Boca Raton, Fla.: CRC Press, 2003.

Scamehorn, John F., and Jeffrey H. Harwell. *Surfactant-Based Separations: Science and Technology.* ACS Symposium Series, no. 740. Washington, D.C.: American Chemical Society, 2000.

Schalley, Christoph A. *Modern Mass Spectrometry.* Topics in Current Chemistry, no. 225. New York: Springer, 2003.

Scheirs, John, and Timothy E. Long. *Modern Polyesters: Chemistry and Technology of Polyesters and Copolyesters.* Wiley Series in Polymer Science. Hoboken, N.J.: Wiley, 2003.

Schmid, R., and V. N. Sapunov. *Non-Formal Kinetics: In Search for Chemical Reaction Pathways.* Vol. 14 of *Monographs in Modern Chemistry.* Deerfield Beach, Fla.: Verlag Chemie, 1982.

Schubert, U., and Nicola Hüsing. *Synthesis of Inorganic Materials.* Weinheim, Germany: Wiley-VCH, 2000.

Schulz, Hedda, and Ursula Georgy. *From Ca to Cas Online: Databases in Chemistry.* 2nd ed. New York: Springer-Verlag, 1994.

Schulz, Horst D., Astrid Hadeler, and Deutsche Forschungsgemeinschaft. *Geochemical Processes in Soil and Groundwater: Measurement-Modelling-Upscaling: Geoproc 2002.* Weinheim, Germany: Wiley-VCH, 2003.

Schulze, E. D. *Global Biogeochemical Cycles in the Climate System.* San Diego, Calif.: Academic Press, 2001.

Schupp, O. E., Edmond S. Perry, and Arnold Weissberger. *Gas Chromatography.* Vol. 13 of *Technique of Organic Chemistry.* New York: Interscience Publishers, 1968.

Schwartz, Anthony M., James W. Perry, and Julian Berch. *Surface Active Agents: Their Chemistry and Technology.* New York: Interscience, 1949.

Schwartz, Steven David, and ebrary Inc. *Theoretical Methods in Condensed Phase Chemistry.* Vol. 5 of *Progress in Theoretical Chemistry and Physics.* Boston: Kluwer Academic, 2000.

Schwarz, James A., and Cristian I. Contescu. *Surfaces of Nanoparticles and Porous Materials.* Surfactant Science Series, vol. 78. New York: Marcel Dekker, 1999.

Schwarz, Robert, Lawrence W. Bass, and Alfred Werner. *The Chemistry of the Inorganic Complex Compounds: An Introduction to Werner's Coordination Theory.* New York: Wiley, 1923.

Schwarzenbach, René P., P. M. Gschwend, and Dieter M. Imboden. *Environmental Organic Chemistry.* 2nd ed. Hoboken, N.J.: Wiley, 2003.

Schweizer, W. *Numerical Quantum Dynamics.* Vol. 9 of *Progress in Theoretical Chemistry and Physics.* Boston: Kluwer Academic, 2001.

Sellergren, Börje. *Molecularly Imprinted Polymers: Man-Made Mimics of Antibodies and Their Applications in Analytical Chemistry.* Vol. 23 of *Techniques and Instrumentation in Analytical Chemistry.* New York: Elsevier, 2001.

Seneci, Pierfausto. *Solid Phase Synthesis and Combinatorial Technologies.* New York: Wiley, 2000.

Sengers, J. V., and International Union of Pure and Applied Chemistry Commission on Thermodynamics. *Equations of State for Fluids and Fluid Mixtures.* Vol. 5 of *Experimental Thermodynamics.* New York: Elsevier, 2000.

Senning, Alexander. *Sulfur in Organic and Inorganic Chemistry.* New York: Marcel Dekker, 1982.

Seymour, Raymond Benedict. *Pioneers in Polymer Science, Chemists and Chemistry.* Boston: Kluwer Academic, 1989.

———. *Introduction to Polymer Chemistry.* New York: McGraw-Hill, 1971.

Seymour, Raymond Benedict, and American Chemical Society Division of the History of Chemistry. *History of Polymer Science and Technology.* New York: Marcel Dekker, 1982.

Seymour, Raymond Benedict, and Charles E. Carraher. *Seymour/Carraher's Polymer Chemistry.* 5th ed. Vol. 14 of *Undergraduate Chemistry.* New York: Marcel Dekker, 2000.

Seymour, Raymond Benedict, G. Allan Stahl, and American Chemical Society. *Macromolecular Solutions: Solvent-Property Relationships in Polymers.* New York: Pergamon, 1982.

Shao, Jiping, and Rensselaer Polytechnic Institute Chemistry Dept. *Linear Liquid Crystal Polymers from Phenylene-Naphthalene Monomers.* New York: Marcel Dekker, 2002.

Shen, Ying, and Rensselaer Polytechnic Institute Chemistry Dept. *Synthesis of Aniline Side Chain Rigid Polymers.* New York: Marcel Dekker, 2000.

Sherrington, D. C., Adrian P. Kybett, and Royal Society of Chemistry. *Supported Catalysts and Their Applications.* Special Publication no. 266. Cambridge, U.K.: Royal Society of Chemistry, 2001.

Shilov, A. E., G. B. Shulpin, and ebrary Inc. *Activation and Catalytic Reactions of Saturated Hydrocarbons in the Presence of Metal Complexes.* Vol. 22 of *Catalysis by Metal Complexes.* Boston: Kluwer Academic, 2000.

Silbey, Robert J., and Robert A. Alberty. *Physical Chemistry.* 3rd ed. New York: Wiley, 2001.

Silva, A. Fernando, and North Atlantic Treaty Organization Scientific Affairs Division. *Trends in Interfacial Electrochemistry.* Boston: Kluwer Academic, 1986.

Silva, J. J. R. Fraústo da, and R. J. P. Williams. *The Biological Chemistry of the Elements: The Inorganic Chemistry of Life.* 2nd ed. New York: Oxford University Press, 2001.

———. *The Biological Chemistry of the Elements: The Inorganic Chemistry of Life.* New York: Oxford University Press, 1991.

Silverman, Louis. *The Determination of Impurities in Nuclear Grade Sodium Metal.* 1st ed. International Series of Monographs in Analytical Chemistry, vol. 44. New York: Pergamon, 1971.

Silverman, Richard B. *The Organic Chemistry of Enzyme-Catalyzed Reactions.* San Diego, Calif.: Academic Press, 2000.

———. *The Organic Chemistry of Drug Design and Drug Action.* San Diego, Calif.: Academic Press, 1992.

Simon, J., and Pierre Bassoul. *Design of Molecular Materials: Supramolecular Engineering.* New York: Wiley, 2000.

Smith, Michael, and Jerry March. *March's Advanced Organic Chemistry: Reactions, Mechanisms, and Structure.* 5th ed. New York: Wiley, 2001.

Sorenson, Wayne R., and Tod W. Campbell. *Preparative Methods of Polymer Chemistry.* New York: Interscience Publishers, 1961.

Sorenson, Wayne R., Wilfred Sweeny, and Tod W. Campbell. *Preparative Methods of Polymer Chemistry.* 3rd ed. New York: Wiley-Interscience, 2001.

Speight, J. G. *Perry's Standard Tables and Formulas for Chemical Engineers.* New York: McGraw-Hill, 2003.

Speight, J. G., and NetLibrary Inc. *The Chemistry and Technology of Petroleum.* 3rd ed. Vol. 76 of *Chemical Industries.* New York: Marcel Dekker, 1999.

Spivey, James J., and ebrary Inc. *Catalysis.* Cambridge, U.K.: Royal Society of Chemistry, 2002.

Steffens, Erhard, Revaz Shanidze, and North Atlantic Treaty Organization Scientific Affairs Division. *Spin Structure of the Nucleon.* Vol. 111 of *Mathematics, Physics, and Chemistry.* NATO Science Series, ser. II. Boston: Kluwer Academic, 2003.

Stegemeyer, H. *Liquid Crystals.* Vol. 3 of *Topics in Physical Chemistry.* New York: Springer, 1994.

Stewart, Alfred W. *Stereochemistry.* Text-Books of Physical Chemistry. New York: Longmans Green, 1907.

Stewart, Jeffrey R. *An Encyclopedia of the Chemical Process Industries.* New York: Chemical Pub. Co., 1956.

Stewart, Kent K., and Richard E. Ebel. *Chemical Measurements in Biological Systems.* Techniques in Analytical Chemistry Series. New York: Wiley-Interscience, 2000.

Stoeppler, M. *Sampling and Sample Preparation: Practical Guide for Analytical Chemists.* New York: Springer, 1997.

Stoeppler, M., Wayne R. Wolf, and Peter J. Jenks. *Reference Materials for Chemical Analysis: Certification, Availability, and Proper Usage.* New York: Wiley-VCH, 2001.

Strathern, Paul. *Mendeleyev's Dream: The Quest for the Elements.* New York: St. Martins Press, 2001.

Stratospheric Aerosol and Gas Experiment II Team (Langley Research Center), and NASA Langley Atmospheric Sciences Data Center. *Stratospheric Aerosol and Gas Experiment II Sage II.* Version 6.0. Hampton, Va.: NASA Langley Research Center Atmospheric Sciences Data Center, 2000.

Sucholeiki, Irving. *High-Throughput Synthesis: Principles and Practices.* New York: Marcel Dekker, 2001.

Suits, Arthur G., and Robert E. Continetti. *Imaging in Chemical Dynamics.* ACS Symposium Series no. 770. Washington, D.C.: American Chemical Society, 2001.

Sun, S. F. *Physical Chemistry of Macromolecules: Basic Principles and Issues.* New York: Wiley, 1994.

Surko, Clifford M., and Franco A. Gianturco. *New Directions in Antimatter Chemistry and Physics.* Boston: Kluwer Academic, 2001.

Sutton, Derek. *Electronic Spectra of Transition Metal Complexes: An Introductory Text.* European Chemistry Series. New York: McGraw-Hill, 1968.

Sutton, Leslie. *Tables of Interatomic Distances and Configuration in Molecules and Ions: Supplement, 1956–1959.* Special Publication, no. 18. London: Chemical Society, 1965.

Sutton, Raul, Bernard W. Rockett, and Peter Swindells. *Chemistry for the Life Sciences.* Lifelines. New York: Taylor & Francis, 2000.

Swadesh, Joel. *HPLC: Practical and Industrial Applications.* 2nd ed. Analytical Chemistry Series. Boca Raton, Fla.: CRC Press, 2001.

Testa, Bernard. *Principles of Organic Stereochemistry.* Vol. 6 of *Studies in Organic Chemistry.* New York: Marcel Dekker, 1979.

Testa, Bernard, and Joachim M. Mayer. *Hydrolysis in Drug and Prodrug Metabolism: Chemistry, Biochemistry, and Enzymology.* Cambridge, U.K.: Wiley-VCH, 2003.

Thomas, Alan F. *Deuterium Labeling in Organic Chemistry.* New York: Appleton-Century-Crofts, 1971.

Thomas, Arthur Waldorf Spittell. *Colloid Chemistry.* 1st ed. International Chemical Series. Edited by J. F. Norris. New York: McGraw-Hill, 1934.

Thomas, David N., and Gerhard Dieckmann. *Sea Ice: An Introduction to Its Physics, Chemistry, Biology, and Geology.* Malden, Mass.: Blackwell Science, 2003.

Thompson, J. M. T., and ebrary Inc. *Visions of the Future Chemistry and Life Science.* New York: Cambridge University Press, 2001.

Tinoco, Ignacio. *Physical Chemistry: Principles and Applications in Biological Sciences.* 4th ed. Upper Saddle River, N.J.: Prentice Hall, 2002.

Tipping, Edward. *Cation Binding by Humic Substances.* 1st ed. Cambridge Environmental Chemistry Series, no. 12. New York: Cambridge University Press, 2002.

Toda, Fumio. *Organic Solid State Reactions.* Boston: Kluwer Academic, 2002.

Todres, Zory V. *Organic Ion Radicals: Chemistry and Applications.* New York: Marcel Dekker, 2003.

Tranter, Roy L. *Design and Analysis in Chemical Research.* Sheffield Analytical Chemistry. Boca Raton, Fla.: CRC Press, 2000.

Trossen, Dirk, Andre Schuppen, Michael Wallbaum, and ebrary Inc. *Shared Workplace for Collaborative Engineering.* Hershey, Penn.: Idea Group, 2002.

Turova, Nataliya Ya, and ebrary Inc. *The Chemistry of Metal Alkoxides.* Boston: Kluwer Academic, 2002.

Turro, Nicholas J. *Molecular Photochemistry.* Frontiers in Chemistry. New York: W. A. Benjamin, 1965.

Van Ness, H. C. *Understanding Thermodynamics. Dover Books on Engineering.* New York: Dover, 1983.

———. *Classical Thermodynamics of Non-Electrolyte Solutions.* New York: Pergamon, 1964.

VCH Publishers. *Ullman's Encyclopedia of Industrial Chemistry.* 6th ed. Weinheim, Germany: Wiley-VCH, 2002.

Vekshin, N. L. *Photonics of Biopolymers.* Biological and Medical Physics Series. New York: Springer, 2002.

Velde, B. *Introduction to Clay Minerals: Chemistry, Origins, Uses, and Environmental Significance.* New York: Chapman & Hall, 1992.

Venables, John. *Introduction to Surface and Thin Film Processes.* New York: Cambridge University Press, 2000.

Visconti, Guido. *Fundamentals of Physics and Chemistry of the Atmosphere.* New York: Springer, 2001.

Vögtle, F., and F. Alfter. *Supramolecular Chemistry: An Introduction.* New York: Wiley, 1991.

Vögtle, F., Christoph A. Schalley, and H. F. Chow. *Dendrimers IV: Metal Coordination, Self Assembly, Catalysis.* Vol. 217 of *Topics in Current Chemistry.* New York: Springer, 2001.

Vögtle, F., J. F. Stoddart, and Masakatsu Shibasaki. *Stimulating Concepts in Chemistry.* New York: Wiley-VCH, 2000.

Vögtle, F., and E. Weber. *Host Guest Complex Chemistry: Macrocycles—Synthesis, Structures, Applications.* New York: Springer, 1985.

Walczak, Beata. *Wavelets in Chemistry.* Vol. 22 of *Data Handling in Science and Technology.* New York: Elsevier Science B.V., 2000.

Waluk, Jacek. *Conformational Analysis of Molecules in Excited States.* Methods in Stereochemical Analysis. New York: Wiley-VCH, 2000.

Washburn, Edward W., National Research Council (U.S.), International Research Council, National Academy of Sciences (U.S.), and Knovel (firm). *International Critical Tables of Numerical Data, Physics, Chemistry, and Technology.* 1st electronic ed. Knovel Scientific and Engineering Databases. Norwich, N.Y.: Knovel, 2003.

Webb, G. A., and ebrary Inc. *Nuclear Magnetic Resonance.* Specialist Periodical Report. London: Royal Society of Chemistry, 2002.

———. *Nuclear Magnetic Resonance.* Specialist Periodical Report. London: Royal Society of Chemistry, 2001.

———. *Nuclear Magnetic Resonance.* Specialist Periodical Report. London: Royal Society of Chemistry, 2000.

———. *Nuclear Magnetic Resonance.* London: Royal Society of Chemistry, 1999.

———. *Nuclear Magnetic Resonance.* Specialist Periodical Report. London: Royal Society of Chemistry, 1998.

Weber, Arthur L., and U.S. National Aeronautics and Space Administration. *Oligoglyceric Acid Synthesis by Autocondensation of Glyceroyl Thioester.* NASA Contractor Report, NASA CR-177130. Washington, D.C.: National Aeronautics and Space Administration, National Technical Information Service distributor, 1986.

———. *Stereoselective Formation of a 2′, 3′-Aminoacyl Ester of a Nucleotide.* NASA Contractor Report, NASA CR-176830. Washington, D.C.: National Aeronautics and Space Administration, National Technical Information Service distributor, 1986.

Weber, Walter J. *Environmental Systems and Processes: Principles, Modeling, and Design.* New York: Wiley-Interscience, 2001.

Weber, Walter J., and Francis A. DiGiano. *Process Dynamics in Environmental Systems.* Environmental Science and Technology. New York: Wiley, 1996.

Weiner, Eugene R. *Applications of Environmental Chemistry: A Practical Guide for Environmental Professionals.* Boca Raton, Fla.: Lewis, 2000.

Wermuth, C. G. *The Practice of Medicinal Chemistry.* 2nd ed. Boston: Academic Press, 2003.

White, Guy K. *Experimental Techniques in Low-Temperature Physics.* 3rd ed. Monographs on the Physics and Chemistry of Materials, no. 43. New York: Oxford University Press, 1987.

Williams, Andrew. *Free Energy Relationships in Organic and Bio-Organic Chemistry.* Cambridge, U.K.: RSC, 2003.

———. *Introduction to the Chemistry of Enzyme Action.* European Chemistry Series. New York: McGraw-Hill, 1969.

Williams, Alan F. *A Theoretical Approach to Inorganic Chemistry.* New York: Springer-Verlag, 1979.

Williams, Alan F., Carlo Floriani, and André E. Merbach. *Perspectives in Coordination Chemistry.* Weinheim, Germany: VCH, 1992.

Williams, Alex, and Helmut Günzler. *Handbook of Analytical Techniques.* New York: Wiley-VCH, 2001.

Williams, Arthur L., Harland D. Embree, and Harold J. DeBey. *Introduction to Chemistry.* Addison-Wesley Series in Chemistry. Reading, Mass.: Addison-Wesley, 1968.

Williams, David A. R., David J. Mowthorpe, and ACOL. *Nuclear Magnetic Resonance Spectroscopy: Analytical Chemistry by Open Learning.* New York: Wiley, 1986.

Williams, David J., and American Chemical Society Division of Polymer Chemistry. *Nonlinear Optical Properties of Organic and Polymeric Materials.* ACS Symposium Series no. 233. Washington, D.C.: American Chemical Society, 1983.

Williams, G. H. *Advances in Free-Radical Chemistry,* vol. 1. New York: Academic Press, 1965.

Williams, Linda D., and ebrary Inc. *Chemistry Demystified.* New York: McGraw-Hill, 2003.

Witczak, Zbigniew J., Kuniaki Tatsuta, American Chemical Society Division of Carbohydrate Chemistry, and American Chemical Society. *Carbohydrate Synthons in Natural Products Chemistry: Synthesis, Functionalization, and Applications.* ACS Symposium Series, no. 841. Washington, D.C.: American Chemical Society, 2003.

Wright, David A., Pamela Welbourn, and ebrary Inc. *Environmental Toxicology.* Cambridge Environmental Chemistry Series, no. 11. New York: Cambridge University Press, 2002.

Wrolstad, Ronald E., and Wiley InterScience. *Current Protocols in Food Analytical Chemistry.* New York: Wiley, 2002.

Wu, Albert M. *The Molecular Immunology of Complex Carbohydrates.* Vol. 491 of *Advances in Experimental Medicine and Biology.* New York: Kluwer, 2001.

Wu, David, and David Chandler. *Solutions Manual for "Introduction to Modern Statistical Mechanics."* New York: Oxford University Press, 1988.

Xiang, Xiao-Dong, and Ichiro Takeuchi. *Combinatorial Materials Synthesis.* New York: Marcel Dekker, 2003.

Yamase, Toshihiro, and Michael Thor Pope. *Polyoxometalate Chemistry for Nano-Composite Design.* Nanostructure Science and Technology. New York: Kluwer Academic/Plenum Publishers, 2002.

Yan, Bing. *Analysis and Purification Methods in Combinatorial Chemistry.* Vol. 163 of *Chemical Analysis.* Hoboken, N.J.: Wiley-Interscience, 2004.

Yan, Bing, and Anthony W. Czarnik. *Optimization of Solid-Phase Combinatorial Synthesis.* New York: Marcel Dekker, 2002.

Young, David C. *Computational Chemistry: A Practical Guide for Applying Techniques to Real World Problems.* New York: Wiley, 2001.

Young, G. J., American Chemical Society Division of Fuel Chemistry, and American Chemical Society Division of

Petroleum Chemistry. *Fuel Cells.* New York: Reinhold, 1960.

Young, Robyn V., and Suzanne Sessine. *World of Chemistry.* Detroit, Mich.: Gale Group, 2000.

Yui, Nobuhiko. *Supramolecular Design for Biological Applications.* Boca Raton, Fla.: CRC Press, 2002.

Zeng, Xian Ming, Gary P. Martin, Christopher Marriott, and ebrary Inc. *Particulate Interactions in Dry Powder Formulations for Inhalation.* London: Taylor & Francis, 2001.

Zerbe, Oliver. *BioNMR in Drug Research.* Vol. 16 of *Methods and Principles in Medicinal Chemistry.* Weinheim, Germany: Wiley-VCH, 2003.

Zielinski, Theresa Julia, and Mary L. Swift. *Using Computers in Chemistry and Chemical Education.* Washington, D.C.: American Chemical Society, 1997.

Zimmerman, Howard E. *Quantum Mechanics for Organic Chemists.* New York: Academic Press, 1975.

Zlatkevich, L. *Luminescence Techniques in Solid-State Polymer Research.* New York: Marcel Dekker, 1989.

Zoebelein, Hans. *Dictionary of Renewable Resources.* New York: VCH, 1997.

Zollinger, Heinrich. *Color Chemistry: Syntheses, Properties, and Applications of Organic Dyes and Pigments.* 2nd rev. ed. New York: VCH, 1991.

———. *Color Chemistry: Syntheses, Properties, and Applications of Organic Dyes and Pigments.* New York: VCH, 1987.

———. *Aromatic Compounds, Organic Chemistry,* ser. 1, vol. 3. Baltimore: University Park Press, 1973.

———. *Azo and Diazo Chemistry; Aliphatic and Aromatic Compounds.* New York: Interscience Publishers, 1961.

Zsigmondy, Richard, Ellwood Barker Spear, and John F. Norton. *The Chemistry of Colloids.* 1st ed. New York: John Wiley & Sons, 1917.

APPENDIX II

CHEMISTRY-RELATED WEB SITES

A sample of Internet sites containing a wealth of chemical data

American Chemical Society. Available online, URL:
http://www.chemistry.org/portal/a/c/s/1/home.html
Accessed April 28, 2004
The official Web site of ACS. Features a molecule of the week, a discussion board, polls, news, and calendar of events. A separate member's section features election results, membership information, resources, and benefits. Registered users can create their own personalized page. An "Educators & Students" page presents suitable-level essays on chemistry-related issues. There is free access to recent articles from *Industrial & Engineering Chemistry Research*, the *Journal of Combinatorial Chemistry*, and the 125 most-cited journals of the ACS publications program.

Chemistry. Available online, URL:
http://chemistry.about.com/mbody.htm
Accessed April 28, 2004
This educational Web site contains articles, dictionaries, science fair ideas, clip art, various databases, and archives of previously published material.

Chemistry Functions. Available online, URL:
http://www.stanford.edu/~glassman/chem
Accessed April 28, 2004
A unique Web site where you can solve a number of chemistry-related problems such as performing molar conversions and balancing equations. The site also has an interactive periodic table and links to other sites.

Chemistry Teaching Resources. Available online, URL:
http://www.anachem.umu.se/eks/pointers.htm
Accessed April 28, 2004
This Web site presents a comprehensive list of chemistry teaching resources on the Internet and features more than 20 categories.

Classic Chemistry. Available online, URL:
http://webserver.lemoyne.edu/faculty/giunta
Accessed April 28, 2004
This Web site presents the full text of classic papers presented by famous chemists, from Arrhenius to Zeeman. There are links to other history of chemistry sites. The site also contains classic calculations from many leading thinkers, but read the teachers notes first. A history-of-chemistry calendar presents a daily rundown of chemistry-related events and major discoveries.

Links for Chemists. Available online, URL:
http://www.liv.ac.uk/Chemistry/Links/links.html
Accessed April 28, 2004
This is a worldwide metasite that contains links to hundreds of chemistry-related sites from companies, research centers, publications, history, and education. This is a good site to start a search for chemistry material.

NIST Chemistry WebBook. Available online, URL:
http://webbook.nist.gov/chemistry
Accessed April 28, 2004
This Web site provides thermochemical, thermophysical, and ion energetics data compiled by NIST under the Standard Reference Data Program. The NIST (National Institute of

Standards and Technology) Chemistry WebBook provides users with easy access to chemical and physical property data for chemical species using the Internet. The data are from collections maintained by the NIST Standard Reference Data Program and outside contributors. Data in the WebBook system are organized by chemical species, and you can search for chemical species by several criteria. Once the desired species has been identified, the system will display data for the species.

Organic Chemistry Resources Worldwide. Available online, URL:
http://www.organicworldwide.net
Accessed April 28, 2004
This Web site was designed for the specific needs of organic chemists, collecting and independently annotating all useful organic chemistry sites. There is no access restriction, as it is available for a worldwide audience, and no registration or user fees are required.

Tutorial for High School Chemistry. Available online, URL:
http://dbhs.wvusd.k12.ca.us/ChemTeamIndex.html
Accessed April 28, 2004
If you need a good place to brush up on those chemistry categories that you are having difficulties with, this site presents several online tutorials that are sure to help. There are some links to other high-school-level chemistry sites.

WebElements. Available online, URL:
http://www.webelements.com
Accessed April 28, 2004
An excellent online interactive periodic table of the elements. Allows you to print your own personal periodic table.

APPENDIX III

CHEMISTRY SOFTWARE SOURCES

Biocatalog

http://www.ebi.ac.uk/biocat/biocat.html
An impressive site that lists software in the biology/biochemistry area categorized in several domains from DNA to servers.

The Chemical Thesaurus

http://www.meta-synthesis.com/Products/ChemThes3.html
A free chemical thesaurus provides useful information about chemical species and chemical interactions and reactions.

Chemistry Software for Linux

http://sal.kachinatech.com/Z/2/
A list of links to chemistry software that runs on the Linux platform.

Computational Chemistry List

http://www.ccl.net/cca/software/
A listing of chemistry-related software programs by computer platforms. Everything from MS-DOS to Unix.

Free Chemistry Programs for Mac Users

http://hyperarchive.lcs.mit.edu/HyperArchive/Abstracts/sci/HyperArchive.html
Contains a number of free chemistry programs for Mac users only. Read the abstracts first to determine if the program meets your needs.

Genamics Softwareseek

http://www.genamics.com/software/index.htm
A large Web site that is a repository and database of free and commercial tools for use in molecular biology and biochemistry, supporting all computer platforms. Currently there are more than 1,200 entries.

Open Science Project

http://www.openscience.org/links.php?section=7
A listing of links to free chemistry programs in more than 20 categories. Contains a search engine.

Software for Chemistry

http://www-rohan.sdsu.edu/staff/drjackm/chemistry/chemlink/teach/teach16.html
This is a metasite that contains links to other sites that list lots of chemistry software.

Tom's Free Chemistry Software

http://allserv.UGent.be/~tkuppens/chem/
The first place to start looking. An alphabetical listing of all free, shareware, and demo software in chemistry. This site is updated regularly by Tom Kuppens in Belgium. The site also has a search engine if you do not want to browse.

WWW Computational Chemistry Resources

http://www.chem.swin.edu.au/chem_ref.html#Software
A listing of many software links for computational chemistry.

APPENDIX IV

NOBEL LAUREATES RELATING TO CHEMISTRY

1901 Jacobus H. van't Hoff
"In recognition of the extraordinary services he has rendered by the discovery of the laws of chemical dynamics and osmotic pressure in solutions."

1902 Hermann Emil Fischer
"In recognition of the extraordinary services he has rendered by his work on sugar and purine syntheses."

1903 Svante August Arrhenius
"In recognition of the extraordinary services he has rendered to the advancement of chemistry by his electrolytic theory of dissociation."

1904 Sir William Ramsay
"In recognition of his services in the discovery of the inert gaseous elements in air, and his determination of their place in the periodic system."

1905 Adolf von Baeyer
"In recognition of his services in the advancement of organic chemistry and the chemical industry, through his work on organic dyes and hydroaromatic compounds."

1906 Henri Moissan
"In recognition of the great services rendered by him in his investigation and isolation of the element fluorine, and for the adoption in the service of science of the electric furnace called after him."

1907 Eduard Buchner
"For his biochemical researches and his discovery of cell-free fermentation."

1908 Ernest Rutherford
"For his investigations into the disintegration of the elements, and the chemistry of radioactive substances."

1909 Wilhelm Ostwald
"In recognition of his work on catalysis and for his investigations into the fundamental principles governing chemical equilibria and rates of reaction."

1910 Otto Wallach
"In recognition of his services to organic chemistry and the chemical industry by his pioneer work in the field of alicyclic compounds."

1911 Marie Curie
"In recognition of her services to the advancement of chemistry by the discovery of the elements radium and polonium, by the isolation of radium

and the study of the nature and compounds of this remarkable element."

1912 Victor Grignard
"For the discovery of the so-called Grignard reagent, which in recent years has greatly advanced the progress of organic chemistry."

Paul Sabatier
"For his method of hydrogenating organic compounds in the presence of finely disintegrated metals whereby the progress of organic chemistry has been greatly advanced in recent years."

1913 Alfred Werner
"In recognition of his work on the linkage of atoms in molecules by which he has thrown new light on earlier investigations and opened up new fields of research especially in inorganic chemistry."

1914 Theodore W. Richards
"In recognition of his accurate determinations of the atomic weight of a large number of chemical elements."

1915 Richard Willstätter
"For his researches on plant pigments, especially chlorophyll."

1916 The prize money was allocated to the special fund of this prize section.

1917 The prize money was allocated to the special fund of this prize section.

1918 Fritz Haber
"For the synthesis of ammonia from its elements."

1919 The prize money was allocated to the special fund of this prize section.

1920 Walther Nernst
"In recognition of his work in thermochemistry."

1921 Frederick Soddy
"For his contributions to our knowledge of the chemistry of radioactive substances, and his investigations into the origin and nature of isotopes."

1922 Francis W. Aston
"For his discovery, by means of his mass spectrograph, of isotopes, in a large number of non-radioactive elements, and for his enunciation of the whole-number rule."

1923 Fritz Pregl
"For his invention of the method of micro-analysis of organic substances."

1924 The prize money was allocated to the special fund of this prize section.

1925 Richard Zsigmondy
"For his demonstration of the heterogeneous nature of colloid solutions and for the methods he used, which have since become fundamental in modern colloid chemistry."

1926 The (Theodor) Svedberg
"For his work on disperse systems."

1927 Heinrich Wieland
"For his investigations of the constitution of the bile acids and related substances."

1928 Adolf Windaus
"For the services rendered through his research into the constitution of the sterols and their connection with the vitamins."

1929 Arthur Harden, Hans von Euler-Chelpin
"For their investigations on the fermentation of sugar and fermentative enzymes."

1930 Hans Fischer
"For his researches into the constitution of haemin and chlorophyll and especially for his synthesis of haemin."

1931 Carl Bosch, Friedrich Bergius
"In recognition of their contributions to the invention and development of chemical high pressure methods."

1932 Irving Langmuir
"For his discoveries and investigations in surface chemistry."

1933 The prize money was allocated with one-third going to the main fund and two-thirds to the special fund of this prize section.

1934 Harold C. Urey
"For his discovery of heavy hydrogen."

1935 Frédéric Joliot, Irène Joliot-Curie
"In recognition of their synthesis of new radioactive elements."

1936 Peter Debye
"For his contributions to our knowledge of molecular structure through his investigations on dipole moments and on the diffraction of X-rays and electrons in gases."

1937 Norman Haworth
"For his investigations on carbohydrates and vitamin C."

Paul Karrer
"For his investigations on carotenoids, flavins and vitamins A and B2."

1938 Richard Kuhn
"For his work on carotenoids and vitamins."

1939 Adolf Butenandt
"For his work on sex hormones."

Leopold Ruzicka
"For his work on polymethylenes and higher terpenes."

1940 The prize money was allocated with one-third going to the main fund and two-thirds to the special fund of this prize section.

1941 The prize money was allocated with one-third going to the main fund and two-thirds to the special fund of this prize section.

1942 The prize money was allocated with one-third going to the main fund and two-thirds to the special fund of this prize section.

1943 George de Hevesy
"For his work on the use of isotopes as tracers in the study of chemical processes."

1944 Otto Hahn
"For his discovery of the fission of heavy nuclei."

1945 Artturi Virtanen
"For his research and inventions in agricultural and nutrition chemistry, especially for his fodder preservation method."

1946 James B. Sumner
"For his discovery that enzymes can be crystallized."

John H. Northrop, Wendell M. Stanley
"For their preparation of enzymes and virus proteins in a pure form."

1947 Sir Robert Robinson
"For his investigations on plant products of biological importance, especially the alkaloids."

1948 Arne Tiselius
"For his research on electrophoresis and adsorption analysis, especially for his discoveries concerning the complex nature of the serum proteins."

1949 William F. Giauque
"For his contributions in the field of chemical thermodynamics, particularly concerning the behavior of substances at extremely low temperatures."

1950 Otto Diels, Kurt Alder
"For their discovery and development of the diene synthesis."

1951 Edwin M. McMillan, Glenn T. Seaborg
"For their discoveries in the chemistry of the transuranium elements."

1952 Archer J.P. Martin, Richard L.M. Synge
"For their invention of partition chromatography."

1953 Hermann Staudinger
"For his discoveries in the field of macromolecular chemistry."

1954 Linus Pauling
"For his research into the nature of the chemical bond and its application to the elucidation of the structure of complex substances."

1955 Vincent du Vigneaud
"For his work on biochemically important sulphur compounds, especially for the first synthesis of a polypeptide hormone."

1956 Sir Cyril Hinshelwood, Nikolay Semenov
"For their researches into the mechanism of chemical reactions."

1957 Lord Todd
"For his work on nucleotides and nucleotide co-enzymes."

1958 Frederick Sanger
"For his work on the structure of proteins, especially that of insulin."

1959 Jaroslav Heyrovsky
"For his discovery and development of the polarographic methods of analysis."

1960 Willard F. Libby
"For his method to use carbon-14 for age determination in archaeology, geology, geophysics, and other branches of science."

1961 Melvin Calvin
"For his research on the carbon dioxide assimilation in plants."

1962 Max F. Perutz, John C. Kendrew
"For their studies of the structures of globular proteins."

1963 Karl Ziegler, Giulio Natta
"For their discoveries in the field of the chemistry and technology of high polymers."

1964 Dorothy Crowfoot Hodgkin
"For her determinations by X-ray techniques of the structures of important biochemical substances."

1965 Robert B. Woodward
"For his outstanding achievements in the art of organic synthesis."

1966 Robert S. Mulliken
"For his fundamental work concerning chemical bonds and the electronic structure of molecules by the molecular orbital method."

1967 Manfred Eigen, Ronald G.W. Norrish, George Porter
"For their studies of extremely fast chemical reactions, effected by disturbing the equilibrium by means of very short pulses of energy."

1968 Lars Onsager
"For the discovery of the reciprocal relations bearing his name, which are fundamental for the thermodynamics of irreversible processes."

1969 Derek Barton, Odd Hassel
"For their contributions to the development of the concept of conformation and its application in chemistry."

1970 Luis Leloir
"For his discovery of sugar nucleotides and their role in the biosynthesis of carbohydrates."

1971 Gerhard Herzberg
"For his contributions to the knowledge of electronic structure and geometry of molecules, particularly free radicals."

1972 Christian Anfinsen
"For his work on ribonuclease, especially concerning the connection between the amino acid sequence and the biologically active conformation."

Stanford Moore, William H. Stein
"For their contribution to the understanding of the connection between chemical structure and catalytic activity of the active center of the ribonuclease molecule."

1973 Ernst Otto Fischer, Geoffrey Wilkinson
"For their pioneering work, performed independently, on the chemistry of the organometallic, so called sandwich compounds."

1974 Paul J. Flory
"For his fundamental achievements, both theoretical and experimental, in the physical chemistry of the macromolecules."

1975 John Cornforth
"For his work on the stereochemistry of enzyme-catalyzed reactions."

Vladimir Prelog
"For his research into the stereochemistry of organic molecules and reactions."

1976 William Lipscomb
"For his studies on the structure of boranes illuminating problems of chemical bonding."

1977 Ilya Prigogine
"For his contributions to non-equilibrium thermodynamics, particularly the theory of dissipative structures."

1978 Peter Mitchell
"For his contribution to the understanding of biological energy transfer through the formulation of the chemiosmotic theory."

1979 Herbert C. Brown, Georg Wittig
"For their development of the use of boron- and phosphorus-containing compounds, respectively, into important reagents in organic synthesis."

1980 Paul Berg
"For his fundamental studies of the biochemistry of nucleic acids, with particular regard to recombinant-DNA."

Walter Gilbert, Frederick Sanger
"For their contributions concerning the determination of base sequences in nucleic acids."

1981 Kenichi Fukui, Roald Hoffmann
"For their theories, developed independently, concerning the course of chemical reactions."

1982 Aaron Klug
"For his development of crystallographic electron microscopy and his structural elucidation of biologically important nucleic acid-protein complexes."

1983 Henry Taube
"For his work on the mechanisms of electron transfer reactions, especially in metal complexes."

1984 Bruce Merrifield
"For his development of methodology for chemical synthesis on a solid matrix."

1985 Herbert A. Hauptman, Jerome Karle
"For their outstanding achievements in the development of direct methods for the determination of crystal structures."

1986 Dudley R. Herschbach, Yuan T. Lee, John C. Polanyi
"For their contributions concerning the dynamics of chemical elementary processes."

1987 Donald J. Cram, Jean-Marie Lehn, Charles J. Pedersen
"For their development and use of molecules with structure-specific interactions of high selectivity."

1988 Johann Deisenhofer, Robert Huber, Hartmut Michel
"For the determination of the three-dimensional structure of a photosynthetic reaction center."

1989 Sidney Altman, Thomas R. Cech
"For their discovery of catalytic properties of RNA."

1990 Elias James Corey
"For his development of the theory and methodology of organic synthesis."

1991 Richard R. Ernst
"For his contributions to the development of the methodology of high resolution nuclear magnetic resonance (NMR) spectroscopy."

1992 Rudolph A. Marcus
"For his contributions to the theory of electron transfer reactions in chemical systems."

1993 The prize was awarded for contributions to the developments of methods within DNA-based chemistry equally between:

Kary B. Mullis
"For his invention of the polymerase chain reaction (PCR) method."

Michael Smith
"For his fundamental contributions to the establishment of oligonucleotide-based, site-directed mutagenesis and its development for protein studies."

1994 George A. Olah
"For his contribution to carbocation chemistry."

1995 Paul J. Crutzen, Mario J. Molina, F. Sherwood Rowland
"For their work in atmospheric chemistry, particularly concerning the formation and decomposition of ozone."

1996 Robert F. Curl, Jr., Sir Harold Kroto, Richard E. Smalley
"For their discovery of fullerenes."

1997 Paul D. Boyer, John E. Walker
"For their elucidation of the enzymatic mechanism underlying the synthesis of adenosine triphosphate (ATP)."

Jens C. Skou
"For the first discovery of an ion-transporting enzyme, Na+, K+ -ATPase."

1998 Walter Kohn
"For his development of the density-functional theory."

John Pople
"For his development of computational methods in quantum chemistry."

1999 Ahmed Zewail
"For his studies of the transition states of chemical reactions using femtosecond spectroscopy."

2000 Alan Heeger, Alan G. MacDiarmid, Hideki Shirakawa
"For the discovery and development of conductive polymers."

2001 William S. Knowles, Ryoji Noyori
"For their work on chirally catalyzed hydrogenation reactions."

K. Barry Sharpless
"For his work on chirally catalyzed oxidation reactions."

2002 For the development of methods for identification and structure analyses of biological macromolecules:

John B. Fenn, Koichi Tanaka
"For their development of soft desorption ionization methods for mass spectrometric analyses of biological macromolecules."

Kurt Wüthrich
"For his development of nuclear magnetic resonance spectroscopy for determining the three-dimensional structure of biological macromolecules in solution."

2003 For discoveries concerning channels in cell membranes:

Peter Agre
"For the discovery of water channels."

Roderick MacKinnon
"For structural and mechanistic studies of ion channels."

2004 Aaron Ciechanover, Avram Hershko, Irwin Rose
"For the discovery of ubiquitin-mediated protein degradation."

APPENDIX V

Periodic Table of Elements

1 — atomic number
H — symbol
1.008 — atomic weight

Numbers in parentheses are the atomic mass numbers of radioactive isotopes.

1 H 1.008																	2 He 4.003
3 Li 6.941	4 Be 9.012											5 B 10.81	6 C 12.01	7 N 14.01	8 O 16.00	9 F 19.00	10 Ne 20.18
11 Na 22.99	12 Mg 24.31											13 Al 26.98	14 Si 28.09	15 P 30.97	16 S 32.07	17 Cl 35.45	18 Ar 39.95
19 K 39.10	20 Ca 40.08	21 Sc 44.96	22 Ti 47.88	23 V 50.94	24 Cr 52.00	25 Mn 54.94	26 Fe 55.85	27 Co 58.93	28 Ni 58.69	29 Cu 63.55	30 Zn 65.39	31 Ga 69.72	32 Ge 72.59	33 As 74.92	34 Se 78.96	35 Br 79.90	36 Kr 83.80
37 Rb 85.47	38 Sr 87.62	39 Y 88.91	40 Zr 91.22	41 Nb 92.91	42 Mo 95.94	43 Tc (98)	44 Ru 101.1	45 Rh 102.9	46 Pd 106.4	47 Ag 107.9	48 Cd 112.4	49 In 114.8	50 Sn 118.7	51 Sb 121.8	52 Te 127.6	53 I 126.9	54 Xe 131.3
55 Cs 132.9	56 Ba 137.3	57-71*	72 Hf 178.5	73 Ta 180.9	74 W 183.9	75 Re 186.2	76 Os 190.2	77 Ir 192.2	78 Pt 195.1	79 Au 197.0	80 Hg 200.6	81 Tl 204.4	82 Pb 207.2	83 Bi 209.0	84 Po (210)	85 At (210)	86 Rn (222)
87 Fr (223)	88 Ra (226)	89-103‡	104 Rf (261)	105 Db (262)	106 Sg (263)	107 Bh (262)	108 Hs (265)	109 Mt (266)	110 Ds (271)	111 Uuu (272)	112 Uub (285)	113 Uut (284)	114 Uuq (289)	115 Uup (288)			

*lanthanide series	57 La 138.9	58 Ce 140.1	59 Pr 140.9	60 Nd 144.2	61 Pm (145)	62 Sm 150.4	63 Eu 152.0	64 Gd 157.3	65 Tb 158.9	66 Dy 162.5	67 Ho 164.9	68 Er 167.3	69 Tm 168.9	70 Yb 173.0	71 Lu 175.0
‡actinide series	89 Ac (227)	90 Th 232.0	91 Pa 231.0	92 U 238.0	93 Np (237)	94 Pu (244)	95 Am (243)	96 Cm (247)	97 Bk (247)	98 Cf (251)	99 Es (252)	100 Fm (257)	101 Md (258)	102 No (259)	103 Lr (260)

The Chemical Elements

element	symbol	a.n.	element	symbol	a.n.	element	symbol	a.n.	element	symbol	a.n.
actinium	Ac	89	erbium	Er	68	molybdenum	Mo	42	selenium	Se	34
aluminum	Al	13	europium	Eu	63	neodymium	Nd	60	silicon	Si	14
americium	Am	95	fermium	Fm	100	neon	Ne	10	silver	Ag	47
antimony	Sb	51	fluorine	F	9	neptunium	Np	93	sodium	Na	11
argon	Ar	18	francium	Fr	87	nickel	Ni	28	strontium	Sr	38
arsenic	As	33	gadolinium	Gd	64	niobium	Nb	41	sulfur	S	16
astatine	At	85	gallium	Ga	31	nitrogen	N	7	tantalum	Ta	73
barium	Ba	56	germanium	Ge	32	nobelium	No	102	technetium	Tc	43
berkelium	Bk	97	gold	Au	79	osmium	Os	76	tellurium	Te	52
beryllium	Be	4	hafnium	Hf	72	oxygen	O	8	terbium	Tb	65
bismuth	Bi	83	hassium	Hs	108	palladium	Pd	46	thallium	Tl	81
bohrium	Bh	107	helium	He	2	phosphorus	P	15	thorium	Th	90
boron	B	5	holmium	Ho	67	platinum	Pt	78	thulium	Tm	69
bromine	Br	35	hydrogen	H	1	plutonium	Pu	94	tin	Sn	50
cadmium	Cd	48	indium	In	49	polonium	Po	84	titanium	Ti	22
calcium	Ca	20	iodine	I	53	potassium	K	19	tungsten	W	74
californium	Cf	98	iridium	Ir	77	praseodymium	Pr	59	ununbium	Uub	112
carbon	C	6	iron	Fe	26	promethium	Pm	61	ununpentium	Uup	115
cerium	Ce	58	krypton	Kr	36	protactinium	Pa	91	ununquadium	Uuq	114
cesium	Cs	55	lanthanum	La	57	radium	Ra	88	ununtrium	Uut	113
chlorine	Cl	17	lawrencium	Lr	103	radon	Rn	86	unununium	Uuu	111
chromium	Cr	24	lead	Pb	82	rhenium	Re	75	uranium	U	92
cobalt	Co	27	lithium	Li	3	rhodium	Rh	45	vanadium	V	23
copper	Cu	29	lutetium	Lu	71	rubidium	Rb	37	xenon	Xe	54
curium	Cm	96	magnesium	Mg	12	ruthenium	Ru	44	ytterbium	Yb	70
darmstadtium	Ds	110	manganese	Mn	25	rutherfordium	Rf	104	yttrium	Y	39
dubnium	Db	105	meitnerium	Mt	109	samarium	Sm	62	zinc	Zn	30
dysprosium	Dy	66	mendelevium	Md	101	scandium	Sc	21	zirconium	Zr	40
einsteinium	Es	99	mercury	Hg	80	seaborgium	Sg	106			

a.n. = atomic number

APPENDIX VI

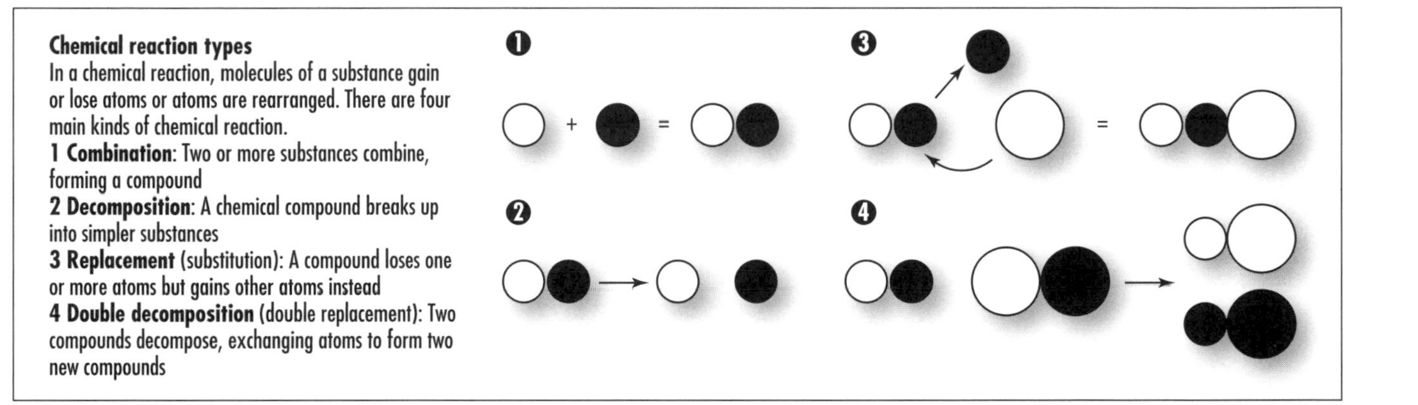
Chemical reaction types
In a chemical reaction, molecules of a substance gain or lose atoms or atoms are rearranged. There are four main kinds of chemical reaction.
1 Combination: Two or more substances combine, forming a compound
2 Decomposition: A chemical compound breaks up into simpler substances
3 Replacement (substitution): A compound loses one or more atoms but gains other atoms instead
4 Double decomposition (double replacement): Two compounds decompose, exchanging atoms to form two new compounds
1
+
=
2
3
=
4

APPENDIX VII

Metals and alloys

All but 40 of the known elements are metals. Metals are elements whose atoms can lose one or more electrons to form electrically positive ions. Most metals are good conductors of heat and electricity. They are malleable (can be beaten or rolled into a new shape) and ductile (can be pulled out into long wires). All metals are shiny, crystalline solids, except mercury, which is a liquid.

Activity series

Some metals form positive ions more easily than others, and so are more chemically active. Sixteen common metals are listed in the order of their activity. Lithium is the most active of all the metals, and gold is the least active.

Native metals

Only four of the least active metals—copper, silver, platinum, and gold—commonly occur in the Earth's crust as native metals (i.e., as free elements). All the others are found in compounds, called ores, that must be chemically treated to obtain the pure element.

Alloys

An alloy is a mixture of two or more metals. Here we list some everyday alloys, the metals from which they are made, and examples of their use.

Alloy	Metals	Examples of use
bronze	copper, tin	"copper" coins
brass	copper, zinc	doorhandles, buttons
cupronickel	copper, nickel	"silver" coins
pewter	tin, lead	tankards
stainless steel	iron, chromium, nickel	cutlery, pots, etc.
sterling silver	silver, copper	jewelry
9, 18, and 22 carat gold	gold, silver, copper	jewelry
dental amalgam	silver, tin, copper, zinc, mercury	filling cavities in teeth
solder	lead, tin	joining metals

Metalloids

These elements are "halfway" between metals and nonmetals. Depending on the way they are treated, they can act as insulators like nonmetals or conduct electricity like metals. This makes several metalloids extremely important as semiconductors in computers and other electronic devices. The eight metalloid elements are boron, silicon, germanium, arsenic, antimony, tellurium, polonium, and astatine.

INDEX

Note: Page numbers in **boldface** indicate main entries; *italic* page numbers indicate photographs and illustrations.

B

D

I

N

T